Lecture Notes in Computer Science 5219

Commenced Publication in 1973
Founding and Former Series Editors:
Gerhard Goos, Juris Hartmanis, and Jan van Leeuwen

Michael D. Harrison   Mark-Alexander Sujan (Eds.)

Computer Safety, Reliability, and Security

27th International Conference, SAFECOMP 2008
Newcastle upon Tyne, UK, September 22-25, 2008
Proceedings

 Springer

Volume Editors

Michael D. Harrison
Newcastle University
School of Computing Science
Claremont Tower, Newcastle upon Tyne, NE1 7RU, UK
E-mail: michael.harrison@ncl.ac.uk

Mark-Alexander Sujan
University of Warwick
Health Sciences Research Institute
Coventry, CV4 7AL, UK
E-mail: m-a.sujan@warwick.ac.uk

Library of Congress Control Number: 2008934760

CR Subject Classification (1998): D.1-4, E.4, C.3, F.3, K.6.5

LNCS Sublibrary: SL 2 – Programming and Software Engineering
ISSN
       0302-9743
ISBN-10   3-540-87697-9 Springer Berlin Heidelberg New York
ISBN-13   978-3-540-87697-7 Springer Berlin Heidelberg New York

Springer is a part of Springer Science+Business Media

springer.com

© Springer-Verlag Berlin Heidelberg 2008
Printed in Germany

Typesetting: Camera-ready by author, data conversion by Scientific Publishing Services, Chennai, India
Printed on acid-free paper   SPIN: 12517558   06/3180   5 4 3 2 1 0

Preface

Safecomp was held in Newcastle upon Tyne, UK in September 2008. The conference was the latest in a long and strong tradition of leading-edge research and practice in Computer Safety, Reliability and Security that began in 1979.

The programme was drawn from a strong international selection of papers from a dozen countries in three continents (32 papers from 115 submissions). Traditional Safecomp themes such as software dependability, software safety arguments and formal methods continued to be represented. This conference also strengthened themes that have been less visible in previous Safecomp conferences, particularly those relating to the complexity and resilience of systems, critical infrastructures and human factors. It broadened the usual domains of application to include, for example, e-Commerce. The programme continued to benefit from strong industrial contributions in safety critical and security critical applications.

We were fortunate to have keynote addresses from Colin O'Halloran (QinetiQ) and Roger Rivett (LandRover) on different complementary industrial experiences in security and reliability. We were also fortunate to have Erik Hollnagel as our opening keynote speaker. Professor Hollnagel's contribution was effective in challenging traditional views, broadening the focus of concern, especially relating to the management and analysis of complexity in large-scale systems.

We would like to express our gratitude and thanks to all those whose effort has made this conference possible: to the submitting authors and to the invited speakers; to the Programme Committee and the external reviewers who helped compile an attractive programme; to the financial and scientific sponsors; and last but not least, to the members of the Organizing Committee who took care of the local arrangements.

We hope that you will find these proceedings of interest and use for your own work.

July 2008

Michael Harrison
Mark-Alexander Sujan

Organization

Safecomp 2008 was sponsored by EWICS TC7.

Organizing Committee

Co-chairs   Michael Harrison (Newcastle University, UK)
                             Mark-Alexander Sujan (Warwick University,
                             UK)
EWICS Chair   U. Voges (Forschungszentrum Karlsruhe, DE)
Organizing Committee   Joan Atkinson (Newcastle University, UK)
                             Massimo Felici (University of Edinburgh, UK)
                             Michael Harrison (Newcastle University, UK)
                             Steve Riddle (Newcastle University, UK)
                             Claire Smith (Newcastle University, UK)
                             Shamus Smith (Durham University, UK)
                             Mark-Alexander Sujan (Warwick University,
                             UK)
                             Christine Wisher (Newcastle University, UK)
                             Nikos Zarboutis (ENERCON Service Hellas,
                             Greece)

Programme Committee

R. Amalberti, France
S. Anderson, UK
T. Anderson, UK
J. Braband, Germany
N. Buth, Germany
S. Cheshire, UK
M. Cooke, UK
P. Daniel, UK
W. Ehrenberger,
  Germany
L. Emmet, UK
C. Fairburn, UK
M. Felici, UK
J. Gorski, Poland
B. Gran, Norway
L. Grunske, Germany
W. Halang, Germany
M. Harrison, UK

M. Heisel, Germany
C. Heitmeyer, USA
A. Hessami, UK
E. Hollnagel, France
C. Johnson, UK
M. Kaaniche, France
K. Kanoun, France
T. Kelly, UK
J. Knight, USA
F. Koornneef,
  The Netherlands
P. Ladkin, Germany
B. Littlewood, UK
J. McDermid, UK
O. Nordland, Norway
P. Palanque, France
A. Pasquini, Italy
M. Pickering, UK

S. Pozzi, Italy
G. Rabe, Germany
F. Redmill, UK
F. Saglietti, Germany
E. Schoitsch, Austria
S. Smith, UK
L. Strigini, UK
M. Sujan, UK
P. Traverse, France
J. Trienekens,
  The Netherlands
M. van der Meulen,
  The Netherlands
U. Voges, Germany
A. Weinert, Germany
S. Wittmann, Belgium
N. Zarboutis, Greece
Z. Zurakowski, Poland

External Reviewers

O. Meyer	I. Wentzlaff	N. Rivire
N. Chozos	T. Santen	M. Roy
T. Storer	H. Schmidt	A. van Moorsel
T. Ma	D. Hatebur	C. Gacek
J. Clark	T. Santen	P. Ryan

Sponsoring Institutions

EWICS TC7

Centre for Software Reliability **CSR**

Newcastle University

Warwick Medical School **Warwick** Medical School

AdaCore AdaCore

ReSIST

Qinetiq *QinetiQ*

Adelard Adelard

TTE-Systems **TTE** Systems

British Computer Society **BCS**

ifip ifip

IFAC IFAC

DECOS **DECOS**

Austrian Computer society

Gesellschaft für Informatik e.V.

Encress

European
Network of
Clubs for
REliability and
Safety of
Software

Table of Contents

Security

Safety Cases

Formal Methods

Dependability Modelling

Security and Dependability

Critical Information Infrastructures: Should Models Represent Structures or Functions?

Erik Hollnagel

École des Mines ParisTech, Crisis and Risk Research Centre
Sophia Antipolis, France
erik.hollnagel@crc.ensmp.fr

Abstract. The common approaches to modeling and analyzing complex socio-technical systems, of which Critical Information Infrastructures is one example, assumes that they can be completely specified. The methods emphasize how systems are composed or structured and how component failures propagate. Since socio-technical systems always are underspecified, they cannot be analyzed in the same way. The alternative is to focus on their functions, and how the variability of functions can combine to create non-linear effects. An example of that is the Functional Resonance Analysis method (FRAM).

Keywords: Complexity, risk, socio-technical systems, functional resonance.

1 Introduction

There is little need to argue that the industrialized societies of today completely depend on information technology, both to carry out individual functions and as a backbone. In 1984 Charles Perrow argued that many systems by then had become so tightly coupled, both within and between domains of activity, that accidents should be accepted as normal. This trend has continued unabated, and Perrow's comment that "(o)n the whole, we have complex systems because we don't know how to produce the output through linear systems" [1, p. 89] is as true today as it was then. One consequence of this is that many information infrastructures become critical for organizations and society, in the sense that the "normal" functioning we have become accustomed to is possible only if the underlying information systems work effectively and reliably. While this usually is so, there is nevertheless an ever growing list of cases where information systems and information infrastructures have failed – both in routine operations or services and in more spectacular ways ranging from the failure of the London Ambulance Service Computer Aided Dispatch [2] to security breaches in petroleum industry information systems [3]. This creates a clear need to be able to prevent such failures from taking place. To do so we need to understand both the nature of the risks and the means of preventing them.

2 The Structural Approach to System Modeling

It is a basic principle of cybernetics that it is necessary to have a model of a system in order to control it [4]. This principle also applies in the case of system safety, since

M.D. Harrison and M.-A. Sujan (Eds.): SAFECOMP 2008, LNCS 5219, pp. 1–4, 2008.

risk identification and safety management clearly are forms of control. The common approaches to modeling critical information infrastructures (CII) are based on the tradition established by the analysis and assessment of technical systems. They emphasize how CIIs are *structured*, using the classical principles of aggregation and decomposition. Although the concern should be the ability of a CII to provide a specific service or functionality, descriptions are invariably given in terms of whether components and/or subsystems work or fail, and whether they are available and reliable.

The common approaches make two fundamental assumptions. The first is that adverse outcomes arise from failures and malfunctions of system parts, whether they be subsystems or components. The focus is therefore on identifying the possible failures modes and on deriving their probability. The second is that the effects of failures or malfunctions propagate linearly. This favors a notation of system components, such as processes, information channels, sensors, evaluation unites, actuators, control loops, etc., and how they are connected, usually in terms of inputs and outputs. While structural models, and the associated methods, have been very successful for purely technical systems where the impact of humans and/or organizations are negligible, they are ill-suited to address the safety issues of socio-technical systems.

Technical systems are basically bimodal, i.e., parts or components either work or fail, and when they fail they are repaired or replaced. Connections among components are defined in advance by the system design and the systems are assumed to be fully describable (tractable). Since technical systems in principle can be completely specified, the quality of their performance depends on how well deviations from the requirements can be prevented.

In socio-technical systems, the performance of parts and of the system as a whole is not bimodal but can vary. While it sometimes is better than normal and sometimes worse, it is exceedingly rare that it fails completely. Socio-technical systems are furthermore outlined rather than designed, and specific connections may therefore develop or disappear as the system adjusts to changing conditions and demands. Since socio-technical systems always are underspecified to some degree, their internal performance must be variable in order to compensate for the incompleteness of the specifications. The quality of their performance therefore depends on how well the variability can be managed in a given situation, rather than on the suppression of variability. For socio-technical systems it is therefore more important to describe their function than their structure.

While it may be convenient to use a conventional linear notation to describe socio-technical systems, it is inadequate both in principle and in practice. The issue for socio-technical systems is not how components can fail but rather how performance variability can combine in an unforeseen manner and thereby lead to unexpected negative, or positive, outcomes. The model used must be able to describe how this happens; otherwise both measurements and their interpretation will be inadequate for monitoring and for effective interventions if and when something goes wrong.

3 The Functional Approach to System Modeling

Even though organizations often are represented as hierarchies or networks, there can be vast differences between the formal and the informal organization. Humans are, of course, even more diverse and cognitive models are rarely more than crude renderings of a few prominent features – despite heroic efforts by cognitive scientists and psychologists. A notation based on components, malfunctions, and linear propagations is therefore unable adequately to represent a socio-technical system.

An adequate notation must be able to describe how the system functions rather than how it is structured. The underlying model and method must furthermore be able to account for how the variability of normal performance can combine in unanticipated ways. The functional resonance analysis method (FRAM; [5]) provides an example of how this may be done. This particular method is based on four principles:

- The principle of equivalence of successes and failures: Successes and failures arise from the same underlying processes or "mechanisms".
- The principle of approximate adjustments: Because systems always are underspecified, people must adjust their performance to match current conditions. These adjustments are approximate because there is a lack of time, resources, and/or information.
- The principle of emergence: This means that it is impossible in practice to trace an effect back to a specific cause or function – or in the case of failures, a malfunction. Causal analyses are therefore often ineffective, and in some cases directly misleading and inappropriate.
- The principle of functional resonance: The variability of a number of functions may sometimes reinforce each other and thereby cause the variability of one function to exceed the expected range. Consequences may spread through tight couplings rather than via identifiable and enumerable cause-effect links. This can be described as a resonance of the variability of normal performance.

The FRAM notation allows couplings and dependencies to be created and mapped as they develop and disappear in real time. An important part of the functional modeling is therefore to determine the variability of each function by characterizing how dependent it is on the environment, i.e., the functioning conditions. For nodes that describe human and organizational functions, one powerful principle is how humans and organizations make sacrificing decisions or trade off efficiency for thoroughness. While this normally is useful and contributes to the overall efficiency of the system, it carries with it the risk of unexpected outcomes.

The FRAM notation offers a set of strong principles by which the psychological, social, and technical couplings in a CII can be modeled. This approach to mapping dynamic dependencies is already used in several other contexts, mostly in relation to industrial safety applications [6, 7]. The step to security applications can make use of this experience to develop a more rigorous approach.

References

1. Perrow, C.: Normal accidents: Living with high risk technologies. Basic Books, Inc., New York (1984)
2. Finkelstein, A., Dowell, J.: A comedy of errors: the London Ambulance Service case study. In: Proc. 8th International Workshop on Software Specification & Design IWSSD-8, pp. 2–4. IEEE Computer Society Press, Los Alamitos (1996)

3. Albrechtsen, E., Hovden, J.: Industrial safety management and information security management: risk characteristics and management approaches. In: Proceedings of ESREL 2007 (2007)
4. Conant, R.C., Ashby, W.R.: Every good regulator of a system must be a model of that system. International Journal of Systems Science 1(2), 89–97 (1970)
5. Hollnagel, E.: Barriers and accident prevention. Ashgate, Aldershot (2004)
6. Mercadier, V.: When incident command system fails. University of Berkeley, Berkely (2007)
7. Nouvel, D., Travadel, S., Hollnagel, E.: Introduction of the concept of functional resonance in the analysis of a near-accident in aviation. In: 33rd ESReDA Seminar: Future challenges of accident investigation, Ispra, Italy, November 13-14 (2007)

Security and Interoperability for MANETs and a Fixed Core

Colin O'Halloran and Andy Bates

QinetiQ
Malvern Technology Park
Worcestershire WR14 3PS
cmohalloran@qinetiq.com

Abstract. The problem of ensuring security and interoperability for mobile ad hoc networks that connect to a valuable fixed core infrastructure is discussed. The problem is broken down into three interdependent research areas of Security versus Risk; Identity Management; Verification, Validation and Certification. The research issues for each area are discussed in detail.

Keywords: NEC, NCO, Security, Interoperability, Risk, Trust, Identity Management, Verification, Validation, Certification.

1 Introduction

A fundamental assumption of Network Enabled Capability[1] (NEC) is that the networks that connect sensors, platforms, troops and commanders, are robust and secure, and the data provided is trustworthy such that reliable and timely information can support local decision-making.

In practice work in the area of security has focused on Confidentiality and Integrity with Availability relegated to a background concern. Availability is a central concern of NEC because interoperability is a prime driver for NEC, hence interoperability is used explicitly in conjunction with Confidentiality and Integrity.

As NEC is implemented, the transition from a traditional largely static core to include a dynamic wireless edge creates significant challenges for security and interoperability. The increasing operational trend to rapidly form military coalitions with foreign nations, or provide humanitarian aid in association with Other Government Departments (OGDs) and Non-Government Organizations (NGOs) drives the need to ensure that the benefits conferred by access is balanced with the risk of loss of confidentiality and data integrity.

Advances in Information and Computing Technologies (ICT) are therefore required which can satisfy the dynamic balance of security versus risk for ad hoc mobile networks, and deal with the real time dynamics of forming and reforming such systems which can be Verified, Validated and Certified (VV&C) and scaled to address very large numbers of wireless nodes without degrading network performance.

---

[1] And the related concept of Network-Centric Operations (NCO).

M.D. Harrison and M.-A. Sujan (Eds.): SAFECOMP 2008, LNCS 5219, pp. 5–11, 2008.
© Springer-Verlag Berlin Heidelberg 2008

The subject of this paper is to propose a technical approach that incorporates near and future ICT developments to bridge the gap between the current situation, via a near term intermediate position, to a full future NEC position.

The research requirements can be split into three inter-dependent research areas:

- Security versus Risk;
- Identity Management;
- Verification, Validation and Certification.

These three research areas need to be addressed within a technical architectural framework for ad hoc secure networks such as that shown in Figure 1. This will ensure a common approach and understanding to assumptions and dependencies when researching different aspects of the ad hoc network security problem.

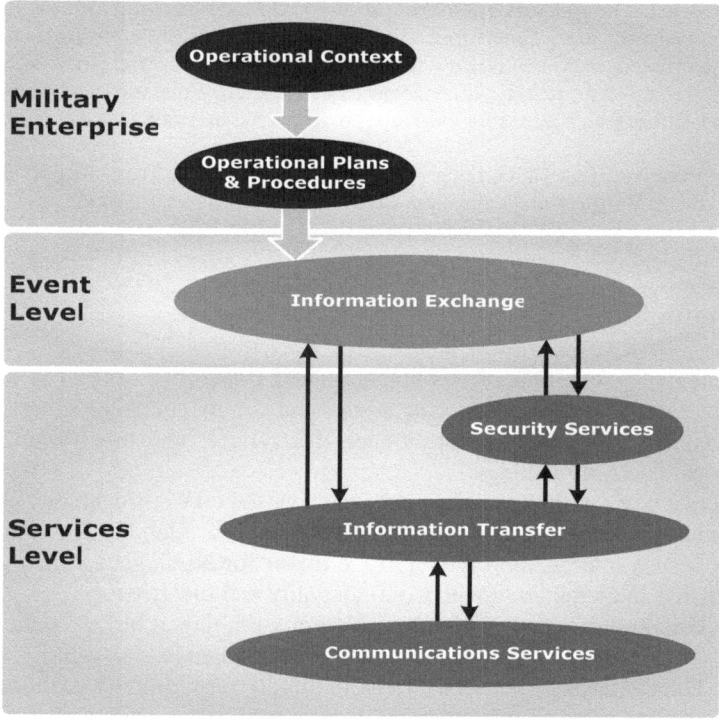

Fig. 1. Technical Architectural Framework for ad hoc Secure Networks Research and Development - this framework will guide how emerging technologies can fit together

2 Security Versus Risk in Mobile Ad-Hoc Networks (MANETs)

This is concerned with the ability of mobile communications and computing nodes to discover one another and create operationally relevant and secure networks. In pre-planned networks, the likely deployment of the network and the risks to it, are generally known about in advance, and judgments are made about the security mechanisms required to protect the network and information. In ad hoc networks, the membership, the connectivity, and the security risks are not known about in advance. Thus, a dynamic risk management approach needs to be adopted.

2.1 Dynamic Risk Management

There are a number of different dimensions to dynamic risk management.

Low –> High Risk Appetite Spectrum

Within a military operation, the Commander's appetite for risk may change, depending on the potential benefits to the operation. For example, allowing NGO nodes to connect to the military ad hoc network will typically increase the security risks to the network, but the operational benefits may justify these increased risks. In response to this increased risk, additional security measures may need to be put in place.

Simple –> Complex Risk Management Spectrum

Depending upon the level of risk appetite partners are willing to accept, the ad hoc network can adopt a spectrum of risk management strategies. At one end of the spectrum the network can adopt very rigid pre-defined rules of network connection (e.g. only allowing a connection if a common security key is held), or at the other end of the spectrum the network may be self-synchronizing, self-organizing, self healing, self-destructing, and self-accrediting. The latter is potentially more flexible to the changing military operation, but the increased complexity will generally increase the security risks resulting from the difficulty of being able to gain assurance about the overall system security behavior.

Human –> Automated Spectrum

The risk management strategy can also vary in the degree a human is involved in the security decision process. Again, there is a spectrum of possibilities. These range from a human being solely involved (as is the case for static systems); to the semi-automated case where some aspects of the decision process are automated, but others are referred to a human for a decision; through to a fully autonomous approach which may operate under very simple or very complex rules of behavior.

Risk versus Trust

The risk management strategy can also vary in the degree it demands of being able to manage previous, current and future security risks. In an ad hoc network fully under the control of a single organization, then it is possible for that organization to put in place dynamic security measures to manage the security risks. In an ad hoc network involving different organizations (e.g. coalition partners and NGOs) it is less likely that the MoD will have full, or even any, control of the security management of those parts of the ad hoc network not owned by MoD. Thus, connecting the MoD parts of the ad hoc network to non-MoD parts will potentially increase the security risks to the MoD part. However, if by some means (directly or indirectly) there is a belief (which may be evidence based) on how security risks are managed within the non-MoD part of the ad hoc network, then the resulting risks can be re-assessed.

We are using here the following generally acceptable definition of trust:

An evolving, contextual and composite belief that one principal (trustor) has that another principal (trustee) will perform certain actions with certain expected results, when not all information about those actions is available.

There will be differing degrees of trust (which will be influenced by the level of evidence available and the strength of the belief). It is also likely that different parts of the ad hoc network will be trusted by MoD to different levels.

Boundary Security

In forming an ad hoc network involving different organizationally owned nodes, there will be boundaries at which these nodes connect. These will be at different levels and places within the ad hoc network architecture (e.g. communications, information transfer, security services, applications etc). At these security domain boundaries, it will be possible to construct security mechanisms (e.g. filters) to control connections and information across the boundary as part of a cross security domain solution. Thus, MoD can help mitigate the resulting security risks of connecting to an NGO node by managing the connection and information flow across that security boundary. Ideally, these security mechanisms should be variable to allow them to be changed as a result of variations in the external risks and the degree of risk appetite the MoD Commander is willing to take.

2.2   Node and Service Discovery in Ad Hoc Networks

Within an ad hoc network, one cannot assume that centralized services are available to support ad hoc nodes. Consequently, alternative models need to be adopted. Peer-to-Peer (P2P) technologies have emerged in recent years from the Internet community. In such networks, information is held on peer nodes, rather than central servers.

Unfortunately, most of the current implementations of P2P are not 'pure' in that although information is held on peers, the discovery on those peers is still supported by central servers, which are used to identify which peers hold relevant information.

In a 'pure' P2P, there is a need for services which support the discovery of relevant other nodes, services and information, and which do not utilize central servers. There are number of ways of achieving this requirement. These include:

- P2P Service Publication – In this model, a peer tells other peers what information or services they have to offer.
- P2P Service Request – In this model, a peer makes a request either to other individual peers, or via a broadcast mechanism.
- Federated Service Publication & Request – In this model, peers build up information on other peers to support the discovery of services and information.

The security challenges to be addressed in ad hoc network discovery are determining:

- The identity, the confidence in the identity integrity and its association with the node and/or service;
- The security properties associated with the node / service;

- The security risks that results from connecting to that node or service;
- The security protection mechanisms that need to be put in place to create that connection with the service and/or node;
- The security implications of publishing, requesting and federated storing of service information.

2.3 MANET and Fixed-Core Interoperability

The previous discussion has focused on the security implications of the ad hoc dynamic MANET. However, there is also a need for nodes and services within the MANET to be able to interoperate with nodes and services within the "Fixed-core". As a consequence, there will be different approaches to risk management adopted in different parts of the network.

As MANETs need to be able to accommodate different risk-management approaches for the reasons stated previously, then the fixed-core part is just a particular type of risk management. However, within the fixed-core there may not be any pre-planning of how to manage the security risks associated with dynamic MANETs. This will particularly be the case for the early states of NEC. Consequently, the Boundary Security approach to risk mitigation will be required at the MANET and Wired-core boundary.

3 Identity Management

Identity management involves the management of the identities of people, nodes, services, organizations, and communities, and the provision of facilities to enable them to identify and authenticate one another when communicating. With a fixed or pre-planned mobile networks, there is some knowledge of the identity of the people, nodes and services within the network, which may be disseminated at system set-up. For example, the identity signatures and security properties of all nodes within the network can be distributed in advance to each node. Alternatively, it is possible to establish a centralized identity management service, which is only distributed a set-up time.

With an ad hoc network, it must be assumed that this is not the case, and that there is a need to dynamically:

- Determine the identity, to validate the integrity of the identity and its association with the node, person or service;
- Determine and validate the current security properties of that identity.

Having achieved the above, it is then possible to start to reason about the security risks of connecting to each node, person or service.

To achieve dynamic identity management there are a number of possible approaches:

- Use of pre set-up information;
- Use of centralized identity management services supported by local storage of identity management information for when the central services are not available;

- Use of a federated identity management service (e.g. using P2P technology to share identity management information);
- Use of multi-channel, multi-event identity information exchanges to bootstrap increasing levels of assurance in identities;
- Identity management based upon trust models (e.g. the US Army tells the British Army that a US UAV belongs to the US Army and has specified security properties);
- Identity management based upon security property composition, rather than absolute identity (e.g. a person has the required security properties rather than confirming their identity and then having to determine their security properties).

4 Verification, Validation and Certification (VV&C)

In fixed networks, it is possible for a human to VV&C the security properties of the network at system design, integration and test. In pre-planned mobile networks, it is possible for a human to VV&C the network, often using support tools, at system design, network planning and set-up.

In ad hoc networks, the following security aspects may not be known in advance:

- Nodes involved;
- ad hoc network construction and connectivity between the nodes;
- Security properties of the nodes;
- Security assurance levels associated with the security mechanisms;
- The ownership of the ad hoc nodes;
- The threats and risks that might emerge during the life of the ad hoc network.

Consequently, in order to achieve VV&C of ad hoc networks, we need to consider the security properties of the ad hoc network, which are known prior to deployment. These may include for example the following:

- The security mechanisms and assurance levels in the owned ad hoc network nodes;
- The level of trust one organization may have in nodes owned by another;
- The security rules of behavior, that an organization's ad hoc nodes will follow given a security event. The security event may be triggered by an interaction with an alien node, or an environmental security event has occurred;
- The agreed rules of secure connection and information exchange between coalition partners.

We do not have to assume that the above security mechanisms, assurance levels, behavioral rules and trust levels remain static during the operation of the ad hoc network, but we must have a security assured means to change these parameters and this means must also be capable of being VV&C prior to deployment of the ad hoc network.

Thus, VV&C in ad hoc networks is achieved, not by VV&C of the overall ad hoc network, but by the following:

- VV&C of security properties, and security behavioral rules prior to deployment;
- VV&C of the any proposed changes to the security properties and behaviors of ad hoc nodes during deployment;
- VV&C of a mechanism to be able to distribute changes and to update the security properties and behaviors in ad hoc nodes during deployment.

If we have the above capability, then it is possible to reason about the security behavior of ad hoc networks, yet also be able to change that behavior in response to operational and environmental developments.

5 Conclusions

The problem of ensuring security and interoperability for mobile ad hoc networks that connect to a valuable fixed core infrastructure has been discussed. The problem has been broken down into three interdependent research areas of

- Security versus Risk;
- Identity Management;
- Verification, Validation and Certification.

The research issues for each area have been discussed in detail.

Technology, Society and Risk

Roger Rivett

Functional Safety Technical Specialist
Land Rover (UK)
`rrivett@landrover.com`

Abstract. There remains a healthy debate among those working in the functional safety field over issues that appear to be fundamental to the discipline. Coming from an industry that is a relative newcomer to this discipline I look to the more established industries to give a lead. Not only are they in debate about key issues, the approaches taken do not always transfer easily to a mass market product, developed within very tight business constraints. Key issues that are debated include:

- What is meant by risk, what is acceptable risk and who does the accepting?
- How do we justify that an acceptable risk has been, or will be, achieved?
- What role does the development process play?
- What is meant by the concept of a Safety Integrity Level?

In this talk I will air some views on these questions based on my experience of deve-loping automotive systems and authoring industry sector guidelines and standards in the hope that this will provoke informed discussion.

M.D. Harrison and M.-A. Sujan (Eds.): SAFECOMP 2008, LNCS 5219, p. 12, 2008.
© Springer-Verlag Berlin Heidelberg 2008

Panel: Complexity and Resilience*

Aad van Moorsel**

School of Computing
Newcastle University
aad.vanmoorsel@newcastle.ac.uk

Complexity and Resilience

The complexity of modern-day information systems creates large dependability challenges. As described in the ReSIST (Resilience for Survivability in IST) working programme [1], "current state-of-knowledge and state-of-the-art reasonably enable the construction and operation of critical systems, be they safety-critical (e.g., avionics, railway signalling, nuclear control) or availability-critical (e.g., back-end servers for transaction processing). However, the situation drastically worsens when considering large, networked, evolving, systems either fixed or mobile, with demanding requirements driven by their domain of application, i.e., ubiquitous systems. There is statistical evidence that these emerging systems suffer from a significant drop in dependability and security in comparison with the former systems. There is thus a *dependability and security gap* opening in front of us that, if not filled, will endanger the very basis and advent of information systems."

The ReSIST working programme [1] further points out that "Two main drivers of the creation and widening of the gap are complexity and cost pressure. Complexity growth under cost pressure results from (drastic) changes that can be functional, environmental and technological. Examples of such changes are: a) growth of systems as demand increases, b) merging of systems in company acquisitions or coupling of systems in military coalitions, c) interactions between systems of differing natures (e.g., large-scale information infrastructure on the one hand and networks of sensors on the other), d) dynamically changing systems (e.g., spontaneous, or ad-hoc, networks of mobile nodes and sensors), e) the ever-evolving and growing problem of attacks both by amateur hackers and by professional criminals."

This panel discusses how to accommodate complexity-induced challenges in resilience, considering resilience with respect to accidental failures as well as malicious attacks. It will address among others the following questions:

- what are the dominating research questions that need to be addressed to provide resilience in highly complex information systems?
- what can we learn from other disciplines regarding resilience in complex systems?

---

* This panel is supported by: EU network of excellence 026764 ('ReSIST: Resilience for Survivability in IST').

** Supported in part by: EU network of excellence 026764 ('ReSIST: Resilience for Survivability in IST') and EU coordination action 216295 ('AMBER: Assessing, Measuring, and Benchmarking Resilience').

M.D. Harrison and M.-A. Sujan (Eds.): SAFECOMP 2008, LNCS 5219, pp. 13–15, 2008.

- can emerging applications be Internet-based, or are other network and communication paradigms required to provide resilience?
- can system resilience be understood and researched outside the context of the application and its users?
- what kind of tools and methodologies must be developed or adapted to improve information system resilience?

Panel Participants

Ann Blandford is Professor of Human Computer Interaction and Director of the UCL Interaction Centre, University College London. The main focus of her research is on ways of answering the question "How well does your system fit?", in terms of cognition, interaction and work. Broad themes include human capabilities and human error, interactions with technology and models, methods and theories for evaluating interactive systems.

Erik Hollnagel is Professor and Industrial Safety Chair at École des Mines de Paris, and author of the book "Resilience Engineering: Concepts and Precept". He is an internationally recognised specialist in the fields of resilience engineering, system safety, human reliability analysis, cognitive systems engineering, and intelligent man-machine systems. He is the author of more than 300 publications including thirteen books, articles from recognised journals, conference papers, and reports.

Marcus Kaiser is RCUK Academic Fellow for Complex Neural Systems. He has worked on organization, development, and robustness of cortical and neuronal as well as of metabolic and protein-protein interaction networks. He is initiator and deputy director of the Wellcome Trust 4-year PhD programme "Systems Neuroscience: From Networks to Behaviour" and member of the Institute of Neuroscience management board. In addition, he is member of the EPSRC funded network Mathematical Neuroscience and member of the editorial board of the journal Frontiers in Neuroinformatics.

Jean-Claude Laprie is "Directeur de Recherche" of CNRS, the French National Organization for Scientific Research. His research has focused on dependable computing since 1973, especially on fault tolerance, on dependability evaluation, and on terminology issues. He currently leads the EU FP6 Network of Excellence ReSIST (Resilience for Survivability in Information Society Technologies), integrating leading researchers active in the multidisciplinary domains of dependability, security, and human factors.

Moderator **Aad van Moorsel** is Reader (Associate Professor) in Distributed Systems at Newcastle University. He worked in industry from 1996 until 2003, first as a researcher at Bell Labs/Lucent Technologies in Murray Hill and then as a research manager at Hewlett-Packard Labs in Palo Alto, both in the United States. He has worked in a variety of areas, from performance modelling to systems management, web services and grid computing. In his last position in industry, he was responsible for HP's research in web and grid services, and worked on the software strategy of the company.

Reference

1. Laprie, J.-C., et al.: working programme for ReSIST: Resilience for Survivability in IST, network of excellence, contract number 026764,
   http://www.resist-noe.org/overview/summary.html

The Effectiveness of T-Way Test Data Generation

Michael Ellims[1], Darrel Ince[2], and Marian Petre[2]

[1] Pi-Shurlok, Milton Hall, Cambridge, UK
[2] Dept. of Computing, Open University
Walton Hall, Milton Keynes, UK
mike.ellims@pi-shurlok.com, {d.c.ince,m.petre}@open.ac.uk

Abstract. This paper reports the results of a study comparing the effectiveness of automatically generated tests constructed using random and *t*-way combinatorial techniques on safety related industrial code using mutation adequacy criteria. A reference point is provided by hand generated test vectors constructed during development to establish minimum acceptance criteria. The study shows that 2-way testing is not adequate measured by mutants kill rate compared with hand generated test set of similar size, but that higher factor *t*-way test sets can perform at least as well. To reduce the computation overhead of testing large numbers of vectors over large numbers of mutants a staged optimising approach to applying *t*-way tests is proposed and evaluated which shows improvements in execution time and final test set size.

Keywords: Software testing, random testing, automated test generation, unit test, combinatorial design, pairwise testing, t-way testing, mutation.

1 Introduction

How to generate test sets automatically has been the subject of much research and a wide range of techniques have been proposed ad investigated. These including random generation [1], search techniques such as generic algorithms [2] and combinatorial techniques [3] based on statistical design of experiments [4] used to identify and isolate the effects of interactions between factors of interest.

For unit testing the factors of interest are the input variables of the function under test and the interactions between different values of those variables and how they effect the outcome of running the code. If we generate vectors that cover all 2-way (pairwise) interactions between n input variables v_1 to v_n then there will be a vector in the test set such that for every value that the variable v_i is allowed to take it will be paired with each value the variable v_j is allowed to take for all i and j where $i \neq j$.

An important consideration is which values each variable will be allowed to take on. In general the tester will select data points for each input variable that are of "interest" based on criteria such as data input ranges, domain partitioning and other heuristic rules. Selection all values is impossible except where only a small number of values are allowed such as for enumerations.

To make this more concrete consider a function with three input variables, v_1, v_2 and v_3 that take on the values a_1, a_2, a_3 and b_1, b_2 and c_1, c_2 respectively. Then a 2-way adequate test set that ensures that a vector exits that contains all values of v_1 paired with all values of v_2 and all values of v_3 and all pairs of v_2 and v_3 is shown in Figure 1.

M.D. Harrison and M.-A. Sujan (Eds.): SAFECOMP 2008, LNCS 5219, pp.16–29, 2008.
© Springer-Verlag Berlin Heidelberg 2008

$$a_1 \ a_2 \ a_3 \ a_2 \ a_1 \ a_3 \ a_1$$
$$b_2 \ b_1 \ b_1 \ b_2 \ b_2 \ b_2 \ b_1$$
$$c_1 \ c_2 \ c_1 \ c_1 \ c_2 \ c_2 \ c_1$$

Fig. 1. An example seven vector, 2-way adequate test set for 3 variables

Larger values of t can be used, for example $t = 3$ would involve matching sets of three variables and $t = 4$, four variables in the same way. The advantage of taking this approach is that far fewer vectors are required to construct a t-way adequate test set than would be required for a test set that contained all combination of values. The work presented in this paper is an investigation of the utility of t-way test generation for unit testing and in particular to determine its suitability for safety-related software.

1.1 Contributions of This Work

The work presented here makes the following contributions;

- it provides a direct comparison with t-way adequate test sets against human generated test sets.
- it provides a practical method of incorporating high factor t-way testing and mutation analysis into a development process which can avoid much of the computational overhead that may otherwise be encountered.

2 Related Work

2.1 Combinatorial Techniques

The original work on using combinatorial techniques for testing was presented by Mandl [5] who used orthogonal arrays to select sets of constructs for testing an Ada compiler. Sherwood [6] developed the Constraint Array Test System (CATS) to generate test sets algorithmically. This work was extended by Cohen *et al.* [3] as the automatic efficient test generator (AETG) system and this algorithm has been the focus of much later work.

The literature on combinatorial testing can be divided into two major classes, first algorithms for generating t-way adequate test sets and second, work that evaluates the technique. The latter in turn falls into two main categories: reports of the tools in field use, and a small body of experimental work.

The studies from field use have examined real systems and on the whole report on the detection of additional errors using combinational techniques. Brownlie *et al.* [7] applied the technique to testing of an email system. The effectiveness claimed for the technique is related to the saving in the number of test cases required and not on a direct comparison of faults found by applying this and any other technique. Cohen *et al.* [8] present information on the AETG tool on two releases of software where it found more faults than standard test techniques, however what the standard techniques are is not stated. Dalal *et al.* [9], [10] report briefly on the use of the AETG tool on a number of systems and the fact that more faults were discovered with its use than without. Smith et al. [11] discusses the use of the technique (2-way) on spaceflight software and compare the number of faults found using the 2-way test sets vs. test sets constructed by other means. Here 2-way adequate test sets did not fair as well as expected.

One further set of field studies is of special interest, these looked at variable inter-actions leading to the activation of faults in several systems. Wallace and Kuhn [12] looked at software failure modes in data collect by the Federal Drug Administration (FDA) involving the recall of medical equipment. Kuhn and Reilly [13] examined the Mozilla and Apache open source projects using their bug tracking databases to deter-mine the number of conditions required to trigger the fault. Finally Kuhn et al. [14] looked at a large distributed system being developed at NASA. This work suggests that in practice small t factor of between four and six was required to reveal all faults reported.

Given the period of time in which combinatorial testing using covering arrays has been in use there are surprisingly few controlled experimental studies. There are five major studies: [15] which addressed coverage, [16], [17], [18]and [19] which ad-dressed effectiveness at detecting seeded faults.

Dunietz et al. [15] compared the code coverage of random designs without re-placement vs. the coverage obtained from systematic designs (i.e. t-way adequate test sets) with the same number of vectors. They concluded that for block coverage low factor t-way designs could be effective.

Nair et al. [16] investigated random testing without replacement and no partition-ing vs. partition based testing, and showed that, in general partition testing should be more effective. The particular case of partition testing was an application of experi-mental design (t-way) and it showed that the probability of detecting the failure for simple random testing is significantly lower than partition based techniques.

Kobayashi *et al.* [17] examined the fault detecting ability of specification based, random, anti-random [20] and t-way techniques applied to the testing logic predicates against mutations of those predicates. The authors concluded that 4-way tests were nearly as effective as specification techniques and better than both random and anti-random.

Grindal et al. [19] examined the fault detecting power of a number of different combinatorial strategies including 1-way (each choice), base choice (a single factor experiment), pairwise AETG and orthogonal arrays. Work was performed on code with hand seeded faults and data reported for branch coverage is consistent with other experimental results. However after examining the data the authors concluded that code coverage methods may also need to be employed. As in [11] it was found that the base choice technique performed as well as orthogonal arrays and 2-way in 3 out of 5 problems. However no technique detected fewer than 90% of the detectable faults.

Schroeder et al. [18] examined effectiveness in terms of code coverage for t-way vs. random selection with replacement on code with hand seeded faults. While this produced results that broadly support the results from other experimental work, it was found that higher values of t were required to reveal some faults. They also concluded that t-way test sets were no more effective that test sets constructed random selection for sets of the same size.

Our conclusion is that the literature indicates that there is no overwhelming con-sensus as to the utility of combinatorial techniques.

2.2 Code Mutation

Much of the empirical work that evaluates the effectiveness of the *t*-way testing technique is constrained by two main limitations. First the reliance on hand seeded faults. Second by the inability of common metrics such as code coverage to distinguish between test sets that reach code but do not stress the code sufficiently to reveal errors and test sets that do.

Code mutation as proposed by Hamlet [21] and DeMillo *et al.* [22] has been used previously in studies to compare test effectiveness [23], [24], [25]. It also has the advantage that it subsumes conditional coverage techniques [26].

Mutation has recently been applied to evaluating random testing with C programs [27] with the aim of determining whether faults inserted using the mutation are representative of real faults. The conclusion is that they are but that they are also possibly more difficult to detect.

3 The Experimental Study

3.1 The Data Set

The functions that were used in this study were drawn from a system that controls a large industrial engine currently employed in safety-critical applications (Wallace). The system was developed in a manner consistent with IEC 61508 [28] and code has been subjected to review, unit, integration and system testing.

Hand generated unit test sets were developed using standard techniques such as boundary value analysis and equivalency partitioning. They also took into account the structure of conditional statements and attempt to ensure that all clauses are tested for both TRUE and FALSE. All sets of test vectors are statement, branch coverage adequate and most are also LCSAJ adequate ([29]). Therefore we have some confidence that the hand generated set of test vectors are of high quality. A full description of how the unit test process is given in [30].

3.2 Procedure Employed

The procedure employed in this experiment consisted of the following steps:

- A simple mutation tool was developed that produced operator, variable name, constant and statement removal mutations [31].
- A set of functions from Wallace was selected with a range of complexities from 12 to 62 executable statements (i.e. excluding comments, blank lines and braces).
- The hand-generated vectors for each function was extracted along with input domain information from the detailed designs and data dictionary.
- Each mutant was run on each vector for the automatically and hand generated test sets. For each complete set of mutants vs. test set executed the number mutants left alive was recorded.

In previous work [31] we employed our AETG based tool, however the tool is inherently inefficient as it performs a liner search to match *t*-way tuples generated in candidate vectors with tuples remaining to be covered. Lei et al. [32] reference the jenney tool [33] and compared the performance of their tool FireEye against other available tool sets. In terms of execution time jenny is far more efficient that our own tool and replaces it in this work.

3.3 Code Selected

Table 1 summarizes the functions examined. Three were selected on the criteria that they contained known errors discovered running the unit tests (vs. designing tests). The remainder were selected based on their complexity, e.g. _gov_gen_ffd_rpm was selected as it contains a large number of conditional statements (eleven).

Table 1. Summary of properties of code used in this study

Function Name	Lines	Valid Mutants	Nesting Factor	Cond'n Factor	if's	Inputs
_dip_debounce	12	81	2	2	2	17
_aip_median_filter	25	217	1	1	4	3
_sdc_fuel_control	17	213	2	2	5	9
aip_spike_filter	22	178	3	1	4	7
_thc_decide_state	16	386	7	2	7	9
_thc_autocal	33	669	5	2	8	6
_aip_apply_filters	30	311	2	2	4	8
_gov_rpm_err	22	783	2	1	5	9
_sdc_pre_start	51	1297	3	1	8	3
_gov_gen_ffd_rpm	62	1227	4	2	11	16

Properties for each of the functions are shown in Table 2 as follows, the first column is the function name and the second is the number of executable statements in the function. Column three gives the number of valid mutants that would actually compile (ignoring warnings for divide by zero etc). The fourth and fifth columns are the nesting factor (maximum depth of nesting) and the condition factor (maximum number of comparisons in a predicate) as used in [34]. The sixth column is a count of the number of if statements in the code with each case of a switch statement being counted as one. The final column is the number of inputs to the function. The function _dip_debounce stands out here, but this is because the underlying data structure is a set of arrays and the original test set contained data values for the first, middle and last elements of those arrays.

3.4 Experiment 1

3.4.1 Aims

The aims of this experiment are two fold. First to evaluate the effectiveness of t-way adequate test sets relative to a set of high quality human generated tests. Second to compare them with other automatic generation techniques that require a comparable level of analysis to allow data to be generated.

3.4.2 Procedure
The Procedure Employed in this Experiment Consisted of the Following Steps for Each Function:

- Generate a t-way adequate test set sets for $t = 2$ to $t = 5$. For numeric variables the minimum, median and maximum values in the range were used. For enumeration variables we used all valid values and one out of range value to exercise the default statement in the code. For Boolean variables TRUE and FALSE were used.
- Generate a test set of the same size as the t-way test sets using random selection from the same set of values with replacement using the same values as for t-way.
- Generate a test set of the same size purely random tests . Numeric values were drawn from the whole range with equal probability and replacement. Enumerations and Boolean values were selected as above. The generator described in [35] was used to ensure long sequences.
- For each function one or more sets of "base choice" [36] test vectors were generated. Base choice is where a base vector is selected, perhaps based on expected or normal use. Additional vectors are generated from this base by changing a single value of one variable in each new vector until all values have been used for all variables.
- For each function, execute each of the valid mutants on each test vector and for each test set recorded the number of mutants that were killed.

3.4.3 Results
Are shown in Table 2. The first column, gives the function name and the second states the information given in the next four rows as follows. For each function the first row (vectors) is the number of test vectors in the set determined by the size of t-way test vectors. The second row (t-way) is the number of mutants killed by t-way vectors for $t = 2$ to 5.

The final two columns (base, hand) gives the number of vectors in the base choice and hand generated test sets with the number of mutants left alive below it. For each function the smallest test set that had the best performance is highlighted in bold.

Table 3 gives indicative information on the amount of time in seconds that it takes to run each set of t-way adequate test sets data for each function.

The primary concern is which of the techniques is best at killing mutants in the selected functions. One approach is to look at which technique kills the most mutants for each function. The results are summaries as follows;

- t-way test vectors win or draw in six of the ten cases.
- Test vectors generated via random selection win or draw in half the cases.
- Random data generation wins or draws in four of the ten cases but notably only has a single win in the second half of the table.

The selection of "a winner" here is arbitrary in that it is the test set that killed the most mutants regardless of the number of vectors required and for some code only small numbers of vectors are required. Another way to approach is to examine the number of cases where a method failed to achieve a result comparable with the hand generated tests. Here there is one failure for t-way and random selection plus a near miss (_sdc_fuel_control by one) and four failures for random testing.

Table 2. Number of mutants killed for each of the sets of test vectors applied

Function Name	Process	2-way	3-way	4-way	5-way	Base	Hand
_dip_debounce	vectors	19	60	205	634	25	18
	t-way	**9**	9	9	9	28	12
	rand sel	14	9	9	9		
	random	11	10	10	9		
_aip_median_filter	vectors	12	28	54		7	27
	t-way	49	40	40		56	41
	rand sel	46	43	40			
	random	**40**	40	40			
_sdc_fuel_control	vectors	17	57	174	504	17	15
	t-way	101	49	25	22	36	21
	rand sel	126	31	24	22		
	random	84	58	25	**18**		
aip_spike_filter	vectors	16	49	146	400	14	40
	t-way	42	23	23	23	80	18
	rand sel	66	37	32	23		
	random	82	82	66	**16**		
_thc_decide_state	vectors	73	271	972	2883	28	17
	t-way	228	206	100	**57**	313	60
	rand sel	182	146	63	**57**		
	random	348	346	307	232		
_thc_autocal	vectors	20	70	181	377	14	6
	t-way	333	188	**187**	187	270	197
	rand sel	407	299	264	189		
	random	410	335	299	221		
_aip_apply_filters	vectors	34	142	562	1949	23	68
	t-way	47	46	46	46	64	64
	rand sel	**46**	46	46	46		
	random	**46**	46	46	46		
_gov_rpm_err	vectors	17	62	208	662	17	17
	t-way	**443**	443	443	443	444	446
	rand sel	**443**	443	443	443		
	random	465	462	462	460		
_sdc_pre_start	vectors	22	79	228	573	13	14
	t-way	736	**673**	673	673	965	675
	rand sel	700	**673**	673	673		
	random	742	742	742	742		
_gov_gen_ffd_rpm	vectors	21	81	299	1040	29	14
	t-way	701	190	158	**140**	785	152
	rand sel	663	270	148	**140**		
	random	502	265	152	152		

We can also calculate the mean number of vectors required to kill each mutant. Here the number of vectors required achieve the best result is used and we find that *t*-way requires 2.62 vectors per mutant, random selection 2.71 and random 3.70.

In no case was base choice the best performing technique and in only two cases was its performance comparable with the hand generated tests. These results were surprising given that previous work as [11], [19] found the technique to perform rather better.

Table 3. Execution times for the *t*-way adiquate test sets

Function Name	Valid Mutants	2-way	3-way	4-way	5-way	Max (hours)
_dip_debounce	81	76	210	743	1649	0.46
_aip_median_filter	217	64	127	248		0.07
_sdc_fuel_control	213	132	362	808	3667	1.02
aip_spike_filter	178	109	433	858	1665	0.46
_thc_decide_state	311	707	2723	8156	43451	12.07
_thc_autocal	386	139	582	2313	4253	1.18
_aip_apply_filters	669	198	420	675	2788	0.77
_gov_rpm_err	783	212	851	3239	8563	2.34
_sdc_pre_start	1237	906	1506	5083	16,231	4.51
_gov_gen_ffd_rpm	1227	972	2612	17,758	33,653	9.35

3.5 Experiment 2

3.5.1 Aims

There are two obvious issues with the data presented above. First that the execution times are long for some functions compared with the time it takes to generate the tests by hand. Timesheet data gives an average of 5.6 hours for AIP functions, 5.7 hours for DIP and 1.9 hours for SDC function. Second, the number of vectors that would have to be examined to determine if a test passed or failed is infeasibly large. In practice a large part of the problem with generating tests by hand is determining whether the output is correct. Given the volume of tests generated automatically, determining whether the code passes or fails places an unacceptable burden on the tester and significantly reduces the utility of any automatic generation technique.

Therefore this experiment has two aims. First to investigate the potential of reducing the amount of time required to exercise all the mutants. Second to determine if a minimal test set can be extracted from the process to reduce the oracle problem to a manageable level.

3.5.2 Procedure

For this experiment we modified the test driver to record which vectors killed which mutants for each set of test vectors. After all vectors had been run over all mutants the optimisation routine determines which vector killed the most mutants and it is selected to be retained, mutants it killed are removed from further consideration. This is repeated until there are no new vectors that kill more than one mutant left.

The run with the next set of vectors excludes from consideration those mutants that were previously killed by all preceding test sets but otherwise the optimisation process is identical. This continues until the final set of vectors is run when the restriction on not selecting vectors that only kill a single vector is removed.

Other procedures have been made to reduce the number of vectors that need to be considered. A suggestion by Offutt [37] was to simply ignore vectors that do not kill any mutants. However these experiments suggest that savings may not be great as large number of vectors kill at least one mutant which is why we delay selecting these until the final pass. Offutt et al. [38] suggest mechanism for selecting minimal sets of vectors that again removes mutants as they are killed but runs the set of vectors in different orders.

3.5.3 Results

From experiment 2 are shown in Table4 which for each function reports the time to run the largest t-way test set (max), the time using the optimisation procedure outlined above (min) and the percentage time saving for the optimisation (gain). Information on vectors given is the number of hand generated vectors (hand), the size largest single t-way adequate test set (max) and the size of the optimised test set (min). For reference the t value of the test set that first resulted in the maximum number of mutants killed is shown in the second column headed t.

Table 4 shows that in terms of time saved the optimisation procedure can deliver significant saving for possibly the majority of functions, with an average saving of close to 53%. However it is also clear that for functions that show no increase in mutants killed at higher values of t the process can be counter productive e.g. _dip_debounce but that it is not always the case e.g. _aip_apply_filters. The benefits where high t values do show improvement are more supportive of the idea that the optimisation scheme trialled here is worth while.

Table 4. Summary data for t-way optimisation runs

Function Name	t	Time (seconds)			Vectors		
		max	min	gain	hand	max	min
_dip_debounce	2	1649	2029	123 %	18	634	6
_aip_median_filter	3	248	67	27 %	27	54	9
_sdc_fuel_control	5	3667	1144	31 %	15	504	12
aip_spike_filter	2	1665	628	37 %	40	400	9
_thc_decide_state	5	43451	6942	16 %	17	2883	13
_thc_autocal	4	4253	1276	30 %	6	377	13
_aip_apply_filters	2	2788	2029	73 %	68	1949	7
_gov_rpm_err	2	8563	6118	71 %	17	662	4
_sdc_pre_start	2	16231	18212	112 %	14	573	12
_gov_gen_ffd_rpm	5	33653	5767	17 %	14	1040	22

Results for the size of the test sets from the optimisation routine are less ambiguous, in eight of the ten cases the test sets are smaller than the hand generated test sets. In the remaining two cases they are not significantly larger in terms of total tests required.

There is however one down side, as reported in [11] vectors that were selected by the optimisation procedure were not very user friendly. That is, it takes a significant effort to understand what is being tested. Here none of the test cases contained tests that would be obvious to an engineer producing the test cases by hand (the first author was the engineer in charge of Wallace) and many, especially those for the function _aip_apply_filters contained data that in practice would not be used and would be disallowed by the tool that vets the engine control unit calibration data.

3.6 Investigations

There are a small number of interesting features present in Table 2 as follows;

- why is it so difficult to obtain a good kill rate for the _sdc_pre_start function?
- is the fault detecting ability of random testing really static for _sdc_pre_start?
- can we improve on the results for _gov_gen_ffd_rpm if we use more random tests?

Examination of live mutants _sdc_pre_start code revels the fact that the majority of live mutants are connected with manipulating variables that have Boolean values. As has been noted in other work [31] and in a large amount of research on searched based test data generation [34], [39]. Boolean data appears to be intrinsically difficult to deal with.

The _sdc_pre_start code was executed with a number of different randomly generated test sets using different seed for 288, 573, 1200 and 2400 values. While some of the vector sets showed some improvement the best result returned was only 717 killed mutants and all data sets showed the same flat pattern as shown in Table 2.

Code for _gov_gen_ffd_rpm was run with a test set of 2000 and 5000 vectors taking 12 and 32.4 hours to execute. The test set of size 2000 showed no improvement while the test set of 5000 vectors killed only an additional 2 mutants.

4 Threats to Validity

Threats to external validity are that code being tested may not be representative of other code though a variant of aip_median_filter has been used by other researchers [40, 41] and the function itself in [42]. This however is a general problem in testing research and code from different application domains is likely to have different. The code used here is thought to be representative of fixed point integer code real-time embedded applications domain.

A novel threat is that as the code development process was strongly controlled that code actually may be easier to detect faults in than more typical code. The implication is that the results presented here are possibly optimistic. The only approach is to use other data sets, however often these do not have the necessary hand generated test vectors available. Another threat is that code mutation may not be representative of real faults. Results in [27] strongly suggest that test sets that are adequate for mutation will also be effective for real faults.

The major threat to internal validity comes from the way that the data points were used in the *t*-way and random selection data sets, being limited to minimum, median and maximum values. This is simplistic however it should tend to bias the results against success, resulting in a false negative. However the data selection process does follows examples in books such as [43] which will possibly provide the primary source of information on combinatorial techniques for practitioners.

The tool used to insert faults into the code may also presents a risk, while it avoids the bias associated with hand seeded faults it is a relatively simple tool and is not capable of introducing mutants over multiple lines. Analysis by one of the authors [44] however suggests that the majority of effective operators have been implemented.

5   Conclusions

The results of these experiments have been surprising. At the start of this study we all thought that 2-way techniques offered a valid way of testing critical software. However our results show that:

- 2-way (pairwise) combinatorial techniques using simple selection criteria for selection data points are not adequate with respect to hand generated tests for the more complex functions as measured by mutants generated, nesting level and number of condition statements.
- Test sets that involve higher values of t-way adequate tests appear to be as effective as hand generated tests at killing mutants. However this statement holds only relative to being able to distinguish mutated from original code. We have not assessed the relationship with "real" faults. However as noted above results from [27] suggest that a test set for one will be effective on the other.
- Random testing can be surprisingly effective but is not reliable in the sense that it may often provide good results, but this cannot be counted on.

6   Future Work

There are some obvious avenues of work that the authors are either currently pursuing or intend to pursue in the near term. Firstly some initial work has been done using a small number of hand generated vectors as the first step in the optimisation process. This has been done by drawing small random samples from the existing hand generated tests. Initial results suggest that while the number of mutants killed is only minimally affected there may be further savings to be made in execution time.

The Wallace code base contains functions with higher level of complexity than those involved in this study. However these have as inputs large arrays of one or more dimensions and it is not clear how to effectively deal with these data structures. Does one treat them as a collection of individual variables or as a complete unit?

As noted above the data selection model used is possibly too simplistic. Previous work [45] shows that there can be an advantage in using more complete data models. This work should be repeated with higher t-way test sets.

The unit tests for the Boar project reported in [30] have been extracted from the project archive and these may provide an interesting comparison. The unit testing for this project was outsourced and it is known that there is a significant difference between Wallace and Boar in what activity in the unit test process (test design vs. test run) errors were revealed.

One area of interest is the effect that the mutant comparison function has on the ability to detect faults. The current comparison functions are derived directly from the hand generated tests and compares not only the output values but in most cases the majority of other input data to check for invalid modification. It would be interesting to determine what effect changing these functions has on the ability of vectors to kill mutants.

Acknowledgments

Our thanks to J. H. Andrews for making his mutation tool available for evaluation.

References

1. Duran, J., Ntafos, S.: An Evaluation of Random Testing. IEEE Trans. Softw. Eng. 10(4), 438–444 (1984)
2. Gallagher, M.J., Narasimhan, V.L.: ADTEST: A Test Data Generation Suite for Ada Software Systems. IEEE Trans. Softw. Eng. 23(8), 473–484 (1997)
3. Cohen, D.M., et al.: The AETG System: An Approach to Testing Based on Combinatorial Design. IEEE Trans. Softw. Eng. 23(7), 437–444 (1997)
4. Diamond, W.J.: Practical Experiment Design For Engineers and Scientists. John Wiley & Sons, New York (2001)
5. Mandl, R.: Orthogonal Latin Squares: an Application of Experiment Design to Compiler Testing. Commun. ACM 28(10), 1054–1058 (1985)
6. Sherwood, G.: Effective Testing of Factor Combinations. In: Third Int'l Conf. Software Testing, Analysis and Review, Software Quality Eng. pp. 151–166 (1994)
7. Brownlie, R., Prowse, J., Phadke, M.S.: Robust Testing of AT&T PMX/StarMAIL Using Oats. AT&T Technical Journal 71(3), 41–47 (1992)
8. Cohen, D.M., et al.: The Automatic Efficient Test Generator (AETG) System. In: Proceedings 5th International Symposium on Software Reliability Engineering, pp. 303–309. IEEE Computer Society, Los Alamitos (1994)
9. Dalal, S., et al.: Model-based Testing of a Highly Programmable System. In: Proc. of the Ninth International Symposium on Software Reliability Engineering. IEEE Computer Society, Los Alamitos (1998)
10. Dalal, S.R., et al.: Model-based Testing in Practice. In: Proc. of the 21st Int'l Conf. on Software Engineering, pp. 285–294. IEEE Computer Society, Los Alamitos (1999)
11. Smith, B., Feather, M.S., Muscettola, N.: Challenges and Methods in Testing the Remote Agent Planner. In: Proceedings of the Fifth International Conference on Artificial Intelligence Planning Systems, pp. 254–263. AAAI Press, Menlo Park (2000)
12. Wallace, D.R., Kuhn, D.R.: Failure Modes in medical device software: an analysis of 15 years of recall data. International Journal of Reliability, Quality and Safety Engineering 8(4), 351–371 (2001)
13. Kuhn, D.R., Reilly, M.J.: An Investigation of the Applicability of Design of Experiments to Software Testing. In: Proceedings of the 27th Annual NASA Goddard Software Engineering Workshop (SEW-27 2002). IEEE Computer Society, Los Alamitos (2002)
14. Kuhn, D.R., Wallace, D.R., Gallo, A.M.: Software Fault Interactions and Implications for Software Testing. IEEE Trans. Softw. Eng. 30(6), 418–421 (2004)
15. Dunietz, I.S., et al.: Applying Design of Experiments to Software Testing: Experience Report. In: Proc.of the 19th Int'l Conf. on Software Eng., pp. 205–215. ACM Press, New York (1997)
16. Nair, V.N., et al.: A Statistical Assessment of some Software Testing Strategies and Application of Experimental Design Techniques. Statistica Sinica 8, 165–184 (1998)
17. Kobayashi, N., Tsuchiya, T., Kikuno, T.: Non-Specification-Based Approaches to Logic Testing for Software. Information and Software Technology 44(2), 113–121 (2002)

18. Schroeder, P.J., Bolaki, P., Gopu, V.: Comparing the Fault Detection Effectiveness of N-way and Random Test Suites. In: ISESE 2004: Proceedings of the 2004 International Symposium on Empirical Software Engineering, pp. 49–59. IEEE Computer Society, Los Alamitos (2004)

19. Grindal, M., et al.: An Evaluation of Combination Strategies for Test Case Selection, in Technical Report, Department of Computer Science, University of Skövde (2003)

20. Malaiya, Y.K.: Antirandom testing: getting the most out of black-box testing. In: Proceedings, Sixth International Symposium on Software Reliability Engineering, pp. 86–95 (1995)

21. Hamlet, R.G.: Testing Programs with the Aid of a Compiler. IEEE Trans. Softw. Eng. 3(4), 279–290 (1977)

22. DeMillo, R.A., Lipton, R.J., Sayward, F.G.: Hints on Test Data Selection: Help for the Practising Programmer. Computer, 34–41 (1978)

23. Daran, M., Thevenod-Fosse, P.: Software Error Analysis: a Real Case Study Involving Real Faults and Mutations. SIGSOFT Softw. Eng. Notes 21(3), 158–171 (1996)

24. Frankl, P.G., Weiss, S.N., and Hu, C.: All-uses vs. mutation testing: an experimental comparison of effectiveness. J. Syst. Softw. 38(3), 235–253 (1997)

25. Zhan, Y., Clark, J.A.: Search-Based Mutation Testing for Simulink Models. In: Proc. of the 2005 Conference on Genetic and Evolutionary Computation, pp. 1061–1068. ACM Press, New York (2005)

26. Offutt, A.J., Voas, J.M.: Subsumption of Condition Coverage Techniques by Mutation Testing, in Tech. Report, Dept. of Information and Software Systems Engineering, George Mason Univ., Fairfax, Va (1996)

27. Andrews, J.H., Briand, L.C., Labiche, Y.: Is Mutation an Appropriate Tool for Test Experiments? In: Proc. of the 27th Int'l Conf. on Software Engineering, pp. 402–411. ACM Press, New York (2005)

28. Anon.: Functional Safety of Electrical/Electronic/Programmable electronic safety-related systems, Part 1: General Requirements, BS EN 61508-1:2002, British Standards (2002)

29. Woodward, M.R., Hedley, D., Hennel, M.A.: Experience with Path Analysis and Testing of Programs. IEEE Trans. Softw. Eng. 6(6), 228–278 (1980)

30. Ellims, M., Bridges, J., Ince, D.C.: The Economics of Unit Testing. Empirical Softw. Eng. 11(1), 5–31 (2006)

31. Ellims, M., Ince, D., Petre, M.: The Csaw C Mutation Tool: Initial Results. In: Mutation 2007. IEEE Computer Society, Los Alamitos (2007)

32. Lei, Y., et al.: IPOG: A General Strategy for T-Way Software Testing. In: 14th Annual IEEE Int'l Conf. and Workshops on the Engineering of Computer-Based Systems (ECBS 2007), pp. 549–556. IEEE Computer Society, Los Alamitos (2007)

33. Jenny (accessed June 2007), http://www.burtleburtle.net/bob/math

34. Michael, C.C., McGraw, G., Schatz, M.A.: Generating Software Test Data by Evolution. IEEE Trans. Softw. Eng. 27(12), 1085–1110 (2001)

35. Wichmann, B.A., Hill, I.D.: Generating Good Pseudo-Random Numbers. Computational Statistics & Data Analysis 51(3), 1614–1622 (2006)

36. Ammann, P.E., Offutt, J.: Using Formal Methods to Derive Test Frames in Category-Partition Testing. In: Proc. of 9th Annual Conf. on Computer Assurance (COMPASS 1994), pp. 824–830. IEEE Computer Society, Los Alamitos (1994)

37. Offutt, A.J.: A Practical System for Mutation Testing: Help for the Common Programmer. In: Proc. of the IEEE Int'l Test Conference on TEST: The Next 25 Years, pp. 824–830. IEEE Computer Society, Los Alamitos (1994)

38. Offutt, J.A., Pan, J., Voas, J.M.: Procedures for Reducing the Size of Coverage Based Test Sets. In: Twelfth Int. Conf. on Testing Computer Software, pp. 111–123 (1995)
39. Bottaci, L.: Instrumenting Programs with Flag Variables for Test Data Search by Genetic Algorithms. In: Proc. of the Genetic and Evolutionary Computation Conference. Morgan Kaufmann Publishers, San Francisco (2002)
40. Gotlieb, A.: Exploiting Symmetries to Test Programs. In: Proceedings of the 14th International Symposium on Software Reliability Engineering, p. 365. IEEE Computer Society, Los Alamitos (2003)
41. Offutt, A.J., et al.: An Experimental Determination of Sufficient Mutant Operators. ACM Trans. Softw. Eng. Methodol. 5(2), 99–118 (1996)
42. Dillon, E., Meudec, C.: Automatic Test Data Generation from Embedded C Code. In: Heisel, M., Liggesmeyer, P., Wittmann, S. (eds.) SAFECOMP 2004. LNCS, vol. 3219, pp. 180–194. Springer, Heidelberg (2004)
43. Copeland, L.: A Practitioner's Guide to Software Test Design. Artech House Publishers, Boston (2004)
44. Ellims, M.: The Csaw Mutation Tool Users Guide, in Technical Report, Department of Computer Science, Open University (2007)
45. Ellims, M., Ince, D., Petre, M.: AETG vs. Man: an Assessment of the Effectiveness of Combinatorial Test Data Generation, in Technical Report, Department of Computer Science, Open University (2007)

Towards Agile Engineering of High-Integrity Systems

Richard F. Paige[1], Ramon Charalambous[1], Xiaocheng Ge[1], and Phillip J. Brooke[2]

[1] Department of Computer Science, University of York, Heslington, York, YO10 5DD
{paige,xchge}@cs.york.ac.uk, ramon.charalambous@gmail.com
[2] School of Computing, University of Teesside, Middlesbrough, UK
pjb@scm.tees.ac.uk

Abstract. We describe the results of a pilot study on the application of an agile process to building a high-integrity software system. The challenges in applying an agile process in this domain are outlined, and potential solutions for dealing with issues of communication, scalability, and system complexity are proposed. We report on the safety process, argumentation generated to support the process, and the technology and tools used to strengthen the agile process in terms of support for verification and validation.

1 Introduction

Critical software systems development is typically approached through use of rigorous processes, through careful consideration of risks and mitigation, and via substantial planning. The so-called *plan-driven processes* that are invariably preferred for high-integrity software (HIS) development have evolved to support rigorous development of products. However, the emergence of new technologies, requirements volatility, the desire to achieve incremental certification, and undesired documentation costs cause difficulties for many organisations in applying plan-driven processes.

Agile processes (APs) are iterative and incremental, and aim to cope with volatile requirements while improving the flexibility of the development process, through a number of concrete technical practices. APs, such as Extreme Programming (XP) [3], pose both opportunities and challenges in the domain of HIS engineering: the flexibility and volatility problems tackled by APs are exactly those experienced by HIS development; and the mechanisms through which APs achieve success are difficult to combine with verification, validation, and certification requirements for HISs.

For a HIS to be deployed, it needs to be certified as acceptably safe. Certification is normally process-based and requires substantial evidence, often in the form of rigorously presented documentation. But APs can deprecate the production of documentation in favour of greater focus on the production of working code. In this paper we report on an experiment to assess the applicability of an AP to the development of high quality software with safety requirements. We also aim to identify areas in which APs require modification in order to become fully compatible with HIS development. The feasibility of applying APs to HIS will be assessed with an attempt to develop a simple avionics application and an accompanying simulator using XP.

M.D. Harrison and M.-A. Sujan (Eds.): SAFECOMP 2008, LNCS 5219, pp.30–43, 2008.

2 Background and Related Work

2.1 Plan-Driven Processes

High-integrity systems – whether safety critical or safety related – are traditionally viewed as best developed via *plan-driven* processes. Plan-driven processes are based on the idea of software undergoing various transformations from specification to code, based on decisions made early in the project's lifecycle. Plans involve estimating costs and resource demands, identifying risks associated with a project and finding ways to eliminate or minimize them, and tracking progress throughout the project's lifecycle. Thorough documentation allows developers to gather information for assessing the quality and effectiveness of their work. Plan-driven methods' main strength is that they provide predictability and repeatability of results [6]. Expert consensus is that the most effective way of facilitating communication between the various engineering fields involved in HIS development is through inherent *documentation* [4,9,17,18]. Documentation also serves as a way of capturing evidence of compliance to standards. Documentation is seen as necessary and unavoidable in HIS development, although the degree to which it is conducted varies according to project.

A potential source of problems in plan-driven processes is reliance on contracts as procurement mechanisms. Contracts often result in delays in schedules and can prove damaging to supplier-customer relations due to ambiguity of terms, which leads to mistrust. [6] identify contracts as the main conduit for interaction between developers and customers, and as a potential stress point in customer-developer relations.

2.2 Agile Processes

The basic concept of all APs is the *Observe-Orient-Decide-Act (OODA)* loop [20] (Fig. 1). Humans make decisions in a cyclic fashion, based on information they collect through *observation* of their environment. That information is used to *orient* one's perspective on the situation, make *decisions* regarding the *actions* one will take, and carry out those actions. If one can reduce the time it takes from observation to action, one can gain advantage over the opposition.

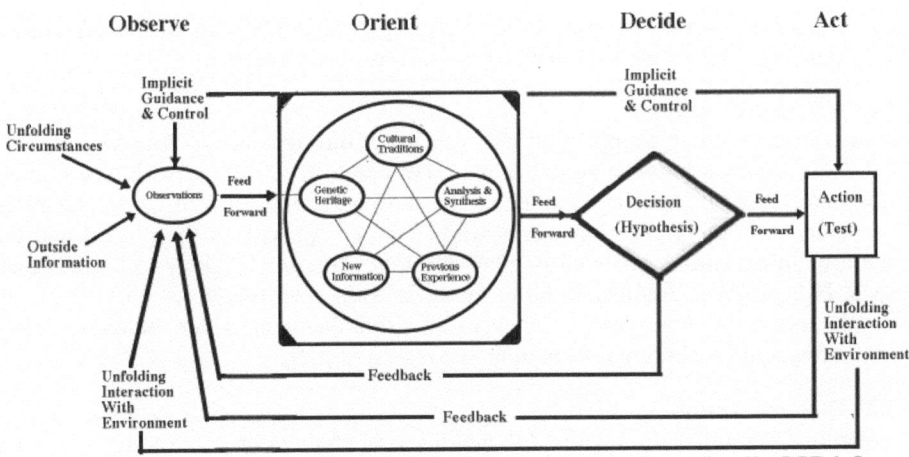

Fig. 1. The OODA Loop [20]

The OODA loop concept is fundamental to an AP. The key idea behind the way APs operate is that by keeping release cycles short, development can rapidly adapt to changes in the environment and requirements through decision making and acting, based on the most relevant and up-to-date information. APs are based on the principles defined in the Manifesto for Agile Software Development [15]. The principles can be summarized as: focus on customer satisfaction through frequent delivery of high quality software; empower the people involved in the development; favor face-to-face communication over comprehensive documentation; simplicity of design; and make the customer an active member of the team.

The focus of APs on software development varies; some APs are built around programming activities, while others consider the entire software lifecycle. Despite their differences, all APs share the same vision of how agile software development should be and they all directly support the principles of agile software development found in the Manifesto. [1] mentions nine of the most prevailing techniques today. These include Adaptive Software Development, Agile Modelling, the Crystal family, Dynamic Systems Development Method, XP, Feature-Driven Development, and Scrum.

APs have been almost exclusively used for in-house development or dedicated development environments. In the HIS domain this assumption rarely holds, as HI (software) systems development typically involves "critical mass" capabilities through monolithic and inflexible requirements, and re-engineering or interfacing with legacy systems built using traditional development methods [6].

2.3 Applying APs to HIS

If the HIS industry desires to accelerate release cycles in order to cope with an increasingly evolving business and technological environment, then APs may be of value. Despite the limited literature on the use of APs in the HIS domain, the scarcity of empirical data, and the fact that research in this field is at an early stage, there are indications that AP are applicable to critical systems development [4,6,7,10,14,16]. AP practices also appear to be partially aligned with standards such as the RTCA DO-178B [9,21] and ISO/IEC 12207-1995 [19]. APs are compatible with frameworks such as the CMMI [5]. The consensus of these studies is that APs will have to undergo extensive tailoring in order to become fully compatible with HIS development. We summarise some of the key challenges that must be overcome.

2.3.1 Communication
APs facilitate communication in an informal way, through face-to-face verbal communication and tacit knowledge [6,8,9,16]. The face-to-face manner with which interaction takes place in APs, is regarded as the most effective form of communication. This manner of communication works well if the complexity of the project and the amount of information conveyed is relatively low. This is not likely to be the usual case in HIS projects. Another limiting factor is team size: the larger a development team, the more communication channels exist within the team. Again, human capacity for retaining and processing information becomes a barrier.

2.3.2 Documentation
Reconciling documentation with AP practices is challenging. Avoiding *excessive* documentation is not only one of the main principles of the agile philosophy, but also

one of the main sources of agility. The misconception that APs reject documentation practices altogether is a result of the effectiveness with which internal communication is achieved. The lack of an explicit documentation mechanism could limit APs' applicability to HIS development, as certification requirements for documentation may prove overwhelming. Artefacts such as source code or unit tests may prove inadequate for certification purposes if they are used as the primary documentation means.

2.3.3 Customer Participation

APs rely heavily on customer participation for project success. As [16] points out there is no single, clear role for customers in HI systems development. Customers can come from a number of industries, business and technical backgrounds, each with their own agendas and perspectives of the system under development. Certification bodies may also be viewed as customers, whose participation requires independence when it comes to verifying that a system meets certification requirements. In APs, customers are responsible for planning iterations, identifying risks and developing acceptance tests. Without customers, teams are left to make these decisions on their own using monolithic requirements as drivers for decision-making. The frequent customer feedback achieved through short development cycles is an important input to APs. The techniques used for HIS procurement isolate customers from the development process to the point where coarse feedback of the type "the system works or the system does not work" is given on an infrequent basis.

2.3.4 Multiple-Domain Engineering

The development of HIS requires knowledge and experience from a variety of engineering backgrounds. This can create friction between the software teams involved in the project and teams from other engineering backgrounds [9].

2.3.5 Testing

Testing has a prominent position among agile practices, especially in the case of XP. In XP, testing is done constantly throughout the lifecycle, through the development of unit tests written prior to the implementation of system features. Test Driven Development (TDD), one of the cornerstones of XP, is found to be compatible with standards such as DO-178B [2,9]. Besides black-box testing, HI software development also involves white-box testing techniques such as static verification and coverage testing. White-box testing is not addressed by APs and this is a potential area in which improvements can be made. However, this is difficult as these techniques are time-consuming, expensive and some of them are difficult to carry out incrementally. *Acceptance testing* of critical systems is expensive and the opportunity to perform it does not appear very often. The iterative nature of APs may require that such tests be performed several times over, to determine the reliability of subsequent increments; this is clearly unfeasible.

2.3.6 Incrementality

HI systems are characterized by large numbers of often inflexible requirements of high priority. APs rely on the assumption that low-priority requirements that hold little value for the customer can be deferred to a future increment. This is an area in which friction between HI software development and the fundamentals of APs start to appear. [16] also highlights the issue of incremental certification; specifically, they question whether a safe but incomplete design can be incrementally certified. Defining what constitutes an increment, as well as the number of increments required to reach a critical mass of functionality adequate for certification, is one of the main concerns with the application of APs to HIS.

3   Adapting APs to HIS Development

The previous section identified a number of challenges that must be overcome in order to attempt to apply an AP for building a high-integrity system. This section provides suggestions for adapting APs to HIS development. In particular, we present suggestions on how APs can be "cross pollinated" with plan-driven approaches, in order to achieve both a degree of agility and the rigor necessary for HIS development.

3.1   Agility Across the HIS Development Spectrum

HI software development should not be performed in isolation from the rest of the development activities. [9] points out that HIS development can benefit from agility when it is applied "across the software/systems boundary". Aligning systems engineering activities with APs can potentially resolve incremental development and interfacing issues between software and system engineering activities, resulting in a more unified relationship between the various engineering disciplines involved in HIS development. The use of tools (such as Simulink) with automatic code generation appears promising in homogenizing software and systems engineering activities [16,19].

Leveraging the benefits of Pair Programming can help align software and systems engineer perspectives [16]. By pairing software and systems engineers, the communication gap between software teams and systems teams is abridged. This practice is likely to face problems in terms of acclimatization of systems engineers with the agile culture. It is also unclear how oversight requirements can be satisfied. However, the benefits of increased design quality through constant evaluation and direct communication are a strong incentive for the application of this practice to HIS development.

3.2   Using Risk as a Driver for Planning and Design

A critical component of plan-driven approaches is risk management. [6] proposed that a balance between agile and plan-based approaches can be achieved with risk management as a catalyst. They propose a hybrid method that uses risk as a driver for determining the mix of agility and planning a software development process should possess, and offer guidelines on applying the method. The concept of risk management as a basis for agility is also discussed in [5]. [9] advocates that risk-driven planning can reduce the dangers of incremental development to an acceptable and manageable level. However, they stress that in order for risk management to be used effectively, a certain level of maturity and experience in applying it are required.

In terms of design, risk management can assist in designing the system so that anticipated changes are easily absorbed [6,9]. Identifying the risks associated with volatile requirements can enable developers to create change-resilient designs. However, there needs to be a balance between simplicity and modifiability of designs.

3.3 Documentation

An approach to compensating for the lack of intra-team documentation in APs is the generation of documentation artefacts from code [19,21]. [19] suggests that APs can be made compatible with documentation requirements of standards such as the ISO/IEC 12007 through the use of software tools. [21] proposes that *agile documentation techniques* be used for documents that cannot be automatically generated. APs limit the need for documentation-based communication, through the use of face-to-face communication. However, in the case of external communication, documentation directed at stakeholders is needed. [19] suggests the introduction of a model based on Brooks' *"Surgical team"*, using a documentation subteam. Information capturing can be enhanced by tools that extract documentation from source code and test suites.

3.4 Incremental Development Process

The incremental nature of APs poses substantial challenges for HIS development. The idea of always delivering a "working system" after each iteration is difficult to reconcile with the reality of building (often embedded) HIS that must also be certified. [9] suggests that a "pipelined iterations" model be used to schedule development activities on increments at different engineering levels throughout the iterations phase of the project lifecycle. Fig. 2 depicts the engineering activities performed on given requirements sets simultaneously. Interactions between the different engineering roles may yield derived safety requirements which further complicate the process.

The authors address the issue of incremental certification by introducing *minor* and *major* iterations to the above model, using DO-178B as the certifiability criterion of an increment [9]. Minor iterations aim at designing, assessing and implementing a given set of requirements across engineering roles, while major iterations aim at preparing releases for acceptance testing. Releases are considered as *conditionally* safe, since acceptance test results may reveal the need for changes. In addition, several acceptance tests will be deferred until the final release, whereupon expensive, one-off tests will be performed.

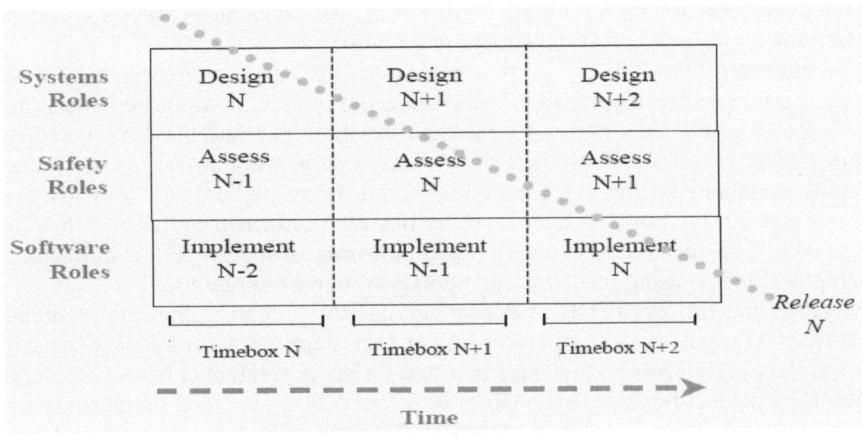

Fig. 2. Pipelined iterations process

3.5 Testing

While APs generally emphasise testing, it is usually acceptance and unit testing that are applied. Incorporating white-box testing, static analysis and verification techniques and MC/DC testing is challenging: these are intensive activities whose execution reduces agility and impedes rapid feedback. [4] mentions that static analysis testing activities for security critical systems can be automated through the use of tools that support the application of such testing activities within an agile context. They also propose the introduction of dynamic analysis tools - such as fault injection - that similarly reduce overheads. [16] propose the use of simulators and testing rigs with which expensive acceptance tests can be accommodated at the end of iterations.

3.6 Customising an Agile Process for HIS

XP is not fully compatible with the needs of critical systems development. If XP is to yield certifiable software within schedule and budget constraints, it has to undergo modifications. The process we have developed is what we term an *evenly weighted* variant of XP, tailored to the needs of the particular case study that we aim to build. XP intended to be tailored to the needs of the team and the project. While defining the tailored process, we drew from practices proven by software engineers for building · HIS. We also included techniques currently used in plan-driven methods for HI software development that traditional XP lacks, while remaining fundamentally agile.

The flexibility of XP allows for its augmentation through the addition of risk management, safety analysis and safety case development techniques. The proposed process retains the original XP practices, with the exception of the Metaphor and Collective Ownership practices. The project workforce should be segmented into domain teams, assigning each with distinct and clear development responsibilities, and supporting them with technical service teams. It is suggested that teams are kept small (up to 12 members) in order to be responsive with regards to communication. Technical writing teams are required, tasked with collecting and documenting information on the various aspects of development that are taken into consideration when the software is submitted for certification. The necessity for the inclusion of technical service teams comprising of domain experts (e.g., aerodynamics experts) will depend on the complexity and size of the system to be built.

The process we have developed resembles pure XP, and comprises a series of steps that are performed iteratively; Fig. 3 illustrates the process. During iteration *n,* Business prepares and selects stories for the next iteration *n+1* and the next increment is defined. TDD is conducted on the current increment *n*, while a team is verifying the previous increment *n-1* through simulations and test runs on hardware testbeds. Safety analysis and safety case development activities are performed on the *n-2* increment. At the end of an increment, evaluation and adjusting of the process is performed by development teams using feedback and metrics from past iterations.

Much of the process (TDD, V&V, planning, story engineering, and evaluation) come from XP. A new and important part of each iteration is *exploration*. An exploration spike ranging from a few days to a few weeks is conducted before the iterative stage of the project begins. This enables developers to understand the purpose of the system to be developed, define an initial architecture which evolves throughout the lifecycle, and serves as a substitute to the Metaphor practice. The need for exploration should diminish with time, as the workforce gains knowledge.

Story engineering and Planning	Plan N	Plan N+1	Plan N+2
TDD and integration	Develop N-1	Develop N	Develop N+1
V&V	Verify N-2	Verify N-1	Verify N
Safety analysis	Assess N-2	Assess N-1	Assess N
Safety case development	Argue N-3	Argue N-2	Argue N-1
Evaluate and adjust	Feedback from N-2	Feedback from N-1	Feedback from N
Iteration	N-1	N	N+1

Fig. 3. The HIS-XP development process with pipelined iterations

4 Case Study

We now report briefly on a case study we carried out to assess the applicability of the HIS-XP process for building safety-related software systems. The *Integrated Altitude Data Display System* (IADDS) is responsible for providing pilots with altitude data during flight. It is also responsible for issuing warnings to pilots whenever altitude limits are reached. Pilots are able to display altitude readings based on the *Sensor Of Interest* (SOI), i.e., barometric or radar equipment. IADDS is in general part of a larger avionics suite consisting of communications, navigational, weather monitoring and flight control running on a common IMA platform. In addition to IADDS, we developed a simple simulator for the operational environment, which served as a test-bed. Development of the simulator application was in parallel to the Altimeter application. IADDS possesses a number of safety properties and constraints, e.g.,

- IADDS altitude must equal simulator's calibrated altitude.
- The software must always issue audible (but suppressible) and visible (non-suppressible) warnings whenever the minimal altitude limit is exceeded while landing gear is retracted.
- Automatic data mode switching function must transparently and without user intervention switch from barometric mode to radar mode whenever the current SOI's indicated altitude equals 1500 feet while the aircraft is descending.

(Typically, such safety properties would include requirements for detection and correction of errors due to temperature lapse; we made simplifying assumptions.) The abstract architecture of IADDS is illustrated in Fig. 4.

The simulator serves a dual purpose; it can be used as a means of verifying the correctness of the Altimeter's behaviour, and during acceptance testing. The simulator software provides users with a GUI through which the various parameters (MSL altitude, local QNH) that determine atmospheric pressure can be set. The simulator will enable users to determine whether the expected altitude is correctly estimated by IADDS, by displaying the calibrated altitude for the currently defined area pressure.

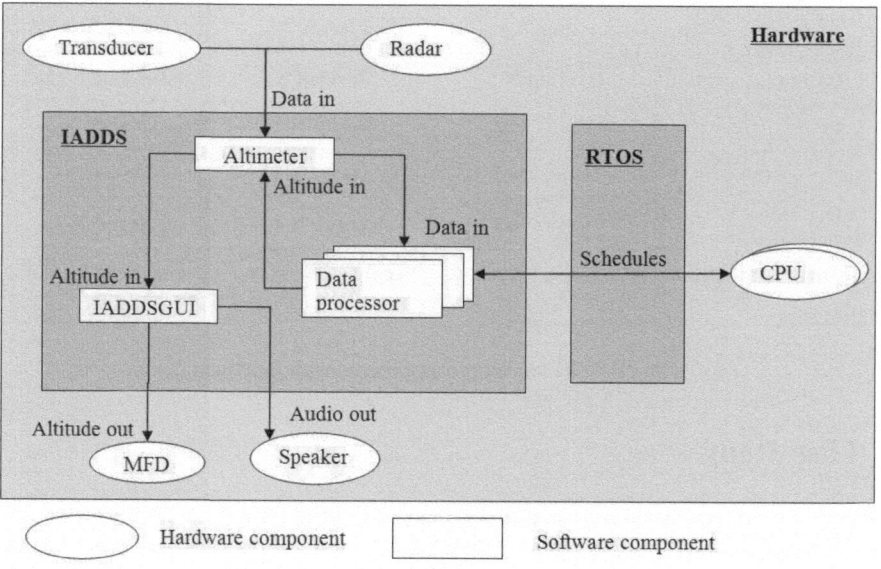

Fig. 4. Abstract Architecture of IADDS

XP heavily relies on tools that automate parts of the development process. Any tool used in XP development must be able to automate the process it is designed to support (e.g., unit testing) as much as possible to make its use worthwhile. For this experiment, we applied Microsoft's Spec# static verification tool, NUnit for TDD, Visual Studio 2005 as an IDE, NCover for coverage testing, UWG3 for creating fault trees, Adelard's ASCE for safety case development, and Enterprise Architect for documentation. We now briefly outline how the HIS-XP process was applied for building IADDS and its simulator, touching on the key phases of the process.

4.1 User Stories

Stories for the IADDS and Simulator applications were developed with the help of an airline pilot, who assumed the roles of customer and domain expert. Information about altitude measuring technologies was provided by the customer. Throughout the story engineering stage, domain knowledge was readily supplied. An interesting challenge encountered during the story engineering stage was educating the customer in story development. Expressivity and clarity of the stories produced improved gradually over time. The main difference between the stories used for this experiment and conventional XP stories is the inclusion of an additional field called 'Fitness Criteria' [18]. Fitness criteria explicitly define any safety properties and constraints that a story must satisfy. Such constraints are normally accounted for by test cases; however the inclusion of fitness criteria in a story makes safety case development activities easier by readily identifying evidence associated with a particular feature.

4.2 Safety Stories

Safety stories represent safety requirements discovered during the safety analysis stage of an iteration. They resemble regular user stories, although their purpose is to capture information related to hazards that exist within implemented user stories. Fig. 5 shows a safety story from the IADDS application. The purpose of distinguishing between user and safety stories is to provide means for documenting the outputs of the safety engineering steps of the process, and ensuring that product-related evidence is captured, as well as a way of prioritizing and estimating changes to existing code.

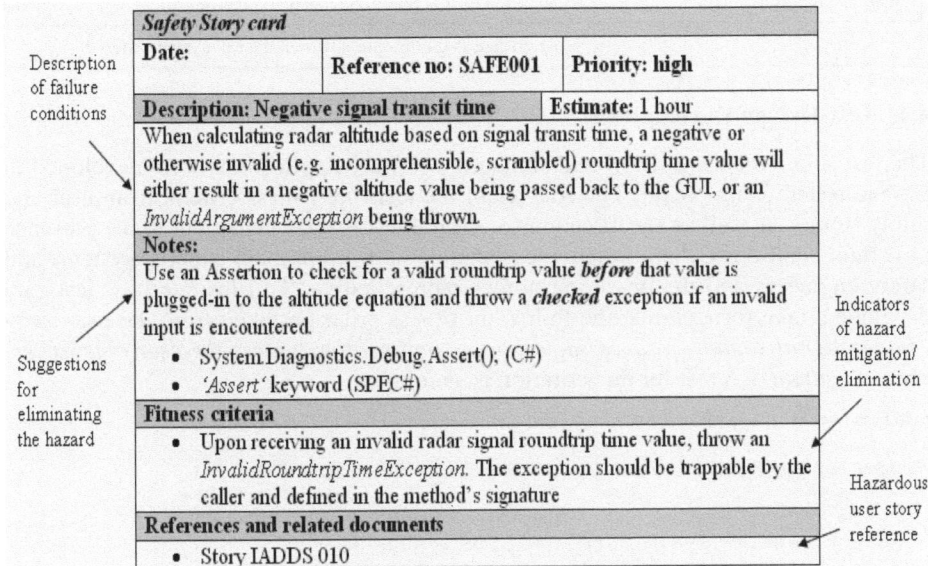

Fig. 5. Safety story card for IADDS

4.3 Planning and Risk Management

The initial release plan was divided into three iterations, each culminating in a working version of the IADDS software. A fourth iteration was added for detecting and correcting residual defects. The customer was only involved in IADDS story engineering. Requirements for the simulation tool were extracted from IADDS system requirements. The average ideal engineering time (IET) per iteration was estimated to 6 days out of 14. The actual average IET per iteration was 5.5 days. Total actual IET for both IADDS and Simulator applications was 16.5 days.

During risk management activities for IADDS, risk-defining variables were taken into consideration for each story. A series of questions representing about the likelihood of each variable (see below) were answered, and the answer to each question produced the value assigned to the variable for a particular story. Unanswered questions were given a high risk value, until variables can be assigned a value by confidently answering the corresponding questions. The number and types of risk variables can vary according to the type of project undertaken. The number of risk variables used for the stories in this project was deliberately kept low, although in a real-world project the number of variables would be much higher and would likely involve different risks for each increment.

Risk variable	Question asked
Technical know-how	Do we know how to develop this feature?
Skill and technology	Do we have the tools necessary to develop this feature?
Story volatility	How likely is it for this feature to change in the near future?
Scale of change	How much of this feature is likely to change?
Criticality	How critical is the feature's role in overall system safety?

4.4 Test-Driven Development

The test suites for the IADDS and simulator software applications were developed in C# and tested using NUnit. The sources of the tests are fitness criteria from user and safety stories, as well as specifications of desirable software behaviour in the presence of certain conditions. Additionally, tests can come from domain experts, systems and safety engineers and the on-site customer representative. To illustrate how tests are developed, take for example the following fitness criterion taken from the user story *"Radar signal roundtrip times should never be negative, unless the user opts for erroneous output"*. A test for this criterion is as follows.

```
public void RadarSignalTimeGreaterThanZero()
{
    Simulation mySim = new Simulation(new decimal(1000),
                             new decimal(1013.25));
    mySim.TerrainElevation = 100;
    Assert.IsFalse(mySim.CurrentAltitude < 0 &&
          mySim.CurrentAltitude >= mySim.TerrainElevation);
    mySim.AllowErroneousOutput = false;
    Assert.IsTrue(mySim.GetRadarSignalRoundtripTime() > 0);
}
```

The above test aims at verifying that the software will always produce a positive roundtrip value for a given radar signal *S*, provided that the necessary environmental conditions hold when the method is called.

The role of tests as evidence that can be used to argue a system's safety raises the question of tool reliability and correctness. Tools such as NUnit classify as verification tools from a DO-178B viewpoint. Although many certified systems have been successfully developed with unqualified tools, software vendors still need to ensure that any TDD tools used have the ability to detect all errors that a piece of software may potentially contain, and does not insert any errors in the software being verified.

4.5 Safety Process

Incremental hazard and safety analysis activities resulted in safety stories, which in some cases adversely affected iteration plans and development activities. It soon became apparent that incremental safety activities are difficult to carry out, as safety requirements discovered may invalidate design decisions made in earlier iterations.

During safety analysis of the software artefacts of the project, tools were used for producing and analyzing items such as fault trees. It became clear that automatic tool support for safety activities is helpful, but it is only partially addresses the issues associated with safety activities. What is needed is true integration of safety analysis activities with APs; we conducted only a partial integration of these activities.

Unlike plan-driven approaches, the incremental nature of Agile development does not allow for complete hazard identification and assessment to be conducted up front, before implementation of the design begins. Hazard management should therefore be incremental in order to be successfully applied to XP projects.

The discovered hazards associated with the stories allocated to a given iteration were recorded in a Hazard Log, along with a textual analysis of each hazard. Details for each hazard include a severity value, a likelihood of occurrence value, and a risk exposure value. The process for identifying hazards was rather simple, consisting of inspections of the user stories allocated to the increment under scrutiny and of any initial design ideas. The conditions under which operation of the software could potentially jeopardize system safety were then assessed. Although less sophisticated than techniques such as SHARD or LISA [17], this proved effective.

Safety analysis was conducted using the extended FTA technique proposed by Kaiser et al [12]. The tool UGW3 was used to construct a set of fault trees for IADDS and to calculate the failure probability of IADDS components. The extended approach to FTA presented in [44] was selected due to its modular approach to safety analysis. The technique allows for analysis to take place on a partial design, and facilitates the reusability of CEG components. The CEG approach can be conducted in an incremental manner, resulting in smaller, understandable and more manageable models.

Within an XP context, the safety process used to assess a system's safety needs to result in rapid feedback in order to be effective. Independence requirements isolate safety teams from development teams, although some participation of software engineers in the safety process is necessary. The boundary between safety and development teams imposed by independence requirements creates overheads in the digestion of safety requirements. Batch delivery of safety requirements can adversely influence the progress of an iteration, causing breakage of tests and code, creating the need for rework to bring already integrated code up to date with safety requirements. Stories may be deferred to later iterations in order for teams to cope with safety requirements implementation, although this suggests a departure from plans and budget estimates.

Safety case development was conducted as an external activity to software development by a group of safety engineers who collected evidence to support claims about the software process' ability to yield HIS products, and a system's suitability for operation within a specific context, by constantly refining a set of contingent preliminary arguments. This was done until a comprehensive safety case emerged.

Although it is preferable that a safety case for a system built with XP evolves incrementally alongside the system itself, safety case development is not in any way constrained to an incremental model. An initial safety case that evolves through evidence collected with every iteration is ideal; however the necessary evidence may not become available until the later stages of a project. If that is the case, safety arguments can be produced towards the end of the lifecycle.

The approach used to argue the safety of IADDS was product-based. Evidence is primarily associated with the software product, although process-related arguments are vital to successfully arguing IADDS safety. The method selected for arguing IADDS' safety was GSN, as the notation and six step method [13] are suitable for incrementally developing maintainable and highly expressive arguments. Arguments were developed for the IADDS application using Adelard's ASCE tool. The

IADDS safety case was relatively small. The strategies adopted in constructing the "implementation safe" argument were based on *safety requirements satisfaction* and *identified hazard omission*. The goals resulting from the hazard omission strategy were supported with fault tree, hardware component failure rates and cutset evidence. Goals regarding safety requirements satisfaction were solved using evidence such as unit testing results and simulation results. The "implementation correct" argument was structured by approaching it with strategies for arguing over *satisfactory test results* and *completeness of implementation*. Process arguments were used to support product arguments for IADDS. These were included in the product arguments as references to process-related arguments. This approach is based on "Away Goals" [11], an extension to existing GSN notation. Away goals are "placeholders" for process-related arguments which defend the provenance of product evidence, as well as strategy context. The tool used to develop the safety case for IADDS does not support away goals; GSN notes were used instead to refer to the supporting arguments.

5 Conclusions

Although XP and APs in general were not designed with safety-critical systems development in mind, they can be adapted to that sort of development. The exact capacity of the XP variant proposed in this report for achieving a given software level (as defined in DO-178B) is not known; it is rather unlikely that level A software can be produced in the near future with the modifications made to the process so far. The current aim of Agile Methods should not be to replace plan-driven approaches, but to complement their use by applying them to the areas where plan-driven methods do not perform as well as their Agile counterparts.

In the future, we should work at developing a framework of procedures and techniques with which the hybrid process used in our experiment can be adapted to the needs of any project size and of complexity. The core parts of the process will remain unchanged; the framework will allow developers to tailor the method to their needs and managers to plan the lifecycle and orchestrate resources, using risk-based techniques. Also of interest should be developing high-level architectural plans that enable developers gain a clear understanding of the system and its purpose, by substituting or supplementing the Metaphor practice. The framework should define guidelines with which such artefacts can be developed and evolved throughout the lifecycle.

Acknowledgments. We thank the referees for their helpful suggestions. The research presented in this paper was supported by the Engineering and Physical Sciences Research Council, as part of the Large-Scale Complex IT Systems research programme, and research grant EP/F001096/1.

References

1. Abrahamsson, P., Wasta, J., Siponen, M.T., Ronkainein, J.: New directions on Agile Methods. In: Dillon, L., Tichy, W. (eds.) Proc. ICSE 2003, pp. 244–254. ACM Press, New York (2003)
2. Amey, P., Chapman, R.: Static verification and Extreme Programming. In: Sward, R. (ed.) Proc. SigADA 2004, pp. 4–9 (2004)

3. Beck, K.: Extreme Programming explained. Addison-Wesley, Reading (2000)
4. Beznosov, K., Kruchten, P.: Towards agile security assurance. In: Sekar, R., McHugh, J. (eds.) Proc. 2004 Workshop on New security paradigms, pp. 47–54 (2004)
5. Boehm, B.: Get Ready for Agile Methods, with Care. IEEE Computer 35, 64–69 (2002)
6. Boehm, B., Turner, R.: Balancing agility and discipline. Pearson, London (2003)
7. Boström, G., Wäyrynen, J., Bodén, M., Beznosov, K., Kruchten, P.: Extending XP practices to support security requirements engineering. In: Bruschi, D., De Win, B. (eds.) Proc. Workshop on Software engineering for secure systems, ICSE 2006, pp.11–18. ACM Press, New York (2006)
8. Eckstein, J.: Agile Software Development in the Large. Dorset House (2004)
9. Galloway, A., Paige, R.F.: On the use of Agile Methods for High-Integrity Real-Time Systems, DARP Technical Report DARP-TR-2006-5 (2006)
10. Grenning, J., Peeters, J., Behring, C.: Agile development for embedded software. In: Zannier, C., Erdogmus, H., Lindstrom, L. (eds.) XP/Agile Universe 2004. LNCS, vol. 3134, pp. 194–195. Springer, Heidelberg (2004)
11. Habli, I., Kelly, T.: Process and product certification arguments. ACM SIGBED Review 3(4), 1–8 (2006)
12. Kaiser, B., Liggesmeyer, P., Mäckel, O.: A new component concept for fault trees. In: Cant, T. (ed.) Proc. 8th Australian Workshop on Safety critical systems and software, pp. 37–46. Australian Computer Society (2003)
13. Kelly, T.P.: Arguing safety – A systematic approach to managing safety cases (PhD thesis), University of York (1998)
14. Manhart, P., Schneider, K.: Breaking the ice for Agile Development of Embedded software: an industry experience report. In: Estublier, J., Rosenblum, D. (eds.) Proc. ICSE 2004, pp. 378–386. ACM Press, New York (2004)
15. Manifesto for Agile Software Development (2007), http://agilemanifesto.org/
16. Paige, R.F., Chivers, H., McDermid, J.A., Stephenson, Z.R.: High-Integrity Extreme Programming. In: Omicini, A. (ed.) Proc. SAC 2005, pp. 1518–1523. ACM Press, New York (2005)
17. Pumfrey, D.J.: The principled design of computer system safety analyses (PhD thesis), University of York (1999)
18. Robertson, S., Robertson, J.: Mastering the requirements process, AWL (2006)
19. Theunissen, W.H.M., Kourie, D.G., Watson, B.W.: Standards and agile software development. In: Eloff, J., et al. (eds.) Proc. Enablement through technology, vol. 47, pp. 178–188 (2003)
20. Wikipedia contributors, OODA Loop, Wikipedia, The Free Encyclopedia, http://en.wikipedia.org/w/index.php?title=OODA_Loop&oldid=154056152
21. Wils, A., Van Baelen, S.: Towards an Agile avionics process (2007), http://www.agile-itea.org/public/deliverables/ITEA-AGILE-D2.12_v1.0.pdf

SafeSpection – A Systematic Customization Approach for Software Hazard Identification

Christian Denger, Mario Trapp, and Peter Liggesmeyer

Fraunhofer Institute Experimental Software Engineering, Fraunhofer-Platz 1,
67663 Kaiserslautern, Germany
{Christian.Denger,Mario.Trapp,
Peter.Liggesmeyer}@iese.fraunhofer.de

Abstract. Software is an integral part of many technical systems and responsible for the realization of safety-critical features contained therein. Consequently, software has to be carefully considered in safety analysis efforts to ensure that it does not cause any system hazards. Safety engineering approaches borrowed from systems engineering, like Failure Mode and Effect Analysis, Fault Tree Analysis, or Hazard and Operability Studies, have been applied on software-intensive systems. However, in order to be successful, tailoring is needed to the characteristics of software and the concrete application context. Furthermore, due to the manual and expert-dependent nature of these techniques, the results are often not repeatable and address mainly syntactic issues. This paper presents the concepts of a customization framework to support the definition and implementation of project-specific software hazard identification approaches. The key-concepts of the approach, generic guide-phrases, and tailoring concepts to create objective, project-specific support to detect safety-weaknesses of software-intensive systems are introduced.

Keywords: Software Safety, Guide-Phrases, SafeSpection, Software FMEA, Software FTA, Software HAZOP.

1 Introduction

Over the last decades, embedded systems have become an integral part of our daily lives. Especially in the automotive domain, software-intensive systems execute and control a variety of functions and safety measures. Without software, many innovative functions and features would be hard or even impossible to realize.

As a part of a safety-critical system, i.e., a system whose failure might endanger human life, cause extensive environmental damage, or lead to substantial economic loss [1], the software itself must be perceived as safety-critical. In other words, as part of the system the software has the potential of putting the overall system into a hazardous situation. In that sense, Leveson defines safety critical software as any software that can contribute to the occurrence of a hazardous system state either directly or indirectly [2]. As an example of the safety criticality of software, General Motors had to recall almost one million cars due to problems with their airbag system. On paved roads under normal conditions the software interpreted the unstable movement

M.D. Harrison and M.-A. Sujan (Eds.): SAFECOMP 2008, LNCS 5219, pp. 44–57, 2008.

of the cars as a crash and activated the airbag. Similar examples can be found that demonstrate the importance of including software in system safety analysis activities executed during the development life-cycle.

Hence, in the automotive domain, recent standards (e.g., IEC 61508, ISO/WD 26262, MISRA Safety Analysis Guide) request a thorough software safety analysis. The standards require the application of safety engineering techniques on the functional concept and on the software architecture [3], [4]. In order to fulfill this, companies typically apply safety analysis techniques like Failure Mode and Effect Analysis (FMEA), Hazard and Operability Studies (HAZOP), and Fault Tree Analyses (FTA) to identify potential system hazards caused by the software. However, the applied techniques are often not customized to the characteristics of software and consequently do not support the identification of conceptual software faults. Mainly, the standard processes of system-FMEA, -FTA, and -HAZOP techniques are used to analyze the software work products. These processes are of a manual nature without concrete guidance on how to identify software faults. The results rely on the experience of the moderator and the participating experts and hence the analysis is not repeatable, subjective and results cannot be compared between different development teams. During the last decades, some approaches have evolved on how safety analysis techniques can be applied to software. As Section 2 demonstrates, the efficient application of these techniques still remains unclear. The main reason for this is that many approaches focus the analysts on very detailed, low-level software causes of hazards like uninitialized variables, too late or too early execution of algorithms, and wrong data models. Additionally, the guidance provided by existing approaches is often on a general-purpose level not tailored to the specific context characteristics of a (software-) project. Hence, only general aspects like correctness and completeness issues, and syntactic aspects are captured. Conceptual software faults, i.e., faults in the logic of the software models, are rarely in the scope of existing approaches. In consequence, what is missing today is an approach that provides systematic guidance on how to customize safety-analysis techniques to the characteristics of a software development context. This approach should support the identification of conceptual, semantic software faults that might cause system hazards. The identification should be performed during the early development phases to provide real added value. Safe-Spection has been developed to close this gap.

Section 2 provides a detailed overview of the state of the art regarding software safety analyses and motivates the approach. Section 3 introduces the core concept of SafeSpection: guide-phrases and a grammar to systematically derive project-specific guidance on detecting conceptual software faults causing system hazards. Section 4 outlines the results of an initial feasibility study in an industrial setting. Section 5 concludes the paper and provides some future research topics.

2   Existing Software Safety Analysis Approaches

Even though the idea of software safety analysis techniques has been around for several decades, the number of publications regarding this topic remains quite small [6]. The following subsections categorize the existing approaches and provide a critical review of these regarding their repeatability, customizability, and focus.

2.1 A General View on Existing Software Safety Approaches

According to Fenelon et al. [5], software safety analysis approaches are categorized according to the direction of the search for software causes of hazards. Explorative, inductive, deductive, and descriptive approaches are distinguished. This classification is based on the categorization of safety analysis techniques used in systems engineering. In order to provide a more intuitive, software-related classification we rephrased and extended this existing scheme.

On an abstract level software safety analysis approaches are classified according to the underlying systems engineering techniques they are based upon. Thus, *HAZOP-like approaches*, *FMEA-like approaches*, *Inspection-like approaches,* and *Formal-approaches* are distinguished indicating that the identification of software causes of hazards is based on standard FMEA, HAZOP, inspection and formal approach, respectively. FTA-like approaches are not considered in this scheme as these require software hazards as an input and therefore do not provide concepts to identify these. In addition to this abstract categorization each approach is classified according to the scheme illustrated in Fig. 1

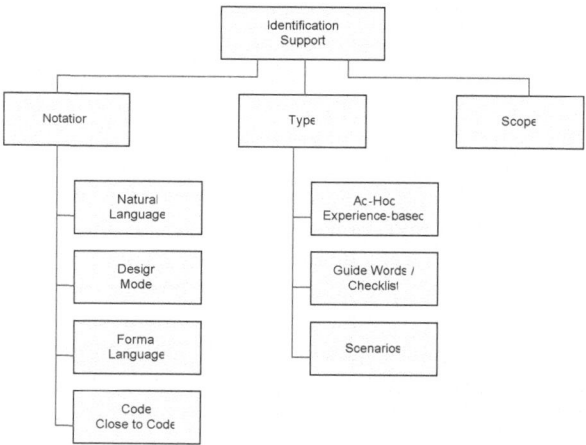

Fig. 1. Categorization Scheme for Characteristics of Software Hazard Analysis Approaches

Notation classifies the approaches according to the software development notations to which they can be applied. The sub-classes of *Notation* are not orthogonal, that is, it is possible that a technique is applicable to different notations. *Type* characterizes the support provided by the approach in detecting software causes of hazards. The subtype "ad-hoc/experience-based" indicates that no explicit support is provided; the sub-type "guide-words/checklists" indicates that some triggers are provided that point analysts to potential software faults; "scenarios" represent a special kind of support that gives procedural guidance to the analysts. *Scope* provides a classification of the types of software issues that are addressed (e.g., communication issues, correctness issues, completeness aspects). Additionally, the approaches are classified according to the software life-cycle phases they are designed for (i.e., requirements analysis, architecture definition, detailed design, and code).

We classified 60 references according to this scheme. 33 approaches explicitly mention the use of FMEA principles to analyze software caused hazards, including approaches that use a combination of FMEA and FTA. 20 approaches are based on HAZOP ideas and five on inspection ideas. This result indicates that FMEA is the technique applied most frequently for identifying and analyzing software causes of hazards. Classifying the approaches according to Fig. 1 shows that most of the approaches are defined for lower-level development phases, i.e., detailed design and code. This finding seems to contradict the finding that natural language is the notation to which most of the approaches are applied. However, a close analysis of the approaches shows that they operate on detailed design specifications of code modules written in natural language, technical models like state-charts, and variable definitions. Analyzing the scope of the existing techniques shows that due to the provided guidance, i.e., guidewords such as commission, omission, early, late, and the application of these on assets like services, variables, data rates, and signals mainly the detection of syntactic faults and correctness issues is supported. Software faults on a more subtle, logical level are often not detectable using these approaches.

2.2 Detailed Discussion of Selected Approaches

In the class of HAZOP-like approaches, the most prominent one is the SHARD approach defined by [9]. The underlying idea of the approach is the suggestion of potential failure modes of the software by means of guidewords. The focus is on the interfaces of major software components and on the data- and control-flow between them. SHARD provides the guidewords service commission, service omission, service timing (early, late), and service value (incorrect) to support the detection of potential deviations of the software behavior during the requirement phase. Lisagor et al. extended the SHARD approach for architecture evaluations [12]. Similar approaches are SoftwareHAZOP [10, 11] which applies standard HAZOP guidewords (more, less, part-of, other-than, before, after, etc.) to different software notations (data-flow diagrams, state-charts, class-diagrams). A formal variant of HAZOP-like techniques is defined by Reese et al. [13]: the software deviation analysis. The idea is that based on pre-defined software deviations, it is possible to derive deviation scenarios from a formal model of the software.

Regarding FMEA-like approaches the most prominent ones are the HiPHOPS approach [14, 15] and the Bidirectional Analysis [8]. Both approaches are a combination of FMEA and FTA approaches during the software requirements and design phase. The idea is to investigate the impact of software failure modes on the software and system level using a FMEA. Then, the identified hazards that are most critical are analyzed in detail by means of an FTA to decide whether or not the hazard really can occur. In case of [15], the FTA can be automatically derived from the formal representation of the FMEA results. More recently, the SoftCare approach has been defined [16]. This is also a combination of FMEA and FTA but some more guidance is provided on how to identify initial software failure modes. For this purpose, the guidewords (commission, omission, service timing, service value) are applied to software-related constructs (data, procedures, variables). The resulting list of potential software failures, however, contains a huge list of items pointing mainly to syntactic issues (e.g., wrong data value, late procedure call).

Summarizing the analysis of the state of the art of existing approaches, the following open issues get evident: 1) Even though requested by many authors (e.g., [2], [6], [11]), there is no systematic approach on how to perform a customization of the analyses to a given project context. 2) Many approaches assume that the software failure modes, i.e., the software causes of a hazard, are already known when the analysis starts. In practice, this is not the case and the identification of the failure modes is dependent on the experience and the knowledge of the analysts. This results in subjective, non-repeatable, hard-to-compare results. 3) Even in the case that guidance is provided for detecting potential software failure modes this guidance mainly focuses on correctness and completeness issues of software work products. However, such aspects are also addressed by standard software inspection approaches and it is important to carefully analyze the overlap of software inspections and the proposed software safety analysis approaches. Independent of that is the fact that systematic guidance on how to detect conceptual faults that have an impact on software-safety are missing. Remember the airbag example given in the introduction. The software was correct and complete but contained a conceptual fault. The algorithms used in the software to detect the system state "crash" contained a conceptual fault as they were too shock-sensitive in certain driving situations.

3   The SafeSpection Framework

The overall objective of SafeSpection is the systematization of the detection of software caused hazards, i.e., software failure modes. In that sense, SafeSpection supports the customization and execution of software FMEA and software FTA analyses by guiding the analysts in the identification of software causes of hazards. Hence, Safe-Spection must not be perceived as a substitution but as an add-on to these approaches to overcome the issues related to their execution. A framework for customizing software safety guide-phrases to a specific project context is the core of the SafeSpection approach that realizes the systematization.

3.1 Approach to Systematization

In order to efficiently and effectively detect software failure modes, it is essential to provide systematic guidance that focuses the analysts not only on syntactic issues but mainly on conceptual software faults. The following example taken from a real-world accident illustrates the importance of focusing on conceptual software faults: The system specification requires that certain functions of an electronic control unit of an aircraft are executable if and only if the plane is "on ground". The system-state "on ground" is realized by the software as:

aircraft is on ground if signal_wheels_turning == true and signal_pressure_wheels >= x lb.

Typically, the moderator of a FMEA or a HAZOP analysis is responsible for triggering the analysis team with suitable questions that point to potentially unsafe behavior. Using HAZOP guidewords, one would trigger the team with questions like "Is the signal_wheels_turning correct?", "Is the signal_wheels_turning late?", "Is the signal_pressure_wheels too early or omitted?" To answer these questions the software is analyzed in detail to determine whether or not these events can occur. This is mainly a syntactic check. What is missing is the check of whether or not the software realization represents a safe solution. An experienced moderator might ask additional questions pointing at conceptual faults like: "Is it possible that the aircraft is on ground but the wheels are not turning?" or "Is a situation possible where the wheels are not turning or pressure is < x lb but the aircraft is on ground?" Asking these questions reveals that the software realization is correct but not safe: in case of aqua-planning the wheels are not turning but the aircraft is on ground!.

In order to overcome the reliance on expert experiences, SafeSpection provides an abstract framework that allows the flexible customization of guide-phrases to a specific project and product context. The guide-phrases are defined in such a way that they support the detection of **conceptual** software faults that cause system level hazards. The underlying idea of the framework is the "formalization" of the provided support in terms of guide-phrase patterns that are derived from a guide-phrase meta-model. Based on the meta-model and the patterns, it is possible to instantiate project-specific guide-phrases that point to conceptual faults. Both the definition of the patterns and the instantiation of the guide-phrases are supported by SafeSpection guidelines. This results in more specific guidance on the detection of software-caused hazards, reduces the overlap of software-safety analysis and standard quality assurance by focusing on conceptual faults rather than syntactic issues (which are addressed by standard quality assurance activities like software inspections), and makes the results of the analysis repeatable, i.e., less dependent on individual experts, and easier to compare between teams. The core elements of the SafeSpection framework and their application are outlined in the following sections.

3.2   The SafeSpection Framework Concepts and Their Application

SafeSpection differentiates between three abstraction layers of guide-phrases (cf. Fig. 3). On the highest level, meta-meta-questions define the building blocks of a guide-phrase. The meta-meta-questions are the fundamental element for defining systematic and repeatable guidance for the detection of conceptual software faults, as they prescribe the structure of a general guide-phrase. According to the SafeSpection approach, a guide-phrase comprises two main parts, a Trigger-Part and an Effect-Part (cf. Fig. 2). The Trigger-Part is a sentence that represents a question pointing to elements in the functional specification of the software that might contain conceptual faults. The Trigger-Part element comprises three sub-elements: The *Object* represents elements of a software specification in the focus of the analysis for potential faults (e.g., a function, service or component). The *Influence Factor* describes issues that can have a potentially negative impact on the *Object*. Finally, *Interference* describes the type of impact that is imposed by the *Influence Factor* on the *Object*. Each trigger-part of a guide-phrase has one object, one influence factor, and one interference. The *Effect-Part* is either a closed question asking about the possibility that an already known hazard is caused by the question described in the trigger-part, or it is an open question asking about the possible / thinkable consequences or impacts if the question described by the Trigger Part becomes true.

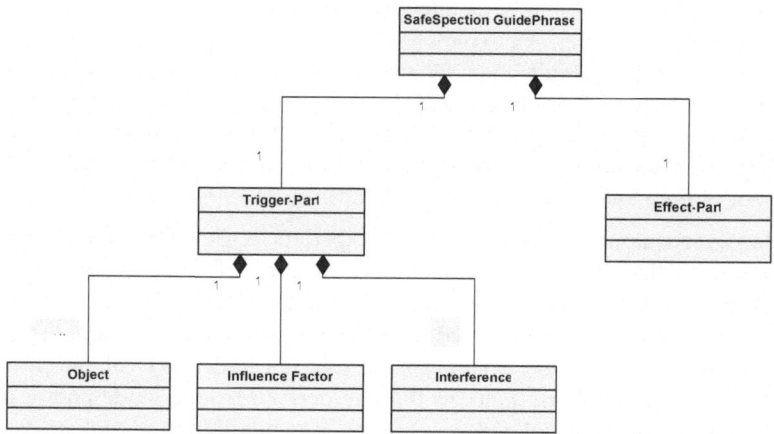

Fig. 2. The Meta-Meta-Questions Defining the Structure of SafeSpection Guide Phrases

Having introduced the basic building blocks, the SafeSpection framework defines meta-questions that represent domain-specific instantiations of the concepts *Objects* and *Influence Factors*.

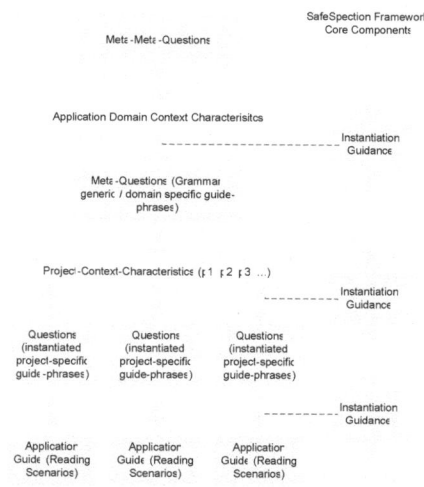

Fig. 3. The Hierarchies for the SafeSpection MetaModel

Consequently, these meta-questions represent generic guide-phrases that are applicable in a certain domain and that are an intermediate abstraction layer allowing the systematic customization of feasible, project-specific guidance.

In order to use the SafeSpection framework efficiently it is the responsibility of a safety manager and a software development leader to identify the domain-specific instantiations of the meta-meta-questions. This activity has to be done in close cooperation between software development and safety management to gather the

Nr.	Question
1	How is the behavior of the application software typically described in the domain (in terms of functions, services, processes)?
2	Can software functions within this domain be characterized with respect to time constraints, pre-conditions, post-conditions?
3	Which modeling elements / components are used to describe the software function in the domain (e.g., sate-machines, data-flow models)
…	……
12	Are assumptions regarding the realization of functions to be considered?
13	What environmental conditions can have an impact on the software behavior (e.g., weather conditions, road-conditions)?
14	Is the software behavior dependent on operational modes (e.g., power-up, power-down)?

Fig. 4. SafeSpection Interview-Guide to Identify Guide-Phrase Objects and Influence Factors

knowledge and experience of both worlds. The concepts need to be identified for the given domain of the systems developed (e.g., electronic control units for a car). Safe-Spection provides an interview guide that supports the identification of relevant objects and influence factors (see an excerpt in Fig. 4).

The result of this activity is a set of generic guide-phrases that comprise the domain-specific objects and influence factors. The results of the interviews should be backed up with a comprehensive study of existing functional specifications of ECUs in the company to identify additional objects and influence factors. In the aircraft example, one could identify software realizations of external conditions as an object, i.e., the formula for "on ground". Examples of influence factors in this domain are weather conditions, flight situations like landing, take-of, and so on. Hence, the identification of a complete set of object and influence factors is a crucial success factor of the SafeSpection approach.

We elicited domain-specific objects and influence factors in the context of functional specifications of electronic control units of cars. In this domain, typical objects are the *functions/services* and their characteristics (pre-conditions, timing constraints, realization, accuracy, assumptions), the *interfaces* of the functions/services to other functions (i.e., exchanged signals, their syntax, their semantic, and timing); and the *interactions* the functions are involved in. The influence factors that can have a negative impact on these objects are in the SafeSpection approach: *environmental conditions* (e.g., weather conditions, road conditions), *operational situations* (e.g., high-speed driving, urban driving), *technical constraints* (e.g., latency of actuators, frequency of sensor polling), *realization assumptions* (e.g., algorithm xyz is used to approximate vehicle speed), *operational modes* (e.g., power-up, power-down, diagnosis), and the *change of technical constraints* (e.g., reusing software in another hardware environment, change of sensor characteristics due to aging).

In order to standardize the definition of the generic guide-phrases, SafeSpection provides a grammar. This grammar defines rules on how objects and influence factors are combined into a SafeSpection guide-phrase. The core structure of each guide-phrase follows the rule: $S \rightarrow Intro \bullet Influence \bullet interference \bullet Object$?, where Intro is a phrase introducing a question, like "Does the…", "Is it possible that…". Influence and Object are the identified domain-specific objects and influence factors, and interference defines the type of impact on the object. In our aircraft example applying SafeSpection leads to the following generic guide-phrase: "Does the <<weather

condition>> invalidate the <<software realization>> of <<system condition>>. The words in <<..>> are the generic objects and influence factors that need to be identified by the experts using the SafeSpection interview guidelines.

According to the combination of objects, interferences, and influence factors, the SafeSpection approach predefines the following types of guide-phrases that address certain types of software-caused hazards in the context of an ECU.

Name	Scope
1. Overall assumptions	Supports the identification of software-caused hazards that stem from a violation of system-wide constraints, pre-requisites and assumptions by the software realization.
2. External Influence	Supports the identification of software-caused hazards that stem from an inappropriate consideration of special characteristics of driving situation, operational modes, and environmental conditions in the software.
3. Changed Environment	Supports the identification of software-caused hazards that stem from changes in the software environment (like changed technical constraints, changed application context, changed sensor characteristics) that are not properly mapped / considered in the software realization.
4. Communication	Supports the identification of software-caused hazards that stem from wrong or inappropriate interactions of software elements and software-realized functions / services / processes.
5. Functional Realization	Supports the identification of software-caused hazards that stem from an improper realization and an insufficient consideration of influences on the software behavior (like the fulfillment of assumptions, prerequisites constraints that are not given in certain operation of modes).
6. Special Functions	Supports the identification of software-caused hazards that stem from the implementation of degradation scenarios that are not properly integrated in the overall functional concept.

Fig. 5. Types of Guide-Phrase Patterns in the SafeSpection Framework

For each type, one or more generic guide-phrase is provided. With respect to the analysis of functional specifications of ECUs, we defined a set of generic guide-phrases for the types defined above. The following questions represent guide-phrases of the type external influence and changed environments:

Does the <<characteristic>> of <<driving situation>> invalidate the
<<pre-condition>> || <<post-condition>> of <<function>>?
Does the change of <<characteristics>> of <<sensor>> || <<actuator>>
violate the timing-constraints of <<function>>?

In the Appendix of thi paper an excerpt of the full set of generic guide-phrases supporting the detection of conceptual faults is listed. The advantage of the guide-phrases is their generic nature aimed at conceptual faults compared to the syntactic guidance provided by existing HAZOP guidewords. Moreover, the guide-phrases are already tailored to the application domain and hence more specific than general-purpose guidewords. The following comparison clarifies this advantage: the checklist questions defined by Leveson [2] typically aim at completeness issues, e.g.,: "A trigger involving the nonexistence of an input must be fully bounded in time". Guide-phrases defined with the SafeSpection approach would perceive the definition of such a trigger and its time bounds as objects of the specification, i.e., SafeSpection takes these as inputs. These objects are combined with influence factors to check whether or not the time bound can be violated for example by external conditions or whether or not the time bound contradicts realization assumptions underlying the software.

Finally, the generic guide-phrases are instantiated to concrete guide-phrases that are applicable in a certain project. That is, the generic meta-questions defined for the application domain are instantiated with concrete objects and influence factors of a software project. In our aircraft example, the generic guide-phrase is instantiated with the concrete objects and influence factors: "Does rainy weather invalidate the software realization plane on ground if the wheels are turning and the pressure is >= x lb?" The person responsible for this activity is typically a quality assurance person of the project team whose functional specification is analyzed. The resulting guide-phrases are used by the analysis team to identify conceptual faults in the functional specification. It is most important to identify those generic guide-phrases that are relevant for the specific project context. Again, the SafeSpection framework provides guidelines on how to perform this instantiation in terms of expert interviews. The project-specific guide-phrases result in systematically tailored guidance addressing the real safety needs in a project context. The detection of conceptual software faults in the project becomes a repeatable activity and focused on the project-characteristics rather than on providing general-purpose guidance. The identified guide-phrases can be used as a stand-alone technique similar to an inspection approach, using the guide-phrases as checklist questions or as part of the software FMEA and software FTA activities where they guide the analysis team in detecting software failure modes and software causes of hazards.

4 SafeSpection Application

In order to validate the applicability of the SafeSpection framework, we validated its core concepts in an industrial project. The objective of the project was the development of a complex, distributed system to realize new functionality in a car. Due to confidentiality reasons it is not possible to show details of the software system or its architecture, but on an abstract basis the project can be described.

4.1 The Application Context

The software system in this project realizes an innovative feature of a future car. The overall software system comprises 8 sub-systems interconnected by a network. Each sub-system is responsible for the realization of one or more features of the functionality. By applying SafeSpection, the manufacturer of the system wanted to ensure that the software system does not impact the overall value-adding processes in an unacceptable way. Hence, in this project, safety was not defined in the common way, i.e., loss of life, or injury to people, but as the loss of an immense amount of money due to such potential negative influences caused by the software system. The SafeSpection approach was used to support the identification of conceptual faults in the general functional specification of the software system and its conceptual architecture. The analysis was performed at the end of the requirements analysis step and after the conceptual architecture of the system had been defined. The manufacturer had already performed an analysis of the potential risks caused by the software system but without systematic guidance.

4.2 The Application Process

The execution of the software safety analysis was organized in 3 full-day workshops. Based on the already identified catastrophic influences of the software system, a fault-tree analysis was performed to analyze the software causes of the unwanted events. In order to support this step, i.e., the identification of conceptual software faults causing the top-event, the SafeSpection framework was used to identify and apply supporting guide-phrases. As outlined in the last section, the first step of the SafeSpection approach is the definition of generic guide-phrases that combine objects and potential influence factors. The analysis of the 500-page software specification written mainly in natural language and the conceptual overview of the software architecture resulted in the following generic objects: processes, components, interactions, pre- and post-conditions, assumptions, and constraints; and in the generic influence factors operational mode, system assumptions, technical and environmental constraints. Based on these concepts generic guide-phrases could be created.

In order to identify potential software causes for the unwanted events, the concrete instances of the identified generic phrases needed to be identified. This was done as part of the FTA workshop. Starting from the unwanted event, those concrete system processes influenced by the software were identified that directly contribute to the unwanted event. Then the components realizing the identified processes as well as the interaction of these components were identified together with the customer's experts. This was done using the customization questions defined in the SafeSpection framework and the results of this step were documented by extended sequence charts showing all concrete objects of realizing the selected processes (see Fig. 6).

The swim-lanes show the concrete components that participate in the identified processes. The grey-boxes represent the objects, pre-conditions, post-conditions, constraints, and assumptions. These were also identified as part of the workshop in cooperation with the customer's experts. For example, the component Pre-Processing 1 requires as a pre-condition the availability of a certain data-item (xyz) and that the initialization has been performed successfully. The component Pre-Processing 2 must fulfill the constraint that the processing of data is completed within 5 ms. The component Data-buffer contains the implicit assumption that not more than 25 requests are sent within one second. Finally, as a post-condition of the whole process the plausible data are presented at the software interface as a output.

The negative form of the post-condition of the whole process represents the unwanted event, i.e., the top-event of the fault tree. Now, the selected guide-phrase patterns guide the identification of the causes of the unwanted top-event. In other words, the guide-phrases were used to systematically identify potential software causes of the unwanted top-events. As it was not possible to derive explicit influence factors prior to the workshop (due to time limitations in the project) the guide-phrases were used as open questions. That is, the guide-phrase patterns were modified in such a way that they ask for potential influence factors that invalidate the object under discussion. The following list shows an excerpt of instantiated guide-phrase patterns derived for analyzing the objects in the sequence chart.

Is it possible that the realization of pre-processing 1 violates the timing constraint "needs to finish in 5 ms" of pre-processing 2?

- Which characteristic of the operational mode contradicts the realization of pre-processing 1?
- Which external condition invalidates the realization of pre-processing 2?
- Which change of characteristics of external components interacting with the application software violate the pre-conditions of pre-processing 1?

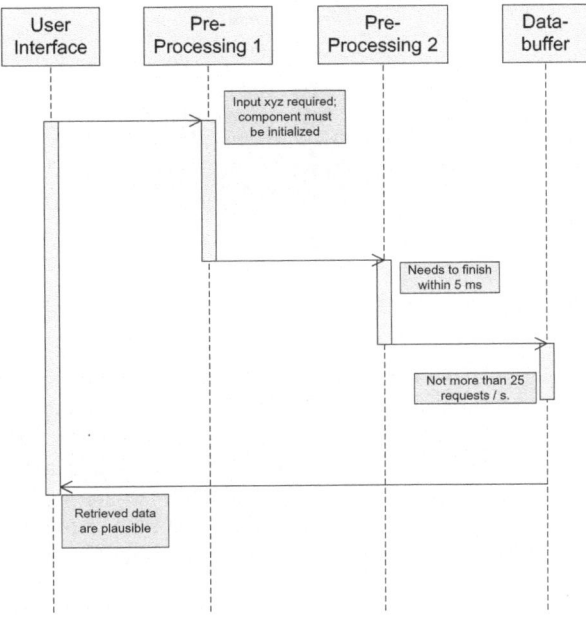

Fig. 6. Application of the Guide-Phrases

- Which change of characteristics of external components interacting with the application software violate the assumptions of the data buffer?
- Is it possible that the semantic of messages is different for pre-processing 1 and pre-processing 2?

The analysis starts from the unwanted top-event and asks whether or not an intermediate event that is described by the guide-phrase triggers the top-event. If this is the case, the event is added to the fault tree, if not, the event described by the next guide-phrase is investigated until all guide-phrases have been considered. The workshop leaders (two of the authors of this paper) derived the guide-phrases, asked the related questions, and modeled the results as extensions of the fault tree.

4.3 The Application Results

Using the guide-phrases defined by the SafeSpection approach resulted in a systematic and easy to apply refinement of the fault tree top-events. The developers involved in the analysis perceived the fault tree technique and the systematic consideration of potential causes as a highly valuable technique to detect conceptual faults in their functional software specification. The application of SafeSpection resulted in project-specific guidance, which could be quickly derived during the FTA-meeting. The management perceived the approach as a success, as the results provided additional conceptual weaknesses in the software specification.

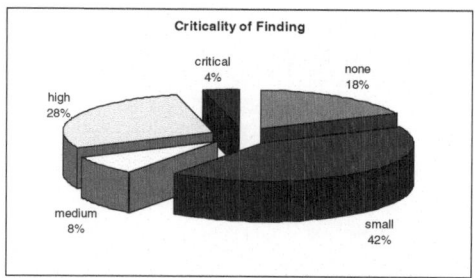

Fig. 7. Criticality of the Software Faults detected with SafeSpection

For the highest prioritized top-event, for example, we could identify 50 additional software faults that could cause the top-event. For the 32 findings rated as high and critical a careful re-consideration of the software specification was performed and mitigation strategies needed to be defined. These results show that the SafeSpection approach created customized guide-phrases that identify so far undetected conceptual software faults quickly and in a feasible way during the FTA-meeting. We could detect and resolve several faults that might have caused catastrophic events for the company.

5   Conclusion

The SafeSpection framework introduced here provides a feasible approach to identify customized and project-specific guidance for the detection of conceptual software faults that have the potential of causing safety-critical system events. We demonstrated the feasibility of the core concept of the approach (a grammar for defining generic guide-phrases) in an industrial case study. The customized guide-phrases supported the identification of 50 additional software faults; 32 of them required the definition of suitable mitigation strategies to prevent a catastrophic top-event.

In future steps, it is important to validate the applicability of the customization approach in more detail. First, it is important to validate the completeness of the provided support on identifying objects and influence factors of the guide-phrases. Second, the resulting guide-phrases will be compared in an empirical study with standard software-safety analysis techniques (like FMEA or FTA) with respect to the type of detected software faults (is it possible to detect more conceptual faults) and the repeatability and comparability of the results.

References

1. Knight, J.C.: Safety Critical Systems: Challenges and Directions. In: 24th International Conference on Software Engineering (ICSE 2002), pp. 547–550. ACM, New York (2002)
2. Leveson, N.: Safeware – System Safety and Computers. Addison Wesley Publishers, Boston (1995)
3. IEC 61508: Institute of Electrical and Electronics Engineers. Functional Safety of electrical/electronic/programmable electronic safety-related systems Part 3 Requirements on Software (1999)

4. ISOWD 26262, Road vehicles, Functional Safety Part 6: Product development software. Working draft (2006)
5. Fenelon, P., McDermid, J.A., Pumfrey, D.J., Nicholson, M.: Towards Integrated Safety Analysis and Design. ACM Computing Reviews 2(1), 21–32 (1994)
6. McDermid, J.A.: Software Hazard and Safety Analysis. In: Lecture Notes in Computer Science, vol. 2469, pp. 23–34 (2002)
7. Papadopoulos, Y., et al.: A Method and Tool Support for Model-based Semi-automated Failure Modes and Effects Analysis for Engineering Designs. In: 9th Australian Workshop on Safety Related Programmable Systems (SCS 2004), pp. 89–95. Australian Computer Society (2004)
8. Lutz, R.R., Woodhouse, R.M.: Bi-directional Analysis for Certification of Safety-Critical Software. In: The proceedings of the International Software Assurance Certification Conference (ISACC 1999), pp. 1–9. Springer, Heidelberg (1999)
9. Pumfrey, D.J.: The Principled Design of Computer System Safety Analysis. PhD thesis. Department of Computer Science, University of York, UK (1999)
10. Chudleigh, M.: Hazard analysis using HAZOP: A case study. In: 12th International Conference on Computer Safety, Reliability and Security (SAFECOMP 1993), pp. 99–108. Springer, Heidelberg (1993)
11. Redmill, F., Chudleigh, M., Catmur, J.: System Safety: HAZOP and Software HAZOP, p. 248. John Wiley & Sons Ltd., Chichester (1999)
12. Lisagor, O., et al.: Safety Analysis of Software Architectures – Lightweight PSSA. In: The proceedings of the 22nd International System Safety Conference (ISSC 2004). IEEE Computer Society, Los Alamitos (2004)
13. Reese, J.D., Leveson, N.G.: Software Deviation Analysis. In: 19th International Conference on Software Engineering (ICSE), pp. 250–260. IEEE, Los Alamitos (1997)
14. Papadoupoulos, Y., et al.: Hierarchically Performed Hazard Origin and Propagation Studies. In: Felici, M., Kanoun, K., Pasquini, A. (eds.) SAFECOMP 1999. LNCS, vol. 1698, pp. 139–152. Springer, Heidelberg (1999)
15. Papadopoulos, Y., et al.: Automating the Failure Mode and Effects Analysis of Safety Critical Systems. In: The proceedings of the 8th International Symposium on High Assurance Systems Engineering (HASE 2004), pp. 310–311 (2004)
16. Rodriguez-Dapena, R.: Software safety verification in critical software intensive systems. Phd Thesis, Eindhoven Technical University, University Printing Office (2002)

Integrating Safety Analyses and Component-Based Design

Dominik Domis and Mario Trapp

Fraunhofer Institute for Experimental Software Engineering, Fraunhofer-Platz 1,
67663 Kaiserslautern, Germany
{dominik.domis,mario.trapp}@iese.fraunhofer.de

Abstract. In recent years, awareness of how software impacts safety has increased rapidly. Instead of regarding software as a black box, more and more standards demand safety analyses of software architectures and software design. Due to the complexity of software-intensive embedded systems, safety analyses easily become very complex, time consuming, and error prone. To overcome these problems, safety analyses have to be integrated into the complete development process as tightly as possible. This paper introduces an approach to integrating safety analyses into a component-oriented, model-based software engineering approach. The reasons for this are twofold: First, component- and model-based development have already been proven in practical use to handle complexity and reduce effort. Second, they easily support the integration of functional and non-functional properties into design, which can be used to integrate safety analyses.

1 Introduction

Today, we are surrounded everywhere by embedded systems. For example, cars have more than 80 microcontrollers, which control, e.g., multimedia systems, comfort functions, and driver assisting functions. A lot of these systems are safety-critical, i.e., a failure of one or more of these systems can lead to accidents involving injury or loss of life. Therefore, standards for the development of safety-critical systems highly recommend considering safety during the complete development process [1]. Safety analyses are intended to be used as part of the constructive development process. They are very valuable for designing safe systems from the very beginning and for having a systematic means for assessing which parts of the system have which impact on safety. This is essential for the cost-efficient development of safety-critical systems. Nonetheless, safety analyses are very time-consuming and, as practical experience shows, are thus often performed only once very late in the development cycle, sometimes even for documentation purposes only. Applied in late phases, however, the analysis results have no direct impact on the development of the system, and their benefit as a constructive means for developing safe systems is thus not recognized. Consequently, from a project manager's or developer's point of view, these analyses become even less important and helpful. Yet, the application of safety analyses is indisputably of crucial importance for the development of safety-critical systems.

M.D. Harrison and M.-A. Sujan (Eds.): SAFECOMP 2008, LNCS 5219, pp. 58–71, 2008.
© Springer-Verlag Berlin Heidelberg 2008

This problem is even more severe for software. In practice, safety analyses are most often limited to hardware, and software has only been regarded as a black box. This is also true for the automotive industry. But automotive software realizes more and more safety-critical functions that can harm people, such as X-by-wire and driving dynamic control systems. It cannot be assumed that these complex embedded systems have zero faults or that their safety can be guaranteed by intensive testing. Besides this, mitigating weak points late in the development process is one of the biggest cost factors in the development of software. So, safety analyses of software architecture and design are as valuable as on the system or hardware level for the constructive development process of safety-critical systems. Furthermore, safety analyses are process-spanning activities, including the system, software, and hardware levels, which cannot be analyzed in isolation. Thus, safety analyses of software are particularly necessary for identifying how failure modes are propagated through or caused by the software and for finding Common Cause Failures that violate the safety assumptions on system level. Because of this, in recent years, the awareness of how software impacts safety has increased rapidly. For example, the working draft of the ISO 26262 and the MISRA safety analysis guidelines [2] recommend safety analyses to also be performed on software. But in order to make safety analyses applicable in the constructive software design process and tap their full potential for the cost-efficient development of safety-critical systems, a significant reduction in complexity and effort is essential. To achieve this, in this paper, we integrate three mature approaches into one design methodology for the design of safety-critical software:

1. Standard safety analyses, i.e., Failure Mode and Effect Analysis (**FMEA**) and Fault Tree Analysis (**FTA**), because they are most intuitively applicable, widespread, and accepted.
2. Semi-automatic safety analyses of model-based design, because their tool support and automation reduces effort, supports the efficient evolution of models, and facilitates consistency between safety and design models.
3. Component-based software engineering, because it uses best software engineering principles and supports reuse.

Chapters 2 to 4 discuss the advantages and disadvantages of the three approaches and the related work in these fields, respectively. All three use different, mutually complementary ways to handle complexity and reduce effort. In order to benefit from all advantages and compensate for disadvantages in the constructive design of safety-critical software, these approaches have to be tightly integrated into one design method that uses and coordinates their activities in an optimal way. This integrated design methodology is presented in chapter 5. The current status and future work are discussed in chapter 6, and chapter 7 gives a short summary and conclusion.

2 Safety Analyses

Safety analyses aim at identifying failure modes, their causes, and those effects that can have an impact on system safety. Their primary goal is not to uncover design faults or prove that an implementation is correct. Safety analyses uncover safety-critical weak points that are theoretically possible failures that may cause hazards and argue whether such hazards are sufficiently improbable in the current system design or not. On the one hand, sufficiently improbable means the actual failure probability of random hardware failures and, on the other hand, the application of appropriate measures and methods for avoiding and mitigating random and systematic faults. Because of this, common verification and validation techniques cannot replace safety analyses, but are prerequisites for developing safe systems.

Most standards and guidelines as well as many experts recommend the combination of an inductive safety analysis, such as FMEA, with a deductive one, such as FTA, to identify and analyze hazards. FMEA identifies failure modes and searches bottom-up for their effects. FTA takes a top event and searches for its causes. In contrast to FMEA, FTA also determines how failure modes are related to each other combinatorially. Both techniques are intuitively applicable, and are the most widely spread and accepted ones. However, the immense effort required to apply FMEA or FTA to complex, software-intensive systems very often impedes their application. While FMEA is accepted and commonly used for hardware and mechanical systems, software is mainly regarded as a black box. Particularly in the automotive industry, FTA is still the exception rather than the rule.

Besides these two, there exist many other techniques that are not widespread or accepted. Many of them are mathematically more powerful, but they are less intuitive, more complex, and therefore less applicable in industry, like Markov chains, Petri nets, or formal models. These can be used to complement more intuitive safety analyses.

3 Automated Safety Analyses of Model-Based Design

Model-based development uses design models, such as Matlab/Simulink, ASCET, or UML, to visually represent software on high levels of design, simulating its behavior and generating code from them. Because of this, model-based development directly helps to handle complexity and reduce effort by using tool support and automation. To support safety analyses, these models can also be annotated with appropriate safety-related information. Based on this, they can be automatically transformed into safety analysis models, or they can be analyzed directly. In this way, information that was already specified in the models is also used in safety analyses to reduce effort. Most of these approaches are based on data flow models, like Matlab/Simulink, ASCET or SCADE, and can be divided into Failure Injection (FI) and Failure Logic Modeling (FLM) [3].

FI injects failure modes into a (formal) model and uses symbolic model checking to identify counterexamples that violate safety requirements [4]. A counterexample is equivalent to a cut set of a fault tree. A cut set is a set of basic events or failure modes that causes the top event of a fault tree. The minimal cut sets of a fault tree are the sets of basic events, where every event must be true for the top event to become true. Because of this, if all counterexamples are identified by the symbolic model checker, they can be used to derive a fault tree whose top event is the disjunction of all minimal cut sets. This is called a minimal cut set tree and its disadvantage is that the tree is completely flat, i.e., it does not show the system structure and therefore, it does not show the failure propagation traces through the system. This, however, is necessary for finding the appropriate places in the system where safety measures need to be implemented. Because of this, additional tools are needed for finding these error traces [5].

The advantage of FI is that by using only formal models, the safety analyses are correct. But this requires the use of a formal design model. If the formal model has to be derived, this is a new source of faults. Additionally, the injection of failure modes is less intuitive than the application of standard safety analyses, resulting in another source of faults. This makes the application of FI more difficult in industrial practice.

Failure Logic Modeling is the second kind of approach to automating the safety analyses of data flow models. Failure Logic Modeling models the local failure flow of modules or components on the lowest hierarchy level, i.e., it analyzes and models the failure modes of the inputs and outputs as well as the components themselves and their causal relationships. For this purpose, logic expressions [6] or finite state machines are used most often. However, there is also one approach that directly uses fault trees to model the local failure flow [7][8]. Based on the models of the local failure flow and the structure of the data flow models, fault trees are automatically generated by most approaches.

One problem and disadvantage of FLM compared to FI that can be found very often in the literature is correctness [3], because FLM is modeled manually. Of course, this can be a source of errors, but it is the strength of safety analyses to use expert knowledge and human intuition to also find problems that are not specified. So, the problem in automating safety analyses is to find the right ratio between human tasks and automation. At least the process of using safety analyses constructively in the top-down development process requires a lot of human thinking. This is why FLM has to be preferred to FI for this purpose. However, FI can be used to verify the correctness of the FLM later in the development process.

Another problem of FLM is the lack of abstraction and refinement [3], which are of major importance in a top-down development process. In FLM, only one hierarchy level can be analyzed, and no relations are defined between the safety analyses of different hierarchy levels. So, most of the time, the entire current system is considered and it is hard to focus only on one hierarchy level. Because of this, solutions for handling complexity in this dimension also have to be found.

4   Component-Based Software Engineering

Abstraction and refinement are inherent parts of component-based software engineering (**CBSE**), which is already a mature approach to handling complexity in the development of IT systems. But for embedded systems, and particularly for safety-critical systems, only proprietary approaches exist until now. Most of them rather address safety-related, non-functional properties, such as real-time behavior and correctness. For example, the Prediction-Enabled Component Technology (**PECT**) of the Predictable Assembly of Certified Components (**PACC**) [9] provides analytical interfaces, which can be used by model checkers to verify properties related to the safety of the system. This use of model checking is equivalent to the Fault Injection mentioned above and also proposed in the Rich Component Model (RCM) [10]. Of course, correct real-time behavior and the formal verification of safety-relevant properties are necessary to guarantee safety, but they are not sufficient for developing safe systems.

However, CBSE is highly likely to further reduce the complexity of constructive safety analyses during the development of safety-critical systems. The main reason for this is *separation of concerns,* which is the basic principle of CBSE [11] and which is applied in three dimensions:

1. Divide and conquer.
2. Rigorous separation between specification and realization.
3. Separation of different functional and non-functional properties by views.

The first two dimensions are illustrated in Figure 1a. Every box is a component, consisting of a specification and a realization. The component specification specifies the black-box behavior of the component, i.e., all externally visible properties or the requirements on the component. This includes the interfaces of the component as well as all externally visible functional and non-functional properties. In contrast to this, the realization shows the component as a gray box, i.e., it shows the black-box specifications of the subcomponents the component consists of and their collaboration. For example, the top component in Figure 1a consists of three subcomponents and the realization of the top component only knows the specifications of the subcomponents and specifies their collaboration. In a top-down development process, this means that the specifications of the subcomponents are derived from the specification of the component based on the component realization. When doing so, a complex system or component is recursively divided into subcomponents until the components are simple enough to be implemented directly. Additionally, the realization of every component is simple, because only the collaboration of appropriate subcomponents has to be defined based on the specifications of the subcomponents. Their inner details are hidden in their realizations. In this way, the development of system and software is recursively separated into many simple and controllable tasks.

The third dimension is illustrated in Figure 1b, which shows the Safe Component Model, an adaptation of the KobrA component model [11]. Both the specification and the realization of a component consist of views. Each view describes another functional or non-functional property of the component. The advantage of the view concept is that different properties of the components are considered separately and clear internal interfaces between the different views are defined. In this way, CBSE not only helps to focus only on one system element on one hierarchy level at any one point in time, but, additionally, on only one property of this element. In this way, the complexity of systems and software is controlled by separating the system into different views of hierarchical components. This is possible because of two reasons: First, for every view, composition rules, which specify how views of different components can be connected with each other, are defined by domain experts. Second, rules for abstraction and refinement between the views of the specification and the realization are defined. Thus, the system is not only divided into components and views, but is also composed of these. Finished components and all their views can be reused, which again reduces effort.

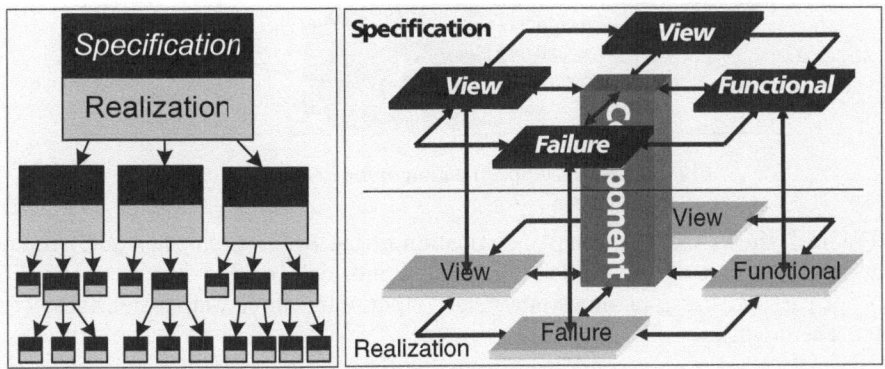

Fig. 1. a) System structure of hierarchical components. b) Safe Component Model.

5   Safe Component Model

In the Safe Component Model (**SCM**), the principle of CBSE is realized and adapted to the model-based design of safety-critical embedded systems. For model-based design, the basic views are data flow models. These are the functional views of specification and realization in Figure 1b. In order to make safety analyses efficiently and constructively applicable during top-down development, safety analysis views and appropriate automations are needed. Besides these, views for other non-functional properties can be used dependent on the application domain. But in this paper, we are focusing on the functional and safety views.

For the top-down design of safety-critical software, the functional views and the safety views have to be tightly integrated. The first step is to specify the intended functional behavior of the component in the *functional specification*. Based on this, the failure behavior of the component has to be assessed directly and the results are modeled in the *failure specification*. After this, the component specification can already be used in analyses on superordinate levels. The specification is described in section 5.1.

In the next step, which is described in section 5.2, the specification is realized by collaborating subcomponents. The subcomponents used and their collaboration are specified in the *functional realization*. Because every subcomponent has a failure specification, Failure Logic Modeling can be used to semi-automatically generate the *failure realization*. In this top-down process, the specification can be seen as the requirements on the realization. Because of this, it has to be checked whether the realization fulfills the specification or not. The relationship between failure specification and realization is described in section 5.3.

5.1   Specification

A functional specification is a simple functional block with input and output interfaces. Therefore, the functional specification is equivalent to a SubSystem in Matlab/Simulink and many other model-based development approaches. Additionally, the syntax and semantic of the interfaces have to be specified in order to describe the functionality of the block, make it reusable and analyzable.

Fig. 2. Functional Specification of the Brake Controller

Figure 2 shows the functional specification of the *Brake Controller* (**BC**) component, which is part of the traction control and anti-lock braking system of the IESE concept car. BC has four input interfaces (inputs) and three output interfaces (outputs). The inputs are:

- I1 *steering_angle_driverInput*, which has the type integer, a value range from 0 to 180, and describes the steering angle in degrees that is set by the driver.
- I2 *v_carRef*, which is of the type double with a value range between 0 and 100 describing the speed of the car in kilometers per hour.
- I3 *brake_driverInput*, which is an integer with the value range 0-100 that specifies the braking power that is set by the driver.
- I4 *v_yaw*, which is of the type int with a value range from 0-360 and describes the rotation of the car in degrees per second around its y-axis (vertical axis).

All four inputs are used to calculate the braking power at the individual wheels that is required to maintain the driving stability of the car and to steer and decelerate it as intended by the driver. Both rear wheels are affected by the same brake, which is controlled by the output O3 *corrected_rearBrake*. O1 *corrected_brake_FL* controls the brake at the front left wheel and O2 *corrected_brake_FR* that at the front right wheel. All three outputs are integer values between 0-100 and describe the power at the brakes. Additional information includes the exact physical functions and assumptions, when the component can be used, and how. This syntactical and semantic information can be described, e.g., by type systems, invariants, preconditions, or post-conditions. But for this simplified example, it is unimportant how the information is specified and what the exact functionality is. The important point is that SCM can be applied on any model that makes minimal use of interfaces in this way.

After specifying the functional requirements of the component in the functional specification, its failure behavior, which might have an impact on safety, has to be assessed. This early safety assessment can be used in the safety analyses of a super component and as safety requirements on the realization of the component. In this way, the failure specification makes safety-critical components reusable, because the safety model is part of the component specification. Besides this, by analyzing safety immediately, no information is lost, because the engineer still remembers what he/she assumed.

Because SCM uses standard safety analyses, the analysis is very intuitive and guided by mature techniques. In the first step, an Interface Focused-FMEA (IF-FMEA) [6] is applied on the functional specification. First, the IF-FMEA searches for failure modes at the inputs and outputs of the component as well as those of the component itself. To identify failure modes, the concept of HAZOP guidewords is used, which have also been adapted for software [12]. This method is called SHARD and proposes the following guidewords: Omission, Comission, Value, Early, Late. A standard set of guidewords is a useful basis, but has to be adapted to the domain and application being analyzed. For this purpose, an object-oriented *Failure Type System* [13] is used (Figure 3). Each failure mode of an input, output, and component gets an unambiguous *failure type*, such as *FM_Omission* or *FM_Value*. The Failure Type System can be adapted to every application; failure modes can be defined and attributes can be used to refine the semantics of the failure modes. For example, for the failure type *FM_High_Deviation* in Figure 3, it has to be specified which deviation is tolerable before it is considered a failure.

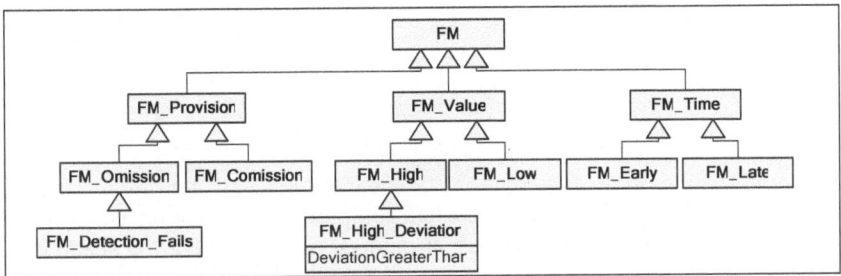

Fig. 3. Example of a Failure Type System

In the second step, the IF-FMEA searches for causes and effects of failure modes. Causes of output failure modes may be internal failure modes of the components, or failure modes of inputs. Vice versa, the effects of input failure modes and internal failure modes are output failure modes. These relationships are defined during the second step of the IF-FMEA. In the next step, the information of the IF-FMEA is refined by a Component Fault Tree (**CFT**) [14], i.e., the combinatorial relationships between output, input, and internal failure modes are investigated and further failure modes are identified. The output failure modes thus become the top events of the CFT, the internal and input failure modes become basic events. CFTs directly support the component concept by enabling the definition of output and input events. The output failure modes can thus be defined as output events (filled triangles in Figure 4) and the input failure modes as input events (triangles, open at the bottom). In this way, the CFT can be easily used in the FTA of a superordinate component.

In the BC example, the input *BC.I1* has the failure types *FM_Low*, *BC.I2 FM_High*, and *BC.I3* as well as *BC.I4 FM_Value* in the CFT of Figure 4. BC itself can have an internal FM, *BC.Int1 FM*, and the failure detection of the input signals can fail, *BC.Int2 FM_Detection_Fails*. These failure modes are part of the specification, because their effects are externally visible, they are requirements on the realization, and they do not show inner details of the component. Because of this, information hiding is also guaranteed in the failure specification. All input and internal failure modes can cause the corrected brake value at the output *BC.O3* to be wrong, which is represented by the failure types *FM_Value* . All input FMs except *BC.I3 FM_Value* can only cause *FM_Value* if the internal failure detection of BC fails. Because of this, *BC.Int2 FM_Detection_Fails* is combined with these input failure modes by an AND-gate. Additionally, *BC.Int1 FM* can delay the corrected rearBrake, which is represented by *BC.O3 FM_Late*. *BC.Int1 FM* and *BC.I3 FM_Value* are single points of failures.

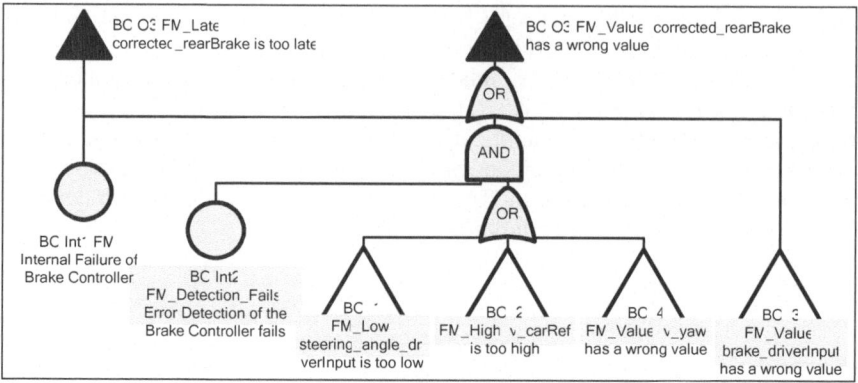

Fig. 4. Failure Specification of the Brake Controller

In this way, all failure modes of the current component specification are easily identified and the failure behavior is assessed. If the analysis shows that the specification is not suitable for achieving safety in the context of the supercomponent, the specification can be changed immediately and alternatives can be compared. For example, one may decide that further inputs are needed to increase the efficiency of error detection and handling. For this purpose, no quantitative analyses are needed, only qualitative and sensitivity analyses assessing the impact of events. In this way, the functional and safety requirements of the component are derived from the super component and are directly considered in the subsequent realization of the component.

5.2 Realization

The functional realization is a gray box specification of the component, which defines or reuses appropriate subcomponent specifications to realize the requirements specified in the component specification. This is done by trial and error or by using expert knowledge or design patterns. Many model-based approaches have similar hierarchical model elements, like the definition of a SubSystem in Matlab/Simulink [15]. So, both the specification and the realization of SCM can be applied to most model-based approaches.

Figure 5 shows the final functional realization of BC. BC consists of the *steering angle delimiter* (**SAD**) and the *yaw rate corrector* (**YRC**). SAD uses the inputs *BC.I1* and *BC.I2* of BC. With these inputs, the YAC calculates the *delimited_steering_angle* at the output *SAD.O1*, which is connected with *YRC.I1*, because YRC requires this input. The other two inputs of YRC, *YRC.I2* and *YRC.I3*, require *v_yaw* and *brake_driverInput*, which are the other two inputs of BC. All three outputs of YRC are directly connected with the corresponding outputs of BC because they provide the necessary signals.

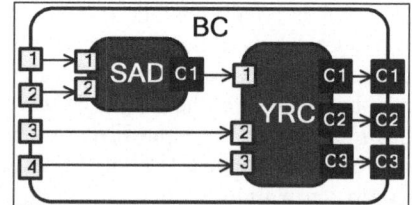

Fig. 5. Functional Realization of the Brake Controller

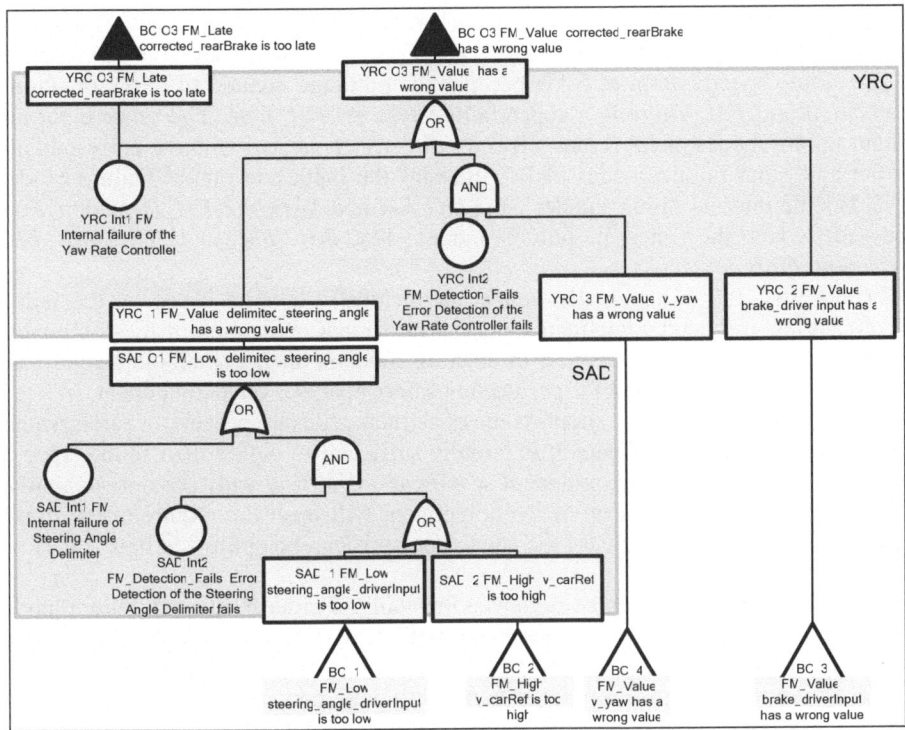

Fig. 6. Failure Realization of the Brake Controller

Thus, BC is composed of YRC and SAD based on their specifications. In order to avoid interface problems or the composition of unsuitable components, interfaces can only be connected with each other if they are syntactically and semantically compatible. For this purpose, appropriate composition rules have to be defined. For SCM, it is only important to mention that these composition rules do not only include the functional views, but also the failure views. This reduces effort and helps to define or identify appropriate subcomponents. It is automatically checked whether two component interfaces can be connected with each other. For this purpose, in addition to the functional syntax and semantic, it is checked whether the failure types of each interface are compatible: The ports of components and subcomponents can only be connected if they have the same failure type or the failure type of the required interface is a super failure type of the provided interface. In case the failure types have attributes, these also have to be considered in the compatibility check. If many provided interfaces have to be connected with a single required interface, an OR-gate has to be used. This guarantees that the failure realization contains a properly connected CFT and corresponds to the semi-automatic safety analyses done by Failure Logic Modeling.

The CFT of the failure realization of BC is shown in Figure 6. Equivalent to the functional realization, the failure realization is composed of the specifications of its subcomponents, but here the failure specifications are used instead of the functional realizations. So, the gray box in the lower left part of the picture, which is labeled SAD, shows the instantiated failure specification of the SAD.

The top event of the SAD failure specification is *SAD.O1 FM_LOW*. This is connected with the input failure mode *YRC.I1 FM_Value*, which is an input failure mode of the failure specification of SAD because the ports are connected in the functional realization and *FM_Value* is a super failure type of *FM_Low*. The other input and output failure modes of SAD and YRC are directly connected with the corresponding output and input failure modes of BC. Besides the input and output Failure Modes, SAD has the internal failure modes *SAD.Int1 FM* and *SAD.Int2 FM_Detection_Fails* and YRC has the internal failure modes *YRC.Int1 FM* and *YRC.Int2 FM_Detection_Fails*.

In this way, the failure realization is automatically derived based on the failure specifications of the subcomponents and the functional realization of the component. In a bottom-up approach, the reuse of existing subcomponents results in a significant decrease in the effort needed for performing safety analyses on a component. In a top-down approach, this directly supports the constructive design process of safe systems. For the defined subcomponents, it is initially sufficient to define their failure specifications. Based on these subcomponent specifications, it is already possible to analyze whether the current realization of the component will meet the requirements. In this way, the failure specifications of the subcomponents can be optimized before they are realized.

The failure realization can be reviewed manually in order to identify failure modes that have not been considered in the model until now. These have to be added manually at the appropriate point, but the failure specification of the subcomponent and the instance used in the realization are automatically kept consistent. Thus, manual steps are also necessary in the failure realization, but because most steps are automated, the effort is low.

5.3 Relation between Specification and Realization

In this top-down design process, the functional realization is used to derive the safety specification of the subcomponents from the safety or failure specification of the component. After the specification and the realization of the component are finished, it has to be checked whether the realization fulfills the specification or not. For this purpose, appropriate and application-specific rules have to be defined. This includes, for example, that both have the same input and output failure modes and that the output failure modes have the same MCS. The internal failure modes of the specification can summarize internal failure modes of the realization. For example, the failure realization of BC has six MCS: *YRC.Int1 FM, SAD.Int1 FM, SAD.Int2 FM_Detection_Fails & BC.I1 FM_Low, SAD.Int2 FM_Detection_Fails & BC.I2 FM_High, YRC.Int2 FM_Detection_Fails & BC.I4 FM_Value, and BC.I3 FM_Value*. When *YRC.Int1 FM* and *SAD.Int1 FM* are summarized to *BC.Int1 FM* and *SAD.Int2 FM_Detection_Fails* is summarized together with *YRC.Int2 FM_Detection_Fails* to *BC.Int2 FM_Detection_Fails*, they constitute all MCS of the BC failure specification. Here, all internal failure modes with the same failure type were summarized, but other rules may also be defined. Failure probabilities might also be used for this purpose when this seems suitable, but the results of qualitative analyses should be preferred in the analysis of software.

The same ideas can be used to automatically derive the failure specification from the failure realization when a preliminary failure specification should be substituted by one that is closer to the realization or when no failure specification exists. In general, the important point of the failure specification is that only externally visible properties of the component are shown and all inner details are hidden. Because of this, the BC failure specification does not show any information about YRC or SAD. All internal failure modes of BC are failure modes of BC itself. So, when a failure specification is generated from a failure realization, all internal events must be renamed. Additionally, they can be summarized and the tree is transformed into an MCS tree, which does not show any details about the inner structure of BC. In this way, no inner details of a component are betrayed by the failure specification. First, this helps to abstract from details and to focus on the relevant things on the current component level. Second, this information hiding is of major importance for the protection of intellectual properties in a distributed development. If a supplier delivers a component that has to fulfill a safety-critical functionality, this component has to be considered in the safety analysis of the system, but without betraying any intellectual properties regarding the component. Because of this, rigorous information hiding is necessary in the failure specification.

6 Current Status

The SCM has already been implemented as part of the ComposeR tool for the component-oriented, model-based development of safety-critical embedded systems. The tool makes it possible to extend SubSystems in Matlab/Simulink with complete component specifications and realizations and to analyze the extended Simulink models. For safety, this includes connectability and safety analyses. The safety analyses can be performed with ESSaRel [16] or Fault Tree + [17]. ESSaRel is easier to use, since it directly supports CFTs, but Fault Tree + is one of the most widely used fault tree tools in industry. Because of this, ComposeR also supports the generation of fault trees of the entire system in Fault Tree+ for certification purposes. Besides safety, ComposeR already supports views for graceful degradation/adaptation and further views are currently being implemented. Moreover, the INProVe tool was developed based on ComposeR for the architectural analysis of dataflow models. The results of this tool are used to support model-based safety analyses.

   The SCM methodology and the ComposeR tool were used in the development of the traction control and anti-lock braking system of the IESE concept car. This is a radio controlled model car with a combustion engine equipped with sensors, actuators, and ECUs for implementing the intended functionality. Thus, it is a real practical example. In the next step, we will validate the methodology and the tool in an industrial case study and develop them further based on the results.

7 Summary and Conclusion

This paper has explained that safety analyses should be used as part of the constructive development process of safety-critical systems and software, in order to develop safe systems and avoid the costs of late analyses and changes. Particularly for software, however, safety analyses are too complex for many companies to apply. To better handle the complexity of software safety analyses, we developed a method for tightly integrating standard safety analyses, like FMEA and FTA, into a component-oriented, model-based software design method. In this way, the safety analyses benefit from the separation of concerns provided by component-based software engineering. The system is divided into controllable subcomponents and the safety analyses either focus on the specification or the realization of the current component. Safety analyses on higher component levels abstract from details that are refined on lower component levels. So, there is a clearly defined scope for every step of the analysis. The impact on safety of every component is automatically analyzable at each component level. The refinement is absolutely traceable across the different component levels and particularly includes the safety analyses. Moreover, through the rigorous separation between specification and realization, information hiding and protection of intellectual properties are guaranteed in distributed development between different companies. Besides this, the method actively supports the reuse of components, because the safety analysis model becomes an inherent part of the component model. The approach is tool-supported and applicable to model-based designs like Matlab/Simulink. Because of this, SCM helps to handle the complexity of safety analyses and makes them constructively applicable during the software design process, where they achieve the greatest benefit.

References

1. IEC 61508: Functional safety of electrical/electronic/programmable electronic safety-related systems, International Electrotechnical Commission (1999)
2. MISRA: Guidelines for safety analysis of vehicle based programmable systems, MIRA Limited, Warwickshire (2007)
3. Lisagor, O., McDermid, J.A., Pumfrey, D.J.: Towards a Practicable Process for Automated Safety Analysis. In: 24th International System Safety Conference, pp. 596–607 (2006)
4. Bozzano, M., Villafiorita, A.: ESACS: An Integrated Methodology for Design and Safety Analysis of Complex Systems. In: 14th European Safety and Reliability Conference, pp. 237–245. Balkema Publishers, Maastricht (2003)
5. Bretschneider, M., Holberg, H.-J., Peikenkamp, T., Böde, E., Brückner, I., Spenke, H.: Model-based Safety Analysis of a Flap Control System. In: Proceedings of the INCOSE 2004 – 14th Annual International Symposium, Toulouse (2004)
6. Papadopoulos, Y., McDermid, J.A.: Hierarchically Performed Hazard Origin and Propagation Studies. In: Felici, M., Kanoun, K., Pasquini, A. (eds.) 18th International Conference on Computer Safety, Reliability and Security. LNCS, vol. 1608, pp. 139–152. Springer, Heidelberg (1999)
7. Grunske, L., Kaiser, B.: Automatic Generation of Analyzable Failure Propagation Models from Component-Level Failure Annotations. In: 5th IEEE International Conference on Quality Software, pp. 117–123. IEEE Computer Society Press, New York (2005)
8. Grunske, L.: Towards an Integration of Standard Component-Based Safety Evaluation Techniques with SaveCCM. In: Hofmeister, C., Crnković, I., Reussner, R. (eds.) QoSA 2006. LNCS, vol. 4214, pp. 199–213. Springer, Heidelberg (2006)
9. Wallnau, K.C.: Volume III: A Technology for Predictable Assembly from Certifiable Components (PACC). Technical report CMU/SEI-2003-TR-009, Pittsburg, PA: Software Engineering Institute, Carnegie Mellon University (2003)

10. Damm, W., Votintseva, A., Metzner, A., Josko, B., Peikenkamp, T., Böde, E.: Boosting Re-use of Embedded Automotive Applications Through Rich Components. In: Proceedings of the Foundation of Interface Technology Workshop. Elsevier Science, Amsterdam (2005)
11. Atkinson, C., Bayer, J., Bunse, C., Kamsties, E., Laitenberger, O., Laqua, R., Muthig, D., Peach, B., Wüst, J., Zettel, J.: Component-based Product Line Engineering with UML. Addison-Wesley, London (2001)
12. Pumfrey, D.J.: The Principled Design of Computer System Safety Analyses, DPhil Thesis, University of York (1999)
13. Giese, H., Tichy, M., Schilling, D.: Compositional Hazard Analysis of UML Component and Deployment Models. In: Heisel, M., Liggesmeyer, P., Wittmann, S. (eds.) SAFECOMP 2004. LNCS, vol. 3219, pp. 166–179. Springer, Heidelberg (2004)
14. Kaiser, B., Liggesmeyer, P., Mäckel, O.: A New Component Concept for Fault Trees. In: Lindsay, P., Cant, T. (eds.) Proceedings of the 8th Australian workshop on Safety critical systems and software, Canberra, vol. 33, pp. 37–46. Australian Computer Society (to be published, 2003); Conferences in Research and Practice in Information Technology Series
15. MathWorks, Simulink: Simulation and Model-Based Design, http://www.mathworks.com
16. Embedded Systems Safety and Reliability Analyser (ESSaRel), http://www.essarel.de
17. Isograph: Fault Tree Analysis Software - FaultTree, http://www.isograph.com

Modelling Support for Design of Safety-Critical Automotive Embedded Systems

DeJiu Chen[1], Rolf Johansson[2], Henrik Lönn[3], Yiannis Papadopoulos[4],
Anders Sandberg[5], Fredrik Törner[6], and Martin Törngren[1]

[1] Royal Institute of Technology, SE-10044 Stockholm, Sweden
{chen,martin}@md.kth.se
[2] Mentor Graphics Corp., SE-41755 Gothenburg, Sweden
rolf_johansson@mentor.com
[3] Volvo Technology Corp., SE-40508 Gothenburg, Sweden
Henrik.lonn@volvo.com
[4] University of Hull, Hull HU6 7RX, UK
Y.I.Papadopoulos@hull.ac.uk
[5] Mecel AB, SE-400 20 Gothenburg, Sweden
anders.sandberg@mecel.se
[6] Volvo Car Corp., SE-40531 Gothenburg, Sweden
ftorner@volvocars.com

Abstract. This paper describes and demonstrates an approach that promises to bridge the gap between model-based systems engineering and the safety process of automotive embedded systems. The basis for this is the integration of safety analysis techniques, a method for developing and managing Safety Cases, and a systematic approach to model-based engineering – the EAST-ADL2 architecture description language. Three areas are highlighted: (1) System model development on different levels of abstraction. This enables fulfilling many requirements on software development as specified by ISO-CD-26262; (2) Safety Case development in close connection to the system model; (3) Analysis of mal-functional behaviour that may cause hazards, by modelling of errors and error propagation in a (complex and hierarchical) system model.

Keywords: Automotive Embedded Systems, Dependability, Model-Based Development, Safety Analysis, Safety Case.

1 Introduction

Safety is posing an increasing challenge for the developers of automotive embedded systems, also referred to as automotive Electrical/Electronic (E/E) systems. While accounting for a large portion of the innovations and flexibility, the underlying computer software and hardware also results in growing product complexity. The last decade has indeed shown that an increasing number of vehicle failures stem from errors related to embedded systems.

Currently, ISO is developing a standard on Functional Safety for Road vehicles (ISO-CD-26262) [1]. As pointed out in its introduction, with the high complexity growth there is an increasing risk of failures in automotive embedded systems. This

M.D. Harrison and M.-A. Sujan (Eds.): SAFECOMP 2008, LNCS 5219, pp. 72–85, 2008.
© Springer-Verlag Berlin Heidelberg 2008

makes safety a key issue in future automobile development. ISO-CD-26262 requires that a complete Safety Case is developed, presenting evidence that the system is safe. It also specifies the requirements on product development at software level.

While state-of-the-art safety analysis techniques [2] provide support for deriving the causes and consequences of errors, the difficulties remain in capturing and maintaining plausible errors, safety requirements, and other related information along with design refinement, changes and evolution, and in providing the safety argument. Such analysis techniques in turn rely on system modelling and management support, as well as the alignment with tools, processes, and standards. One challenge with current methods for automotive E/E systems development is the lack of systematic approaches to information management, architecting and verification. Solutions relying on social and traditional text-based communication do not scale for handling advanced embedded systems. Software architectures and/or exchange format standards such as AUTOSAR [3] offer a significant improvement of the state of practice. However, experience tells us that advanced and complex systems also require model-based engineering encompassing appropriate abstractions and views for both cost-efficiency and development effectiveness. Over the years, the demand for additional levels of abstractions and views has been continuously raised [6, 12].

System modelling based on an architecture description language (ADL) is a way to keep the engineering information in a well-defined information structure. In this paper we present how the architecture description language EAST-ADL2, complementary to AUTOSAR, provides a basis for systematic development of safety-critical automotive systems. As a language for architecture description, the EAST-ADL2 captures the domain knowledge for automotive embedded systems and provides the modelling means for keeping various engineering information, e.g., across multiple levels of abstraction and concerns, within one infrastructure. Three important areas of EAST-ADL2 will be highlighted in this paper: (1) System development based on models on different levels of abstraction. This enables fulfilling many requirements on software development as specified by ISO-CD-26262; (2) Safety Case development in close connection to the system design; and (3) Analysis of hazardous failures by modelling of errors and the propagations in a hierarchical system model. The integration of these aspects provides structured information handling of requirements, design, safety analysis, other verification and validation information, and design decisions. The approach supports reuse, consistency between models, automated handling of dependencies, view generation, transformations and analysis.

The paper is organised as follows: We first give an overview of EAST-ADL2 showing its capabilities for model-based development, and how it is complementary to AUTOSAR. Then we describe the modelling support for a Safety Case. In the following section we describe error modelling and modelling of error propagation, and the link to the HiP-HOPS safety analysis tool. Finally, we illustrate the approach with an industrial case study on one ECL (Electronic Column Lock) system.

2 Overview of EAST-ADL2

EAST-ADL2 is developed in the ATESST project (www.atesst.org), further extending and refining the EAST-ADL language from the EAST-EEA project (www.east-eea.org). It is a domain-specific architecture description language aiming to adequately meet the engineers' needs regarding information management and practical methods in the development of advanced automotive embedded systems. The language provides an ontology for all the related engineering information and a set of well-defined constructs for the capturing and structuring of such information in a standard format. The covered system aspects include requirements, vehicle features, functions, variability, software and hardware design, and environment, as well as the related structures and behaviours. For the purpose of early quality assessment and verification, the language also supports the capturing of other necessary non-functional properties and thereby enables the reasoning of system timing and failure modes. Through its constructs for traceability, the EAST-ADL2 allows the modelling of dependencies across requirements, structural items and V&V information.

2.1 Hierarchies and Levels of Abstraction

While stipulating the abstractions and viewpoints that are of particular importance in the development of automotive embedded systems, the EAST-ADL2 further enforces separation-of-concerns and complexity-control through a multi-viewed and hierarchical modelling language. The core concept is to structure the solution architecture into five levels of abstraction: *VehicleLevel*, *AnalysisLevel*, *DesignLevel*, *ImplementationLevel*, and *OperationalLevel*. See also Figure 1. The levels correspond to system views that can be used to support a variety of processes (from top-down to bottom-up), including the typical scenario for platform-based product families, where a new function is added to an existing system. The architectural solution at each abstraction level is self-contained in the sense that it constitutes a complete model of the system under consideration from a particular viewpoint. Within each of these architectural solutions, hierarchies of composition are supported by dedicated constructs to describe the part-whole relations of functions/components.

The models at the *VehicleLevel* provide a top-level view of the E/E system of a vehicle where the intended electronic features are described and elaborated in respect to the related product-line organizations. One view that captures the realizations of such

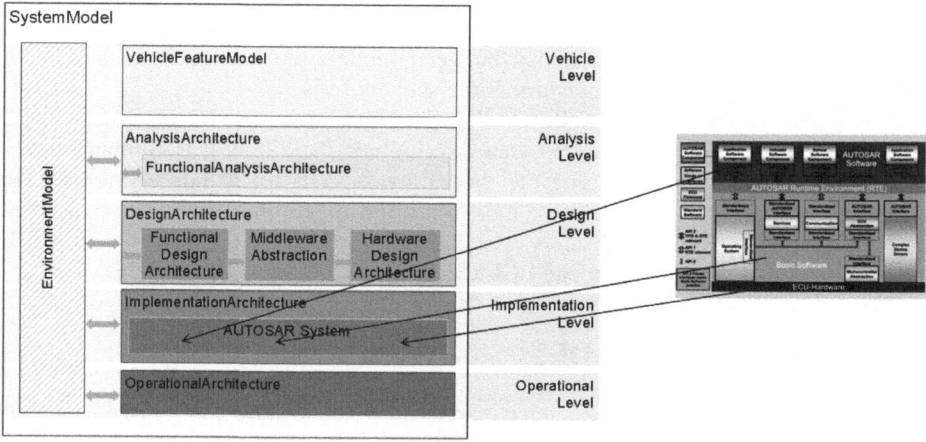

Fig. 1. EAST-ADL2 abstraction layers and its relation to AUTOSAR [9]

electronic features in terms of logical functions and principal interfaces is given at the *AnalysisLevel*. A further refined view is provided at the *DesignLevel* where more implementation-oriented aspects are taken into consideration, such as alignment with intended software decomposition and the target platform, fault tolerance, sensor and actuator interfacing, etc. The support by EAST-ADL2 at this level includes the functional design architecture for application software, the middleware abstraction for platform software (e.g., middleware, RTOS etc.), and the hardware architecture for target platform (e.g. I/O, sensor, actuator, power, ECU, topology and electrical wiring including communication bus). It allows the reasoning of partitioning and allocation of functions as well as the verification of the preliminary design either by simulation or analysis techniques. The overall structure at the *DesignLevel* is such that one or several entities can be later realized by AUTOSAR entities captured at the *ImplementationLevel* [9]. Full traceability is possible from function definitions at the vehicle level to AUTOSAR entities. The *OperationalLevel* is hidden by AUTOSAR concepts via deployment on the AUTOSAR RTE (Run-Time Environment), representing the E/E system as it is realized in the manufactured products.

One example of this hierarchical multi-viewed modelling approach is illustrated in Figure 2, with an electronic feature *Brake* (denoted by the *EFeature* construct), models of more detailed solutions, and the final implementation. The solutions at the *AnalysisLevel* include the logic function *BrakeCtrl* (denoted by the *ADLFunction* construct) and the abstract interfaces *BrakePedal* and *BrakeMotor* for the interactions with the vehicle environment (denoted by the *FunctionalDevice* construct). The corresponding software and hardware design solutions are shown at the *DesignLevel*. While the logic function *BrakeCtrl* is realized by the software function *BrakeCtrl*, the abstract interfaces are represented by hardware devices (denoted by the *DeviceIF* construct) and software components for signal transformation (denoted by the *LocalDeviceManager* construct). The implementation of the design is given by AUTOSAR concepts at the *ImplemenationLevel* (e.g., an elementary *ADLFunction* is mapped to a *RunnableEntity* of an *AtomicSoftwareComponent*).

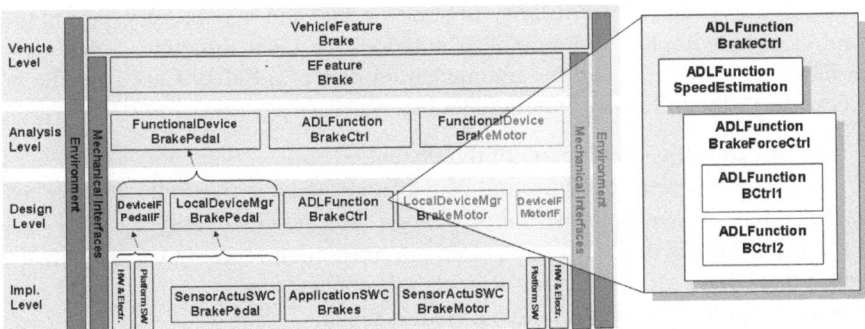

Fig. 2. An example showing the electronic feature (Brake) and its representations with the EAST-ADL2 abstraction levels

2.2 Requirements and Traceability Support

EAST-ADL2 provides explicit support for requirement specification and management in the development of advanced embedded systems. It differentiates between functional requirements, which typically focus on some part of the "normal" functionality that the system has to provide (e.g. "ABS shall control brake force via wheel slip control"), and quality requirements, which typically focus on some external property of the system seen as a whole (e.g. "ABS shall have an MTTF of 10,000 hours"). To allow integration of external requirements tools, EAST-ADL2 provides supports for the mapping of Requirements Interchange Format (RIF) [7] concepts.

The language treats requirements as separate entities and provides specific constructs to support the traceability by extending and adapting related principles from SysML [8]. Typically, based on requirements on the higher abstraction levels of EAST-ADL2, more detailed requirements are derived along with the refinements and decompositions. Specific associations are introduced to relate requirements to their target elements (through the *ADLSatisfy* construct). EAST-ADL2 introduces the notion of *Verification&Validation Case* (denoted by the *VVCase* construct) in order to show how a certain requirement is verified in a particular system context as well as to support the planning, tracking, and updating of V&V efforts,. While linking certain requirements and target entities, each *VVCase* provides a description of the related evaluation information and activities.

3 Safety Case Support in EAST-ADL2

A Safety Case provides structure to the qualitative argumentation about why a system is safe enough. Hence, the Safety Case is dependent on referencing and aggregating information of different types related to the systems functionality and realization. Therefore, integration with an ADL is useful for system development. Currently, there are no requirements for Safety Cases in the automotive industry, but in the upcoming automotive safety standard ISO-CD-26262 [1] such requirements are raised. This section will provide the safety case metamodel which was implemented in EAST-ADL2. A more detailed description is presented in [10].

A Safety Case can consist of large amounts of data and may be very hard to grasp. To mitigate the complexity, a graphical notation, the Goal Structure Notation, have been introduced by Kelly for the argumentation part of a Safety Case [4]. The notation consists of the following building blocks:

- *Goal* – A claim about a property of the system.
- *Strategy* – A description of how and why a Goal can be derived into other Goals.
- *Justification* – Provides further rationale for a selected GSN entity.
- *Evidence* – This is the set of leafs of an argumentation representing the actual evidence that shows satisfaction of the goals it is connected to.
- *Context* – Defining in what context a Goal is given.

A safety case metamodel is shown in Figure 3. It is based on a description of GSN and shows how the GSN entities relate to each other. The safety case entity itself is the top level of a safety case, and it consists of the GSN argument entities. The safety case can also consist of several other safety cases, in a hierarchical structure. In order to maximize the traceability of the design data, each GSN class can be associated to any EAST-ADL2 entity. This will provide support for consistency of the data as well as support for the change management process.

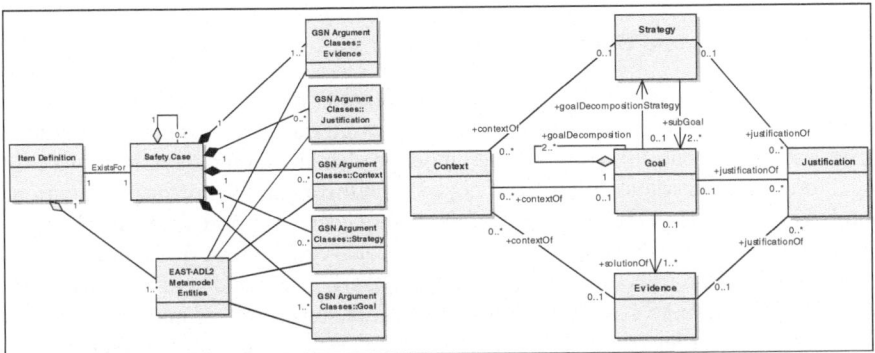

Fig. 3. The safety case metamodel based on GSN

A *Safety Case* is valid for a system or function, and this scope needs to be defined. As shown in Figure 3, the safety case scope is defined by the *ItemDefinition* class that is a collection of EAST-ADL2 entities, i.e. all available specifications entities for the given item. The metamodel also contains the relations between the internal elements. As shown in Figure 4, the *Goal* is the centre of the safety case structure. It can be decomposed directly, or through the usage of a strategy, into two or more *Goals*. Each *Goal* shall have a solution relation to at least one *Evidence*, also known as *Solution* in the GSN notation. The *Evidence* entity can have several specializations, ranging from protocols of V&V activities to design decisions. Each element in the GSN structure can also be related to a *Context* entity indicating that a description of the context can be provided. Similarly, a *Justification* for increased clarity can be provided for each GSN entity, by a justification relation.

The *Evidence* represents any information that supports or, in its ultimate form, proves that the *Goal* it is connected to is achieved. As such, the information can be of many types, e.g. analysis reports, design specifications, requirements, protocols from V&V activities, etc. The system model of EAST-ADL2 captures most of these entities in suitable packages, e.g. a package focusing on verification and validation, the V&V package. In [10], the described safety case metamodel clearly visualizes the GSN entities and interdependencies which has the advantage of facilitating the comprehension and easing the training effort.

4   Error Modelling Support in EAST-ADL2

As an overall system property, safety is concerned with the anomalies (in terms faults, errors, and failures) and their consequences under given certain environmental conditions. Functional safety represents the part of safety that depends on the correctness of a system operating in its context [11] and addresses the hazardous anomalies of a system in its operation (e.g., component errors and their propagations). The objective of the EAST-ADL2 error modelling is to allow an explicit reasoning of functional safety and thereby to facilitate safety engineering along with an architecture design or maintenance process.

4.1 Key Concepts and Domain Model

EAST-ADL2 facilitates safety engineering in regards to the modelling and information management. While supporting the safety design through its intrinsic architecture description and traceability support, the language also allows the developers to explicitly capture the error logics in terms of component errors and the error propagations in an architecture error model through its error modelling support (see also Figure 4). The error modelling is treated as a separated analytical view. It is not embedded within a nominal architecture model but seamlessly integrated with the architecture model through the EAST-ADL2 meta-model. This separation of concerns is considered necessary in order to avoid some undesired effects appearing when error modelling and nominal design is mixed, during comprehension and management of nominal design, reuse of models, and system synthesis (e.g., code generation).

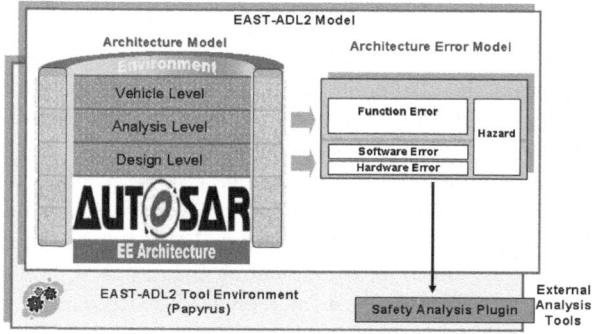

Fig. 4. EAST-ADL2 error modelling extends the nominal architecture in a separate view and provides analysis leverage through external tools

The EAST-ADL2 error modelling package extends a nominal architecture model, typically at the *AnalysisLevel* and *DesignLevel*, with the information of failure semantics and error propagations. The failure semantics can be provided in terms of logical or temporal expressions, depending on the analysis techniques and tools of interest. Such analytical information, together with environmental conditions, forms the basis for identifying the likely hazards, reasoning about the causes and consequences, and thereby deriving the safety requirements. The relationships of local error behaviours are captured by means of explicit error propagation ports and connections. Due to these artefacts, EAST-ADL2 allows advanced properties of error propagations, such as the logical and temporal relationships of source and target errors, the conditions of propagations, and the synchronizations of propagation paths. Hazards or hazardous events are characterized by attributes for severity, exposure and controllability according to [1]. A hazardous event may be further detailed by e.g. use cases, sequence or activity diagrams. In an architecture specification, an error is allowed to propagate via design specific architectural relationships when such relationships also imply behavioural or operational dependencies (e.g., between software and hardware). Fig. 5 shows the domain model definitions of constructs for the error modelling. The key concepts include:

- *ErrorBehavior*: the definitions of possible failure behaviours of an *ADLEntity* (i.e., an abstract function or component).
- *ErrorModel*: the container for the usages or instantiations of particular *errorBehaviors* in a particular architecture context.
- *propagationPort*: ports through which the faulty events defined in an ErrorBehavior propagate to other *ErrorBehaviors* or result in *Hazards*.
- *ErrorPropagation*: abstractions for error propagations that in turn relies on particular instances of *ADLEntity* (e.g. communication connectors) or the explicit or implicit dependencies between them (e.g., allocations described by the *ADLRealization* construct).
- *ErrorToHazard*: a link between *errorBehaviors and* their effects on the system.

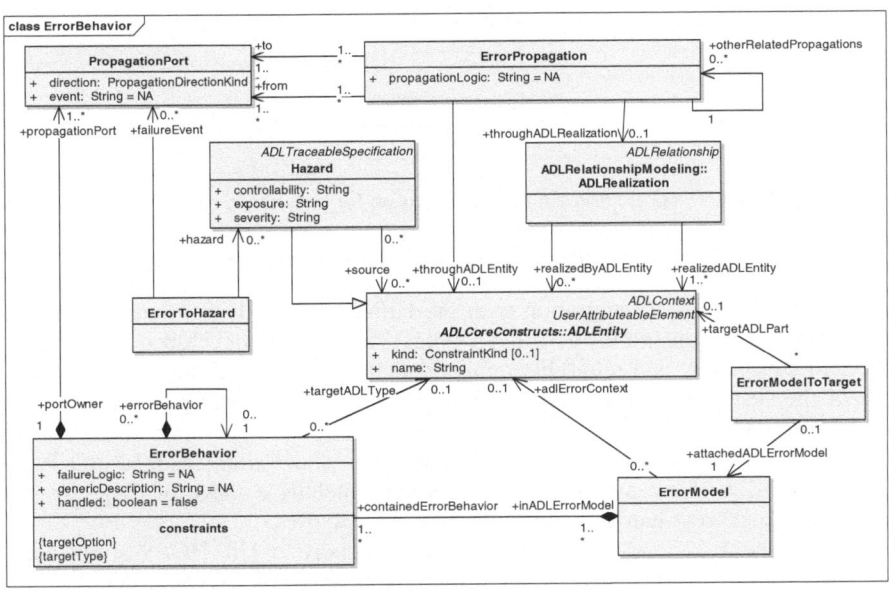

Fig. 5. The EAST-ADL2 domain model definitions for error modelling

In EAST-ADL2, the support for safety requirements and analysis is specifically addressed. The safety requirements, which are specialized to define the safety goals to be met, have attributes and related entities to define the related functional and non-functional requirements and the hazards to be mitigated. Hazards or hazardous events are associated with both errors of abstract functions/components and the environment model and characterized by attributes for severity, exposure and controllability. See

Fig. 6. for the domain model definition. This concept is in line with [1] where each hazard is related to an *Item*, which is defined as "E/E system (i.e. a product which can include mechanical components of different technologies) or a function which is in the scope of the development according to this standard". When modelling a system in EAST-ADL2 this means that for each level of abstraction a complete set of *Items* is identified. The hazardous event may be further detailed by e.g. use cases, sequence or activity diagrams. A safety requirement specifies the necessary safety functions and their effectiveness (i.e., ASIL levels [1]). It can be traced all the way to its derived requirements and thereby to the subsequent hardware and software solutions as well as the needed V&V efforts.

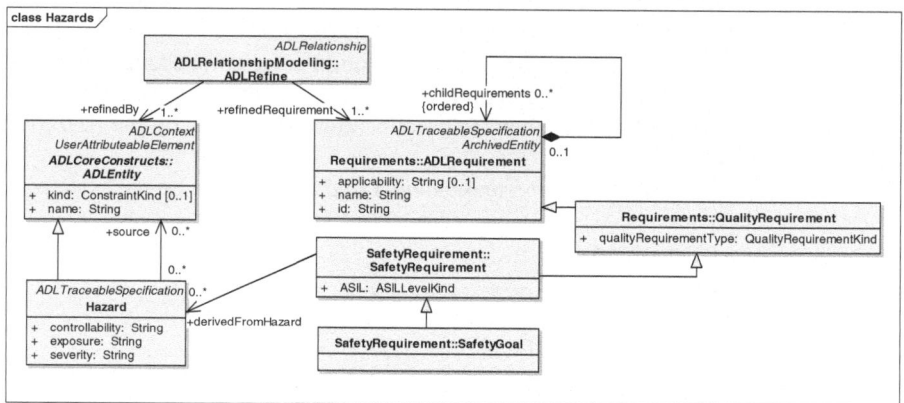

Fig. 6. The EAST-ADL2 domain model definition for hazard and safety requirements

4.2   Analysis Leverage and Tool Support through HipHOPS Method

A proof-of-concept tool integration with the HiP-HOPS method (Hierarchically Performed Hazard Origin and Propagation Studies) [5] has been developed. HiP-HOPS is a model-based safety and reliability analysis technique in which topological descriptions of the system (hierarchically composed if required to manage complexity) that are annotated with formalised logical descriptions of component failures, are used as a basis for the automatic construction of fault trees and Failure Modes and Effects Analyses (FMEA) for a system. Suitable models include a range of diagrams commonly used to express hardware and software architectures.

Through the EAST-ADL2 error modelling support, a HiP-HOPS study can be performed on the abstract models at the *AnalysisLevel* or on the more detailed architecture models at the *DesignLevel*. This creates opportunities for systematic identification of safety related requirements, re-use of earlier analysis, and the ability to achieve a consistent and continuous assessment in the centre of which lies the design of the system itself. Given an EAST-ADL2 model which contains descriptions of *ErrorBehaviours*, a global view of system failure can be captured via HiP-HOPS in a set of system fault trees which are automatically constructed as expressions that describe local fault propagation are being evaluated during the traversal. The synthesised fault trees are interconnected and form a directed acyclic graph sharing branches and basic events that arise from error propagations defined in the model. Classical

Boolean reduction techniques and recent algorithms for fault tree analysis that employ Binary Decision Diagrams (BDDs) are applicable on this graph. Thus, qualitative analysis (e.g. of abstract functional models) or quantitative analysis (e.g. calculation of system-level failure rates from known probabilistic component data) can be automatically performed on the graph to establish whether the system meets the desired safety or reliability. The logic in the graph can also be automatically transformed into a simple table which is equivalent to a multiple failure mode system.

5   Example Case Study: Electronic Column Lock

Steering column lock is a security function for preventing any steering wheel movement without an authorized starter key. Traditional solutions use the position of physical starter key as the securing and unlocking mechanism. For the reasons of user-friendliness, as well as crash safety and vehicle security, keyless engine start solutions with the immobilizer transponder and start button have been increasingly adopted, allowing advanced cryptography for authentication control prior to engine start. As a physical starter key is no longer present, there is a need to replace the traditional steering column lock principle. With electric steering column lock (ECL), a logical key position rather than the physical key position is used to enable and disable steering. The implementation normally consists of a mechanical lock placed on the steering column as the actuation element, and a control unit for reading the immobilizer transponder code and vehicle state and for controlling the mechanical lock. Fig. 7. depicts the modelling coverage by EAST-ADL2 for an ECL system.

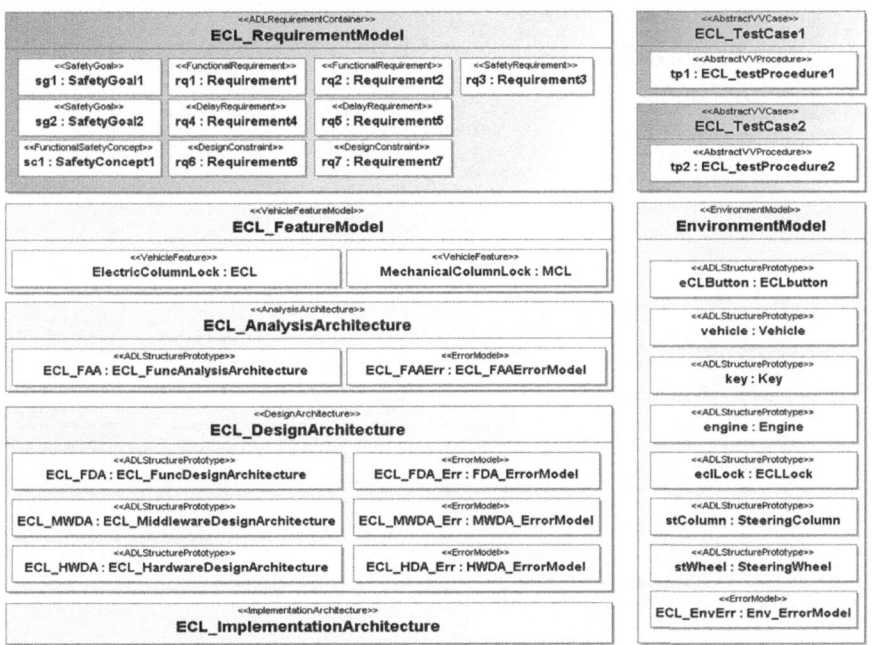

Fig. 7. The modeling coverage of EAST-ADL2 for an ECL system

Taken from the legislation, 95/56/EC annex IV, some of the basic requirements for the security function with relevance for safety are: (1) Devices to prevent unauthorized use shall be such as to exclude any risk of accidental operating failure while the engine is running, particularly in the case of blockage likely to compromise safety; (2) Locking shall only be possible after making one operation to stop the engine and then a second operation designed to lock the column.

The major top level hazard is: *Steering is disabled while driving.* This top level hazard must be controlled to a risk level where the risk of the hazard is kept sufficiently low. To model this hazard we use the mechanisms described in Fig 3. The initial Safety Case for ECL will consists of a root 'Item', with one 'Hazard' and two 'ADL Functions', One being the user function and the other the Environmental Model entity for the steering wheel. The safety assessment; in this case this is an function with the highest safety integrity level (ASIL D according to ISO 26262), is required to enable the correct actions on technical and process elements on the function development. The safety analysis for the function requires that we make an architecture defining which outputs and inputs are needed in order to perform the user function. Fig 8. shows the abstract functional definition of the ECL.

Sensors for vehicle speed, engine speed and key-position are required for ECL operation. For unlocking there are requirements on using an approved key, this is not fully considered in this example. There is also the case of retry strategies that can use

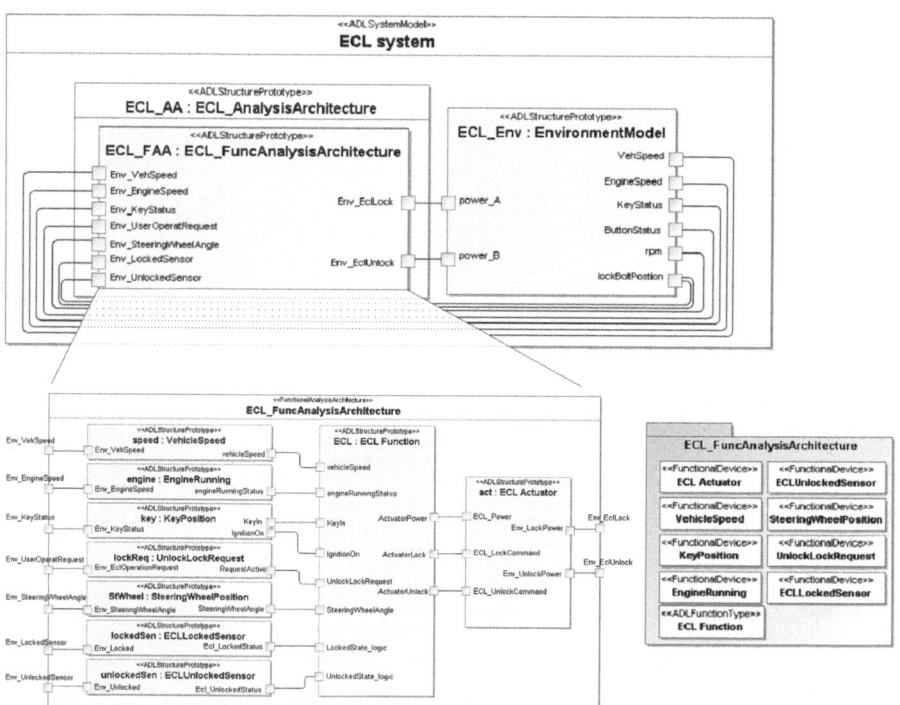

Fig. 8. Functional Analysis Architecture with Electrical Column Lock function, where the components from the package "ECL FuncAnalysis Architecture" act as parts

movement of the steering wheel as a trigger. The results of the safety analysis are the basic safety requirements and how these must be implemented. In this example, the outcome is that if the ECL lock is unpowered when the vehicle is moving and/or the engine is running, the Hazard cannot occur (unless there are electrical hardware or mechanical failures). The following architecture requirements (shown in Fig. 7) capture this and should be implemented in the design of the ECL function:

- **SafetyGoal1** – The ECL must not be powered when vehicle is moving and ignition is on.
- **SafetyGoal2** – The ECL must not be powered when engine is running and ignition is on.

As we rely on vehicle speed, engine running and ignition state on an ASIL D class function, we need to consider the need for redundant decision making on this set of data. As illustrated in the example case, the *ADLFunctions* on the highest level are similar between the analysis and the design level respectively. When we realize the *AnalysisArchitecture* (AA) on the analysis level with the *DesignArchitecture* (DA) on the design level, we can still hide a number of details on the highest level of the chosen ADLFunction hierarchy. Please observe the difference between the *realization* that is done when going from the more abstract analysis level to the design level (relation between "Architectures" in Figure 7), and the unpacking of details that is done when going down an ADLFunction hierarchy (the composition relation between contained and containing functions in Figure 8).

To perform a HiP-HOPS based hazard analysis, the first step is the establishment of *ErrorBehaviors* of components (i.e. functions, hardware or software elements) as failure expressions which show how output failures of each component can be caused by internal malfunctions and deviations of the component inputs. A variant of Hazard and Operability Studies (HAZOP) can be used to identify plausible output failures such as the omission, commission, value (hi, low) or timing (early, late) failure of each output and then to determine the local causes of such events as combinations of internal component malfunctions and similar types of input failures. Thereafter, the structure for error propagations in a particular architecture context is determined using *ErrorModels* for instantiating the predefined *ErrorBehaviors* and *ErrorPropagation* for the propagations. The system effects of failure events in terms of *Hazards* are captured with the *ErrorToHazard* construct. This global view in terms of fault trees are the generated with the HiP-HOPS plug-in.

Our experience from case studies suggests two useful design patterns that can be derived from this type of analysis: (a) when the analysis indicates that the omission of a function has only marginal effects while commission and value failures have catastrophic effects, a design recommendation should be made to design the function in a way that it "fails silent"; (b) on the other hand, when all potential failure modes of a function are shown to have catastrophic effects on the system then a design recommendation should be made to allocate the function to a fault tolerant architecture. With the help of these results, the abstract functions are allocated to appropriate hardware and software architectures in which case HiP-HOPS studies can become much more detailed and quantitative in nature making use of available information about component failure modes and failure rates.

ErrorBehaviours are now extended to include real failure modes and probabilistic component failure data. Such failure modes include electrical and mechanical component failures caused by wear and environmental conditions or, in the case of programmable components, statistically observed functional failures caused by unspecified random or systematic faults. Note that credible probabilistic failure data (often not available) is not essential for producing useful results. Qualitative application of the technique can still produce useful results. The logical reduction of fault trees into minimal cut-sets and FMEA, for instance, can indicate single points of failure in the system and point out potential design weaknesses. Clearly, the ability to iterate fast this process ultimately also defines the ability to manage effectively the evolution of an EAST-ADL2 design.

We should note that there is a range of other emerging techniques that are aiming at automation of system safety analysis (Altarica [13] and FSAP-NuSMV [14] among others). Most use model-checking and simulation as means of inferring the effects of component failures in a system. However, the analysis of individual failure modes via simulation or model-checking is computationally expensive and the inductive nature of the analysis (from causes to effects) creates difficulties, especially when combinations of failures need to be considered. Assuming that there are N possible component failures in a system, assessment of combinations of M of those failures requires that the analysis is repeated N!/((N-M)! x M!) times. For a system that has 1.000 failure modes, assessment of the effects of combinations of 2 failure modes requires that the analysis is repeated approximately half a million times. In HiP-HOPS, the analysis of propagation of failures is deductive (from effects to causes) and therefore the technique always synthesises fault trees in linear time not determined by the highest order cutset (i.e. the maximum number of failure modes considered in combination which is defined only by the positioning and nesting of AND gates in the error propagation model). The fast algorithms of HiP-HOPS have not only enabled its application on large systems but also its combination with computationally greedy heuristics such as Genetic Algorithms for the purpose of architectural optimisation with respect to dependability and cost [15] - a capability which is unique in HiP-HOPS. Moreover, HiP-HOPS has advanced capabilities for probabilistic analysis which include Poisson, Binomial and Weibull calculation models, as well as capabilities for common cause and zonal analyses, while most formal techniques tend to focus on functional safety analysis only. Clearly there is often a need in software design to consider the probability of failure of components.

6 Conclusions

In this paper we have presented how the architecture description language EAST-ADL2 supports the development of safety-critical automotive E/E systems. The integration of the safety case metamodel, safety analysis, and the system modelling will achieve several benefits. The Safety Case development will be eased by the systematic development that is supported by the EAST-ADL2, including structured information handling, support for reuse, consistency between the models, etc. The EAST-ADL2 language will benefit by having support for the Safety Case approach, an important technique in safety relevant system development. Further, the safety case metamodel will provide support for motivating why certain design decisions are needed and provide means for connecting the argumentation and design information. The connection between error modelling and system modelling supports quick safety design iterations, the creation of views, and structured information management.

We believe this approach presents an important step in making the design and safety processes more efficient and effective. Future work will concentrate on further evaluation of the approach, developing systematic support for integrating several relevant analysis techniques, and considering optimization with respect to safety properties. Another direction is to assess how the proposed approach, biased by automotive specifics and standards, is applicable to other domains.

Acknowledgments. This work was supported by contribution of all the partners of the ATESST project consortium funded by the European Commission. We wish to acknowledge feedback from the anonymous reviewers.

References

1. International Organization for Standardization: Draft 26262. ISO Committee (2008)
2. Chen, D.J., Törgren, M., Lönn, H.: Elicitation of relevant analysis and V&V techniques. D2.2.1. ATESST EC FP6 (2007), http://www.atesst.org
3. AUTOSAR Development Partnership, http://www.autosar.org
4. Kelly, T.P.: Arguing Safety - A Systematic Approach to Managing Safety Cases. PhD Thesis. University of York (1998)
5. Papadopoulos, Y., McDermid, J.A.: Hierarchically Performed Hazard Origin and Propagation Studies. In: Felici, M., Kanoun, K., Pasquini, A. (eds.) SAFECOMP 1999. LNCS, vol. 1698, pp. 139–152. Springer, Heidelberg (1999)
6. Sangiovanni-Vincentelli, A., Di Natale, M.: Embedded System Design for Automotive Applications. IEEE Computer 40(10), 42–51 (2007)
7. HIS Members and Partners: Specification Requirements Interchange Format (RIF). v1.1a (2007), http://www.automotive-his.de
8. SysML Partners: Systems Modeling Language (SysML). Open Source Specification Project, http://www.sysml.org
9. Cuenot, P., Frey, P., Johansson, R., Lönn, H., Reiser, M.-O., Servat, D., Tavakoli Kolagari, R., Chen, D.J.: Developing Automotive Products Using the EAST-ADL2, an AUTOSAR Compliant Architecture Description Language. Ingéniurs de l'Automobile 793, 58–64 (2008)
10. Törner, F., Chen, D.J., Johansson, R., Lönn, H., Törngren, M.: Supporting an Automotive Safety Case through Systematic Model Based Development - the EAST-ADL2 Approach. Technical Paper Series, 2008-01-0127. SAE (2008)
11. International Electrotechnical Commission: Functional safety of electrical/electronic/programmable electronic safety-related systems – Part 0: Functional safety and IEC 61508 (2005)
12. Martin, T., Chen, D.J., Malvius, D., Axelsson, J.: Chapter - Model based development of automotive embedded systems. In: Navet, N., Simonot-Lion, F. (eds.) Automotive Embedded Systems Handbook. Industrial Information Technology. Taylor and Francis CRC Press, Abington (2008)
13. Arnold, A., Griffault, A., Point, G., Rauzy, A.: The Altarica formalism for describing concurrent systems. Fundamenta Informaticae 40, 109–124 (2000)
14. Bozzano, M., Villafiorita, A., et al.: ESACS: an integrated methodology for design and safety analysis of complex systems. In: ESREL European Safety and Reliability Conference, Balkema, pp. 237–245 (2003)
15. Papadopoulos, Y., Grante, C.: Evolving car designs using model-based automated safety analysis and optimization techniques. Journal of Systems and Software 76(1), 77–89 (2005)

Resilience in the Aviation System

Antonio Chialastri[1] and Simone Pozzi[2,3]

[1] Aviation Lab, Rome, Italy
[2] Deep Blue srl, Rome, Italy
[3] Sapienza University of Rome, Department of Psychology of Social and Developmental
Processes, Rome, Italy
anto.chialastri@tiscali.it, simone.pozzi@gmail.com

Abstract. This paper presents an overview of the main characteristics of the civil aviation domain and their relation with concepts coming from the approach of resilience engineering. Our objective is to first outline the structural properties of the aviation domain (i.e. regulations, standards, relationships among the various actors, system dynamics), to then present some example processes that bear an effect on the system resilience. We will in particular reason on training and on the role of automation, to discuss how and to what extent they impact on system resilience. We contend that, in a complex system like aviation, resilience engineering is not a matter of simple technical upgrades, rather is about facing contradictory tensions and dynamic system changes. This paper contains a pilot's first-hand reflections, so it aims to stimulate discussion on some issues that are still open, rather than providing solutions.

1 Introduction

Given the unbearable human, economical and legal impact of an air disaster, safety has always been the main concern for airline management. However, technological innovation like the introduction of the so-called glass cockpits in the beginning of the Nineties has questioned well-established safety management methods, calling for the adoption of new safety models. For instance, Leveson [1] mentions how reductionist approaches, which derive the whole system safety from ensuring that each single component is safe, fail to appreciate the systemic dimension of safety. Traditional Probabilistic Risk Assessment focuses on functional failures, i.e. on the non-performance or inability of specific components to perform their intended functions. However the more complex safety critical systems have become, the more accidents have been determined by so-called *dysfunctional interactions*. *Dysfunctional interactions* take place when system elements perform as they are expected (i.e. as specified by requirements) but still the overall system behaviour is unsafe. The increasing role of human and software in supervisory control addresses this issue, as it is quite common to have situations in which a component satisfies its specified requirements, even though the requirements may include behaviour that is undesirable from a larger system context.

A coherent approach with the points raised by Leveson comes from Resilience Engineering [2, 3]. Whereas conventional risk management approaches are based on

M.D. Harrison and M.-A. Sujan (Eds.): SAFECOMP 2008, LNCS 5219, pp. 86–98, 2008.

hindsight and emphasize failure probabilities, Resilience Engineering aims to enhance the ability of organizations to create processes that are robust yet flexible, that can use resources proactively to accommodate for external disruptions or internal ones (e.g. production pressures, human errors). In Resilience Engineering, failures do not stand for a breakdown or malfunctioning of normal system functions, but rather represent failure to adapt to the real world complexity. Resilience engineering focuses on the capabilities on all levels of a system to respond to regular and irregular threats in a robust yet flexible manner, and to anticipate the consequences of disruptions. However, all systems to some extent adapt to changes, even if this adaptation might be slow or not apparent. Robustness is provided by specified structures inside the organizations that should respond to intentional attacks or unintentional mishaps, while flexibility is achieved by stretching normal behaviors to cope with situation not previously codified.

Resilience engineering refers to a broader definition of adaptation, whether the system can handle variations that fall outside of the co-called design envelop, that is the variance amplitude as defined in that system. The system should be "designed-for-uncertainties, which defines a 'textbook' performance envelope and how a system recognizes when situations challenge or fall outside that envelope – unanticipated variability or perturbations" [2].

Individuals and organizations must always adjust their performance to the current conditions; and because resources and time are finite it is inevitable that such adjustments are approximate. Success has been ascribed to the ability of groups, individuals, and organizations to anticipate the changing shape of risk before damage occurs; failure is simply the temporary or permanent absence of that.

Given these definitions of resilience engineering, some problems arise regarding the scope of their applicability in aviation.

According to Erik Hollnagel: *"Safety is something a system or an organisation does, rather than something a system or an organisation has. [...] This creates the dilemma that safety is shown more by the absence of certain events – namely accidents – than by the presence of something. Indeed, the occurrence of an unwanted event need not mean that safety as such has failed, but could equally well be due to the fact that safety is never complete or absolute"* [2]. Which begs the question of which is the correct approach to safey in a system such as aviation. The answer depends on how we see the entire system. Fifty years ago, when a reductionist paradigm was dominant, the answer was that safety could be achieved via the engineering approach, by improving onboard technologies. Everything was measurable, predictable, modelled in different shapes to fit for the special field of application. During the eighties, the answer to the same question shifted from engineering to psychology. Following several accidents, due to poor human interaction, the goal was to improve the "liveware" part of the system. Technology was considered safe, while man was not. Today, we recognize that in complex systems we cannot isolate single causes, since every element is interconnected with the other elements.

The approach we propose in this paper is to move away from reductionism and take a philosophical perspective on system dynamics and to address one of the key contradictions at the core of the resilience engineering approach. On the one hand, most of the authors acknowledge that a complex system cannot be reduced to the sum of single components and are aware of the role played by emergent properties. On the other hand, we need to have better engineering principles, that can be applied by industries. Resilience and engineering do not match. Numbers, graphics and models may give more confidence in the manageability of the system and may reduce the uncertainty given by complexity, but they still cannot address emergent properties. In our opinion, this is the key challenge that resilience engineering should tackle.

2   What Is Resilience in Aviation?

Given this explanation of resilience, we must clarify some concepts that could be misleading for the discussion. From a system's theory point of view, accidents are considered as an unexpected combination of events rather than a single failure or action leading to disaster. In a similar manner resilience is the ability to cope with *unexpected* circumstances that could put in jeopardy the whole system. Accidents in aviation, likewise other domains, show very similar and recurrent patterns of events. That is why we are able to categorize types of accidents by their dynamics and by their shared characteristics, e.g. "controlled flight into terrain" or "loss of control". To overemphasise the point for clarity's sake, in most of the accidents we already know every step leading to the negative outcome before they actually start unfolding. Otherwise even with the benefit of hindsight we would not be able to identify the single links of the event's chain.

To recap, aviation resilience is itself a problematic notion to be analysed deeply, not a simple solution to fix organizational latent failures. This leaves us with some open issues. First, how to define aviation resilience. Ability to cope with unexpected events? Robustness towards ambiguity of information? Functional plasticity and structure remodelling, in order to achieve the same result, namely safety? What do we mean by saying that something is unexpected in aviation? Second, which is the appropriate system level for improving resilience? Who are the stakeholders? Shall we concentrate on the final operator (i.e. pilots, air traffic controllers, etc) or on the organisational level or on international institutions (e.g. ICAO, IATA, etc.)

To further our reflection on resilience engineering and the aviation domain, this paper will present some of the characteristics of the aviation domain, to then describe first-line processes and the way they might impact on the system's resilience.

3   The International Nature of Air Transport: Rules and Regulatory Bodies

The aviation industry has been among the first to go global. Its workplace is the world, so it deeply needs international rules to be enforced worldwide. International organisms and national regulators emit a set of rules regarding the air transport. An airline must comply with the "Airworthiness of operation certificate" criteria. Another institution that sets worldwide rules is the ICAO (International Civil Aviation Organization) agency of the United Nations, who emits, among others, regulations regarding flight procedures (i.e. setting the criteria on the design of instrument approach). The IATA (International Airline Transport Association) is responsible of the rules for passengers and good transportation.

We must mention as well the international agreement signed following some international conference as in Chicago (1944), where the States issued an agreement to create ICAO, together with a series of documents also know as *technical annex*. At the moment, 18 annexes have been issued regulating several aspects of international flight.

From then on, other conventions took place, in Tokyo, in Montreal, and so on.

The United States, the cradle of the aviation industry and the commercial flight on a large scale, often set the pace of air regulation, regulation that is later adopted worldwide. The FAA (Federal Aviation Administration) emitted in the early days of flight a set of rules regarding airplane's manufacturers, pilot training, hiring and scheduling, maintenance action and so forth, in order to guarantee the system reliability to customers and workers in every country adopting those rules. Today, another super-national regulator, the JAA (Joint Aviation Authority) issues its own rules for an European Standard to be applied to airplanes and aircrew flying in Europe.

National regulators should comply with international rules and should implement also other safety measures to ensure safe, smooth and orderly flight operations, They should also take the role of the "system watchdog" whenever required. For this reason, some flight rules are still derived from national legislations, which may sometimes be outdated. For instance, as recently as in 1995, in Italy flight was regulated by the old "Navigation Code", issued some sixty years ago (30/3/1942), thus applying the same measures to ships and airplanes. National legislations is sometimes outdated compared to international rules, because aviation still defines most of its rules and standards at a transnational level. This requires national legislations to quickly comply with international standards, which is not always easily done in the appropriate time frame.

3.1 Main Actors

Having covered the regulatory bodies that set worldwide rules for air transport, we can now move to the core business's actors of air industry and how they interact. Main actors are the manufacturers (i.e. Boeing, Airbus, etc), the operators (airlines, charter companies, cargo, and so on), crews, and auxiliary services. Each actor faces its own peculiar safety challenges and has its own responsibilities.

The *manufacturer* builds a new airplane, according to the rules, and after the flight tests, it sells the aircraft with the relative operation manual to the airline (the operator). The manufacturer usually provides the following information:

- system's limitation
- check list for normal, abnormal and emergency situations
- conditional procedures (a non-routine, but non-dangerous procedures)
- special operation (operation with degraded performance depending either on systems or environment)
- performance tables, including engine(s) out performance
- loading
- MEL (minimum equipment list) enabling the crew to fly with inoperative devices until home base where the repair is made according to a schedule (this deviation must be previously approved by national regulator)

– runways tables that indicate the performances the aircraft can develop on the specified runway: i.e. the maximum weight allowed during takeoff or the flight path to be followed in case of engine failure soon after take off.

The *operator* buys the airplane and plans its operating schedule. In addition to the manufacturer's manual, every major airline provides the crew with rules of conduct either on ground or in flight. This is called *General basic*, and specifies almost every aspects of the crew members' working life.

The *crew* flies according to the national and international rules and laws, it must comply with the procedures specified in the operation manual (i.e. the manufacturer's manual) and the guidelines set by the operator (i.e. airline).

Auxiliary services to air travel include: air traffic management, airport services (catering, fuelling, etc.), maintenance, marketing services. Even if each of these bears a significant effect on aviation resilience, in this paper we will only briefly mention the role of maintenance. Airplane overhauling is regulated by international standards and by strict national rules. Every aircraft should be checked every day, and cyclically after a determined number of flight hours. On the average, airplanes are brought to an hangar every three months to have a complete overhauling, in order to check every system and guarantee safe operations. If the crew experiences any system malfunction during a flight, they file a report. The next flight cannot depart unless the problem has been fixed. Documents proving that the maintenance action has been made are quickly sent to the national regulator who has the right (and duty) of supervision on every repair.

According to Amalberti theory on ultra-safe systems [4] (less than one accident every million take offs), we can point out significant difference in safety records between military and civil flights, and also among civil flights: airline, charter and private flights.

Military flight is made in variable environment that does not allow a strict regulation, leaving room for the pilot to arrange his flight in order to be "combat ready". Often the airplane is flown to its limits, with erosion of the risk margins. Sometimes the enemy is inside the cockpit.

Civil aviation is made of different kind of subjects: airline, charter, private. Airlines are structured in a very organised model that relies on detailed procedures to carry on its activity. Accident rate have been estimated in one accident per ten million take-offs. Crews are trained to comply strictly with these procedures. Charter companies are instead driven by profit in a more aggressive way, so economical pressure on crews could be stronger than in the airlines (estimate rate of accident one per 10^5). Keeping accidents at bay is a serious concern for managers and there is a concrete risk of misperception by employees about the management's real priorities. Private flight are less keen on procedures and mainly relies on pilot's experience, but on the other side, two elements could be critical (the estimated rate of accident is one per 10^4). First, pilots often lack a professional community with whom to share their experience, thus hampering effective proactive learning. Second, maintenance is not often carried out by expert engineers, as maintenance people is hired from big companies on a temporary basis.

4 Resilience between Automation and Training

This section will present some of the processes that the aviation domain has established to increase the resilience of its operations. In the first paragraph, we will outline how pilots are trained to perform with the primary objective of safety. In the second part, we will focus on the role of automation in modern aviation, highlighting a progressive shift in the underlying design philosophy. These two examples will show how resilience engineering is about facing contradictory tensions and dynamic system changes.

4.1 Building a "Safe Crew"

Aviation is a socio-technical system made of men and a variable environment. Everyone working around an airplane plays his role in assuring the final target: safety. In doing so, everyone should be strongly committed to ensure the best performance s/he can in order to avoid a deterioration of safety margins. According to the so-called hologrammatic principle (i.e. every single particle contains the properties of the whole in which it is embedded, e.g. a cell in the human body), in air industry every operator should share the basic approach to safety, since any of her/his action could affect the final result. It is thus crucial to review means and processes that make sure that every operator shares the same approach to safety. Among the main drivers, we may mention training programmes, but also organisational culture and the force of examples.

To tackle this issue, the airline industry adopts a knowledge-based approach to safety, where the system resilience is ensured by appropriate performance at the single operator's level. The result is a bottom up approach to safety in which everyone is strongly committed to safety because s/he shares the same value of the entire organisation. Every area has its own principles and varies from role to role. Selection is very important in hiring pilots, less for ramp agents. Teaching is very important for ramp agents, given their sensitive role in assuring flight balance, less for pilots already hired with a valid license. Given the author's experience, this paper will focus only on the pilot role. The main processes put in place by the aviation industry to ensure a knowledge-based approach are selection, training and checking. In the next part of the paper, we will particularly focus on training.

Selection
Selection is the first "filter" of candidates and it is structured taking into consideration several factors: attitude towards the job, reliability, cognitive skills, social abilities, etc. The performances of the would-be pilots are evaluated by a team made of psychologists, old pilots, managers. The desired profile is set in advance, so that only the suitable candidates are enrolled into the flying school. The other key turning point in a pilot's career is the upgrading to the rank of commander. To achieve this rank, the candidate pilot should be positively assessed by many instructors and check pilots.

Training
Training is a lifelong process that endures till a pilot's retirement. It is based on a series of competences and knowledge pieces, ranging from flying skills, to flight management, to role attitude. Each element is required in the pilot profile. Piloting is

not the sole skill required to be an airline pilot. A pilot also needs to make crucial decisions on the basis of theoretical knowledge, previous experiences, current flight data (which include the present situation, the aircraft status, meteorological conditions). Just like a surgeon, pilots have a strong theoretical knowledge, but they also need experience, which sometimes comes paired with mistakes. Both surgeons and pilots always focus on the same object of operation, a human being for the surgeon and a flight for the pilot, but this object everyday changes in subtle or sharp ways. All the recent technological improvements provide help to carry out simple and repetitive tasks, but in the end a good pilot or a good surgeon are required to exercise their sound judgement to evaluate complex situations, whenever they arise. The core of their profession is the "artfulness of the intelligent worker", that reads reality and puts into connection the single, unique, situations they are living with a set of theoretical tenets.

For instance, if we take the case of procedures, we cannot simply claim they positively contribute to safety. They certainly provide support in routine jobs for smooth, clear, precise operations. However, it is impossible to get a procedure for every aspect of a pilot's job. In a complex environment, threats are too many to be foreseen in advance, thus the safest way to cope with unexpected situation is to provide pilots with the appropriate resources to cope with these variations. Usually a pilot working for an airline is taught to fly well within the safety margins. The safety margins protect the system from technical failures, unexpected circumstances or human errors. But while pilots see the margin area as a buffer over risk, managers tend to see buffers as inefficiencies. While pilots sometimes face the trade-off between money and safety to comply with the company's goals, the managers are oriented to maximize the performances pushing the costs at their lowest edge. To make sure pilots possess the right resources in the right situation, the aviation community has identified four resource categories, also known as the "4 P" approach: Philosophy, Policies, Procedures and Practices.

Philosophy is the guiding principle of airline business. The philosophy of airlines should be *safety first*. Every organizational policy, procedure or practice should be implemented according to this basic principle.

Policies are issued by the management to reach the operational target. They are guidelines concerning a determined area. For instance fuel consumption, given the actual oil cost, is object of a common policy in most of the airline. To minimize fuel consumption a series of measures could be adopted, from avoiding of carrying extra fuel, to requesting air traffic control for higher cruising altitudes, etc.

Procedures details the flow chart required to carry out the user's task. They are designed by the operator (airline) to comply with policies and regulations. They take into consideration several aspects: manufacturer's recommendations on the airplane's management, regulator's criteria on crew composition. In the routine job they ensure a safe and smooth flow of operation.

Practices are what people really do to bridge the gap between procedures and reality demands. There is of course a well known potential mismatch between procedures and practices. Whenever it is impossible to comply with the procedure, the captain has the responsibility to deviate in order to ensure a higher level of safety, in accordance with the philosophy of operations. In these situations, it is essential to evaluate the attitude towards risk, variable from pilot to pilot according to multiple factors. Such attitude is commonly described as follows:

- risk expectancy: what is the real chance that something happens?
- risk sensitivity: in case something actually happens, which are my effective resources to cope with this new situation?
- risk penalty: which will be the possible consequences in case something actually happens?

For this reason, training programmes include flying skills, flight management skills, role attitudes. Flying skills are the ability to fly an airplane according to basic flight principles, with or without autopilot. They represent the "knowing how-to", cognitive-physical skills on execution tasks requiring coordination of external input perception with actions. This area has a key prominence for a novice pilot, who has to develop familiarity with locating the airplane position in a three-dimensional space and planning/controlling corresponding movements. Flight management skills refer to the ability of managing aircraft systems in order to perform at the requested level of safety. These skills are developed by internalising operating rules and procedures to understand the rationale behind them. In this training phase, pilots should move beyond the mere knowledge of rules to understand how to use rules as resources to ease work and make it safer. Rule should become "safety resources", so that any violation can only be justified if it is clearly required for safety reasons. Role attitudes cover interpersonal skills, like assertiveness, critique, communicativeness, etc.. These are required to perform in coordination with all the crew members, to develop and maintain a shared view on the objectives, to manage the available resources, to handle interpersonal conflicts that might disrupt the team performance. While the former qualities are named technical skills, the latter is a non-technical skill. Leadership, communication, and other non-technical skills can play a major role in many accidents. For instance, a Controlled Flight Into Terrain accident (CFIT) is caused by pilot's misbehaviour or misconduct, as a perfectly efficient airplanes hits an obstacle or overruns the runway end.

Another core area of training programmes deals with error management. Since errors are unavoidable, this area is still an important one, even though state-of-the-art safety literature [5-9] has deeply questioned the assumption that human error causes more than 90% of the accidents. Anyway, pilots are trained in order to be aware of human behaviour in flight. To improve error management, we articulate the training in three levels of error's awareness: avoid, detect, mitigate.

- Avoid: the ability to develop one's own safety net that ensure a smooth, quick and safe way to operate the system. It includes also flight discipline, intended more as a shared value, rather than a rule to comply with.
- Detect: ability in the perception of something deviating from the natural course of action and from intentional input to the system. A key risk area can be found whenever perception does not match the user's expectation, as sometimes expectations can normalise very deviant perception.
- Mitigate: once the deviation is manifest, a quick return to a desired path is a pilot's "must".

Detection can be particularly tricky, as pilots may underestimate a risk on the basis of the lack of negative outcomes in their experience. This phenomenon is commonly known as "drift to danger" [10]. It is an incremental, slow and pervasive attitude toward risk that drives the sharp-end operator to pursue targets even beyond the managers' will. There is, basically, a misperception of the real margin of risk that the organization, as a whole, is ready to accept. This dynamic can be exemplified with a discussion on the fuel policy.[1] Due to the oil price soaring, many airlines are trying to save money, by reducing the fuel consumption. To reach this target, pilots are invited by staff manager to uplift just the minimum fuel required for the flight. Many pilots complied with this policy to eventually find out that they have significantly eroded safety margins, even to a larger extent than they intended to. Recently, there have been several "lack of fuel" emergencies in the United States and in Europe. The CAA (the English regulator) emitted some years ago a recommendation to all crews flying in UK, to consider the right amount of fuel to carry onboard to avoid distress on passengers and special requests to Air Traffic Control units [11]. Even though not every fuel policy critical event is properly detected and reported by crews, there are clear evidences that this area of concern is spreading worldwide, and pilots are struggling between production demands and protection needs. Furthermore, declaring emergency leads to the fear of inter-peers judgement and blaming. A declared emergency with a good functioning aircraft is an ambiguous event, that could be regarded either as a lack of professionalism or as sound judgment. This is a clear example of how economic pressures, organizational climate, raising expectations could impair pilots' day-to-day choices, making the organization unintentionally drift towards the risk area.

Learning in a professional community

The last point we should mention on pilot training is sometimes disregarded, even if it plays a major role in lifelong learning. A pilot should become aware that s/he is part of a professional community, with which s/he can share experiences and discuss problematic issues. Pilots learn from their mistakes, and no pilot can live long enough to commit all the mistakes by her/himself. Pilots see one flight at a time, which does not ensure that they possess an appropriate perception of flight risks. How can a single crew assess if the mistake it has just done is due to poor training, to poor system design or to a coincidence? There are currently no better means to conduct this assessment than by ensuring that the community can openly discuss these events and can share a common interpretation. In this case resilience comes from the cohesion of a community, and not by dynamics strictly related to flight. Though in aviation the informal communication is seen as potentially unsafe, we should point out the paramount importance of peer-to-peer experience sharing, since it provides a valid resource to cope with unexpected events.

Checking programmes

Checking programmes are set by the national and international regulators to define the minimum requirements for licence validation. Big airlines check their pilots on national regulator's behalf.

---

[1] See "Fuel policy and resilience" by Antonio Chialastri, unpublished manuscript, 2008.

4.2 The Role of Automation: the Tension between Under-Redundancy and Over-Redundancy

About twenty years ago, the air industry, looking for more redundancy in the avionics systems, started introducing automation in flight management. Autopilot and other automatic devices had already been present for some fifty years ago, but that kind of innovation was still guided by the pilots, in the end the final user. The new conception of automation was to provide a whole set of system's redundancies, able to calculate every aspect either of lateral navigation or vertical performances. The pilot was then moved to a monitoring position, rather than being the flight manager. That approach raised questions about the erosion of competence in a pilot (you must know WHAT, not HOW), since the pilot was no longer required to understand the logic of what s/he was using. The basic message was: just use it.

This historical shift in the automation philosophy can be described as a movement from a tactical approach to a strategic one. In the past, every input given to the flight automation (e.g. Flight director, Autopilot, Autothrottle), was immediately visible on a display and pilot's awareness about the mode of automatism was reasonably high. This is called "tactical approach" because input and output were always clear and displayed in cockpit. Nowadays, following the introduction of the Flight Management System (a system that manages and computes several flight aspects in order to minimize pilot's input and provide a protection against flying skills issues, e.g. stall, bank, etc.), a strategic approach is in place. A data (e.g. route deviation, flight level change, etc) inserted now in the computer might be processed hours later, without any displayed information at pilot's reach. If the pilots wants to know which will be the airplane behaviour s/he should review the flight Management System Computer pages.

In the automation case, redundancy is achieved by improving and adding systems in the cockpit, but new risks may arise, as these additional resources contribute to the system resilience only by interacting with human resources, which cannot be considered as a neutral factor. Each situation is exposed to its own peculiar threats. An airplane with few systems (under redundancy) keeps the pilot under stress, fatigue, distraction, information overload, so that workload management is the main area of concern. Such a situation was usual in the middle of 20th century before the introduction of the autopilot, which gave support to pilots during extended operation. Risks were due to flying skills failures, often induced by a too high workload. As automation increasingly supported the pilot's flying skills, the main safety concerns have simply moved to another place. Flight management has become the main risk factor. Over-redundancy has kept the pilot at bay so that s/he lost the basic ability to take over control when needed. Sometimes, pilots cannot understand the system's logic, they lose resources to "fly ahead" of the airplane. Pilots should be a step ahead of the automated flight management system, but as soon as the pilot loses such situation awareness, it should always be possible to revert to basic mode and put back pilot in the position of actually flying the airplane and not merely monitoring automated systems. Nowadays, the primary source of accident has become the *loss of control*, that is the pilot not being able anymore to keep the airplane in a safe flight path. So over-speed, excessive bank angles, stalls, etc, started to show the negative effects of excessive onboard automation. The "erosion of pilot's competence" resulted in a lack of airmanship, caused by excessive confidence in the flight automation system as the primary resource of flight path and performance management. As a result, few years ago, FAA issued a recommendation to airlines to train pilots *back to basics*, in order to develop the ability to fly regardless of the automated systems.

This discussion shows how in aviation automation often does not increase the margin over risk, instead it keeps the risk ratio constant, allowing the crew (or the system as a whole) to work at the maximum capacity. A similar point comes from the analysis of the development of the instrumental approach to an airport. Ground facilities and onboard receivers allow the pilot to identify the runway to land safely. Before landing s/he must be sure that conditions warrant for a safe approach. It is common, at the operational level, to establish a *decision height* where the crew must positively identify by visual contact the runway and decide if landing is safe or not. If the airplane reaches the *decision height* without getting the runway in sight, the approach must be discontinued. When ground facilities and onboard instruments were not so accurate (i.e. non directional beacon - NDB), the *decision height* was set, say, at 1000 ft above ground and the minimum required visibility was four kilometres. As the technologies improved and the VOR (Very high frequency Omnidirectional Radio) was introduced, the decision height was lowered, say, to 500 ft above ground and the minimum visibility to two kilometres. With the implementation of the Instrument Landing System (ILS – a system that provides the pilot with the correct glide path), the relevant *decision height* was further decreased to 200 ft and minimum visibility required to 600 metres. Nowadays the ILS has been improved to a greater accuracy and the crew may wait 20 ft over the runway before making a decision. That is to say: two eye blinks and you land in the middle of a foggy day with visibility of 125 metres. As we see, gradual introduction of new technologies made the airport operable in almost all weather conditions, but it did not increase safety margins. Safety remained constant, while productivity (operability) of the entire system boosted.

We might discuss other examples, like the introduction of reduced vertical separation minima (RVSM), implemented few years ago, that allowed aircraft flying at cruising level to be spaced vertically of 1000 feet, instead of 2000 feet as before. Even here we see that the system is not safer, but more flight levels become available to let more traffic flow.

This brief excursus shows how resilience engineering is not a matter of simple technical upgrades. We might argue that the introduction of automation has made the 1950 aviation system more resilient (at least under certain conditions), but we would miss the point that automation has also caused the system to change, thus making it more vulnerable to other threats. In a complex system like aviation, resilience engineering is not about increasing the safety level by "solving some issues", nor by introducing specific technical solutions, rather it should focus on managing changes and studying a problem from various aspects. It should provide the system view, to counter balance excessive specialisation and reductionism.

5 Conclusions

This paper has analysed the aviation system according to the complexity paradigm approach. In doing so, we should drop the old habit used in aviation as far as twenty years ago to analyse the accident causes: a linear, pre-programmed, highly codified system, made of sub-systems accurately designed by engineers, able to cope with a foreseeable environment. According to this approach, human behaviour is the unique variable, single source of malfunctions leading to disaster. The complexity paradigm invests discipline, from biology to general system theory, to cybernetics, and prompts us not to oversimplify living systems or organization as a whole. It rejects the "standard view" approach based on predictability, verification, measurability, theory of meaning as correspondence, neutral observation, distinction between data and theory. Instead, data are intertwined with theory, observation is never neutral, depending on the observer's light on facts; confutation has replaced verification, and so on.

However, common sense has not followed fully such a paradigm shift. We still see organization as a machine that can be designed, built and checked in every detail, according to the principles of mechanics. For example, in the air industry, quantification is still seen as the main base for decision making. A continuous monitoring activity based on collecting numbers (a huge amount of numbers), followed by scarce analysis and even less synthesis. The loop is not closed with the domain experience, so data remain separated from an overall framework of knowledge.

We have shown in this paper how the concrete mechanisms put in place by the aviation domain to increase its resilience are by far more complex than simple mechanics, as they are multi-faceted, containing inner contradictions and tensions, always developing and subtly changing. Even if we have kept separated the discussion on training and on automation, we eventually have to study the interactions between the two, thus adding further complexity. The lesson we would like to draw from our first-hand experience is that resilience engineering should be a dynamic approach to safety, a never-ending monitoring of the flying activity, which accepts the probable negative outcome and studies all the means to exploit to try and avoid such outcome. An improvement action does not simply fix a safety problem, it also triggers adaptations and interactions. Resilience engineering should be about heightened monitoring of system's changes.

Is it possible to create a model to do it "a priori"? Or should we be satisfied with post-accident analysis that teaches us what went wrong? At the moment, the only sensible answer in aviation is to spread knowledge in order to make people aware of their own behaviour as a single element of the system and as an emergent property, a unique feature, which can contribute to the whole safety. The final question is how to enhance safety via a feed back system that, starting from managers' inputs, collects all the relevant deviation from an ideal centreline accepted as safe. Spontaneous report made by the front line actors (crews, engineers, ramp agents and so on) is vital to detect such a gap between reality and theory. A "no penalty policy" is often endorsed by major airlines in order to encourage people to show their own mistakes, failures in their line operations. At moment this is the only valid approach able to avoid a hidden, and highly dangerous, mismatch between the intended outcome and the actual one.

Acknowledgments. The authors gratefully acknowledge the support provided to this work by the EU project "ReSIST: Resilience for Survivability in IST".

References

1. Leveson, N.G.: A New Accident Model for Engineering Safer Systems. Safety Science 42(4), 237–270 (2004)
2. Hollnagel, E., Woods, D.D., Leveson, N.: Resilience engineering: concepts and precepts. Ashgate, Burlington (2006)
3. Hollnagel, E.: Resilience-The challenge of the unstable. Resilience engineering: concepts and precepts. Ashgate, Aldershot (2006)
4. Amalberti, R.: The paradoxes of almost totally safe transportation systems. Safety Science 37(2-3), 109–126 (2001)
5. Reason, J.T.: Human error. Cambridge University Press, Cambridge (1990)
6. Reason, J.T.: Managing the risks of organizational accidents. Ashgate Publishing Limited, Hampshire (1997)
7. Leveson, N.G.: Safeware. System safety and computers. Addison Wesley Publishing Company, Reading (1995)
8. Dekker, S.: The re-invention of human error. Human Factors and Aerospace Safety 1(3), 247–265 (2001)
9. Dekker, S.: Ten Questions About Human Error: A New View Of Human Factors And System Safety. Lawrence Erlbaum Associates, Mahwah (2005)
10. Dekker, S.: Why we need new accident models. Journal of Human Factors and Aerospace Safety, 2 4(1), 1–18 (in press, 2004)
11. Sindall, T.: Special Objectives Check on air Operator's Fuel Planning Policies. FOCUS on Commercial Aviation Safety 42 (summer, 2000)

Resilience Markers for Safer Systems and Organisations

Jonathan Back[1], Dominic Furniss[1], Michael Hildebrandt[2], and Ann Blandford[1]

[1] University College London Interaction Centre
{j.back,d.furniss,a.blandford}@ucl.ac.uk
[2] OECD Halden Reactor Project, Industrial Psychology Division
michael.hildebrandt@hrp.no

Abstract. If computer systems are to be designed to foster resilient performance it is important to be able to identify contributors to resilience. The emerging practice of Resilience Engineering has identified that people are still a
primary source of resilience, and that the design of distributed systems should
provide ways of helping people and organisations to cope with complexity. Although resilience has been identified as a desired property, researchers and
practitioners do not have a clear understanding of what manifestations of resilience look like. This paper discusses some examples of strategies that people
can adopt that improve the resilience of a system. Critically, analysis reveals
that the generation of these strategies is only possible if the system facilitates
them. As an example, this paper discusses practices, such as reflection, that are
known to encourage resilient behavior in people. Reflection allows systems to
better prepare for oncoming demands. We show that contributors to the practice
of reflection manifest themselves at different levels of abstraction: from individual strategies to practices in, for example, control room environments. The
analysis of interaction at these levels enables resilient properties of a system to
be 'seen', so that systems can be designed to explicitly support them. We then
present an analysis of resilience at an organisational level within the nuclear
domain. This highlights some of the challenges facing the Resilience Engineering approach and the need for using a collective language to articulate knowledge of resilient practices across domains.

Keywords: Human error, distributed cognition, control rooms, nuclear domain.

1 Introduction

In this paper we analyse manifestations of resilient practice at different levels of abstraction from the individual working with simple artefacts to more complex team
working situations. Resilience markers can be any system feature or procedure that
enables resilient practice to manifest. Identifying these markers may provide useful
performance indicators, and allow the resilient characteristics of a system to be communicated, so that existing features or procedures can be augmented in a way that
increases the capacity for resilience beyond that which is already present.

Resilience markers specify the conditions that need to hold for a system to perform
resiliently. In addition to enabling the detection of error-prone or non-resilient computer
systems, our approach provides a means of reasoning about resilience. This allows us to

M.D. Harrison and M.-A. Sujan (Eds.): SAFECOMP 2008, LNCS 5219, pp. 99–112, 2008.

look at distributed systems from a new perspective. Resilience engineering takes the view that resilience is a characteristic of a system. This implies that a holistic perspective is required to develop an understanding. We are aware that the levels of granularity presented here are interrelated and so they should be considered collectively. However, much more work is needed to integrate these different levels. Indeed it could be argued that the nature of resilience goes against a level-based composition, however, our central focus is on finding evidence for resilience in the behaviour we observe, and identifying what type of behaviour we would classify as resilient. The aim of this approach is to develop an understanding of the system attributes that encourage people to engage in resilient activities (*see* Sections 3 and 4). We also discuss the difficulties of understanding resilience issues at an organisational level by presenting a case study from the nuclear domain (*see* Section 5). The examples presented in this paper should not be considered a full set of resilient behaviours that need to be supported: they have been selected as being representative of different levels of granularity that researchers and practitioners need to consider when designing systems that foster resilient performance (*see* Table 1).

Table 1. Levels of Granularity

Granularity	Examples of Vulnerabilities	Resilient Manifestations	Resilient Markers
Individual Level (*see* Section 3)	Errors in procedural routine	1. Reflection 2. Cue creation	Providing an opportunity for meta-cognitive activities.
Small Team Level (*see* Section 4)	Coping with increased demand	1. Buffering 2. Work shadowing 3. Artefact use	Optimised flow of information and physical layout. An understanding of artefact use, social conditions.
Operational Level (see Section 5)	High complexity	Error recovery	Symptom-based emergency procedures, automatic safety systems, strategic crew leadership.
Plant Level (*see* Section 5)	Plant shut downs or failures to start up, major accidents	1. Plant safety record 2. Response to major disturbances	Maintenance regime, plant upgrades, risk analysis, training programs.
Industry Level (*see* Section 5)	Political and regulatory intervention	Performance necessity and availability of alternatives	Regulatory compliance, public/political perception, cost-benefit ratio, competitiveness.

2 Background

Making a system safer involves coupling the capabilities of humans with the technology they work with so that they can stay in control. A resilient system is able to recognise, adapt to and absorb disturbances so that it remains safe by being flexible to new demands [1]. We report on work using experimental microworlds that enable cognitive strategies to be understood, as well as studies of team working situations using distributed cognition modelling. We also look at how the design of computer systems in control room environments explicitly supports resilient practice.

Historically, there has been much more focus on why things go wrong than on why they work well. Conventional engineering approaches to ensuring safety attribute failure to a system component (human or technological) rather than the system as a whole. When systems fail, the cause is often attributed to 'human error' or to a technical problem associated with a control process. Attributing blame to a faulty component offers a pragmatic solution; the component can simply be replaced, fixed, or retrained. The traditional view of managing safety involves attempts to reduce the complexity of a system so that humans can maintain control under stress [2]. For example, one technique is to try and design systems that minimise the number of procedures by automating subsidiary interactions and leaving only the main parameters for the operators to worry about. Ostensibly, this decreases the system complexity from a human-computer interaction (HCI) perspective. However, Perrow's account of high-risk technologies highlights that it is not complexity per se that causes accidents [3]. The existence of many system components is not a problem for either system designers or operators if their interactions are expected. Based on the analysis of case studies and foundational empirical work, we found that dealing with unexpected or hidden events is facilitated by: designs that provide operators with an opportunity to engage in reflection [4]; expanding the variability of actions operators can take [2]; supporting the use of artifacts (such as dynamically generated checklists) that augment the capabilities of human cognition [5, 6]. These types of interactions allow a system to maintain control by anticipating new demands. We classify them as being resilient interactions.

The performance of cognitive systems, ranging from the individual to a team, has been found to be sensitive to external factors such as time constraints and workload which erode control [1, 7]. However, experts are able to generate strategies that support resilient practice (e.g., [8]). Understanding how these strategies are generated will enable the development of computer systems that explicitly support resilient activities. Our approach is about understanding how systems can support the cognitive and communicative capabilities of humans. This enables the socio-technical system as a whole to adapt to oncoming demands. Work suggests that the process of managing demands is influenced by task structures and team roles [9], external cognitive artefacts and computer system design [10]. We suggest that these factors shape the potential for resilient interactions rather than simply attributing resilience to the capabilities of individuals themselves.

An opportunity to think about oncoming demands is essential for individuals, teams, and organisations to reason about ways that performance can be better supported, enabling future strategies to be formulated. For example, an opportunity to reflect can enable an individual to offload workload, allowing them to maintain levels of performance under stress or high load situations. For example when anticipating being in a rush to leave home for work, positioning your bag by the door reduces the likelihood of forgetting to take it with you. Reflecting in a team setting can allow for interruption management [11], task collaboration and temporal coordination [12]. Foundational work suggests reflection at an organisational level is unlikely to take place during routine operation. Nathanael and Marmas's Repetitions-Distinctions-Descriptions model [13] suggests that encountering abnormal or different scenarios forces 'distinctions' from the normal routine to be made. These 'distinctions' trigger reflection-in-action to alter practice; this altered practice can then be absorbed back into normal routine if appropriate. The ability of a socio-technical system, in which computer systems are an integral part, to prepare for oncoming demands is an important aspect of resilience. However, it is by no means the only one. Other aspects are discussed in Hollnagel and Woods [14].

3 Cognitive Resilience at the Individual Level

The first level of granularity to be considered is cognitive resilience. In safety-critical domains operators frequently perform routine tasks. Research on procedural routine has demonstrated that under increased workload individuals are more prone to slips [15]. Although the consequences of a slip do not necessarily move a system towards failure, the ability of an operator to perform effectively is influenced, since some control over the processes they are trying to manage has been lost. While most day-to-day slips result in minor annoyances, those that occur in safety-critical situations (such as in the aviation domain) can be catastrophic. Slip errors can occur systematically even when individuals have the required 'expert' procedural knowledge to perform a task correctly. Manifestations include omission errors (e.g. forgetting to collect the original document after making photocopies), and mode errors (e.g. typing with the Caps Lock mode activated). Slips cannot be eliminated through practice or increased motivation [16] but they can be reduced by adopting a resilient strategy (such as leaving your bag by the door). We hypothesised that reflection can support performance during HCI, allowing slip errors to be mitigated. To test this hypothesis, an understanding of under what conditions individuals are able to engage in reflection was needed. In order to address the question of how an individual's resilient cognitive activities emerge, a 'Fire Engine Dispatch Centre' microworld was developed [6]. The development of a microworld to study how individuals avoid slips improves understanding of what factors shape performance.

The overall objective of the microworld experiment was to send navigational information to fire engines enabling the fastest possible incident response times. When a call was processed the location of the nearest fire engine and the location of the incident were displayed automatically as waypoints on a map. Participants had three minutes to identify the best route based on information displayed on a traffic information ticker. Training trials were used to ensure that participants became familiar with the sequence of actions. After performing two 'error free' training trials consecutively, a participant was allowed to move on to twelve experimental trials. Two error-prone task steps, outlined below, were built into the design: an initialisation step and a mode selection step. The emergence of resilient strategies associated with these steps provides concrete examples of cognitive resilience.

Initialization Step. When commencing a new trial an individual had to decide which call to prioritize before clicking on the 'Start next call' button (*see* Figure 1).

For each trial there was only one correct call prioritization selection. Participants were trained to know that incidents in poor fire engine coverage areas should be selected before incidents in good coverage areas. They also knew that high priority calls took precedence over normal priority calls irrespective of fire engine coverage. The first step in the process of setting call priority involved clicking on the radio button that was located alongside the required call ID. For example, in Figure 1 a participant is required to select ID 4. Clicking on 'Confirm priority change' is the second procedural step. Participants were instructed that the 'Start next call' button should only be clicked when both the new call ID has been selected and the 'Confirm priority change' button has been clicked.

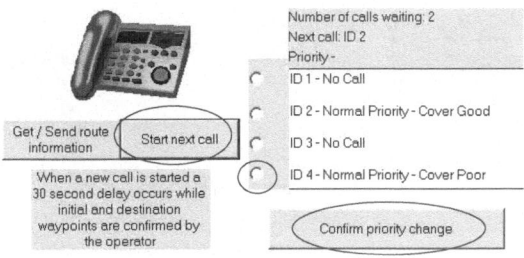

Fig. 1. Inititalisation Step

When a routine task is learned task, steps become associatively linked, i.e. action x (e.g. inserting a DVD) becomes a procedural cue for action y (e.g. locating the remote control). The initialization step could not be procedurally cued, since there was no preceding step, making it highly error-prone. The error occurred when participants omitted the initialisation step, which involved prioritizing calls to the dispatch centre, and instead clicked on the start next call button. The start button captured attention away from the correct procedure since it moved a participant towards starting the primary task of routing fire engines. Experimentation revealed that initialization errors were more avoidable if participants were given the opportunity to reflect on task requirements. The number of initialisation errors made by participants in Condition A, where the system encouraged reflection by displaying the control interface during a trial resumption delay, was compared with the number of errors made by participants in Conditions B, where participants were presented with a blank screen. The mean error rate when display cues were present was 6.09% compared to 23.12% when cues were absent (*Mann-Whitney U* = 40.2, *Wilcoxon W* = 158.5, Z = -2.605, $p < .01$, across 24 participants). Providing users with an opportunity to rehearse procedural steps allows for reflection. System designers can modify the task environment to ensure that rehearsal is possible and in some cases, where problematic interactions have been identified in the past, is actively encouraged (by enforcing delays). Providing a window-of-opportunity as a means of facilitating reflection is a useful marker for resilient design.

Mode Selection Step. After identifying a route, a participant had to select the required route construction mode.

When a participant commenced the route construction procedure (after clicking on the start button) the first requirement was to identify the most appropriate route on the map. Participants had to select the best route based on traffic information (i.e. they had to ensure a proposed route did not run through an accident or heavy traffic area). The device provided a signal that informed participants of the required method of route construction (located above the telephone image, *see* Figure 2). This signal was available after 35-45 seconds from pressing the start button. Participants were required to attend to this signal so that they could determine what type of route information was needed. If GPS was available then the centrally located menu could be used. Clicking on this menu enabled one of the automatically generated routes to be selected. The drop-down menu located below and to the left of the automatic route selection menu was used for manual route construction. A mode error occurred when a participant used the wrong route construction method.

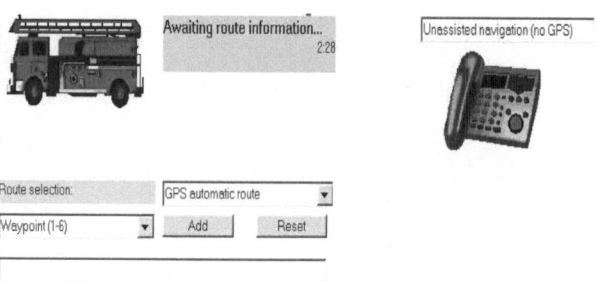

Fig. 2. Mode Selection Step

Attending to the mode selection step indicator required an attentional shift away from the main problem-solving task, making it highly error prone. A post-hoc analysis revealed the generation of a tractable resilient cognitive strategy. If participants placed the mouse cursor close to the signal status display (above the telephone in Figure 2) before the signal status appeared, they were less likely to forget to attend to the display before selecting the appropriate route construction method. When the mouse cursor was placed < 2cm from the display, participants were significantly less likely to make a mode error (Wilcoxon $Z = -1.870$, $p < .05$, two related samples test, across forty-eight participants). Positioning the mouse cursor enables the creation of a sensory cue. If the cursor is attended to then it may indicate that the display should be attended to when route identification is complete.

Further experimentation revealed that the generation of this resilient strategy was significantly more likely under a mixed workload condition. The complexity of the routing task was manipulated so that half of the participants only performed difficult routing tasks while the other half performed both easy and difficult tasks (mixed workload). In the mixed workload condition 64% of participants adopted the cursor strategy. In the high workload only condition only 27% of participants used the mouse cursor as a candidate cue. Critically, participants in the mixed workload condition who adopted this strategy were able to apply it during easy and difficult tasks (Wilcoxon signed rank test, related samples, *Wilcoxon Z = -1.039*, $p < .05$). Analysis of these findings enables us to identify a further marker for resilience that has implications for the design of computer systems. Personalised cue creation is spontaneous and can be used to minimize the likelihood of error. Allowing users to position markers (like 'Post-it' notes) provides support for attentional control. However, the use of such cues is only likely in situations where distinctions to the normal routine can be made. Mixed workload participants had: the cognitive resources available to think of an appropriate cue to guide attention (when workload was low) and the motivation for doing so, i.e. to support performance during high workload trials. Systems designers need to design scenarios that encourage metacognition during routine performance. It is generally agreed that the metacognitive activity consists of two basic processes occurring simultaneously: monitoring progress, and selecting or generating strategies to support performance [17]. Individuals need to be encouraged by the system to engage in metacognition so that they can develop a repertoire of resilient strategies. Reflection encourages the development of appropriate strategies and so enables levels of performance to be maintained under stress.

4   Resilience at the Small Team Level

As illustrated in the previous section, markers for resilient performance can be 'seen' in the laboratory. However, as previously discussed, manifestations of resilient practice occur at different levels of abstraction; next, we consider more complex team working situations. There are many different things to 'see' in socio-technical contexts, often too many, and so it is helpful to have approaches that can facilitate our perception in the 'noise' of real world contexts. DiCoT (Distributed Cognition in Teamwork) has been developed as an approach to applying distributed cognition to teamwork contexts [18]. Distributed cognition is a theoretical area which maintains the computational vocabulary associated with cognitive psychology but expands its unit of analysis. Hollan et al. [19] suggests three ways in which this expansion occurs:

- "Cognitive processes may be distributed across members of a social group";
- "Cognitive processes may involve coordination between internal and external (material or environmental) structure"; and
- "Processes may be distributed through time in such a way that the products of earlier events can transform the nature of later events."

This expansion has important implications for reflecting on and preparing for oncoming demands. For example: What system are we considering to be receiving these demands e.g. an individual, a team, a department, a company? Who is passing on the information and how? What timeframe and what sort of demands are we talking about e.g. restructuring the company over years or preparing for the next five minutes? How is this information structured internally within individuals? How is it represented externally in procedures? DiCoT encourages a system description which helps engage with these issues. Hollan et al. [19] indicate that what functionally influences the computation of the system is the concern of DC. DiCoT encourages analysts to look at these functional influences through five interdependent models. These look at the structure of information flows in the system, the artefacts which are used, the physical layout of the system, the social structures and factors in the system, and how the system has changed over time. These models, and the way they can be used to reflect on oncoming demands, are introduced below with reference to a London Ambulance Service control room study.

Information flow model. The information flow model concerns itself with the propagation and transformation of information within the system. This model underlies the other models. Firstly, the overall computational function of the system is represented in an input-process-output diagram. For example, the input-process-output diagram of an ambulance dispatch system is shown in Figure 3. After this the make-up of the computational system can be explored. Figure 4 shows the abstract computational structure of an ambulance dispatch team. From this we notice that the structure of the system is designed to cope with the oncoming demands of the system. The raw material from the External Callers is filtered into critical information for the decision hub. The buffers control information to the decision hub considering its workload and the criticality of the information. The filter does not hold up information in this way: it just changes its form for computational purposes. If the flow of information around a system is designed in a way that enables critical performance to be maintained during variability in workload, this can be considered a marker for resilience.

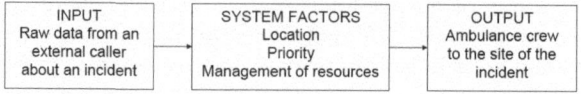

Fig. 3. The input-process-output diagram of an ambulance dispatch system

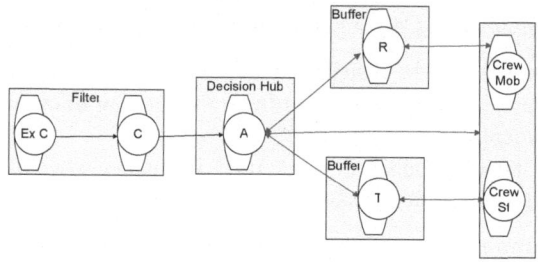

Fig. 4. Overview of main information flow properties the ambulance dispatch system. The Call Taker (C) filters the raw information from the External Caller (Ex C). This structured information is passed to the Allocator (A) who decides which ambulance should attend based on priority, availability and location. Depending on the status of the ambulance the Allocator (A) will channel information to the Telephone Dispatcher (T) or the Radio Operator (R), who will contact an ambulance crew at a station (Crew St) or one which is mobile (Crew Mob). Feedback from the ambulance crews (Crew St and Crew Mob) goes back through the Telephone Dispatcher (T) and the Radio Operator (R) who act as buffers for the decision hub i.e. holding up information when the hub is busy, if it is non-critical and would be disruptive.

Physical Model. The physical model concerns itself with functional influence of the physical layout of the system. For example, at the time of the study, the ambulance dispatch control room in London had seven desks, each of which is responsible for allocating ambulances to a different area of London. The arrangement of the seven desks reflects their geographical location, as adjacent areas will sometimes collaborate on the shared use of resources and attending incidents. This is particularly important with incidents near their shared border. This layout facilitates the oncoming demand of cross-boundary collaboration.

Figure 5 shows the seating arrangement of one of the allocating desks. The Allocator and Radio Operator work closely together, and so are adjacent. This facilitates their collaboration as the Radio Operator is implicitly aware of the Allocator's activities by shadowing them i.e. listening to their communication with others and watching their monitors. This allows the Radio Operator to prepare for oncoming activities before their receipt. This augmented awareness of work demands, through physical co-location, can be considered as a marker for resilience.

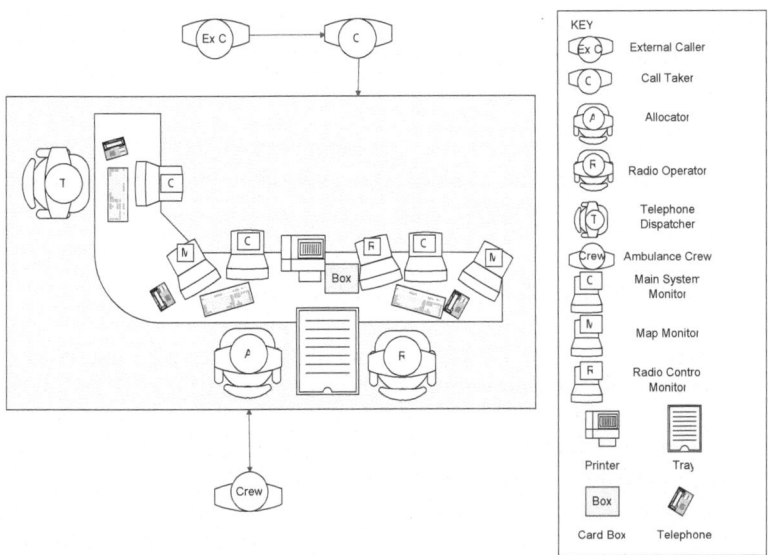

Fig. 5. An allocating desk. The information flow concerned with this sector desk is described in Figure 2.

Artefact Model. The artefact model concerns itself with the influence of the use of artefacts in the system.

Two brief examples of preparing for oncoming demand include: that the sector desks use a computer and card system which prepares them for the eventuality that the computer system might fail; and the computer system will manipulate the colour of incidents to indicate their criticality, facilitating the Allocator's prioritisation of incidents. Also, as soon as the Call Taker has established the location of the incident, the Allocator will have access to the updating details so that they can prepare for the oncoming demand. Redundancy and the support of decision making are important resilience markers.

Social Model. The social model concerns itself with the functional influence of the social structure and factors within the system. An example of inbuilt resilience at this level is that people generally get promoted from Call Taker, to Telephone Dispatcher, to Radio Operator to Allocator: so the more responsibility they have, the more aware they are of the other functions in the system and the way they work. An example of such resilience is that the Allocator may contact the ambulance directly if the Telephone Dispatcher is busy. Effective knowledge and responsibility transfer is a marker for resilience.

Evolutionary Model. The evolutionary model concerns itself with how the computational structure and functions of the system have changed over time. An example of a major change in the ambulance dispatch scenario was the introduction of GPS mapping. This gives Allocators a dynamic visual display of where the incident is and where their ambulances are located. These changes typically happen as a result of a constant pressure to improve processes, to respond to increasing demand from the environment, and to respond to the potentials new technology can offer. Exploiting technological advances to better cope with demands from the environment is a marker for resilience.

DiCoT can be used to understand the computation of the socio-technical system within these five interdependent models. This analysis notices how the system is co-ordinated to cope with oncoming demands.

5   Identifying Resilience in the Nuclear Domain

Two factors make the nuclear industry a particularly interesting context in which to discuss resilience. First, in a high-revenue, high-consequence socio-technical system such as the nuclear industry, significant safety and productivity gains can be expected if the promises of the Resilience Engineering approach can be delivered. Second, the nuclear domain presents an ideal environment for developing, operationalising and testing models of resilience. One of the reasons for this is that analysis of events in the nuclear industry requires a systemic approach. It is virtually impossible to discuss issues at one level of abstraction (operational, plant, industry and regulatory) without recourse to other levels. The degree of interconnectedness becomes clear when elaborating some of the defining characteristics of nuclear operations: information-rich operational environment; stable operations; possibility for severe disturbances; highly trained crews of operators; operational support network; highly proceduralised emergency operations reflecting thorough analysis of design-base accident scenarios; possibility of beyond-design-base incidents (e.g. fire); tight regulatory oversight and reporting regime; high investments and operations cost; high revenue; a variety of stakeholders, including operators, utilities, vendors, politicians and the public.

This section aims to identify manifestations of resilience at different organisational levels in the nuclear industry. The analysis is based on a number of information sources, including results from full-scale simulator experiments and training; incident and event reports; observational, ethnographic and interview studies (e.g. [20]); as well as Performance Shaping Factors that have been found to affect mission success over a range of scenarios in the context of Human Reliability Assessment.

Operational level. Nuclear operations are characterised by a high level of proceduralisation (especially during emergencies), and by a set of automatic safety functions designed to prevent the most severe consequences of accidents (core damage, release of radiation). There is crew-to-crew variability in procedure adherence, but crews are expected to follow the procedures as closely as possible. This system of operators, procedures, control room equipment and automation is expected to perform reliably for design-base incidents, i.e. those scenarios that have been considered during system design and in Probabilistic Risk Assessment. It is the successful interaction between these system components that creates resilience for design-base scenarios. Beyond these systemic properties, a number of factors have been recognized to improve the ability of the system to respond to disturbances. For instance, the move from event-based to symptom-based emergency procedures has allowed a wider range of plant states to be addressed, and provides operators with a simpler and more unified way of responding to complex events [21].

Even with these well-designed and well-tested procedures, plant conditions can arise that challenge the procedures and require knowledge-based situation assessment [22]. To respond successfully to these unanticipated, beyond-design-base events, both instrumentation and crew responses play an important role. Instrumentation helps the crew maintain an overview of the situation and develop an appropriate response plan. Other industries (e.g. petroleum) have already gone further down this path, and the nuclear industry can benefit from developments such as large-screen and information-rich displays, trend displays and ecological interface design. When considering crew responses to beyond-design-base events, a number of characteristics for success have been identified in recent simulator studies [22], including shift supervisors' team leadership style and situation assessment. This suggests that success in nuclear control tasks at a mission level *may* not depend only on success or failure of low-level activities, such as slips, lapses or misidentifications. Given the operational context and time available, such erroneous actions *should* be recovered from without significantly affecting the overall mission. Instead it appears to be crew-level factors, work styles and orientations that are more likely to determine mission-level success or failure. Differences between domains in the significance of low-level failures may be accounted for by the role of time. In domains with acute time pressure such as aviation, it is more likely that low-level erroneous actions can have catastrophic consequences, whereas in the timeframe available to nuclear operators, recovery mechanisms are in place that can compensate for low-level failures. Therefore, available time, and the situational and systemic factors that compensate for failures of individual system components, can be considered resilience markers. Investigating differences between domains as to how these factors influence mission success may provide important insights into markers for resilience.

Plant level. Plants react to outside influences (safety requirements, economics, public opinion) through upgrade programs, training, perseverance, or closure. Several candidates for markers of resilience at this level are available, including performance measures, safety measures (incidents / accidents), and safety culture measures. If and how these indicators measure resilience, in the sense of the plant's ability to respond to and recover from major disturbances, and to adapt to long-term outside forces, is unclear. Analysis of cases where plants have been built but never started up, were shut down well before the end of the designed life cycle, or consistently produce below-expectation power outputs may significantly improve our understanding of resilience. Case studies suggest that the management of organisational change plays an important role, and may constitute a marker for resilience. Organisational factors include conflicts between professional groups within a plant (e.g. operations, maintenance, engineering, managerial), problems of staff recruitment and retention (especially with regards to an aging work force in a so-called 'sunset industry'), and the effects of organisational re-structuring (e.g. mergers, change of ownership). Each of these factors can generate disturbances that compromise the resilience of the plant. A better understanding of these factors is needed as plants prepare for upgrades that will see their lifetimes extend for several decades.

A critical factor for resilience at both the operational and plant level, and a potential marker for resilience, is training. While regular training on well-known initiating events (e.g. steam generator tube rupture) improves response reliability on design-base scenario, training for beyond-design-base operations may require different approaches. More recently, training programs have started to place emphasis on scenarios that challenge procedure support, require knowledge-based diagnosis and planning, involve close crew interactions and communication, and are specifically designed to promote the shift supervisor's situation assessment. Debriefing of simulator training runs is moving from an instructional, failure-based approach towards a crew-guided, reflection-oriented approach.

Industry level. Many of the themes discussed in the previous section re-emerge when considering resilience at the industry level. Judging by the outcome, the nuclear industry possesses remarkable resilience. It recovered from severe accidents and the resulting hostile public opinion. While the survival of the industry was predicated on the organisational changes and safety improvements that followed in the wake of these events, the need for power output and lack of alternatives also played an important role. This suggests that resilience refers not only to the internal quality of a system to adapt to changes in its environment. Instead the environment itself (in this case: politics, the public) is in turn shaped by the perceived value of the products and services provided by the system. From this point of view, resilience markers at an industry level include pricing, demand and competition as well as safety records. Even the sheer size of the industry and the investments made in the infrastructure may contribute to its continued survival (resilience by inertia).

Finally, an important aspect of resilience in the nuclear industry is the role of the regulator. Many aspects of nuclear operations are subject to regulatory oversight. Regulatory practices such as risk-informed decision making have made safety assessment of highly complex systems feasible, while leaving plants some degree of flexibility in implementing and managing their own safety programs. The effect of regulatory oversight on the ability of the industry to adapt and change, the model of performance variability embedded in regulatory practices, and the analysis of outside forces affecting the regulators themselves, are important fields for resilience research.

6 Discussion and Conclusions

The examples presented in this paper are representative of different levels of granularity that researchers and practitioners need to consider when designing computer systems that foster resilient performance. All these examples demonstrate that people are an important source of resilience in creating safety under performance pressure. Our findings are incompatible with the view that erratic people degrade an otherwise safe system, and align with the viewpoint of Cook and Woods [23], who argue that humans need to be supported in a way that helps them cope with complexity. As Rochlin [24] identified, when managing hazardous technical operations, a high level of performance does not flow from eliminating error but rather through anticipating and planning for events and surprises.

At the cognitive level (*see* Section 3) we demonstrated how computer systems can be designed to enable individuals to develop resilient strategies. By allowing individuals to reflect on task requirements, the generation of these strategies becomes spontaneous. The spontaneity of using artefacts in the environment (such as a mouse cursor) to support performance when task demands are increased results in resilient human performance. At the small team level (*see* Section 4) the use of a methodological approach such as DiCoT is able to reveal the hidden complexity of team interactions. DiCoT provides potential to be used as a tool to analyze the performance of the system and recommend improvement in processes, in layout, in technologies, and in social structures within a system's history of change. Being able to represent interactions at a team level is important for understanding resilience, as manifestations, such as the ability to buffer, need to be supported by the way a control room is designed. Computer systems play an increasingly influential role in control rooms so should be considered as an integral component during design. At operational, plant, and industry levels (*see* Section 5) manifestations of resilience are harder to observe. However, the examples presented illustrate that people are still an essential source of resilience, and that the design of complex distributed systems should provide ways of helping people cope with complexity. Computer systems need to support: symptom-based diagnosis of problems at the operational level; flexibility and extendibility at the plant level; and survivability at the industry level.

Resilience markers can aid analyses of simulated scenarios at the individual and team levels, which can be used to evaluate the performance of safety-critical systems. Resilience markers at operational, plant, and industry levels can be used retrospectively. However using markers to predict performance and survivability requires researchers and practitioners to consider the interrelations between all levels collectively. More work needs to be done on understanding their integration.

Acknowledgement. Back, Furniss, and Blandford were supported by EPSRC grant GR/S37494.

References

1. Hollnagel, E., Woods, D.D.: Joint cognitive systems: Foundations of cognitive systems engineering. Taylor & Francis, Boca Raton (2005)
2. Dekker, S.: Failure to adapt or adaptations that fail: contrasting models on procedures and safety. Applied Ergonomics 34(3), 233–238 (2003)
3. Perrow, C.: Normal Accidents: Living with High-Risk Technologies. Basic Books (1999)
4. Back, J., Furniss, D., Blandford, A.: Cognitive Resilience: Reflection-in-action and on-action. In: Proc. Resilience Workshop, pp. 1–6. Linköping University (2007)
5. Masino, G., Zamarian, M.: Information technology artefacts as structuring devices in organizations. Interacting with Computers 15(5), 693–707 (2003)
6. Back, J., Blandford, A., Furniss, D., Curzon, P.: Avoiding Slips. Submitted for journal publication (2008)
7. Wright, P.: The harassed decision maker: Time pressures, distractions, and the use of evidence. Journal of Applied Psychology 59, 555–561 (1974)
8. Klein, G., Orasanu, J., Calderwood, R., Zsambok, C.E.: Decision Making in Action: Models and Methods. Ablex Publishing Co., Norwood (1993)
9. Kirsh, D.: Adapting the environment instead of oneself. Adaptive Behaviour 4(3/4), 415–452 (1996)
10. Spillers, F., Loewus-Deitch, D.: Temporal attributes of shared artifacts in collaborative task environments. In: Proc: HCI 2003 workshop on temporal aspects of tasks (2003)

11. Furniss, D., Blandford, A.: Understanding Emergency Medical Dispatch in terms of Distributed Cognition: a case study. Ergonomics Journal 49(12/13), 1174–1203 (2006)
12. Bardram, J.E.: Temporal coordination: On time and coordination of collaborative activities at a surgical department. Computer Suppoted Cooperated Work 9, 157–187 (2000)
13. Nathanael, D., Marmas, N.: The interplay between work practices and prescription: a key issue for organisational resilience. In: Proc. 2nd Resilience Eng. Symp., pp. 229–237 (2006)
14. Hollnagel, E., Woods, D.D.: Epilogue: Resilience engineering precepts. In: Hollnagel, E., Woods, D.D., Leveson, N. (eds.) Resilience engineering: Concepts and precepts, pp. 347–358. Ashgate (2006)
15. Byrne, M.D., Bovair, S.: A working memory model of a common procedural error. Cognitive Science 21, 31–61 (1997)
16. Back, J., Cheng, W.L., Dann, R., Curzon, P., Blandford, A.: Does being motivated to avoid procedural errors influence their systematicity? In: Proc. HCI 2006, pp. 151–157 (2006)
17. Ertmer, P.A., Newby, T.J.: The expert learner: Strategic, self-regulated, and reflective. Instructional Science 24, 1–24 (1996)
18. Blandford, A., Furniss, D.: DiCoT: A methodology for applying Distributed Cognition to the team working systems. In: Gilroy, S.W., Harrison, M.D. (eds.) DSV-IS 2005. LNCS, vol. 3941, pp. 26–38. Springer, Heidelberg (2006)
19. Hollan, J., Hutchins, E., Kirsh, D.: Distributed cognition: toward a new foundation for human-computer interaction. ACM Trans. Comput.-Hum. Interact. 7(2), 174–196 (2000)
20. Perin, C.: Shouldering Risks. Princeton University Press, Princeton (2004)
21. Ujita, H., Kubota, R., Ikeda, K.: Development and Verification of a Plant Navigation System. Cognition, Technology & Work 3, 22–32 (2001)
22. Halden Work Report 844. The International HRA empirical study – Pilot phase report. OECD Halden Reactor Project. Halden, Norway (2008)
23. Cook, R.I., Woods, D.D.: Operating at the Sharp End: The Complexity of Human Error. In: Bogner, M.S. (ed.) Human Error in Medicine, pp. 255–310. Lawrence Erlbaum, Mahwah (1994)
24. Rochlin, G.: Safe operation as a social construct. Ergonomics 42, 1549–1560 (1999)

Modeling and Analyzing Disaster Recovery Plans as Business Processes

Andrzej Zalewski, Piotr Sztandera, Marcin Ludzia, and Marek Zalewski

Warsaw University of Technology, Institute of Automatic Control and Computational Engineering, Warsaw, Poland
a.zalewski@ia.pw.edu.pl

Abstract. The importance of business continuity and disaster recovery (BC/DR) plans has grown considerably in the recent years, becoming a well-established practice to achieve organization's resiliency. There are several applicable standards, like BS 25999-1:2006, sets of guidelines and best practices in this field. BC/DR plans are typically text documents and exercising is still the main measure used to verify them. On the contrary, to the common practice we suggest to model BC/DR plans as business processes using ARIS methodology and models, which have proven successful in the Enterprise Resource Planning systems projects. This provides uniform representation of BC/DR plans that can be applied across the whole distributed organization, strengthens the efficiency of traditional manual analysis techniques like walk-throughs, helps to achieve completeness, consistency and makes possible computer simulation of BC/DR processes. Timing and dynamic behavior, resource utilization and completeness properties have been also defined. It is possible to analyze them with computer support based on proposed ARIS model of BC/DR plan.

1 Introduction

The catastrophes of last decade, like hurricane Katrina or terrorist's attack on World Trade Center in New York, have shown the importance of organization's resilience against severe disruptions. This caused a rapid development in the genre of Business Continuity, which resulted in:

- the development of a number of standards and recommendation sets – e.g. Business Continuity Management (BCM) standard BS 25999-1:2006 [1], Standard on Disaster/Emergency Management and Business Continuity Programs NFPA 1600 [4], recommendations for contingency planning by NIST, U.S. Department of Commerce [5];
- the inclusion of business continuity practices in IT services management standard ISO 20000 [2] and IT auditing standard COBIT [3];
- numerous books published on the topic of BCM – e.g. [6], [7], [8].

Business Continuity Managements system is implemented within an organization to enable structured, well-organized and timely recovery from severe disruptions. Business Continuity (BC) Plans (including Disaster Recovery Plans)

M.D. Harrison and M.-A. Sujan (Eds.): SAFECOMP 2008, LNCS 5219, pp. 113–125, 2008.

are a key element of this system. As these plans are of vital interest to the organization they should not only be diligently elaborated but also validated and verified (either during the development or during the audits and maintenance).

As it has been shown in section 2 most of BC/DR plans are currently textual documents of different levels of detail and formality. As such they are prone to incompleteness, inconsistency and other imperfections, being at the same time difficult to analyse and verify. To compensate for these disadvantages we advocate an idea of integrating Business Continuity Management with business process modeling to increase the level of formality of BC/DR plans.

BC/DR plans in our approach are treated as a specific kind of business processes activated only in case of severe disruptions. As such, they can be modeled with notations used for business process modeling. In this paper, we use Sheer's Architecture of Integrated Information Systems (ARIS) methodology and notation, which has proved successful in commercial applications, especially Enterprise Resource Planning systems projects. Formalizing one of the disaster recovery plans available from the Internet we show the superiority of formalized diagrammatic representation to traditional textual form of BC/DR plans. Definitions of the important properties of BC/DR plans modeled with ARIS and techniques of their analysis have been provided.

The rest of the paper is organized as follows: the missing parts of BC management are discussed in detail in Section 2, the core concept of the paper i.e. modeling of BC plans with ARIS methodology and models are presented in Section 3, analysis of ARIS models are discussed in Section 4, the results of the paper are discussed in Section 5, future research areas have been suggested in Section 6.

2   The Missing Parts of Business Continuity Management

BS 25999-1:2006 defines how to implement Business Continuity Management within an organization. It defines Business Continuity Management life cycle. The cycle starts from identifying critical services and products, business impact analysis and risk analysis. It is aimed generally at identifying recovery requirements and threats. These in turn lead to the identification of BC Management options and elaboration of appropriate response in the form of incident management, business continuity and disaster recovery plans. These plans play a key role in the resiliency assurance. All these arrangements are subject to exercises, maintenances, audits and self-assessment in the last phase of the BCM life cycle. Similar approaches have been presented in numerous papers (e.g. [13], [11]).

BCM practices seem to be present in the majority of large organizations in the developed economies (see survey for US [9]). The Internet research on the form of BC/DR plans – presented in table 1 reveals that most of the BC/DR plans are just textual documents. The list of analysis/verification techniques for BC/DR plans is rather short – it includes mainly manual methods like deskchecks, walk-troughs, simulations (manual) as well as executions of a part or even entire plan (see BS25999:1 [1]). Only simulations are subject to computer

Table 1. Disaster Recovery/Business Continuity plans level of formalization – survey of the practice

No.	Organization / source	Level of formalization
1	The Australian National Herbarium Canberra – the aim of the plan is to protect and restore the Collection. http://www.anbg.gov.au/cpbr/disaster-plan/	Low – DR plan represented as textual enumeration organized into chapters and subchapters.
2	University of Arkansas – the aim of the plan is to restore all computer operations without loss of any data. http://www.uark.edu/staff/drp/	Low – DR plan represented as textual enumeration organized into chapters and subchapters.
3	University of California – the aim of the plan is to protect and restore the book collection of the general library. http://palimpsest.stanford.edu/bytopic/disasters/plans/ucdaviis_disasterplan2004.pdf	Low-medium – emergency plans represented as textual enumerations, short sentences are used. There is a lot of white space used between each step in printable version to make easy the orientation in the plan.
4	Systems Support Inc. – the aim of the MIS Contingency Plan is to protect corporate resources and employees. http://www.drj.com/articles/drpall.html	Medium – detailed recovery plans presented as textual enumerations, actions are presented in tabular form with explicate naming heading, executing person and action.
5	Massachusetts Institute of Technology – the aim of the plan is to restore critical functions of MIT and the resources required to support them. http://web.mit.edu/security/www/pubplan.htm	Low – recovery processes are presented as textual enumerations. Teams and their emergency actions have been described.
6	University of Arkansas Computing Services Disaster Recovery Plan http://www.uark.edu/staff/drp/	Medium – disaster recovery plans are presented in textual form (including both, actions and resources). Detailed description of roles, actions and resources, but without logical connections between them.
7	NIH Data Center http://datacenter.cit.nih.gov/pdf/disasterplan.pdf	Low – detailed recovery plans are presented in enumerated text form. People have not been explicitly assigned to the recovery actions.
8	Abilene Christion University http://www.acu.edu/technology/is/recovery.html#PCRecovery	Low – recovery action plans are presented in textual form. Disaster recovery teams and their responsibilities are described.

support. The efficiency of manual analysis methods is strongly limited by the textual form of BC/DR plans. Full assessment can be achieved only through real execution of a plan or its part. Apart from the costs of such an execution it is worth noting that there are important cases, in which such experiments are risky themselves and probably would not be accepted by the appropriate authorities: consider case of an art gallery with a collection of precious paintings or sculpture.

The literature on the properties of BC/DR plans and their analysis is rather sparse – the problem has not been so far treated in its entirety – only narrow publications are available e.g. [10], [12].

3   Modeling BC/DR Plans as Business Processes

As a first step to resolve, the issues raised above we present how to model BC/DR plans using Sheer's ARIS methodology and notations – see [14], [15], [16]. The major competitors to ARIS seem to be Business Process Modeling Notation (BPMN) by Object Management Group as well as Unified Modeling Language. Both of them lack models of organization, data (resources) and products while they are focused on the flow of processing and documents (data). This is a major deficiency as all these elements are an integral and important element of every BC/DR plan. ARIS methodology, in turn, defines five views of an organization – organizational, data, function, product/service, process. All the elements comprising BC/DR plans can be assigned to one of those perspectives, which has been shown in table 2.

Table 2. Representation of BC/DR elements in ARIS Methodology

BC/DR element	ARIS view	ARIS model element
Role/Team Responsibility	Function	Function
Critical function	Function	Function
Supporting equipment and supplies	Data	Entity type
BCMS Documentation	Data	Entity type
Organizational structure	Organization	Organizational chart
Groups and Roles	Organization	Organizational chart
Senior Management	Organization	Position/Group
Stakeholders	Organization	Person type
Staff resources	Organization	Internal person
External services and supplies	Organization	External person
Activity	Function	Function
Business Continuity Plan	Process / Control	EPC diagram
Incident management plan	Process / Control	EPC diagram
Incident	Process / Control	Event
Business interruption	Process / Control	Event
Products and services	Product / Service	Product/Service
Business Continuity Management Life cycle	Process / Control	Value Added Chain Diagram

The modeling of BC/DR plans in each of the above perspectives has been presented below in Section 3.1 – 3.5 and illustrated on DR plan for the general library of the University of California [17] (see also table 1, pos. 3).

3.1   Organizational View of BC/DR Plans

The main model of the Organizational View is Organizational Chart. It models the internal structure of the teams engaged in BC/DR plans representing the relations between different members of those teams.

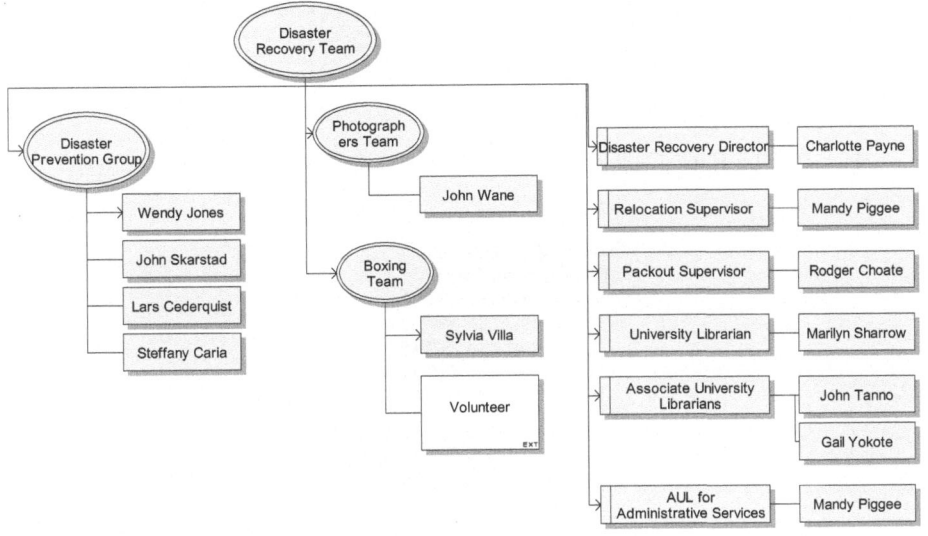

Fig. 1. The Organizational Chart of the Disaster Recovery Team

The Organizational Chart in figure 1 defines Disaster Recovery Team of the University of California. The enclosed diagram shows DR team consists of groups (Boxing Team) and positions like Disaster Recover Director. More information about the teams and their members can be registered as attributes of appropriate objects (see table 2).

3.2   Data View of BC/DR Plans

Data view models resources (excluding human resources) used in BC/DR plan. The relations between them are modeled as Entity-Relationship Model (ERM).

Figure 2 models some of the resources used in DR Plan of University of California, i.e. emergency box consisting of such first aid kit, camera and the other.

3.3   Function View of BC/DR Plans

The function view models functions (i.e. technical tasks or other activities) and their hierarchy. The latter is modeled with Function Tree Diagram. Functions are characterized by the attributes of costs or execution time, which are useful for simulation.

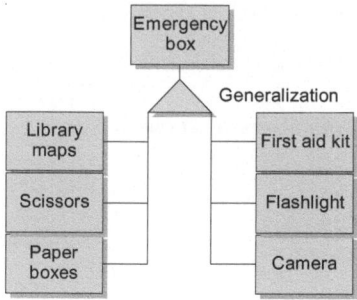

Fig. 2. The Entity-Relationship Model – The content of emergency box

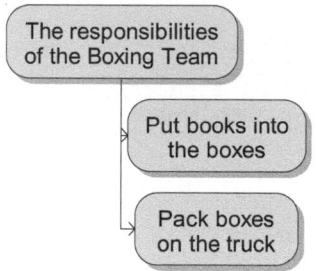

Fig. 3. The Function Tree - Responsibilities of The Boxing Team

The diagram in figure 3 presents the role (function) of Boxing Team in disaster recovery: they are responsible for putting the books into paper boxes and packing them onto the truck.

3.4   Product/Service View of BC/DR Plans

Product or Services are results of the execution of BC/DR plan. They are typically of different levels of abstraction constituting product/service hierarchy – several partial products make an entire higher-level product. This hierarchy is represented by the Product/Service Tree diagram.

The Product Tree diagram in figure 4 shows the partial products comprising "The pack out final report", which is one of the final products of the "Pack out process". It consists of budget, packing report and photographs. The budget is a product of function "Prepare a recovery budget", which is one of the functions in "Pack out process".

3.5   Process/Control View of BC/DR Plans

A Process View consists of two main models: Value-Added Chain Diagram (VACD) and Event-Driven Process Chain (EPC). They have been used to model the processes of BC/DR plans putting the data contained in all the other views into a single, legible model.

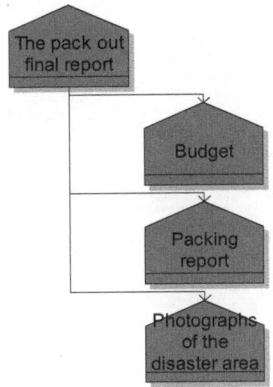

Fig. 4. The Product Tree - The partial products of The pack out final report

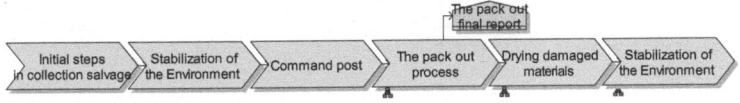

Fig. 5. Value-Added Chain Diagram – the simple processes of DRP

The VACD describes the top-level functions or processes. They usually form a chain illustrating the process of gradual achieving of a higher level goal (product).

Figure 5 shows process of Disaster Recovery Plan of the University of California, which consists of several subprocesses, among them is "The pack out" subprocess modeled below with EPC diagram.

The Event-Driven Process Chain models the procedures of BC/DR plan integrating the information from all the other views:

- resources defined in data views become inputs to the functions;
- products become outputs of the functions;
- elements of the organization view are assigned to the functions (activities) to show the responsibility of the BC/DR teams and/or their members.

The process is event-driven, as every functions is activated with the occurrence of an event and its completion also generates one or more events. Events are graphically represented as hexagons.

The EPC diagram in figure 6 models "The pack out" process. It starts when fire department gives permission to enter the affected area and finishes when the "Final report" is ready. Note that EPC diagram integrates all the information needed to understand and manage the modeled process. It makes possible simulation of a process providing information about cost, time and workload.

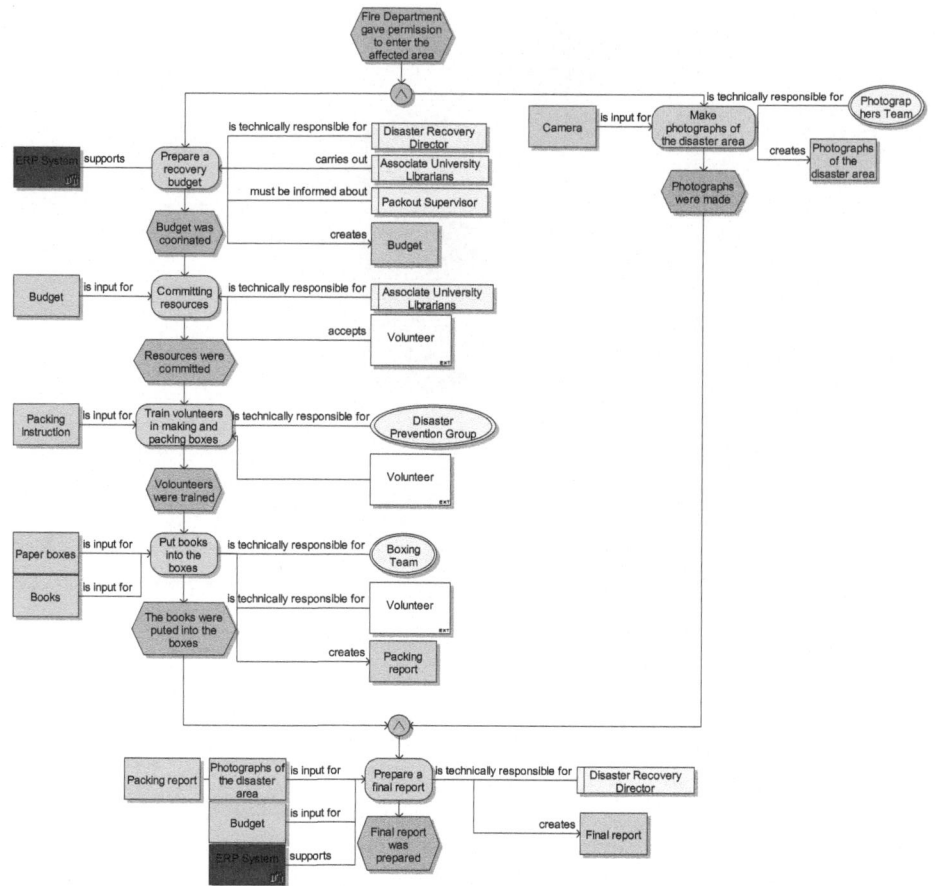

Fig. 6. The Event-Driven Process Chain - The pack out process

4   Analyzing Properties of BC/DR Plans

EPC representation of BC/DR plan makes possible analysis of its timing and dynamic behaviour, completeness and resource utilization properties. This can be achieved by the simulation of EPC models of BC/DR procedures or the analysis of data collected in the ARIS perspectives and their inter-relations. An automated software tools support can be easily developed to support such analyses. At the same time manual analysis techniques like walk-through, manual simulation or desk-checks become more efficient – obviously, it is easier to analyze diagrams than textual documents.

4.1   Simulation

Full formalization of all the ARIS models used to represent BC/DR plans exceeds the scope of this paper. Therefore, the concepts presented below, especially the definitions of the properties of BC/DR plans are semiformal but the ideas behind them are clear and easy to implement in practice.

In our approach BC/DR plan is modeled with a set of EPC diagrams. The execution of a plan can be simulated with discreet event simulation techniques. Necessary prerequisites are:

- Duration times assigned to the functions – these should typically be worst-case durations of the modeled activities;
- Simulation scenarios defined as:
  - times of occurrence of certain events – typically external ones – this typically defines the sequence of process activations due to the occurrence of external events;
  - indicating which choices to select during the simulation in case of conditional constructs – here various strategies can be applied – random choices, selection of either negative or positive choices, user-defined choice.

The simulation can be carried out with computer support – it is possible to use standard modeling tools or develop some of one's own. The result of a simulation of a set of EPC models has been referred to as *event trace*.

Definition 1. *Event Trace of a simulation of a set of EPC diagrams is a sequence of 3-tuples (e, t, p), where e stands for unique identifier of an event, t – time of occurrence of event e measured from the start of the simulation, p – process in which event e has occurred.*

4.2   Timing and Dynamic Behaviour

Timing properties of BC/DR plans are obviously of highest interest to the stakeholders as such plans are usually aimed at bringing the length of the disruption period to a minimum. Analyzing event trace of a given simulation, it is possible to calculate the time between any pair of events that happened during the course of the simulation. This makes possible to estimate the whole duration of BC/DR procedures and relates them to the Business Continuity requirements, expressed in terms of Maximum Tolerable Period of Disruption or Recovery Time Objective. A number of simulations can provide worst-case estimates on the duration of BC/DR procedures.

Definition 2. *If the duration of a given EPC process is predictable, than the directed graph made of EPC diagram limited to events, functions, conditions and logical operators is acyclic. (necessary condition)*

The above definition indicates that if the BC/DR plan contains any conditional construct leading to the functions performed earlier some sequences of function executions can be performed more than once and the number of such repetitions cannot be deduced just from the diagram. Such situations indicate potential errors in BC/DR plan or a risky organizational solution.

4.3   Completeness

Definition 3. *The BC/DR plan is* **complete** *if:*

1. *Each team has at least one team member,*
2. *Each team/team member is assigned to at least one activity (function),*
3. *Each resource has been assigned to at least one activity (function),*
4. *Each product has been assigned as a result of at least one activity,*
5. *Each function is performed at least once in any EPC model comprising BC/DR plan.*

As the data of all the modeling perspectives is strongly interconnected – analysis of these connections can uncover defined but unused resources, teams or team members taking parts in no activities as well as activities defined but not performed during the course of the plan. This indicates potential error in BC/DR plan. Completeness can easily be verified automatically by analyzing the data gathered in each of the ARIS perspectives and its interconnection with appropriate other perspectives – e.g. to verify that all the functions have been utilized it is necessary to compare the set of functions from function view against all the EPC models of process views.

4.4   Technical and Human Resource Utilization

Event traces can be algorithmically transformed to the function (activities) execution traces (basing on the assignment of teams/roles/persons and resources to functions), which model the occupation of given resources during the simulation of BC/DR plan.

Definition 4. *Function Execution Trace of a simulation of EPC diagram D is a sequence of 4-tuples (a, s, f, p), where a stands for unique identifier of an activity (function), s, f – respectively: time of the start and the end of the execution of activity a, p identifies the process containing executed function f.*

Event trace makes it possible to establish:

- The total occupation of a given resource r by all the processes comprising BC/DR plan – it is given by the sum of execution times of functions f to which are assigned resources r;
- The utilization of given resource r – it is the occupation of resources r related to the total duration of BC/DR plan execution;
- The action that possibly conflict on given resource r – such a conflict may take place when two actions use the same resource and their execution periods overlap.
- The timing of the potential resource usage conflicts.

All the above analyses can be automated with appropriately developed software tools.

5   Discussion

Preparing the example illustrating the concepts of modeling BC/DR plans in ARIS approach we tried to represent the DR plan for the library of the University of California using ARIS models. This experiment revealed both drawback of traditional textual forms of BC/DR plan as well as the advantages of modeling such plans with ARIS models.

Although the analyzed plan defines all the necessary components of BC/DR plan, i.e. roles, team member, resources, products, activities and their sequencing, it is very difficult to put all these things together. The connections between activities and teams or team members responsible for performing them, activities and necessary resources and products resulting from these activities are very difficult to locate as all this vital information is spread all over the text document – the references between them are unclear and difficult to maintain. This may lead to incompleteness of BC/DR plans. In fact, we have found the following flaws:

- several activities without any responsible role or person assigned,
- a few activities with undefined resources or incomplete resources assigned,
- resources indicated as needed for a given activity but remaining undefined (the need for rooms for book drying has been specified, however, even potential rooms have not been indicated),
- ambiguous and potentially conflicting roles – e.g. photographing was a duty of the Recovery team, however there is also photographer mentioned in the whole plan whose role does seem to be conflicting with the recovery team unless he is a member of this team, which is not quite clear – the diagram presented in figure 1 is a proposition of resolving this ambiguity,
- one of the persons is probably overloaded with the assigned duties.

All the connections between the components of BC/DR plan, which are so difficult to identify in the textual form of BC/DR plan are explicitly and legibly expressed in ARIS models, especially in EPC diagrams. This makes traditional verification techniques like walkthroughs and desk-checks easier to perform and more efficient, while providing the ability of analyzing properties of BC/DR plan as described in Section 4. Of course full assessment of such a plan is only achievable with its full execution, however precise expression and prior analysis should help to avoid exercising a defective BC/DR plan.

The properties and analysis techniques described in Section 4 provide for basic verification and analysis of BC/DR plan properties. They can help to identify serious flaws in BC/DR plans. The properties of resource utilization, resource conflicts, loops in processes require in-depth analysis, usually requiring more detailed information than defined in our approach. Exemplary issues have been listed below:

- Some resources may be used only exclusively by single person or team at a time. This can force other teams to wait until necessary resource has been released by the other person or team. This situation has not been included in our model. To account for that our model has to be enhanced.

- Conflicts on resources may in extreme cases lead to deadlocks – as they do in case of all parallel systems. To detect such situations it is necessary to convert ARIS model to a fully formal model that makes appropriate analysis possible.
- The resources can also be characterised by their capacity e.g. the capacity of a team is number of man-hours that certain team can work during a unit of time. Again this may be subject to further research.

6   Conclusion and Future Research

There are numerous advantages of modeling BC/DR plans as business processes with ARIS models and methodology:

- It increases preciseness of expression and consistency of BC/DR plan;
- It ensures legible and easily understandable way of documenting and communicating BC/DR plan;
- It increases the efficiency of traditional verification techniques like desk-checks and walkthroughs;
- It makes BC/DR maintenance, on-demand adjustment and audit easier;
- Assessment of BC/DR plan can be performed prior to its execution by means of simulation or using analysis techniques and property definitions described in this paper. The analysis encompasses timing and dynamic behaviour, completeness and resource utilization properties;
- Monitoring and supervision of the execution of BC/DR plan is easier and more efficient when it is modeled as business process with appropriate diagrams;
- It ensures considerable money savings as only plans validated and verified on ARIS models could be exercised in reality;
- It might help to standardize BC/DR plans within a distributed organization.
- ARIS models of BC/DR plans are a common language to be used by all the stakeholders. As it's level of formalism is considerably higher than in the case of a textual form it makes the communication between different stakeholders more precise and unambiguous.

The main directions for the further research are:

- Extension of the model presented in this paper to enable in-depth analysis of resource utilization and resource access conflicts,
- Further formalization of ARIS model – precise expression of the models used for BC/DR modeling in algebraic terms,
- Conversion of ARIS models or its formal form to one of the models of dynamic, parallel systems (like Petri Nets, CSP, Lotos),
- Defining further properties of BC/DR plans that can be subject to analysis,
- Extending analysis techniques with the analysis of dynamic properties (e.g. liveness), resource utilization, conflicts on resource usage.

References

1. BSI: Standard BS 25999-1:2006. Business continuity management. Code of practice, http://www.bsi-global.com
2. ISO/IEC: Information technology – Service management – Part 1: Specification (ISO 20000-1), Part 2: Code of practice (ISO 20000-1). ISO/IEC (2005)
3. ITGI: COBIT 4.1: Control Objectives for Information and related Technology. IT Governance Institute (2007)
4. NFPA: NFPA 1600 – Standard on Disaster/Emergency Management and Business Continuity Programs. National Fire Protection Association (2007)
5. Swanson, M., et al.: Contingency Planning Guide for Information Technology Systems, Recommendations of the National Institute of Standards and Technology, pp. 800–834. NIST Special Publication (June 2002)
6. Snedaker, S.: Business Continuity and Disaster Recovery for IT Professionals. Elsevier, Amsterdam (2007)
7. Barbara, M., et al.: Effective Strategies to Ensure Business Continuity/Disaster Recovery. Dr. Mueller.Verlag
8. Thejendra, B.: Disaster Recovery and Business Continuity. IT Governance Ltd (2007)
9. Nelson, K.: Examining Factors Associated with IT Disaster Preparedness. In: Proceedings of the 39th Hawaii International Conference on System Sciences (HICSS 2006), p. 205b. IEEE, Los Alamitos (2006)
10. Zambon, E., et al.: A Model Supporting Business Continuity Auditing & Planning in Information Systems. In: Second International Conference on Internet Monitoring and Protection (ICIMP 2007), pp. 33–33. IEEE, Los Alamitos (2007)
11. Kepenach, R.: Business Continuity Plan Design. 8 Steps for Getting Started Designing a Plan. In: Second International Conference on Internet Monitoring and Protection (ICIMP 2007), p. 27. IEEE, Los Alamitos (2007)
12. Cloth, L., Haverkort, B.R.: Model Checking for Survivability! In: Proceedings of the Second International Conference on the Quantitative Evaluation of Systems (QEST 2005), pp. 145–154. IEEE, Los Alamitos (2005)
13. Hayes, P., Hammons, A.: Picking up the Pieces: Utilizing Disaster Recovery Project Management to Improve Readiness and Response. In: IEEE Industry Applications Magazine, November/December 2002, pp. 27–36. IEEE, Los Alamitos (2002)
14. Scheer, A.W.: ARIS – Business Process Frameworks. Springer, Heidelberg (1999)
15. Scheer, A.W., et al.: Business Process Automation. Springer, Heidelberg (2004)
16. Weske, M.: Business Process Management: Concepts, Languages, Architectures. Springer, Berlin (2007)
17. University of California: Disaster Prevention, Preparedness and Recovery Plan, http://palimpsest.stanford.edu/bytopic/disasters/plans/ucdaviis-disasterplan2004.pdf

Analysis of Nested CRC with Additional Net Data in Communication

Tina Mattes[1], Frank Schiller[1], Annemarie Mörwald[2], and Thomas Honold[3]

[1] Technische Universität München, Institute of Information Technology in Mechanical Engineering, Boltzmannstr. 15, D-85748 Garching near Munich, Germany
{mattes,schiller}@itm.tum.de
[2] sd&m AG, software design & management, Carl-Wery-Str. 42, D-81739 Munich, Germany
annemarie.moerwald@sdm.de
[3] Zhejiang University, Institute of Information and Communication Engineering, Zheda Road, 310027 Hangzhou, P.R. China
honold@zju.edu.cn

Abstract. Cyclic Redundancy Check (CRC) is an established coding method to ensure a low probability of undetected errors in data transmission. CRC is widely used in industrial field bus systems where communication is often executed through different layers. Some layers have their own CRC and add their own specific data to the net data that is meant to be sent. Up to now, this nesting is not yet included in the safety proof of systems. Hence, additional effort is made to achieve a required degree of safety which was probably on hand but could not be proven. The paper presents an approach to involve the nesting in the calculation of the residual error probability based on methods of coding theory. This approach helps to reduce the number of worst case assumptions in the overall safety proof and finally to reduce the necessary online efforts like the number of parity bits.

Keywords: Cyclic Redundancy Check, Residual error probability, Safety-critical communication.

1 Introduction

Cyclic Redundancy Check (CRC) is a common coding method to detect errors in industrial data transmission. Especially for automated plants in safety-critical applications, the integrity of data (e.g. data that is sent from sensors to processing units or from processing units to actuators) is very important, since undetected errors could lead to dangerous accidents. The goal of safety-critical communication is to detect errors and to initiate the overall process into a safe state, e.g. a state of reduced functionality like low speed or zero voltage. Therefore, data transmission is an essential part of the overall safety proof [1]. A precise measure to quantify the quality of error detection is the residual error probability P_{re}. CRC guarantees a very small residual error probability with a relatively small number of redundant bits of a checksum. That is one reason, why

M.D. Harrison and M.-A. Sujan (Eds.): SAFECOMP 2008, LNCS 5219, pp. 126–138, 2008.
© Springer-Verlag Berlin Heidelberg 2008

CRC is widely used e.g. in industrial field bus systems. There, communication is often executed through different layers according to the ISO/OSI model [2]. Some layers usually have their own CRC and add their own specific data to the net data that is meant to be sent. Since CRC itself is very efficient, it is obvious to analyze this nesting of CRC in order to involve the decrease of P_{re}, which is caused by the nesting, in the safety proof of automated plants where it has not been considered explicitly yet. Usually, worst case assumptions are applied that lead to additional effort and unnecessary equipment costs.

The paper is the sequel to [3]. It is structured as follows. Mathematical principles of CRC are given in the next section including remarks on the calculation of the residual error probability. Then the nesting and the determination of its residual error probability are introduced. Results and examples follow before final conclusions are drawn.

2 Principles of CRC

In this chapter, basic principles of CRC are summarized. For further and detailed information see e.g. [4], [5].

2.1 Functionality of CRC

CRC applies a checksum, FCS (Frame Check Sequence), for error detection. This checksum is calculated in the sender as follows: The sender handles the original data (net data ND, information bits) consisting of m bits as a binary polynomial $nd(x)$. A so called generator polynomial $g(x)$ of degree r has to be chosen. Polynomial $nd(x)$ is first multiplied by x^r, and then divided by $g(x)$. The corresponding bit pattern of the remainder polynomial $fcs(x)$ (see (1)) consisting of r bits is the checksum FCS that is attached to ND:

$$(nd(x) \cdot x^r) \bmod g(x) = fcs(x) \tag{1}$$

For instance, the bit pattern of information bits ND = [1 0 1 0 1 1] and the generator polynomial $g(x) = x^3+x+1$ are given. The bit pattern ND leads to the binary polynomial $nd(x) = 1 \cdot x^5+0 \cdot x^4+1 \cdot x^3+0 \cdot x^2+1 \cdot x^1+1 \cdot x^0 = x^5+x^3+x+1$. The degree of $g(x)$ is $r = 3$. The polynomial counterpart of the checksum FCS is obtained by application of (1): $fcs(x) = ((x^5+x^3+x+1) \cdot x^3) \bmod (x^3+x+1) = x$. That means, the bit pattern FCS = [0 1 0] has to be attached to the original data ND. The resulting bit pattern of length $n = m + r$ is the telegram T = [ND, FCS] that is sent to the receiver. In the example, the telegram consists of nine bits, i.e. T = [1 0 1 0 1 1 0 1 0]. Since the equations (2) hold in the space of binary polynomials,

$$\begin{aligned} t(x) \bmod g(x) &= (nd(x) \cdot x^r + fcs(x)) \bmod g(x) \\ &= nd(x) \cdot x^r \bmod g(x) + fcs(x) \bmod g(x) \\ &= fcs(x) \bmod g(x) + fcs(x) \bmod g(x) \\ &= 0 \end{aligned} \tag{2}$$

It is checked in the receiver if the polynomial counterpart $t'(x)$ of the received telegram T' is divisible by the generator polynomial (cf. (3)):

$$t'(x) \bmod g(x) = 0? \tag{3}$$

If (3) is not true, the received telegram is erroneous and the error is detected; if (3) holds, T is regarded to be transmitted correctly. For instance, as in the example above, T $= [1\,0\,1\,0\,1\,1\,0\,1\,0]$ is sent, $g(x) = x^3 + x + 1$ and the received telegram T' $= [0\,1\,0\,0\,1\,1\,0\,1\,1]$. Since $t'(x) \bmod g(x) = x^2 + x \neq 0$, the falsification is detected. The determination of FCS in the receiver and the check in the sender is often realized by a linear feedback shift register (LFSR). It can also be modeled by a matrix-vector-multiplication which is used to explain the determination of the residual error probability of nested CRC in the following. Let I_m denote the unit matrix of dimension $m \times m$, $nd = (d_{m-1}\ d_{m-2}\ \cdots\ d_0)$ a vector whose coefficients are the bits of ND, and $t = (d_{m-1}\ d_{m-2}\ \cdots\ d_0 fcs_{r-1}\ fcs_{r-2}\ \cdots\ fcs_0)$ a vector consisting of the bits of telegram T. Then t can be calculated by means of a matrix A of dimension $m \times r$ that depends on the generator polynomial $g(x)$ by:

$$t = nd \cdot (I_m \mid A). \tag{4}$$

The matrix $G = (I_m \mid A)$ is called generator matrix. The matrix $H = (A^T \mid I_r)$ is called parity-check matrix and is used for the check in the receiver. This check can be formulated as follows where t' denotes the vector whose coefficients are the bits of the received telegram T':

$$(A^T \mid I_r) \cdot t' = 0? \tag{5}$$

If (5) holds, T is regarded to be transmitted correctly.

Matrix $A = (h^{n-1}, h^{n-2}, \ \cdots\ , h^r)^T$ is assembled by the vector counterparts $h^{n-1}, \ \cdots\ , h^r$ of the result of the modulo-division of the monomials $x^{n-1}, \ \cdots\ , x^r$ and the generator polynomial $g(x)$. Consequently, the i-th row of A consists of the FCS of the net data polynomial $nd(x) = x^{n-i}, i \in \{1, 2, \ \cdots\ , m\}$. The i-th row of the overall generator matrix G consists of the net data ND $= (d_{m-1}\ d_{m-2}\ \cdots\ d_0)$, where $d_{m-i} = 1$ and all other coefficients are 0, and its corresponding FCS.

In the example above with generator polynomial $g(x) = x^3 + x + 1$ and telegram length $n = 9$, the calculation of the modulo-division of the monomials and $g(x)$ leads to: $x^8 \bmod (x^3 + x + 1) = x$, hence $h^8 = (0\ 1\ 0)$, $x^7 \bmod (x^3 + x + 1) = 1$ and $h^7 = (0\ 0\ 1)$, $x^6 \bmod (x^3 + x + 1) = x^2 + 1$, thus $h^6 = (1\ 0\ 1)$; the further vectors are $h^5 = (1\ 1\ 1)$, $h^4 = (1\ 1\ 0)$ and $h^3 = (1\ 1\ 1)$. Therefore matrix A is given by:

$$A = \begin{pmatrix} h^8 \\ h^7 \\ h^6 \\ h^5 \\ h^4 \\ h^3 \end{pmatrix} = \begin{pmatrix} 0\ 1\ 0 \\ 0\ 0\ 1 \\ 1\ 0\ 1 \\ 1\ 1\ 1 \\ 1\ 1\ 0 \\ 0\ 1\ 1 \end{pmatrix}$$

it can easily be calculated by applying (4) to matrix A and $nd = (1\ 0\ 1\ 0\ 1\ 1)$ that $t = (1\ 0\ 1\ 0\ 1\ 1\ 0\ 1\ 0)$ and hence the corresponding telegram is T = [1 0 1 0 1 1 0 1 0] (cf. example above). Further information about the method of matrix-vector multiplication is given in [4], [6] and [7].

2.2 Undetectable Errors

Obviously, CRC cannot detect all errors. If in the example above, T' = [1 0 1 0 1 1 0 1 0] then (3) holds for $t'(x)$ and the falsification is not detectable.

Transmission errors can be modeled by superimposed error patterns F. These patterns have the same length n like T. A bit of F is assigned with value 0, if the corresponding bit in T is transmitted correctly, and a bit of F is allocated by value 1, if the corresponding bit in T is falsified during the transmission. Consequently, T is superimposed by F such that T' = T + F holds[1] . A transmission error is undetectable by CRC if and only if the polynomial corresponding to F, $f(x)$, is divisible by the generator polynomial $g(x)$, since:

$$t'(x) \bmod g(x) = (t(x) + f(x)) \bmod g(x)$$
$$= t(x) \bmod g(x) + f(x) \bmod g(x)$$
$$= f(x) \bmod g(x)$$

holds. Therefore, if $t'(x) \bmod g(x)$ is equal to zero, then $f(x) \bmod g(x)$ is equal to zero and vice versa. These undetectable error patterns including the pattern consisting of only zeros[2] form a linear code.

Since not all transmission errors are detectable it is necessary to define criteria to measure the quality of error detection. One important criterion is the Hamming Distance, which is the number of bits that at least have to be falsified to constitute an undetectable error. This conforms to the minimum number of entries 1 in an error pattern F of all possible error patterns.

More meaningful than Hamming Distance is the residual error probability P_{re}, that is the probability that an erroneous telegram is regarded to be transmitted correctly.

2.3 Calculation of the Residual Error Probabillity

The exact calculation of the residual error probability is usually very complex. There are various methods to calculate P_{re}. One is the *direct code analysis*. There all $2^m - 1$ undetectable error patterns have to be generated explicitly. The numbers A_i of those of i erroneous bits have to be counted ($A_i, i = 1, \ldots, n$ is the so-called weight distribution). Using the weight distribution, P_{re} is calculated by (6), where p denotes the probability that a bit is falsified during transmission[3]

---

[1] Note that '+' stands for exclusive-or in the space of binary polynomials.

[2] The pattern that consists of zeros only is not really an error pattern since each zero stands for the correct transmission of the corresponding bit.

[3] The model of the BSC (binary symmetric channel) is assumed where bits are corrupted independently and the falsification from value 0 to 1 is of the same probability p as the falsification from value 1 to 0.

$$P_{re} = \sum_{i=1}^{n} A_i \cdot p^i \cdot (1-p)^{n-i} \tag{6}$$

Obviously, the generation of all these error patterns leads to complexity of 2^m and the computation becomes feasible only for short telegrams. A more practicable determination of the residual error probability is the *transformed code analysis*. Instead of generating all undetectable error patterns of the original code, a much smaller set of patterns (2^r patterns instead of 2^m) of the corresponding dual code are generated. The weight distribution B_i of this code (so called dual weight distribution) is determined. Based on the dual weight distribution it is either possible to calculate P_{re} directly or to calculate the weight distribution A_i of the original code by means of the MacWilliams Identity (see [6]) which is a numerically more stable alternative. Both options sometimes lead to numerical problems and to inaccurate results (cf. [8]).

One possibility to generate the elements of the dual code is the application of the matrix-vector-multiplication described in Section 2.1. Let $k = (k_{r-1} \ k_{r-2} \ \ldots \ k_0)$ be a vector that comprises all possible vectors element of $\{0;1\}^r$, then all elements of the dual code can be generated by applying (7) to all possible vectors k. Consequently, the parity-check matrix H has to be generated (or the generator matrix G, the parity-check matrix H can be easily derived from).

$$f = k \cdot (A^T \mid I_r) = k \cdot H \tag{7}$$

3   Nested CRC with Additional Net Data

In this section, the nesting of CRC with additional net data is introduced and the calculation of its residual error probability is explained.

3.1   Description of the Nesting

Because of the standardized layer oriented communication according to the ISO/OSI-model (cf. [2]) in typical industrial applications, a nesting of CRC is given (s. Fig. 1).

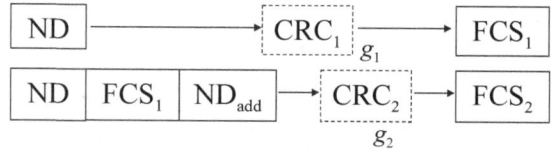

Fig. 1. Nesting of CRC

In an upper layer, e.g. the application layer, the checksum FCS_1 is calculated from the net data ND in a CRC with generator polynomial $g_1(x)$ according to (1) and is attached to ND. ND and FCS_1 build a temporary telegram that is transmitted to the lower layers. One of these layers, e.g. the link layer, attaches its specific net data ND_{add} and a second checksum FCS_2 to the telegram. This FCS_2 is calculated by a CRC with generator polynomial $g_2(x)$ based on a bit pattern consisting of ND, FCS_1, and ND_{add}. The mathematical formulation of the calculation of FCS_2 is given in (8) where madd denotes the number of ND_{add} bits, r_1 the degree of $g_1(x)$, and r_2 the degree of $g_2(x)$.

$$(((nd(x) \cdot x^{r_1} + fcs_1(x)) \cdot x^{m_{add}} + nd_{add}(x)) \cdot x^{r_2}) \bmod g_2(x) = fcs_2(x) \quad (8)$$

The telegram that is finally sent to the receiver consists of ND, FCS_1, ND_{add}, and FCS_2. The receiver checks in the corresponding lower layer if the complete received telegram [ND', FCS_1', ND_{add}',FCS_2'] is divisible by $g_2(x)$ and in the corresponding upper layer if [ND', FCS_1'] is divisible by $g_1(x)$, i.e. it executes the following checks:

$$(nd'(x) \cdot x^r + fcs_1'(x)) \bmod g(x) = 0? \quad (9)$$

$$(((nd'(x) \cdot x^{r_1} + fcs_1'(x)) \cdot x^{m_{add}} + nd_{add}'(x)) \cdot x^{r_2} + fcs_2'(x)) \bmod g_2(x) = 0?$$

Only if both checks of (9) hold, the telegram is regarded to be transmitted correctly.

3.2   Residual Error Probabillity of Nested CRC with Additional Net Data

To compute the residual error probability, the method of transformed code analysis is applied, where the undetectable error patterns are computed by matrix vector multiplication as described in Section 2.3. Basis for this computation is the parity-check matrix H or the generator matrix G, respectively. As demonstrated in Section 2.1, the i-th row of a generator matrix consists of the net data $ND = [d_{m-1}\ d_{m-2}\ \cdots\ d_0]$, where

$$d_{m-j} = \begin{cases} 1, & j = i \\ 0, & j = 1,\ \cdots\ ,m,\ j \neq i \end{cases}$$

and its corresponding FCS. Because of the nesting of CRC, there are two kinds of net data, and ND_{add}. The structure of the generator matrix is partitioned correspondingly. For $i \in \{1, 2,\ \cdots\ ,m\}$ the i-th row of G has the form:

$$G^i = d_{m-1}\ d_{m-2}\ \cdots\ d_0\ \underbrace{fcs_1^i}_{r_1}\ \underbrace{0\ \cdots\ 0}_{m_{add}}\ \underbrace{fcs_2^i}_{r_2} \quad (10)$$

where d_{m-j} is defined as above, fcs_1^i denotes the FCS of $ND = [d_{m-1}\ d_{m-2}\ \cdots\ d_0]$ calculated with generator polynomial $g_1(x)$, and fcs_2^i denotes the FCS of the bit pattern $[d_{m-1}\ d_{m-2}\ \cdots\ d_0\ fcs_1^i\ \underbrace{0\ \cdots\ 0}_{m_{add}}]$ calculated with generator polynomial $g_2(x)$. For $i \in \{m+1,\ \cdots\ ,m + m_{add}\}$, the i-th row of G has the form:

$$G^i = \underbrace{0 \ \ldots \ 0}_{m} \ \underbrace{0 \ \ldots \ 0}_{r_1} \ d^{add}_{m_{add}-1} \ \ldots \ d^{add}_0 \ \underbrace{fcs^i_2}_{r_2} \tag{11}$$

where

$$d^{add}_{m_{add}} = \begin{cases} 1, & j = i \\ 0, & j = 1, \ \ldots \ , m_{add}, \ j \neq i \end{cases}$$

and fcs^i_2 denotes the FCS of the bit pattern $[d^{add}_{m_{add}-1} \ \ldots \ d^{add}_0]$ calculated with generator polynomial $g_2(x)$. To recapitulate (10), (11) with matrix denotation, the generator matrix G has the form:

$$G = \begin{pmatrix} I_m \mid A_1 \mid & 0 & \mid \tilde{A}_2 \\ 0 \mid 0 \mid & I_{m_{add}} & \mid A_2 \end{pmatrix} \tag{12}$$

In (12), A_1 denotes the common generator matrix for a CRC with generator polynomial $g_1(x)$ and m net data bits, A_2 the common generator matrix for a CRC with generator polynomial $g_2(x)$ and m_{add} net data bits, a row of $\tilde{A}_2$ consists of the FCS of the corresponding row of $(I_m \mid A_1 \mid 0)$ calculated with generator polynomial $g_2(x)$. 0 respectively I denote zero matrices and identity matrices, respectively, of appropriate dimensions. The parity-check matrix H is now easily derived from G as shown in (13).

$$H = \begin{pmatrix} A_1^T \mid I_{r_1} \mid & 0 & \mid 0 \\ \tilde{A}_2^T \mid 0 \mid & A_2^T & \mid I_{r_2} \end{pmatrix} \tag{13}$$

The following example illustrates the theory explained above. Given are: ND length $m = 3$, ND_{add} length $m_{add} = 2$, $g_1(x) = x^2 + x + 1$, $g_2(x) = x + 1$, hence $r_1 = 2$, $r_2 = 1$. First, matrices A_1, A_2 are calculated according to the explanations in Section 2.1, because A_1 is needed for the generation of $\tilde{A}_2$: Since $x^4 \bmod x^2 + x + 1 = x$, $x^3 \bmod x^2 + x + 1 = 1$, $x^2 \bmod x^2 + x + 1 = x + 1$ and, analogously, $x^2 \bmod x + 1 = 1$, $x \bmod x + 1 = 1$ the matrices are:

$$A_1 = \begin{pmatrix} 1\,0 \\ 0\,1 \\ 1\,1 \end{pmatrix}, \quad A_2 = \begin{pmatrix} 1 \\ 1 \end{pmatrix}.$$

Matrix $\tilde{A}_2$ in this example is a 3×1 matrix. The first row of $\tilde{A}_2$ is the checksum of the bit pattern $[1\,0\,0\,1\,0\,0\,0]$ calculated with $g_2(x)$ according to (1): $(x^6 + x^3) \cdot x^1 \bmod (x + 1) = 0$. The second row is $(x^5 + x^2) \cdot x^1 \bmod (x + 1) = 0$ and the third row is $(x^4 + x^3 + x^2) \cdot x^1 \bmod (x + 1) = 1$. That leads to the generator and control matrices

$$G = \begin{pmatrix} 1\,0\,0 & 1\,0 & 0\,0 & 0 \\ 0\,1\,0 & 0\,1 & 0\,0 & 0 \\ 0\,0\,1 & 1\,1 & 0\,0 & 1 \\ 0\,0\,0 & 0\,0 & 1\,0 & 1 \\ 0\,0\,0 & 0\,0 & 0\,1 & 1 \end{pmatrix}, H = \begin{pmatrix} 1\,0\,1 & 1\,0 & 0\,0 & 0 \\ 0\,1\,1 & 0\,1 & 0\,0 & 0 \\ 0\,0\,1 & 0\,0 & 1\,1 & 1 \end{pmatrix}.$$

The weight distribution of the dual code can now be obtained by creating all undetectable error pattern according to (7) where k in this case comprises all elements of $\{0; 1\}^{r_1+r_2}$. As described in Section 2.3, the weight distribution of the original code is computed by means of the MacWilliams Identities and the residual error probability is calculated by applying (6). The calculation of the residual error probability of the nesting is also possible by means of stochastic automata [9].

4   Examples and Results

The residual error probability of the introduced nesting will by definition in this section solely refer to the falsification of the net data ND, i.e. only error patterns F that have at least one bit of value 1 in the first m bits are involved in the weight distribution. Therefore, the weight distribution of all undetectable error patterns that have only zeros in the first m bits are subtracted from the weight distribution of the nesting. Since error patterns, that have only zeros in the first m bits, have also bit assigned by zero in the following r_1 bits (because that is the corresponding FCS to the first m bits), the error patterns, whose weights have to be subtracted, have he form as shown in Fig. 2. These error pattern form a code with the same weight distribution like the code generated by $g_2(x)$ for m_{add} ND bits.

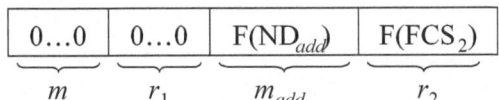

Fig. 2. Error patterns neglected by definition

The calculated residual error probability P_{re} depends on the bit error probability p. An example of $P_{re}(p)$ is given in Fig. 3.

There P_{re} is drawn over p for polynomials $g_1 = 14\text{EABh}^4$ and $g_2 = 1\text{FFh}$ for 256 ND bits and 64 ND_{add} bits. The dashed horizontal line marks the value $2^{-(r_1+r_2)}$, that is the residual error probability in case of uniform distribution, i.e. at $p = 0.5$. This line is drawn for orientation: Whenever the residual error probability is below this line the polynomials are a good choice for the given data lengths, if P_{re} crosses the line, other polynomials should be chosen. A multitude of classified polynomials have been analyzed for different numbers of ND and ND_{add} bits in order to see which classes of polynomials (primitive, reducible, irreducible) are adequate for which data lengths. Note that all results are trends since for every result at least one counter-example can be found. For more detailed information see [10].

---

[4] Note that polynomials are denoted hexadecimal; e.g. 14EABh corresponds to the dual number 1 0100 1110 1010 1011, which corresponds to the polynomial $x^{16} + x^{14} + x^{11} + x^{10} + x^9 + x^7 + x^5 + x^3 + x + 1$.

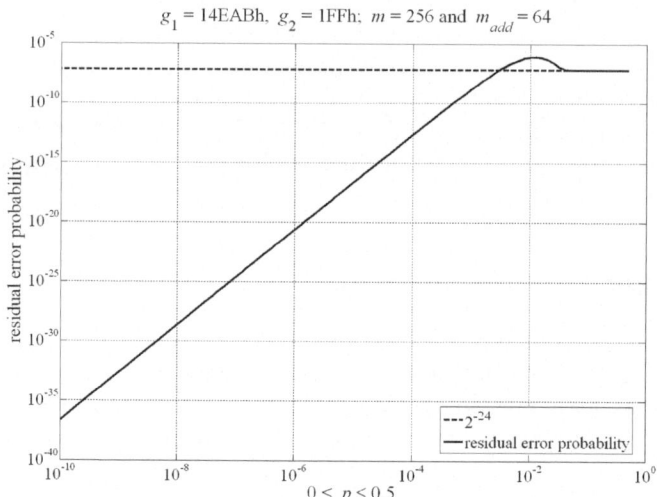

Fig. 3. Example of generated graphs

4.1   Analysis of the Number of ND Bits

Regarding the number of ND bits, m, it can be said, that this number should be relatively small independent of the number of ND_{add} bits and polynomial classes. If for any reasons m has to be large, it turned out, that if m exceeds a certain value further enlargement has no significant effect on the residual error probability. Fig. 4 shows the residual error probability for reducible polynomials $g_1 = 15Dh$, $g_2 = 133h$, 256 ND_{add} bits and various ND bits; the asterisk line

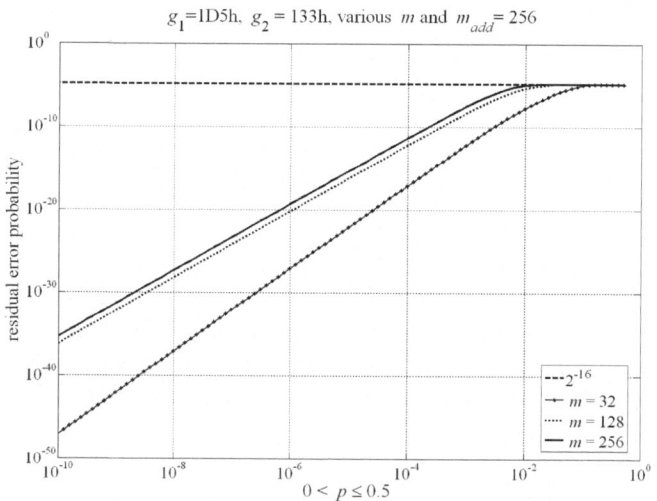

Fig. 4. Example of various ND lengths

marks P_{re} for 8 ND bits, the dotted line for 108 ND bits and the drawn through line for 208 ND bits.

In this example, the increasing of the number of information bits from 128 to 256 has no significant effect.

4.2   Analysis of the Number of ND$_{add}$ Bits

The results of the analysis regarding the number of ND$_{add}$ bits, m_{add}, are very different and therefore not to generalize. For some polynomials, a short number of additional net data bits turn out to be good, in other cases, the number of ND$_{add}$ bits has no remarkable impact on the residual error probability, and there are also combinations where major m_{add} guaranties a smaller residual error probability than smallish m_{add}. Remarkable are the graphs for identical generator polynomials as for example in Fig. 5. There $g_1 = g_2 = $ 1CFh is primitive, $m = 32$ and m_{add} vary.

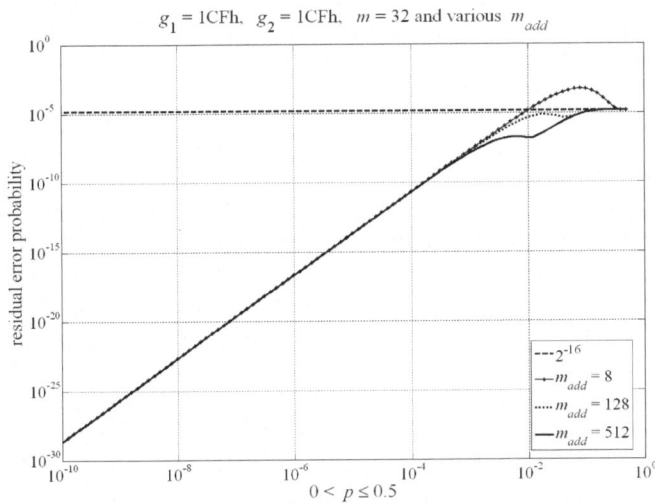

Fig. 5. Example of identical generator polynomials with various m_{add}

There the graphs are identical up to a bit error probability greater than 10^{-4}. Then the graph for greater m_{add} (drawn through line) turns out to be the best, while the one with smallest m_{add} (asterisk line) is the worst alternative. This example illustrates the impact of the bit error probability on the residual error probability.

4.3   Analysis of Polynomials

Concerning the choice of polynomials, it is advisable to choose different generator polynomials since equal generator polynomials lead to a relatively high residual

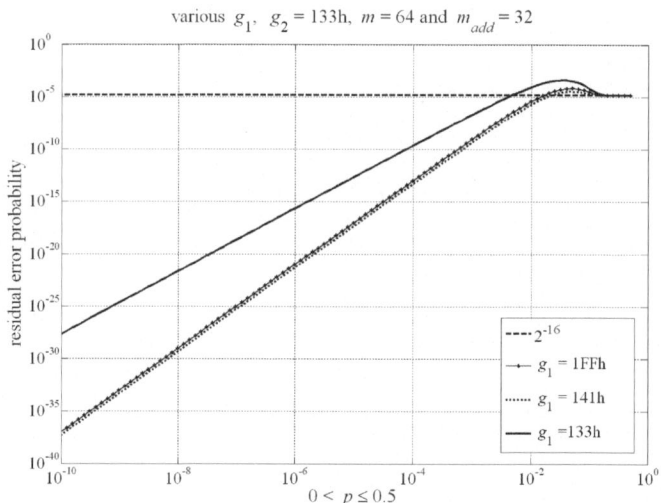

Fig. 6. Example of various first generator polynomials

error probability. Fig. 6, where $g_2 = 133h$, $m = 64$, $m_{add} = 32$ and g_1 varies, stresses this fact, since the residual error probability for equal generator polynomials (drawn through line) is much worse than the others. The second recommendation that can be posed is that the first generator polynomial should not be a multiple of the second polynomial. In most cases, this recommendation holds vice versa but there are a couple of exceptions in which the residual error probability of a nested CRC where the first polynomial is divisible by the second is better than a nested CRC with coprime polynomials. In general, it can be said, that coprime polynomials tend to be a good choice. Additionally, a first generator polynomial that is "good" for given ND length m is advisable. (A polynomial is good for a given data length if the residual error probability of the common CRC with that polynomial remains below the mark of 2^{-r}). Conversely, it has no impact on the quality of nested CRC if the second polynomial is good for length $m + r_1 + m_{add}$.

4.4   Remarks

The presented algorithm has theoretically no limits by the degrees of generator polynomials (solely the computation time increases logarithmically) but for accuracy reasons, the sum of data lengths should be smaller than 1000 bits. The computation of the residual error probability for Fig. 7, where $g_1 = 3h$, g_2 is the Ethernet polynomial of degree 32, $m = 16$ and $m_{add} = 64$, took about 10 days[5]. For data lengths larger than 1000 bits the method of stochastic automata (mentioned at the end of section 3) should be applied.

---

[5] Pentium4 HT, 3,2 Ghz, 1GB RAM Computer.

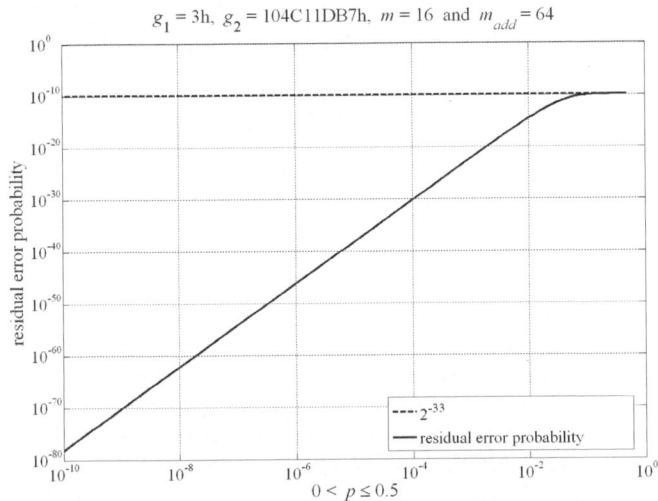

Fig. 7. Ethernet polynomial

5   Conclusion and Future Work

An algorithm for the calculation of the residual error probability has been developed and implemented. Consequently, the residual error probability of the nested CRC with additional net data can be involved in safety proofs of e.g. automated plants and help to reduce the number of worst case assumptions. As the analysis of the nesting showed, unfortunately, it cannot generally be said which polynomials or polynomial classes guarantee the smallest residual error probability for which data lengths. But according to the analysis, some recommendations for the creation of a new protocol can be made regarding the parameters: number of ND bits m, number of ND_{add} bit m_{add}, choice of $g_1(x)$ and $g_2(x)$, (see Table 1).

Table 1. Recommendations for given parameters

Parameter	Recommendation
m	relatively small
m_{add}	relatively small (no significant impact)
$g_1(x)$	high degree coprime to $g_2(x)$), good for regular CRC with m ND bits
$g_2(x)$	high degree, coprime to $g_1(x)$

These recommendations constrain the multitude of variations of the parameters and will therefore help to accelerate the process of finding suitable parameters. But anyhow it is still indispensable to validate ones parameters by a concrete calculation of the residual error probability. Future work will include:

- the identification of the best polynomial $g_1(x)$ for given m, m_{add} and $g_2(x)$,
- the identification of the best polynomial $g_1(x)$ for given m and for unknown m_{add} and $g_2(x)$
- the identification of the best polynomial $g_1(x)$ for given m and for unknown m_{add} and $g_2(x)$, if the rate of detected errors is evaluated online and used in the safety proof (see e.g. [11]).

References

1. International Electronical Comission: Functional Safety of Electrical/Electronic/Programmable Electronic Safety-related Systems. (IEC 61508) (2005)
2. International Organization for Standardization, International Electrotechnical Commission (ISO/IEC): Information Technology - Open Systems Interconnection - Basic Reference Model: Basic Model (ISO/IEC 7498-1) (1996)
3. Mattes, T., Pfahler, J., Schiller, F., Honold, T.: Analysis of Combinations of CRC in Industrial Communication. In: Saglietti, F., Oster, N. (eds.) SAFECOMP 2007. LNCS, vol. 4680, pp. 329–341. Springer, Heidelberg (2007)
4. Peterson, W., Weldon, E.J.: Error Correcting Codes. MIT Press, Cambridge (1996)
5. Schiller, F., Mattes, T.: An Efficient Method to Evaluate CRC-Polynomials for Safety-Critical Communication. Journal of Applied Computer Science 14, 57–80 (2006)
6. Mac Williams, F.J., Sloane, N.J.A.: Theory of Error-Correcting Codes. North-Holland Mathematical Library, Amsterdam (1991)
7. Sweeney, P.: Codierung zur Fehlererkennung und Fehlerkorrektur. MIT Press, Cambridge (1996)
8. Mattes, T.: Untersuchung zur effizienten Bestimmung der Güte von Polynomen für CRC-Codes. University of Trier, Siemens AG, Nuremberg (2004) (in German)
9. Mattes, T.: Analysis of Nested CRC with Additional Net Data by Stochastic Automata. In: 7th IEEE International Workshop on Factory Communication Systems Communication in Automation, Dresden, Germany, May 20-23, pp. 295–304 (2008)
10. Mörwald, A.: Analyse der Verschachtelung von CRC-Verfahren in der industriellen Kommunikation. TU München (2007)
11. Schiller, F., Mattes, T., Büttner, H., Sachs, J.: A New Method to Obtain Sufficient Independency of Nested Cyclic Redundancy Checks. In: 5th International Conference Safety of Industrial Automated Systems, SIAS 2007, Tokyo, Japan, pp. 149–154 (2007)

Symbolic Reliability Analysis of Self-healing Networked Embedded Systems

Michael Glaß, Martin Lukasiewycz, Felix Reimann,
Christian Haubelt, and Jürgen Teich

Hardware/Software Co-Design, Department of Computer Science
University of Erlangen-Nuremberg, Germany
{glass,martin.lukasiewycz,felix.reimann,
haubelt,teich}@cs.fau.de

Abstract. In recent years, several network online algorithms have been studied that exhibit self-x properties such as self-healing or self-adaption. These properties are used to improve systems characteristics like, e.g., fault-tolerance, reliability, or load-balancing.

In this paper, a symbolic reliability analysis of self-healing networked embedded systems that rely on self-reconfiguration and self-routing is presented. The proposed analysis technique respects resource constraints such as the maximum computational load or the maximum memory size, and calculates the achievable reliability of a given system. This analytical approach considers the topology of the system, the properties of the resources, and the executed applications. Moreover, it is independent of the used online algorithms that implement the self-healing properties, but determines the achievable upper bound for the systems reliability. Since this analysis is not tailored to a specific online algorithm, it allows a reasonable decision making on the used algorithm by enabling a rating of different self-healing strategies. Experimental results show the effectiveness of the introduced technique even for large networked embedded systems.

1 Introduction

Systems like, e.g., *automotive* or *avionics electronic control unit (ECU) networks*, networks from the area of industrial control automation, *body-area networks*, or *sensor networks* combine the aspects of both embedded systems and networks. Due to constraints in area consumption, monetary costs, and energy consumption, the used resources exhibit limiting properties in the field of, e.g., computational power or memory size which is typical for embedded systems. On the other hand, the resources are distributed within the systems and, thus, they resemble networks. This distribution is crucial to allow controlling, monitoring, and analysis of the system under the aspect of limited and remote installation spaces. Moreover, these systems have to be optimized with respect to different criteria, ranging from monetary costs, area and power consumption, or throughput to flexibility, reliability and fault-tolerance. Commonly, this system category is referred to as *networked embedded systems*.

For the reliability analysis proposed in this work, several aspects of networked embedded systems are of great importance. Commonly, networked embedded systems are

M.D. Harrison and M.-A. Sujan (Eds.): SAFECOMP 2008, LNCS 5219, pp. 139–152, 2008.

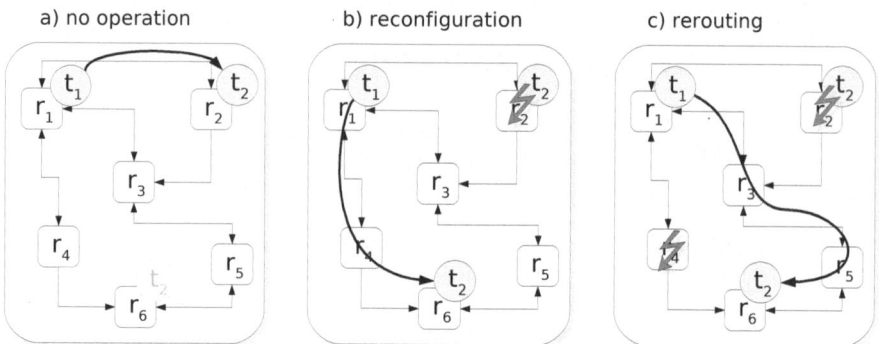

Fig. 1. A self-healing networked embedded system with a data transfer from task t_1 to task t_2. In a), no defect resources are present. In case of resource failures, a reconfiguration activates redundant task instances, cf. b), or reestablishes the communication using a dynamic rerouting, cf. b).

deployed in unattended areas and, thus, administrative tasks and maintenance are expensive and should be avoided or are sometimes even impossible to accomplish. As a matter of fact, resources of the networked embedded systems move from dedicated and protected mounting spaces to installation spaces with destructive agents. These spaces can be found near sensors or within, e.g., an engine or moving parts of vehicles. This trend especially increases the amount of destructive influences on the used hardware, such that permanent failures occur more frequently.

In recent years, systems have been proposed that exhibit so called *self-x* properties. These systems monitor themselves and are able to react autonomously to unwanted system states. Popular properties for self-x systems are *self-adaption* that allows the system to react to different environment conditions or new applications and *self-healing* that allows the system to react to failures and, thus, increases the reliability and fault tolerance of the system, cf. [1]. In this work, we will focus on self-healing networked embedded systems that are based on self-reconfiguration and self-routing. An example of such a system is given in Fig. 1. On the systems architecture, a given set of communicating applications is executed while the communication is implemented via static routes. Hence, there are hardly any dedicated routing resources but the resources perform both the computation of the tasks and the routing using their point-to-point interconnections. In case of a resource failure, the failure has to be detected, cf. [2,3]. After the detection, the redundant instances of the tasks executed on the defect resource that are available in the system are activated using a reconfiguration of the other resources, cf. [4]. For all newly activated task instances and for all communication routes that use the defect resource, a rerouting is performed. Besides the fact that such a system can act autonomously, self-reconfiguration and self-routing allow a resource sharing of active tasks and redundant, inactive tasks. Thus, highly increased costs introduced by static structural redundancy can be avoided.

In this paper, we present a symbolic reliability analysis of such self-healing networked embedded systems. This analysis aims to determine the achievable reliability of the system, independent of the used algorithms to implement the self-healing property. The knowledge of the achievable reliability is important since it allows to quantify the quality of the available online algorithms. Moreover, in the design phase of the system, the achievable reliability allows a coarse grained selection from different system layouts and can be embedded into a design space exploration, since the calculation is much faster than an evaluation of different online algorithms. The presented reliability analysis is based on *Binary Decision Diagrams* (BDDs) [5] and considers both reconfiguration and routing using a symbolic fixed-point iteration. Moreover, even constraints that depend on the dynamic activation of tasks are encoded in the BDD, allowing to respect constraints like, e.g., maximum computational load of a resource.

The remainder of the paper is as follows: Section 2 discusses prior work. In Sec. 3, a formulation of the problem we target in this work is given. Section 4 introduces our symbolic analysis approach for self-healing networked embedded systems. Sec. 5 shows the results of the introduced technique applied to examples where several implementations of self-healing algorithms are available as well as an networked embedded system that corresponds to systems currently used as automotive ECU networks. The paper is concluded in Sec. 6.

2   Related Work

The reliability of self-healing networks has been widely studied, cf. [6,7,8,9]. Hence, all these approaches focus on the network as a communication platform itself while the aspects of nodes of the networked embedded system as both communication and computational resource are neglected. Thus, the approaches are more related to the reliability analysis of classical networks [10] and restricted to given self-healing strategies.

Other approaches can be found in the area of *self-repairing embryonic cells* [11] and *wireless sensor networks* [12,13]. Embryonic cells have a very special architecture that comes with a high degree of spatial redundancy to implement the self-repairing property and, thus, are not appropriate as a networked embedded system model. Wireless sensor networks on the other hand are highly meshed networks and have changing communication possibilities due to their wireless communication medium. Hence, since these

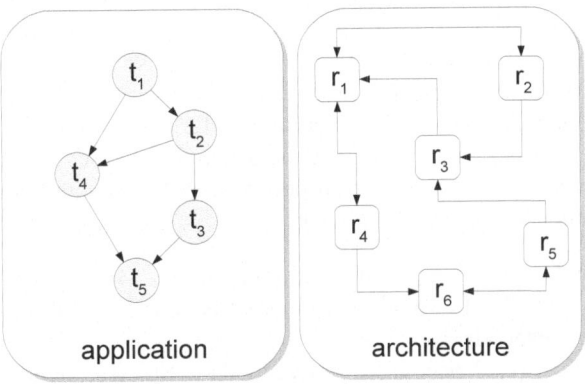

Fig. 2. A system specification with an application and the available resource architecture

sensor networks often include aspects of maintenance as well, simulative approaches are used to quantify their reliability.

The typical networked embedded system is hard wired with maintenance being hardly possible. Thus, the reliability analysis proposed in this work is inspired by the reliability analysis of embedded systems [14] and networked embedded systems without self-healing properties [15]. We will extend these approaches by considering the reconfiguration and dynamic routing that is used by the online algorithm. Moreover, constraints that depend on the online activation of tasks are included in the analysis as well.

3   Problem Description

In this paper, we target the problem of determining the reliability of a self-reconfigurable and self-routing networked embedded system. Hereby, the tasks that are executed on the resources as well as their data dependencies are given. Moreover, the system topology and the reliability attributes of each resource are known. The assumed failure model is permanent failure due to resource failures.

Our formal specification of a *system* consists of the *application*, the system layout or *architecture* and the relation between these two views:

- The application is modeled by a task graph $g_t(V_t, E_t)$ that describes the behavior of the system. The vertices $t_1, ..., t_{|V_t|} \in V_t$ denote tasks whereas the directed edges E_t are data dependencies. Attributes like, e.g., memory usage, computational load are assigned to tasks.
- The architecture is modeled by a graph $g_a(V_a, E_a)$ and represents possible interconnected hardware resources. The vertices $r_1, ..., r_{|V_a|} \in V_a$ represent resources that can be both processing units or communication units like buses or gateways. The edges E_a model available communication links between the resources. Attributes like, e.g., the memory size, maximum computational capacity or reliability are assigned to the resources.

Each resource has the ability to route information to all resources along the directed communication connections, thus, from a routing point of view, all resources can be seen as *network nodes* with point-to-point connections. An example of an application and a given architecture is shown in Fig. 2.

In this model, the execution of a given task is limited to selected resources. This is due to the fact that not each device is generally present on every resource in a heterogeneous system. Therefore, a relation between application and architecture called *mapping* is introduced in the system model:

- The mapping $M : V_t \rightarrow 2^{V_a}$ assigns to each task t a set of possible resources for its execution and $M : V_a \rightarrow 2^{V_t}$ assigns each resource r a set of tasks that can be executed on r, respectively.
- An *instance* $i = (t, r)$ of a task t corresponds to this task being executed on the resource $r \in M(t)$. The set of all available instances of all tasks is defined as $I = \{(t, r) \mid t \in V_t, r \in M(t)\}$.

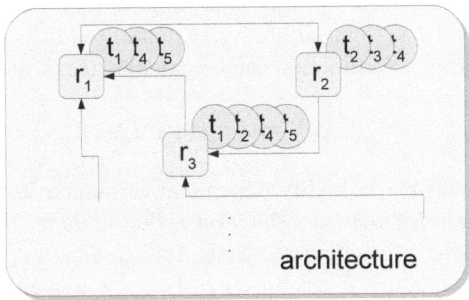

Fig. 3. A part of the application shown in Fig. 2. Depicted are the tasks that are mapped to the resources.

In the online phase of the system, at least one instance of each task has to be *activated*, i.e., has to be executed. An example of a part of the application and its relation to the given architecture is shown in Fig. 3.

The task of the used online algorithms is to keep the networked embedded system *feasible*.

Definition 1. *A system is called* feasible *if the execution of each task of the system's applications and their data dependencies can be correctly carried out by proper operating system resources.*

A task execution can be successfully carried out on a resource if the resource has enough capacity to execute the task and is operating properly, i.e., it is not defect. Data-dependencies can be implemented if there exists a set of properly operating resources that allows to pass the data correctly from the sending to the receiving resource. In this definition, failures that happen at task level like, e.g., soft errors or errors in the task itself, are assumed to be handled at task level using, e.g., task re-execution, cf. [16].

4   Reliability Analysis

In this section, the symbolic reliability analysis is presented. The calculation and representation of the so called *structure function* φ is explained in three steps: First, the requirements for a feasible system are introduced. Afterwards, the representation of the dynamic routing within the structure function is presented. In a final step, the given constraints are integrated directly into φ. Moreover, the evaluation of φ to quantify the systems reliability is explained.

4.1   The Structure Function φ

To model the systems behavior under the influence of failures, the structure function $\varphi : \{0,1\}^{|V_a|} \rightarrow \{0,1\}$ with the Boolean vector $V_a = (r_1, \ldots, r_{|V_a|})$ is calculated, cf. [14]. At this, for each allocated resource $r \in V_a$, a binary variable r is introduced with $r = 1$ indicating a proper operation and $r = 0$ a resource failure, respectively.

This Boolean function indicates a proper operating system, i.e., a feasible system by evaluating to $\varphi = 1$ and a system failure by evaluating to $\varphi = 0$, respectively. For a given system specification, this function can be calculated as follows:

$$\varphi(\boldsymbol{V_a}) = \exists \boldsymbol{I} : \psi(\boldsymbol{V_a}, \boldsymbol{I}) \tag{1}$$

Whether a system is feasible is highly dependent on which instance of each task is activated. Thus, the *extended structure function* ψ that includes both the resources and the available task instances, is calculated first. At this, $\boldsymbol{I} = (i_1, \ldots, i_{|I|})$ is a vector of Boolean variables encoding a task instance being activated. Applying the *exists-operator* $\exists$ to ψ allows to eliminate the I variables by asking if there exists at least one set of task instances that ensures a feasible system.

The requirements for a feasible system are encoded in ψ:

$$\psi(\boldsymbol{V_a}, \boldsymbol{I}) = \bigwedge_{t \in V_t} \left[\bigvee_{i=(t,r) \in I} i \right] \wedge \tag{2a}$$

$$\bigwedge_{i=(t,r) \in I} i \rightarrow r \wedge \tag{2b}$$

$$\bigwedge_{(t,\tilde{t}) \in E_t} \bigwedge_{\substack{i=(t,r), \\ \tilde{i}=(\tilde{t},\tilde{r}) \in I}} i \wedge \tilde{i} \rightarrow R_{r,\tilde{r}}(\boldsymbol{V_a}) \wedge \tag{2c}$$

$$\bigwedge_{r \in V_a} C_r(\boldsymbol{I}) \tag{2d}$$

At least one active instance of each task $t \in V_t$ is needed to allow each application to work properly. This is ensured by Term (2a). Term (2b) states that an activated task instance implies a proper operating resource. Furthermore, if two instances of data dependent tasks are activated, they must be able to communicate and a correct routing has to be possible, cf. Term (2c). At this, the function $R_{r,\tilde{r}}(\boldsymbol{V_a})$ encodes possible routings and, thus, enables to decide whether two data dependent task instances are able to communicate. The calculation of this function is presented in Sec. 4.2. The given constraints that are imposed on the resources are realized by Term (2d) using the function $C_r(\boldsymbol{I})$. The calculation of this function is presented in Sec. 4.3.

4.2   Encoding the Routing

The function $R_{r_s,r_d} : \{0,1\}^{|V_a|} \rightarrow \{0,1\}$ evaluates to 1 if there exists a *route*, i.e., a loop free path, that implements a communication between the data dependent task instances being executed on resource r_s and r_d by passing data over currently proper operating resources only. Thus, the function evaluates to 0 if there is no route that is able to implement the data dependency. In the following, a fixed-point iteration approach formally introduces the calculation of $R_{r_s,r_d}(\boldsymbol{V_a})$. Additionally, a symbolic version of this fixed-point iteration is presented that enables an efficient determination of the desired Boolean function.

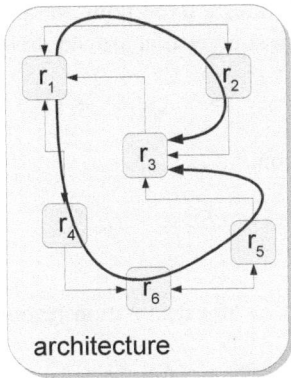

architecture

Fig. 4. All possible routes from resource r_1 to r_3 from the example shown in Fig. 2

Fixed-Point Iteration. A route from a sender resource r_s to a destination resource r_d is carried out by passing the data from one resource to another using the point-to-point connections between the resources.[1] Passing data from one resource to another is called taking a *hop*, thus, one route consists of a ordered sequence of hops starting from the sender and ending at the destination without visiting a resource more than once. Formally, taking a hop between two resources r and $\tilde{r}$ is possible if

$$\exists e = (r, \tilde{r}) \in E_a \tag{3}$$

That means, a hop can only be taken if the resources are able to communicate using a point-to-point connection. The possible routes from the example depicted in Fig. 2 are shown in Fig. 4.

The determination of $R_{r_s,r_d}(V_a)$ has to take all possible routes into account. In the following, the determination of this Boolean function is done by a fixed-point iteration.

The single state of the fixed-point iteration are in the set S with

$$S = V_a \times 2^{V_a}. \tag{4}$$

One state $(r, R) \in S$ consists of a reached resource $r \in V_a$ and the set of resources $R \subseteq V_a$ in form of predecessor resources that have been passed starting from the sender resource to reach the current resource r. The function $\delta : S \rightarrow 2^S$ determines for a given state (r, R) the set of reachable states:

$$\delta((r, R)) = \{(r', R \cup \{r'\}) \mid \text{with } (r, r') \in E_a\} \tag{5}$$

By using δ, for a given set of states $S \subseteq S$ the successor states are calculated by the successor function $SUCC : 2^S \rightarrow 2^S$:

$$SUCC(S) = \{(r', R') \mid \exists (r, R) \in S : (r', R') \in \delta((r, R))\} \tag{6}$$

---

[1] Buses are modeled as a single resource with many point-to-point connections to nodes that are attached to the bus.

Thus, the following function defines a fixed-point iteration that searches all reachable resources with the given set of resources that are required to ensure a route:

$$S_{j+1} = S_j \cup SUCC(S_j) \tag{7}$$

The iteration stops in the iteration k if

$$S_{k+1} = S_k \tag{8}$$

and the fixed-point is reached.

For a given sender resource r_s and destination resource r_d the initial state of the iteration is

$$S_0 = \{(r_s, \{r_s\})\} \tag{9}$$

and the desired states for the fixed-point S_k are those where the current resource equals the destination:

$$\widetilde{S_k} = \{(r_d, R)|(r_d, R) \in S_k\} \tag{10}$$

With the calculated set $\widetilde{S_k}$ the Boolean function that indicates whether a communication between r_s and r_d is possible is as follows:

$$R_{r_s,r_d}(\boldsymbol{V_a}) = \bigvee_{(r,R)\in\widetilde{S_k}} \bigwedge_{\widetilde{r}\in R} \widetilde{r} \tag{11}$$

However, the complexity of this iteration equals the enumeration of all *simple paths* that is known to be #P-complete [17].

Symbolic Approach. The basis of the determination of $R_{r_s,r_d}(\boldsymbol{V_a})$ is #P-complete and, thus, of a high computational complexity. In the following, the set-based fixed-point iteration is done by a symbolic approach using *Binary Decision Diagrams* (BDDs). From the experiences of *Model Checking* a symbolic encoding [18] speeds up a fixed-point iteration by some orders of magnitude.

Preliminary, for each $r \in V_a$ a distinct Boolean function b_r is defined as

$$b_r : X \to \{0, 1\} \text{ with } X = \{0, 1\}^{\lceil ld |V_a| \rceil}, \tag{12}$$

with $x \in X$ being of the form

$$x = \{\boldsymbol{x_0}, \dots, \boldsymbol{x}_{\lceil ld |V_a| \rceil}\}. \tag{13}$$

Hereby, for two resources $r, \widetilde{r} \in V_a$ it holds

$$b_r(x) \neq 0 \tag{14a}$$
$$b_r(x) \wedge b_{\widetilde{r}}(x) = 0 \tag{14b}$$

Thus, the function $b_r(x)$ maps a resource r to a specific binary representation by evaluating to 1 if x is the binary representation of r and evaluating to 0 if x is not the binary representation of r, respectively.

Correspondingly to Eq. (4) a single state or a set of states can be defined in the binary representation and, thus, as a BDD:

$$S : X \times \{0, 1\}^{V_a} \to \{0, 1\} \tag{15}$$

Correspondingly to Eq. (5) the function $\delta : X \times X \times \{0, 1\}^{|V_a|} \to \{0, 1\}$ encodes whether taking a hop is possible represented as a BDD:

$$\delta(x, x', V_a) = \bigvee_{e=(r,\tilde{r}) \in E_a} b_r(x) \wedge b_{\tilde{r}}(x') \wedge \tilde{r} \tag{16}$$

This function evaluates to 1 if the requirements stated in Eq. (3) are fulfilled. Otherwise, δ evaluates to 0, respectively. The required paths in the form of predecessor resources stated in Eq. (4) is incorporated through the V_a variables. This is important since these variables encode the path that is needed for the fixed-point iteration and allow defect resources[2] to falsify the possibility of taking hops at the same time.

Correspondingly to Eq. (6) the successor function is defined as

$$SUCC(S(x, V_a)) = \exists x' : S(x', V_a) \wedge \delta(x', x, V_a). \tag{17}$$

Thus, the fixed-point iteration from Eq. (7) is carried out by

$$S_{j+1}(x, V_a) = S_j(x, V_a) \vee SUCC(S_j(x, V_a)). \tag{18}$$

The iteration stops and the fixed-point is reached if the BDDs for two subsequent iterations are equal, cf. Eq. (8).

For the sender resource r_s and the destination resource r_d the initial state is defined as a BDD correspondingly to Eq. (9):

$$S_0(x, V_a) = b_{r_s}(x) \wedge r_s \tag{19}$$

For the fixed-point $S_k(x, V_a)$ the restricted states to the destination resource are determined correspondingly to Eq. (10) as follows:

$$\widetilde{S_k}(x, V_a) = S_k(x, V_a) \wedge b_{r_d}(x) \tag{20}$$

Thus, correspondingly to Eq. (11) the desired Boolean function or BDD, respectively, is determined as follows:

$$R_{r_s, r_d}(V_a) = \exists x : \widetilde{S_k}(x, V_a) \tag{21}$$

The resulting BDD for the example in Fig. 4 is shown in Fig. 5.

---

[2] Link failures can be seamlessly introduced by adding binary Variables E that encode a proper operation of the communication links E_a.

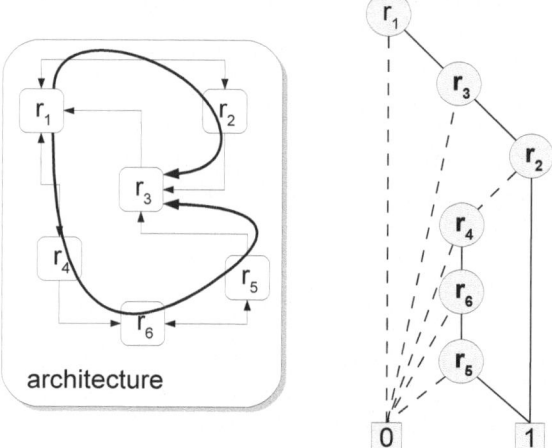

Fig. 5. A BDD encoding the routing possibilities from resource r_1 to r_3 from the example shown in Fig. 2. Edges represent the corresponding variable to be 1 while the dashed edges depict the variable to be 0, respectively.

4.3   Incorporating Constraints

Typical constraints for resources in self-x networked embedded systems are the maximum computational load or the maximum memory capacity of a specific resource. Since the calculation of these objectives can be approximated using linear functions, they can be expressed as linear constraints of the form

$$a^T x \circ b \tag{22}$$

with $a \in \mathbb{Z}^n$, $b \in \mathbb{Z}$ and $\circ \in \{<, \leq, =, \geq, >\}$. A typical constraint for, e.g., the maximum computational load of a resource r has the following form:

$$\sum_{i=(t,r) \in I} l_i \cdot i \leq L_r \tag{23}$$

At this, l_i denotes the computational load arising from activating task t on resource r while the maximum computational load of resource r is denoted as L_r. By incorporating these constraints into ψ, system states that violate a constraint are excluded from the set of feasible system states.

An encoding algorithm for linear constraints as Binary Decision Diagrams has been presented in [19]. Using this algorithm, the function

$$C_r : \{0,1\}^{|I|} \rightarrow \{0,1\} \tag{24}$$

can be realized. The function C_r evaluates to 1 if the task instances that are executed on r do not violate the given constraints and evaluates to 0 if at least one of the encoded constraints is violated, respectively. As an example, resource r_3 from Fig. 2 with an

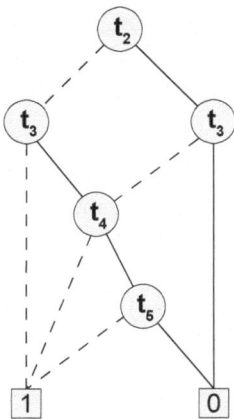

Fig. 6. A BDD encoding the computational load constraint of resource r_3 shown in Fig. 2. Edges represent the corresponding variable to be 1 while the dashed edges depict the variable to be 0, respectively.

L_{r_3} of 5 is to encode. The computational demand of the tasks that are bound to r_3 are $l_{t_2} = l_{t_3} = 3$ and $l_{t_4} = l_{t_5} = 2$. Thus, the constraint can be written as

$$3i_{r_2} + 3i_{r_3} + 2i_{r_4} + 2i_{r_5} \leq 5$$

The resulting BDD for C_{r_3} that is constructed using the Algorithm presented in [19] is shown in Fig. 6.

4.4   Evaluating φ

In the following, we describe how to quantify the reliability of the self-x networked embedded system based on the determined structure function φ. Since the reliability R_r of each resource $r \in V_a$ is given in the system model, e.g., by distribution functions like an *exponential distribution* or a *Weibull distribution*, the reliability of the system at time t can can be calculated through using a modified *Shannon-decomposition* [20] on the BDD representing φ:

$$R(t) = \varphi = R_r(t) \cdot \varphi|_{r=1} + (1 - R_r(t)) \cdot \varphi|_{r=0} \qquad (25)$$

If a *Mission Time* (MT) of the system is given, this decomposition directly quantifies the reliability $R(MT)$ of the system. In our experimental results, the *Mean Time To Failure* (MTTF) $= \int_0^\infty R(t)dt$ is used as the measure of reliability and is determined by a numerical integration of Equation (25). MTTF denotes the expected value for the failure-free time of the system.

5   Experimental Results

In this section, the results of applying our proposed analysis approach to two different examples are presented. First, examples of an available self-healing technique for networked embedded systems known as *ReCoNets* [4] are analyzed. Afterwards, a networked embedded system with a specification inspired by state-of-the-art ECU networks from the automotive area are used to show the applicability of the proposed approach.

ReCoNets. The self-healing technique called ReCoNets implements the self-healing property using a one replication strategy. At this, for each active task, a replica task is created by copying the byte code of the tasks to another resource. Of course, a replica can only be placed at resources that are allowed by the given mappings. There are two strategies available: The *load balancing* (LB) strategy places replicas under load balancing aspects. The strategy that is more focused on lifetime maximization called *BCC* determines *bi-connected components* that can, in case of a failure, lead to a partitioning of the networked embedded system. With this knowledge, the BCC strategy tries to place replicas such that a partitioning does not prevent data dependent tasks from a correct communication. Hence, strategies based on using only one replica can, in general, not achieve the maximum reliability, but reduce the amount of memory needed in every resource since task replicas are created dynamically. With this example, the possibility of quantifying the effectiveness of a self-healing technique using our proposed approach is shown. For this reason, a simple measure for the effectiveness e is used:

$$e = \frac{\text{MTTF}}{\text{MTTF}_{\text{max}}} \tag{26}$$

Table 1 shows the results of the proposed reliability analysis for networked embedded systems where a simulation of the ReCoNets techniques is available. In these testcases, an exponential distribution function was used to model the resource reliability. The size of the networked embedded systems was varied between 10 and 30 resources, each having the same number of tasks. Due to the high vertex degree, these examples can be considered to be complex examples for the analysis. For the small examples, both ReCoNets techniques nearly reach the upper bound of the achievable MTTF. However, the high standard deviation shows that both techniques can also make suboptimal decisions for the replica placement, leading to very early system failures. This is often the result of placing task and corresponding replica in a small subnet that can be isolated from the network by a single link or resource failure. For larger networks, the effectiveness of both ReCoNets techniques decreases, but can still be considered as good. In all testruns carried out, the time consumption of the proposed analysis algorithm and the ReCoNets simulation were nearly equal. Morover, the time consumption of the proposed algorithm is small enough to be applied in design space exploration approaches.

Table 1. Comparison of theoretical upper bound for the MTTF with the ReCoNets self-healing system

testcases	symbolic analysis MTTF$_{\text{max}}$	ReCoNets LB			ReCoNets BCC		
		MTTF	deviation	e	MTTF	deviation	e
small	54.92	50.34	40.04	0.917	51.21	39.44	0.932
medium	66.91	57.20	46.28	0.854	60.25	43.21	0.900
large	176.12	137.18	127.71	0.779	144.64	138.24	0.821

Table 2. Time consumption of a single analysis run for different ECU networks

testcases	specification			time consumption
	#ECUs	#Tasks	#Buses	[s]
small	30	30	2	1.16
medium	50	50	3	3.63
large	70	70	4	6.69

Especially interesting is the relatively small difference in the effectiveness of both ReCoNets methodologies with regards to the known upper bound. In [4], the relative difference between the LB and BCC approach seemed significant. With regards to the upper bound, the designer may choose the load balancing approach as well, since this approach is less than 5% worse than the BCC technique and offers a better load balancing of the resources.

ECU Network. In this section, our proposed methodology is applied to artificial examples inspired by typical *Electronic Control Unit* (ECU) networks from the automotive domain to show its time consumption. In these examples, the resources have a relatively low vertex degree since they are connected via buses. These buses are typically arranged in a star topology. The large example with 70 ECUs corresponds to recent real world automotive networks in premium class automobiles. The experiments were carried out on an Intel Pentium 4 3.20GHz machine with 1GB RAM.

Table 2 shows the results for three different ECU networks. The time consumption per analysis of 1.16 to 6.69 seconds per analysis run shows that the proposed approach is applicable for these kind of networks. Moreover, the time consumption is small enough to be applied in design space exploration approaches where many different network layouts have to be analyzed in order to find the optimum. However, at a certain complexity, the memory consumption of the calculated BDD exceeds the computers main memory and, thus, makes an analysis impossible. In our testcases, this problem arises at networks with about 90 ECUs and 90 tasks.

6   Conclusion

In this paper, a reliability analysis for self-healing networked embedded systems has been proposed. This technique allows to determine an upper bound for the MTTF that can be achieved by self-healing techniques that rely on self-reconfiguration and self-routing. The technique uses a fast and memory-aware symbolic representation and respects given constraints of the system like, e.g., the maximum computational load. Given the proposed technique, the effectiveness of different self-healing techniques for networked embedded systems can be quantified. Moreover, an effective dimensioning of the system in the design phase is enabled. In the future, the proposed technique will be extended to respect a possible maintenance of resources or, more accurately, of subnets of the networked embedded system.

References

1. Dai, Y.S.: Autonomic computing and reliability improvement. In: Proc. of ISORC 2005, pp. 204–206 (2005)
2. Koch, D., Streichert, T., Dittrich, S., Strengert, C., Haubelt, C., Teich, J.: An operating system infrastructure for fault-tolerant reconfigurable networks. In: Grass, W., Sick, B., Waldschmidt, K. (eds.) ARCS 2006. LNCS, vol. 3894, pp. 202–216. Springer, Heidelberg (2006)
3. Garlan, D., Schmerl, B.: Model-based adaptation for self-healing systems. In: Proc. of WOSS 2002, pp. 27–32 (2002)
4. Streichert, T., Glaß, M., Wanka, R., Haubelt, C., Teich, J.: Topology-aware replica placement in fault-tolerant embedded networks. In: Brinkschulte, U., Ungerer, T., Hochberger, C., Spallek, R.G. (eds.) ARCS 2008. LNCS, vol. 4934, pp. 23–37. Springer, Heidelberg (2008)
5. Bryant, R.E.: Graph-based algorithms for boolean function manipulation. IEEE Trans. on Comp. 35(8), 677–691 (1986)
6. Cankay, H.C., Nair, V.S.S.: Reliability and availability evaluation of self-healing sonet mesh networks. In: Proc. of GLOBECOMM 1997, pp. 252–256 (1997)
7. Cankay, H.C., Nair, V.S.S.: Accelerated reliability analysis for self-healing sonet networks. SIGCOMM Comput. Commun. Rev. 28(4), 268–277 (1998)
8. Kawamura, R., Sato, K., Tokizawa, I.: Self-healing atm networks based on virtual path concept. IEEE Journal on Selected Areas in Communications 12(1), 120–127 (1994)
9. Lee, J.: Reliability models of a class of self-healing rings. Microelectronics and Reliability 37(8), 1179–1183 (1997)
10. Politof, T., Satyanarayana, A.: Efficient algorithms for reliability analysis of planar networks - a survey. IEEE Trans. on Reliability 35(3), 252–259 (1986)
11. Ortega, C., Tyrrell, A.: Reliability analysis in self-repairing embryonic systems. In: Proc. of EH 1999, pp. 120–128 (1999)
12. Dressler, F., Dietrich, I.: Lifetime analysis in heterogenous sensor networks. In: Proc. of DSD 2006, pp. 606–616 (2006)
13. Elliot, C., Heile, B.: Self-organizing, self-healing wireless networks. In: Proc. of Aerospace Conference 2000, pp. 149–156 (2000)
14. Glaß, M., Lukasiewycz, M., Streichert, T., Haubelt, C., Teich, J.: Reliability-Aware System Synthesis. In: Proceedings of DATE 2007, pp. 409–414 (2007)
15. Streichert, T., Glaß, M., Haubelt, C., Teich, J.: Design space exploration of reliable networked embedded systems. Journ. on Systems Architecture 53(10), 751–763 (2007)
16. Izosimov, V., Pop, P., Eles, P., Peng, Z.: Synthesis of fault-tolerant schedules with transparency/performance trade-offs for distributed embedded systems. In: Proceedings of DAC 2004, pp. 550–555 (2004)
17. Valiant, L.G.: The complexity of enumeration and reliability problems. SIAM Journal on Computing 8, 410–421 (1979)
18. Burch, J.R., Clarke, E.M., McMillan, K.L., Dill, D.L., Hwang, L.J.: Symbolic model checking: 1020 states and beyond. Inf. Comput. 98(2), 142–170 (1992)
19. Eén, N., Sörensson, N.: Translating Pseudo-Boolean Constraints into SAT. Journal on Satisfiability, Boolean Moelding and Computation 2, 1–25 (2006)
20. Rauzy, A.: New Algorithms for Fault Tree Analysis. Reliability Eng. and System Safety 40, 202–211 (1993)

Investigation and Reduction of Fault Sensitivity in the FlexRay Communication Controller Registers

Yasser Sedaghat and Seyed Ghassem Miremadi

Dependable Systems Laboratory, Sharif University of Technology, Tehran, Iran
y_sedaghat@ce.sharif.edu, miremadi@sharif.edu

Abstract. It is now widely believed that FlexRay communication protocol will become the de-facto standard for distributed safety-critical automotive systems. In this paper, the fault sensitivity of the FlexRay communication controller registers are investigated using transient single bit-flip fault injection. To do this, a FlexRay bus network, composed of four nodes, was modeled. A total of 135,600 transient single bit-flip faults were injected to all 408 accessible single-bit and multiple-bit registers of the communication controller in one node. The results showed that among all 408 accessible registers, 30 registers were immediately affected by the injected faults. The results also showed that 26.2% of injected faults caused at least one error. Based on the fault injection results, the TMR and the Hamming code techniques were applied to the most sensitive parts of the FlexRay protocol. These techniques reduced the fault affection to the registers from 26.2% to 10.3% with only 13% hardware overhead.

Keywords: Safety-critical applications, Distributed embedded systems, FlexRay protocol, Fault injection.

1 Introduction

Today, many safety-critical applications are implemented as distributed embedded systems [13], e.g. X-by-wire applications. These systems are composed of several different types of hardware units (called nodes), e.g., processing units, sensors, and actuators, interconnected by a communication network.

Communication in a distributed architecture can be triggered either dynamically, in response to an event (event-driven), or statically, at predetermined moments in time (time-driven). Examples of event-triggered protocols are Byteflight [1], CAN [2], LonWorks [3], and Profibus [4]. The main drawback of event-triggered protocols is their lack of predictability [5]. Examples of time-triggered protocols are SAFEbus [6], SPIDER [7], and TTP/C [8]. The main drawback of time-triggered protocols is their lack of flexibility [5]. To resolve the drawbacks of both event-triggered and time-triggered protocols, other protocols such as TTCAN [9], FTT-CAN [10], and Flex Ray [11] are introduced that can support both time-triggered and event-triggered transmissions.

Among the latter protocols, the FlexRay protocol is advancing as the predominant protocol and will become the de-facto industry standard for X-by-wire applications [12], [13], [5], [14], [15]; e.g., the next edition of the BMW X5 will use the FlexRay

M.D. Harrison and M.-A. Sujan (Eds.): SAFECOMP 2008, LNCS 5219, pp. 153–166, 2008.

protocol in its electronically controlled dampers [12].The FlexRay protocol was started by an industry consortium with four founding members (BMW, Daimler-Chrysler, Philips, and Freescale) [15]. Three top design objectives were considered in the standardization of the FlexRay protocol: high speed transmission, deterministic communication, and fault-tolerant communication [15].

In safety-critical distributed embedded systems, a fault-tolerant communication between different nodes has a significant impact on the overall system reliability. It has been reported [16], [13] that the overall reliability of a safety-critical distributed embedded system not only depends on the reliability of the nodes, but also on the reliability of the communication network.

This paper investigates the fault sensitivity of all parts of the FlexRay communication controller using fault injection. The most and the least sensitive registers in the FlexRay are characterized. Then, appropriate fault-tolerant techniques are applied to the most sensitive registers, to protect the communication controller against transient faults.

The remainder of the paper is organized as follows: Section 2 introduces the FlexRay protocol briefly. Error models and error handling mechanisms in the FlexRay protocol are presented in Section 3. In Section 4, the experimental environment is presented. Section 5 includes the experimental results and finally, the conclusions are given in Section 6.

2 The FlexRay Protocol

The FlexRay protocol provides key features of synchronization that include scalable data transmission in both synchronous and asynchronous modes. It can support the data rate up to 10Mbit/sec. The protocol itself offers deterministic data transmission, guaranteed message latency and message jitter. The FlexRay supports dual and redundant transmission channels and transmission mechanism is arbitration free. In addition, it has optional support of optical or electrical physical layers. The physical layer will provide support for bus, star, and multiple star topologies [11].

From the dependability point of view, the FlexRay documents [11] specify solely bus guardian mechanism and clock synchronization algorithms. Other features, such as a membership service or mode management facilities, should be implemented in software or hardware layers on top of the FlexRay. This will allow to conceive and to implement exactly the services that are needed with the drawback that correct and efficient implementations might be more difficult to achieve in a layer above the communication controller [16].

One of the main purposes of this paper is to convince developers of the FlexRay communication controller, by the experimental results, how necessary it is to reduce the fault sensitivity of critical registers. This reduction causes to improve the reliability of the FlexRay protocol, noticeably; and it is possible to reduce the need for expensive fault-tolerant techniques, such as bus guardian mechanism or clock synchronization algorithms.

2.1 Protocol Operation

Communications in the FlexRay protocol are based on predetermined interval times which are named communication cycles (bus cycles). These communication cycles are executed periodically. In this protocol a communication cycle is a concatenation of a time-triggered (or static) window, an event-triggered (or dynamic) window, a symbol window and a network idle time (NIT) window. The size of each communication window is set statically at design time. The time-triggered window uses a Time Division Multiple Access (TDMA) [17] mechanism; a node in FlexRay might possess several slots in the time-triggered window, but the size of all the slots is identical. In the event-triggered part of the communication cycle, the mechanism is Flexible TDMA (FTDMA) [18]: time is divided into so-called minislots, each station possesses a given number of minislots (not necessarily consecutive), and it can start the transmission of a frame inside each of its own minislots. A minislot remains idle, if the station has nothing to transmit which actually induces a loss of bandwidth [16]. The symbol window is a communication period in which a symbol can be transmitted on the network. The NIT window is a communication-free period that concludes each communication cycle. Fig. 1 shows an example of communication cycle in the FlexRay protocol.

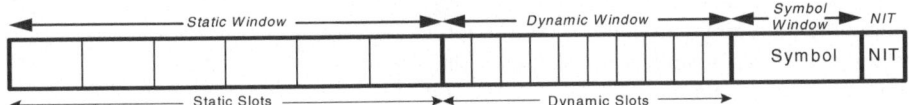

Fig. 1. Communication Cycle in the FlexRay Protocol

The FlexRay frame consists of three parts: the header segment, the payload segment, and the trailer segment. The FlexRay header segment consists of 5 bytes. These bytes contain one reserved bit, payload preamble indicator, null frame indicator, sync frame indicator, startup frame indicator, frame ID, payload length, header CRC, and cycle count.

The payload segment contains 0 to 254 bytes (0 to 127 two-byte words) of data. Because the payload length contains the number of two-byte words, the payload segment contains an even number of bytes. The FlexRay trailer segment contains a single field, a 24-bit CRC for the frame. The Frame CRC field contains a cyclic redundancy check code (CRC) computed over the header segment and the payload segment of the frame. The computation includes all fields in these segments.

In the FlexRay protocol, frames are sent in static slots or dynamic slots of each communication cycle. Fig. 2 shows the frame format in the FlexRay protocol.

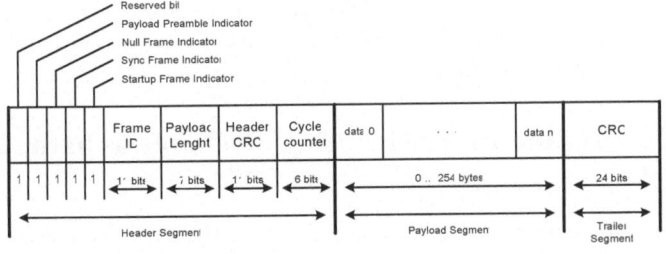

Fig. 2. Frame format in the FlexRay Protocol

2.2 Protocol Structure

The FlexRay communication controller consists of six modules [11]: controller host interface (CHI), protocol operation control (POC), coding and decoding (CODEC), media access control (MAC), frame and symbol processing (FSP), and clock synchronization process (CSP). Fig. 3 shows relation between these modules.

The CHI module, manages data and control flow between the host processor and the FlexRay protocol engine within each node. The CHI module manages all data exchange relevant to the protocol operation and manages all data exchanges relevant to the exchange of messages. Moreover, this module manages protocol configuration data, protocol control data, and protocol status data.

Operational modes of FlexRay modules are adjusted by POC module. Proper protocol behavior can only occur if the mode changes of the core modules are properly coordinated and synchronized. The purpose of the POC is to react to host commands and protocol conditions by triggering coherent changes to core modules in a synchronous manner, and to provide the host with the appropriate status regarding these changes.

The CODEC module is responsible for encoding the communication elements into a bit stream and is responsible for receiving communication elements, making bit streams and investigating correctness of bit streams.

The MAC module controls access to the bus. In the FlexRay protocol, media access control is based on a recurring communication cycle. Within one communication cycle, the FlexRay offers the choice of two media access schemes, i.e., TDMA scheme and FTDMA scheme. The communication cycle is the fundamental element of the media access scheme within FlexRay.

The FSP module checks the correct timing of received frames and received symbols with respect to the TDMA scheme, applies further syntactical tests to received frames, and checks the semantic correctness of received frames.

Finally, the CSP module is responsible for generation of timing units in the FlexRay communication controller, e.g., communication

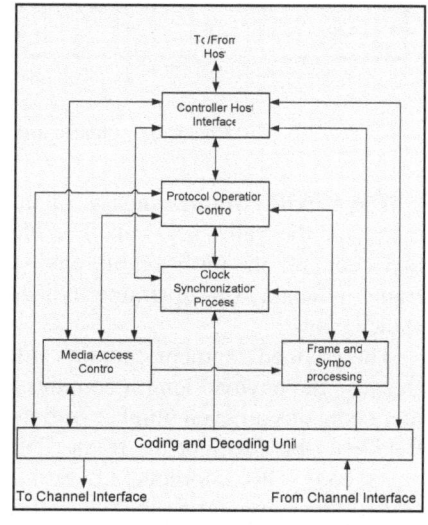

Fig. 3. The FlexRay Structure [11]

cycles. Moreover this module uses a distributed clock synchronization mechanism in which each node individually synchronizes itself to its cluster by observing the timing of transmitted sync frames from other nodes.

3 Error Models and Error Handling Mechanisms in the FlexRay Protocol

Safety-critical applications have to function correctly even in presence of faults. Faults can be permanent (e.g., damaged microcontrollers or communication links), transient (e.g., caused by single event upsets or electromagnetic interferences), or intermittent (appear and disappear repeatedly). The transient faults are the most common, and their number is continuously increasing due to the continuously raising level of integration in semiconductors [19]. These transient single bit-flip errors are more common consequences of transient faults [22].

3.1 Error Models in the FlexRay Protocol

According to the FlexRay protocol, the following three categories of errors are possible [11]:

1) Syntax error
Syntax error denotes the presence of a syntactic error in a time slot, and occurs in following conditions:
- The node starts transmitting while the channel is not in the idle state.
- A decoding error occurs.
- A frame is decoded in the symbol window or in the network idle time.
- A symbol is decoded in the static segment, the dynamic segment, or the network idle time.
- A frame is received within the slot after the reception of a semantically correct frame.
- Two or more symbols are received within the symbol window.

2) Content error
Content error denotes the presence of an error in a received frame, and occurs in following condition:
- In the static segment, the header length the header of the received frame does not match the stored header length in a special register (this register contains globally configured value of the payload length of a static frame).
- In the static segment, the startup frame indicator, contained in the header of the received frame, is set to one while the sync frame indicator is set to zero.
- In the static or in the dynamic segment, the frame ID, contained in the header of the received frame, does not match the current value of the slot counter or the frame ID equals to zero in the dynamic segment.
- In the static or dynamic segment, the cycle count, contained in the header of the received frame, does not match the current value of the cycle counter.
- In the dynamic segment the sync frame indicator, contained in the header of the received frame, is set to one.
- In the dynamic segment the startup frame indicator, contained in the header of the received frame, is set to one.
- In the dynamic segment the null frame indicator, contained in the header of the received frame, is set to zero.

3) Boundary violation error
Boundary violation error denotes whether a boundary violation has occurred at the boundary of the corresponding slot. A boundary violation occurs if the node does not consider the channel to be idle at the boundary of a slot.

3.2 Error Handling Mechanisms in the FlexRay Protocol

In order to respond to errors, two basic mechanisms are provided in the POC module [11]. For significant errors, the POC:halt state is immediately entered. The POC also contains a three-state degradation model for errors that can be endured for a limited period of time. In this case entry to the POC:halt state is deferred, at least temporarily, to support possible recovery from a potentially transient condition.

Errors causing immediate entry to the POC:halt state
There are three general conditions that trigger entry to the POC:halt state:

- Product-specific error conditions such as BIST errors and sanity checks.
- Error conditions detected by the host that result in a FREEZE command being sent to the POC via the CHI.
- Fatal error conditions detected by the POC or one of the core mechanisms.

Product-specific errors are accommodated by the POC, but not described in FlexRay specification. Similarly, host detected error strategies are supported by the POC's ability to respond to a host FREEZE command, but the host-based mechanisms that trigger the command are beyond the scope of this specification, hence they were not considered in this paper.

Errors handled by the degradation model
Integral to the POC is a three-state error handling mechanism referred to as the degradation model. It is designed to react to certain conditions detected by the clock synchronization mechanism that are indicative of a problem, but that may not require immediate action due to the inherent fault tolerance of the clock synchronization mechanism. This makes it possible to avoid immediate transitions to the POC:halt state while assessing the nature and extent of the errors. The degradation model is embodied in three POC states - POC:normal active, POC:normal passive, and POC:halt.

In the POC:normal active state, the node is assumed to be either error free, or at least within error bounds that allow continued "normal operation". Specifically, it is assumed that the node remains adequately time synchronized to the cluster to allow continued frame transmission without disrupting the transmissions of other nodes.

In the POC:normal passive state, it is assumed that synchronization with the remainder of the cluster has degraded to the extent that continued frame transmissions cannot be allowed because collisions with transmissions from other nodes are possible. Frame reception continues in the POC:normal passive state in support of host functionality and in an effort to regain sufficient synchronization to allow a transition back to the POC:normal active state.

If errors persist in the POC:normal passive state or if errors are severe enough, the POC can transit to the POC:halt state. In this state it is assumed that recovery back to the POC:normal active state cannot be achieved, so the POC halts the core mechanisms in preparation for reinitializing the node. The conditions for transitioning between the three states comprising the degradation model are configurable. Furthermore, transitions between the states are communicated to the host allowing the host to react appropriately and to possibly take alternative actions using one of the explicit host commands.

3.3 Error Indicator Registers of the FlexRay Communication Controller

In this protocol, there are some registers that are set in the mentioned error conditions. In this paper, these registers are named "error indicator registers". Table 1 shows these registers and their locations in the FlexRay communication controller.

Activating each of these registers, may result in one or more main error types. Faults, depending on when and where they occur, may change the expected value of some of these registers and cause one or more main error types. In this paper, the type of occurred error is not considered. However, if any of registers in the Table 1, is unexpectedly changed, this change is considered as an error.

Table 1. Error indicator registers (registers of the FlexRay showing the error occurrences) in the FlexRay protocol

Registers	Module	Registers	Modules
decoding_error_on_A		vPOC_Freeze	
TSS_ok		vPOC_CHIHaltRequest	POC
TSS_too_long		vPOC_ErrorMode	
FSS_ok		zSyncCalcResult	CSP
payload_ok	CODEC	Content_error_on_A	
trailer_ok		Fatal_protocol_error	
BSS_ok		T_StatusSlot_ValidFrame	
FES_ok		T_StatusSlot_SyntaxError	FSP
zBssError		T_StatusSlot_ContentError	
Header_Crc_error		T_StatusSlot_TxConflict	
Frame_Crc_error		T_StatusSlot_BViolation	

4   Experimental Environment

The FlexRay communication controller was implemented by hardware description language, Verilog HDL, and specifications of this controller, e.g. timing and configuration, were tested according to the FlexRay protocol conformance test specification [20]. This controller, according to its specifications [11], has six modules to perform its functions: controller host interface (CHI), protocol operation control (POC), clock synchronization process (CSP), frame and symbol process (FSP), media access control (MAC), and coding and decoding (CODEC). A cluster was formed consisting of 4 nodes with single bus topology (Fig. 4). In this topology, a node is composed of a host and a communication controller. The host typically is a hardware unit that generates data to exchange with other nodes through a communication channel.

In the experiments, instead of a real host, a data generator was implemented to generate static frames with fixed length and dynamic frames with variable length at the start of the communication cycles. In this cluster, any node was allowed to send and receive frames on the communication channel.

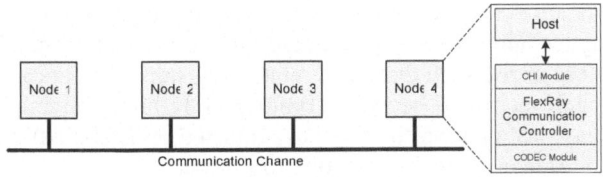

Fig. 4. Experimental setup

To investigate the fault tolerance of the FlexRay communication controller, transient single bit-flip faults were injected in all accessible registers of communication controller modules of the node 2 and their effects on the error indicator registers were observed in node 2 and node 4 (for observing more fault effects); and results were stored.

4.1 Fault Sensitivity Calculation Process

To inject the transient single bit-flip faults at the behavioral level in node 2, the SIN-JECT fault injection tool [21] was used.

A fault sensitivity calculation process of a bit, by using SINJECT tool, consists of four steps:

1- When the given workload is applied, behaviors of the error indicator registers in a fault-free network are simulated and stored.

2- During the second step, to consider fault effects, the given workload is applied again to the network, a single transient bit-flip fault is injected to a bit of a communication controller register of node 2, at a random time, and the behavior of the error indicator registers of node 2 and node 4 are observed.

3- During the third step of the fault sensitivity calculation process, the faulty network behavior is compared with the behavior of the fault-free network, which is gathered at first step, and if there is a mismatch, this injected fault is considered as an activated fault and otherwise, this injected fault is considered as an overwritten fault.

4- To achieve accurate fault sensitivity of a bit, several faults should be injected to this bit (repeating the first three steps). After injecting enough bit-flip faults and determining the number of activated faults (be Equation 1), the fault sensitivity of this bit is calculated by Equation 2:

$$\#\text{injected faults} = \#\text{activated faults} + \#\text{overwritten faults} \tag{1}$$

$$\text{fault sensitivity of a bit} = \frac{\text{Number of activated faults}}{\text{Number of all injceted faults to that bit}} \times 100\% \tag{2}$$

The process was repeated for all bits in all accessible registers in FlexRay communication controller and the fault sensitivity of these registers was determined.

4.2 Fault Tolerance Improvement Strategies

After determining the fault sensitivity of a register, if its sensitivity was more than an acceptable value, a proper fault-tolerant technique would be used to reduce its vulnerability. The Hamming code technique with single bit correction ability and Triple Modular Redundancy (TMR) technique were used for this purpose.

The Hamming technique was implemented on several sets of vulnerable registers. Those sets were organized such that most related registers were encapsulated in a set; and the size of each set varied between 10 bits and 32 bits. This implementation did not incur any delay or limitation to access to protected registers. After changing value of a protected register in a register set, due to protocol operations, Hamming bits of that register set is calculated while other parts of communication controller were allowed to access to that register set.

TMR or Hamming techniques should be used consciously; for example, if there is a highly fault sensitive register which immediately triggers other parts of communication controller by changing its value, the Hamming code technique should not be used to reduce the sensitivity of this register. The main reason is that if a bit-flip fault occurs in this register, the other parts of communication controller react to that fault immediately and some errors may occur in other parts of communication controller; in such situation, if Hamming technique is used, the occurred fault in the register is detected and corrected while other parts of communication controller react to this changing value again. Consequently, a bit-flip fault causes two incorrect reactions in other communication parts. Also, if Hamming technique is implemented such that the accessibility to that register is not allowed until the Hamming bits of this register are calculated, some delay is inserted into the operation of communication controller and this delay may corrupt the timing of FlexRay protocol operations. In this situation, the TMR technique is the better option, but if the imposed delay due to Hamming technique for this type of register does not damage the FlexRay protocol timings, by checking and testing according to the FlexRay protocol conformance test specification [20], it is beneficial to use Hamming technique instead of TMR technique.

On the other hand, if there is a highly fault sensitive register which does not trigger immediately other parts of communication controller by changing its value, the TMR technique should not be used because of its ultra-high hardware overhead (200%). In this condition, the Hamming technique is the better option.

In this paper, with respect to properties of the FlexRay communication controller registers, the TMR technique and the Hamming technique (without incurring any delay) are suggested to improve fault tolerance of this controller.

5   Experimental Results

In this paper, to assess the fault sensitivity of the FlexRay communication controller, the nodes were connected through a passive bus network. The main reason of selecting bus topology is to prevent some error propagations in star coupler of star topology. This prevention results in hiding the fault sensitivity of some communication controller registers.

To simulate the experiments, the ModelSim 5.5 simulation environment was used. The simulation includes four communication cycles; in the first two cycles, single transient bit-flip fault was injected randomly, then simulation was resumed two cycles to guarantee that the injected fault caused an error or overwritten.

5.1   The FlexRay Communication Controller Modules

To reach an accurate fault sensitivity of each register, 50 transient bit-flip faults were injected to each bit of all accessible FlexRay controller registers (according to the fault sensitivity calculation process) and gathered results were investigated. If there existed a register with more than 20% of fault sensitivity, a proper fault-tolerant technique based on properties of this register were used to reduce its vulnerability.

As discussed in the previous section, the TMR technique was only used for registers which were immediately triggered other communication controller parts with their changing values. In this controller, all of these kinds of registers were single-bit register; consequently, for each single-bit register, two redundant flip-flops were added to implement the TMR technique. Furthermore, to implement the Hamming technique, all vulnerable registers which were not improved by TMR technique, were grouped in some sets. These register sets were organized as discussed in the previous section.

The results show that the TMR technique masks all injected faults but the Hamming technique is not able to tolerate all injected faults; because it is probable that faulty registers are used immediately before they are corrected. The experiment results are presented in Table 2, whereas the modeled FlexRay communication controller is not still synthesizable, the estimated hardware overhead is based on the number of accessible flip-flops. Table 3 contains hardware overhead of implemented techniques.

Table 2. Fault injection results

FlexRay Module	# Injected Faults	Standard FlexRay		Improved FlexRay		Improvement (%)
		Activated Faults		Activated Faults		
		#	%	#	%	
POC	5200	2343	45.1	512	9.8	357
CODEC	32300	5396	16.7	3586	11.1	50
MAC	11050	1805	16.3	1181	10.7	52
CSP	47850	16574	34.6	4774	10	246
FSP	6850	1230	18	688	10	80
CHI	32350	8154	25.2	3300	10.2	147
ALL	135600	35502	26.2	14041	10.3	154

Table 3. Hardware overheads

FlexRay Modules	Standard FlexRay		Improved FlexRay	HW Overhead (%)
	# Registers	# Flip-Flops	# Flip-Flops	
POC	28	104	(104 + 32) = 136	30.8
CODEC	104	646	(646 + 46) = 692	7.1
MAC	64	221	(221 + 28) = 249	12.7
CSP	94	957	(957 + 162) = 1119	16.9`
FSP	41	137	(137 + 20) = 157	14.6
CHI	77	647	(647 + 66) = 713	10.2
ALL	408	2712	(2712 + 354) = 3066	13.0

In the modeled FlexRay communication controller, all registers, signals and other components are named based on FlexRay specification document (version 2.1, revision A) [11]; for more details about their responsibilities, readers are referred to [11]. Based on experimental results, Fig. 5 shows the fault sensitivity of all FlexRay communication controller modules in the standard implementation (according to the FlexRay specifications [11]) and in the improved implementation. In this figure, the fault sensitivities of the FlexRay module registers which are more than 20% sensitive to injected faults are presented. For more clarity, fault sensitivities are sorted in a descending order.

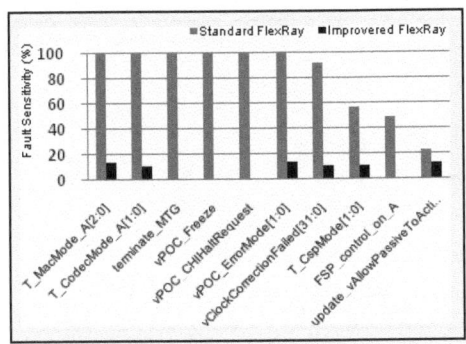

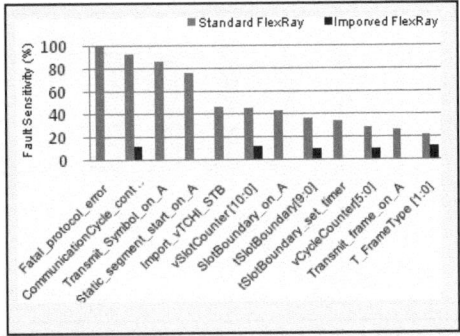

(a) Fault sensitivities of FlexRay Registers (POC Module)

(b) Fault sensitivities of FlexRay Registers (MAC Module)

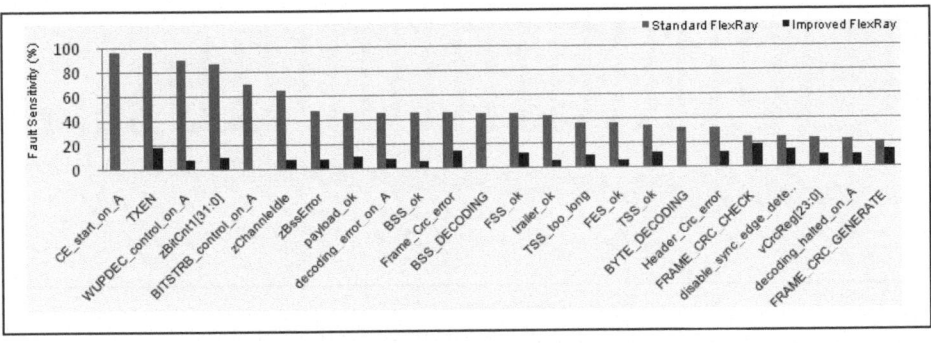

(c) Fault sensitivities of FlexRay Registers (CODEC Module)

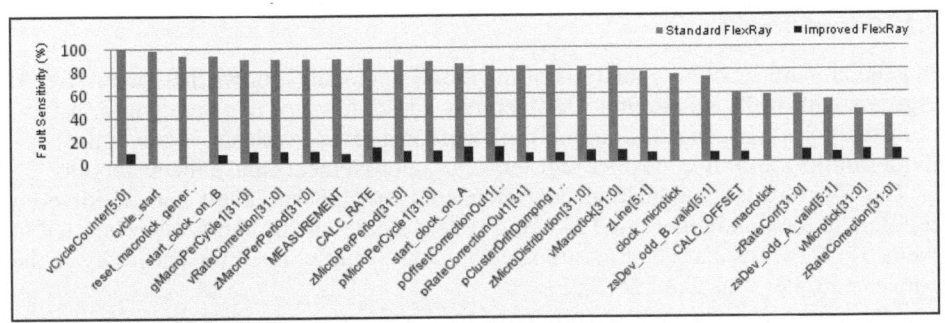

(d) Fault sensitivities of FlexRay Registers (CSP Module)

Fig. 5. Fault sensitivities of FlexRay modules in standard and improved implementations

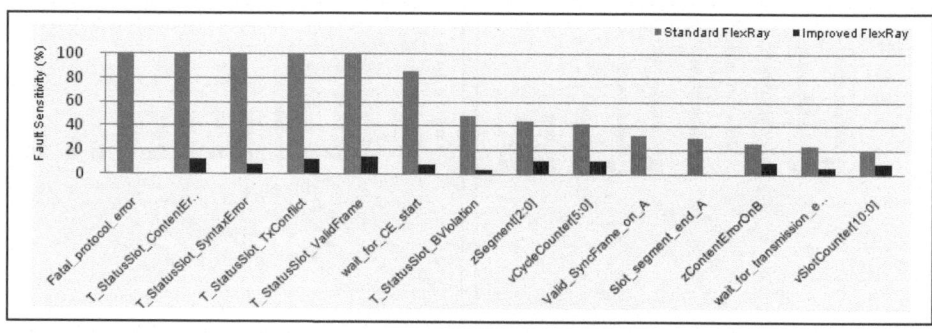

(e) Fault sensitivities of FlexRay Registers (FSP Module)

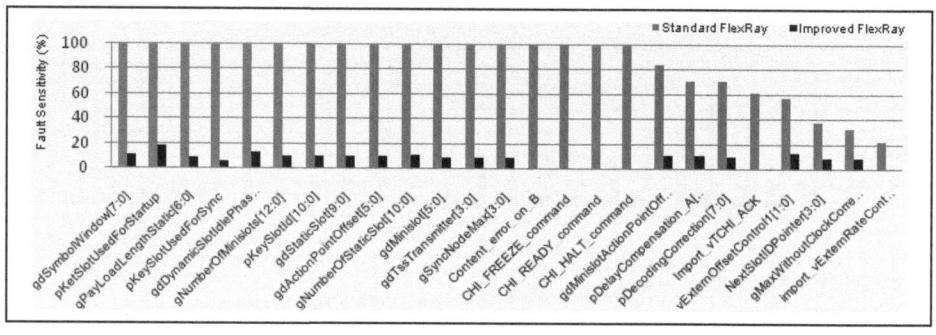

(f) Fault sensitivities of FlexRay Registers (CHI Module)

Fig. 5. (*continued*)

5.2 Overall Results

In general, fault sensitivity analysis of the FlexRay modules shows that there were 30 registers with 100% sensitivity. This fact may question use of this protocol for safety-critical applications. In addition, in Fig. 5 there is a severe variance in the fault sensitivities of the FlexRay controller registers. Our improvements make them smooth.

The FlexRay communication controller contains 408 single-bit and multiple-bit registers in total. A number of 135,600 transient single-bit flip faults were injected to them. 35,502 faults caused at least one error; consequently, the fault sensitivity of the whole controller was about 26.2%.

After improving the fault sensitivity of the FlexRay communication controller, its sensitivity was reduced from 26.2% to 10.3% (about 154% improvement), while adding 354 extra flip-flops costs the controller about 13% flip-flop overhead.

Fig. 6 shows the fault sensitivity of each module in the standard implementation and the improved implementation of FlexRay communication controller. This figure also shows that the POC module is the most sensitive part of FlexRay communication controller and CODEC module is the least sensitive part. Furthermore, our results show that we were able to reduce the sensitivity of FlexRay modules to almost equal

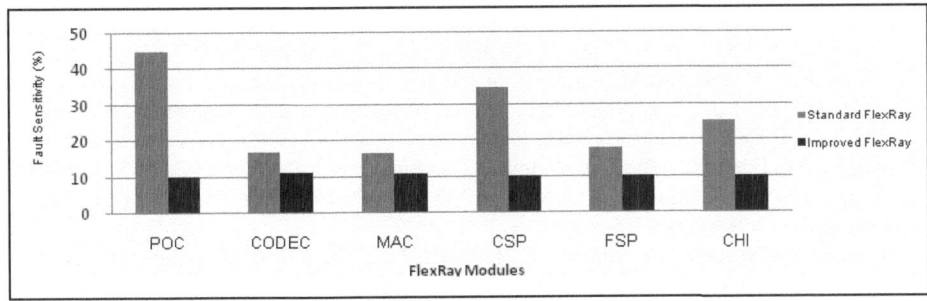

Fig. 6. Fault Sensitivity of FlexRay Communication Controller

values (difference about 1%) as compared with previous values in the standard implementation.

6 Conclusions

Safety-critical automotive control systems are nowadays complex distributed embedded systems and the communication protocol is an essential part of them. The FlexRay communication protocol is now expected to become the future standard for in-vehicle communication.

In this paper, the fault sensitivities and vulnerabilities of FlexRay communication controller registers, based on 135,600 single-bit flip fault injections to all accessible registers, are investigated.

The results show that the fault sensitivities of POC, CODEC, MAC, CSP, FSP, and CHI modules are 45.1%, 16.7%, 16.3%, 34.6%, 18%, and 25.2% respectively. Moreover, according to the fault injection results, among all 408 accessible registers, 30 registers were immediately affected by the injected faults, 84 registers were affected between 20% and 99%, while the remaining (294) registers were affected by less than 20%.

After determining the sensitive registers, proper fault masking and fault-tolerant techniques, based on their properties, are applied to reduce the vulnerability of these registers. This caused, the fault sensitivity of POC, CODEC, MAC, CSP, FSP, and CHI modules to reduce to 9.85%, 11.1%, 10.7%, 10%, 10%, and 10.2% respectively.

In general, the fault sensitivity of FlexRay communication controller was improved more than 2 times and in this improved implementation, none of the registers has more than 20% fault sensitivity.

References

1. Byteflight Specification, http://www.byteflight.com/
2. CAN Specification 2.0, http://www.can-cia.org/
3. LonWorks networks, http://www.echelon.com
4. PROFIBUS DP Specification, http://www.profibus.com
5. Pop, T., Pop, P., Eles, P., Peng, Z.: Bus Access Optimization for FlexRay-based Distributed Embedded Systems. In: Design, Automation & Test in Europe Conference & Exhibition 2007 (DATE 2007), pp. 1–6. EDA Consortium, Nice (2007)

6. Hoyme, K., Driscoll, K.: SAFEbus. In: IEEE Aerospace and Electronic Systems Magazine (ISSN 0885-8985), vol. 8(3), pp. 34–39. IEEE Press, Los Alamitos (1992)
7. Miner, P.S., Malekpour, M., Torres-Pomales, W.: Conceptual design of a Reliable Optical BUS (ROBUS). In: 21st AIAA/IEEE Digital Avionics Systems Conference, pp.13D3-1–13D3-11. IEEE Press, Irvine (2002)
8. Kopetz, H., Bauer, G.: The Time-Triggered Architecture. J. IEEE. 91(1), 112–126 (2003)
9. Road Vehicles—Controller Area Network (CAN)—Part 4: Time-Triggered Communication, ISO 11 898-4 (2000)
10. Ferreira, J., Pedreiras, P., Almeida, L., Fonseca, J.A.: The FTT-CAN protocol for flexibility in safety-critical systems. J. IEEE Micro. (Special Issue on Critical Embedded Automotive Networks) 22(4), 46–55 (2002)
11. FlexRay Communications System - Protocol Specification V2.1 Revision A, http://www.flexray.com
12. Sethna, F., Stipidis, E., Ali, F.H.: What Lessons Can Controller Area Networks Learn From FlexRay. In: Vehicle Power and Propulsion Conference (VPPC 2006), pp. 1–4. IEEE Press, Windsor (2006)
13. Pop, T., Pop, P., Eles, P., Peng, Z., Andrei, A.: Timing Analysis of the FlexRay Communication Protocol. In: 18th Euromicro Conference Real-Time Systems (ECRTS 2006), pp. 203–216. Kluwer Academic Publishers, Dresden (2006)
14. Hagiescu, A., Bordoloi, U.D., Chakraborty, S.: Performance Analysis of FlexRay-based ECU Networks. In: 44th ACM/IEEE Design Automation Conference (DAC 2007), pp. 284–289. ACM, San Diego (2007)
15. Makowitz, R., Temple, C.: FlexRay- A Communication Network for Automotive Control Systems. In: IEEE International Workshop on Factory Communication Systems (WFCS 2006), pp. 207–212. IEEE Press, Torino (2006)
16. Navet, N., Song, Y., Simonot-Lion, F., Wilwert, C.: Trends in Automotive Communication Systems. J. IEEE 93(6), 1204–1223 (2005)
17. Tindell, K., Clark, J.: Holistic Schedulability Analysis for Distributed Hard Real-Time Systems. J. Microprocessing & Microprogramming 40, 117–134 (1994)
18. Cena, G., Valenzano, A.: Performance analysis of byteflight networks. In: IEEE Workshop on Factory Communication Systems (WFCS 2004), pp. 157–166. IEEE Press, Vienna (2004)
19. Izosimov, V., Pop, P., Eles, P., Peng, Z.: Design Optimization of Time- and Cost-Constrained Fault-Tolerant Distributed Embedded Systems. In: Design, Automation and Test in Europe Conference and Exhibition 2005 (DATE 2005), vol. 2, pp. 864–869. IEEE Computer Society, Munich (2005)
20. FlexRay Communications System - Protocol Conformance Test Specification V2.1, http://www.flexray.com
21. Zarandi, H.R., Miremadi, S.G., Ejlali, A.: Dependability Analysis Using a Fault Injection Tool Based on Synthesizability of HDL Models. In: 18th IEEE International Symposium on Defect and Fault Tolerance in VLSI Systems, pp. 485–492. IEEE Press, Boston (2003)
22. Armengaud, E., Rothensteiner, F., Steininger, A., Horauer, M.: A Method for Bit Level Test and Diagnosis of Communication Services. In: IEEE Workshop on Design & Diagnostics of Electronic Circuits & Systems 2005 (DDECS 2005), p. 6. IEEE Press, Hungary (2005)

Secure Interaction Models for the HealthAgents System

Liang Xiao, Paul Lewis, and Srinandan Dasmahapatra

University of Southampton, UK
{lx,phl,sd}@ecs.soton.ac.uk

Abstract. Distributed decision support systems designed for healthcare use can benefit from services and information available across a decentralised environment. The sophisticated nature of collaboration among involved partners who contribute services or sensitive data in this paradigm, however, demands careful attention from the beginning of designing such systems. Apart from the traditional need of secure data transmission across clinical centres, a more important issue arises from the need of consensus for access to system-wide resources by separately managed user groups from each centre. A primary concern is the determination of interactive tasks that should be made available to authorised users, and further the clinical resources that can be populated into interactions in compliance with user clinical roles and policies. To this end, explicit interaction modelling is put forward along with the contextual constraints within interactions that together enforce secure access, the interaction participation being governed by system-wide policies and local resource access being governed by node-wide policies. Clinical security requirements are comprehensively analysed, prior to the design and building of our security model. The application of the approach results in a Multi-Agent System driven by secure interaction models. This is illustrated using a prototype of the HealthAgents system.

Keywords: Clinical Information System, Multi-Agent System, Security Model.

1 Introduction and Motivation

In a distributed collaborative healthcare environment, multiple clinical organisations from geographically different sites may be involved together in the delivery of healthcare services, each having its own users, resources, and access policies. Clinical users, residing in their own sites and doing their specific jobs, often need to access globally available resources and services under locally set constraints.

Such an environment brings challenges to distributed healthcare system infrastructures, especially when security is a concern. Security concerns, either to a conventional system or a distributed system, spread all over the system and differ from one system to another. If they are not taken into account, as early as a system begins to be built, the integrity and usability of the system may be critically compromised. Security challenges for a distributed healthcare system are notable in several aspects. Firstly, no global user repository will be available for distributed authorisation. Clinical centres may join or leave independently. The management and administration of resource access will have to be de-centralised in the network, where each site maintains their own users and resources to be accessed. Secondly, although access control

M.D. Harrison and M.-A. Sujan (Eds.): SAFECOMP 2008, LNCS 5219, pp. 167–180, 2008.

becomes complicated in a distributed environment, we shall bear in mind that unless some degree of open access is promoted where hospitals and users are able to join in freely, the system will not be able to improve clinical decision making by using the knowledge they share. Thirdly, in such an open access condition, healthcare records which contain sensitive private information shall by no means be disclosed, even to collaborative centres and friendly clinicians, except for healthcare purposes. Lastly and more complicatedly, we shall consider the access constraints not only on individual cases, but also on what each of them consists of. Can doctors have access to all patient records in connected hospitals? Can a pathologist have access to complete records or even alter irrelevant reports?

Generally, securing healthcare information systems should authorise users with genuine needs to have access to the services and resource items, in order to perform their job responsibilities. Clinical requirements must be carefully studied in order to understand the constituents of job responsibilities and build the security model.

The rest of the paper is structured as follows. Section 2 analyses clinical security requirements including the principles that need to be supported by the security architecture under development. Section 3 discusses existing security approaches and describes their weaknesses in handling the requirements identified. Section 4 gives an overview of a layered security model and Section 5 illustrates this and the process of building it in detail, using the HealthAgents system. Section 6 concludes the paper.

2 Security Requirements of Healthcare Information Systems

We shall, in the beginning, draw distinctions between the types of threats imposed to healthcare systems and their likelihood. Though eavesdropping or hacking is a major concern to computer network security, it is so expensive that dedicated and capable intruders may consider using a more convenient way. Actually, 10% of GPs (general practitioner) in the UK have experienced their computers being physically stolen [8]. More likely, improper use of the system may lead to privacy leaks, by careless (or malicious) users, extra privileges given by the system incorrectly. A well-designed system should not only protect the communication sites and end users, but also carefully authorise users with genuine needs to have access to selective sharing of information without exposing additional information under protection. This security need has currently not been well addressed in healthcare information systems [4]. In this section, we outline the challenges and common security requirements of healthcare systems in a distributed environment, where preserving privacy and maintaining openness are crucial and information access decisions depend upon role and context.

2.1 The Distributed Environment of Healthcare Information Systems

Aggregating dispersed data into large databases is expensive and practically unfeasible, since geographically different healthcare centres have to have control over their datasets and at the same time maintain a globally consistent data schema. A more important reason to oppose data consolidation is concerned with healthcare data confidentiality. In the UK, the National Health Service (NHS), driven by the motives of easier central administration and better information availability, attempted to build a unified electronic patient record system and give access to extended NHS community. This has been opposed [7] [22] for the reason that such a system, collecting data from existing GP systems but out of their control, is in conflict with the ethical principle that no patient should be identifiable other than to the GP without patient consent [5] and the result from a survey that most patients are unwilling to share their information with NHS [6]. Another objection arises from the overwhelming workload such a centralised system could possibly put upon a security officer responsible for managing the data sharing [4].

A distributed healthcare service infrastructure, however, implies the capability that is required to cope with the administrative burden and the continuous maintenance needs arising from fully functional and networked clinical centres, each of which has its own users, data, access policies, and which assumes that cross-centre access is the norm. A distributed environment and its associated dynamics bring other concerns, such as patient privacy preserving, to the information-sharing healthcare network.

2.2 Preserving Privacy and Confidentiality in Shared Access

The privacy of patient information is an important issue and failure to recognise this will lead to risk of patient safety, loss of public confidence in clinical organisations, and so on [23]. A fundamental ethical principle stated by both the EU and the General Medical Council in the UK is that, patients must consent to data sharing. The British Medical Association [10] advises that clinical professionals, who have access to patient confidential information in order to perform their duties, are responsible for the information they hold under ethical or professional obligations of confidentiality and shall not use or disclose such information for any purpose other than the clinical care of the patient to whom it relates. This means patients shall be assured that they can trust the access of their information, by a care team within their treating hospitals or experts involved from collaborative centres, if any, is safe and accords with their agreement. The moving from a traditional patient-doctor relationship towards a modern patient-healthcare service relationship implies trust to clinical systems must be maintained rather than reliance on doctor responsibilities. The absence of a mechanism or policy framework in the interest of information governance and confidentiality protection, hence, may damage the healthcare services aimed to be delivered, since private information of any individual patient may be made available by systems to people not directly related with the care of that patient. This will give opportunities to potential threats, possibly coming from inside workers, as well as outside hackers. Such threats include ungraceful private information disclosure and abuse or even more risky, incorrect clinical decisions made for vulnerable patients due to clinical data being wrongly altered, accidentally or deliberately. It is worth noting that threats from outside intruding into the network are much rarer than from inside. The security risks tend to increase dramatically, therefore, when an interconnected clinical system network is in place which makes separately stored patient records and clinical information easily accessible and lets a wider range of people have access to them. Appropriate access control to patient records is the fundamental need for patient privacy and information security [23].

2.3 Maintaining an Open Access

Two aspects of openness must be maintained: 1) open for joining the system and not preventing any friendly but previously unknown clinical centre (bringing in its previously unrecognised users) from accessing information available across organisational boundaries; 2) open for information sharing to the network. Conducting healthcare research with more open use of information (identifiable data, etc.) under legitimate constraints and user acceptance, though not related with the clinical care directly, advances medical knowledge and promotes higher quality of healthcare service in the long run and is welcomed by the society. A clinical system can benefit most from clinical data as well as patient-specific data if such information can be machine-analysed and digested. The knowledge accumulated can be useful for later decision makings, particularly for rare but similar cases encountered in the future, confidential information contained in cases not being revealed.

2.4   The Different Access Needs to Data Subsets Due to Distinct Job Nature

The need of distinguishing only the relevant data for sharing among clinical professionals rather than the whole records arises from preserving privacy while maintaining open access. Even if name, address and other privacy information is removed to produce a seemingly anonymised record, a NHS clinician can easily identify a patient by the NHS number and they must be able to do so to perform their jobs. Therefore, it is sensible to grant access permission to particular record parts on the basis of users' expertise. This expertise determines their actual needs of access, to the data parts they routinely work with and by doing so, healthcare roles fulfilled. For example, pathology medical records or reports may be sent to a pathologist involved in a patient's care; prescription sent to a pharmacist; and sensitive parts not sent out at all. A specialist may have more control over their own partitions, e.g. write their reports or order certain tests, but limited permissions to other specialists' partitions or even not at all, e.g. to very sensitive medical test results.

2.5   The Access Policies and Principles Pertinent to Patients as Individuals

It is not rational to allow a professional to have access to all patient records, even if limited to the data subset fitting his/her expertise. Only relevant clinicians who have real life relationships with patients in clinical centres should access their records. This is documented in British Medical Association's security policy principles for clinical information systems [7], and the feasibility of adopting it has been evidenced in [23]. Two major principles are as follows.

Principle of Access: "Each identifiable clinical record shall be marked with an access control list naming the people or groups of people who may read it and append data to it. The system shall prevent anyone not on the access control list from accessing the record in any way."

Principle of Control: "One of the clinicians on the access control list must be marked as being responsible. Only she may alter the access control list, and she may only add other health care professionals to it."

A named responsible clinician, possibly a patient GP, as in the UK or a primary care physician (PCP), as in the US, may set up a workgroup including the specialists who together deliver healthcare to the patient. According to the Principle of Access, it is the members of this group who will be in the patient access control list, as used by

RBAC for files [16], have access to a subset of data they are responsible for, reflecting their job nature. The one who sets up the workgroup will let the system know the group members and their roles in the group, in accord with the Principle of Control. This implies a data ownership. Such a scheme decentralises management burden and increases scalability. The distributed environment and open access requirements suggest that a named doctor may involve specialists from other sites (remote consultants, temporary attending physicians, etc.) into healthcare procedures. For example, a medical opinion requested on a surgical patient may require a medical registrar, from other directorates, to exercise override access to that patient's notes [23]. This is related with delegation [4]. Essentially, a responsible doctor grants access to local or remote users from trusted sites and occasionally, someone acts on their behalf, implying ownership transfer. A triangle relationship is described in [15]: a patient is associated with a workgroup, of which a user is a member, so that a user is permitted access via the workgroup to patient ("self-claimed" or "colleague-granted"/delegation).

3 Existing Security Solutions: Role-Based Access Control and Role Mapping in a Distributed Environment

In Role-Based Access Control (RBAC) [16], permissions that describe operations upon resources are associated with roles. Users are assigned to roles to gain permissions that allow them to perform particular job functions. Privileges may be calculated as follows [2]:

*Privileges = User-Role * Role-Definition + Rules-Function (User-Attributes)*

In addition to the static collection of rights accumulated by roles, a user can dynamically achieve extra rights if they expose certain attributes as defined by rules. This model is efficient when many users require the same set of rights in an organisation but otherwise unmanageable or even useless when roles vary in different conditions under which users act. In a hospital, roles can be defined for a number of classified groups to aggregate permissions, e.g. consultant, radiologist, nurse, who have static job functions. However, dynamic contexts exist in role playing, e.g. patients may be additionally assigned to or removed from a list for which a named doctor is responsible and this influences this doctor's role in caring these patients. RBAC has difficulties to capture such security-relevant contexts as patient, location, and time in healthcare environment [4]. Patient-doctor relationship is identified as a critical clinical security constraint to record access, described in Section 2.

The Community Authorisation Service (CAS) [1] provides a solution to the management of user access control within Virtual Organisations (VOs) spanning over multiple sites in the Grid environment. It breaks the tradition of requiring each resource provider to maintain the mapping of individual users (across VOs) to its local database roles in order to authorise access to its resources. Using CAS, user memberships are instead based on VO roles and local resource providers only need to map these to local database roles. This dramatically reduces the number of mapping entries across resource providers and the duplicated maintenance burden put on them once a new user joins or a current user privilege changes.

Such an approach requires no global user repository. However, a presumption of using the approach, as it is in RBAC, is that a large number of users can be grouped into several role groups requiring certain access levels in involved organisations.

For the same reason that RBAC is infeasible to address the clinical requirement that information access or travelling may alter from patient to patient and user led as stated in the Principle of Access, the CAS is encountered with similar difficulties. Suppose clinicians A and B with the same speciality are from hospitals P and Q respectively. They will be categorised into the same VO role and the same access rights to data in P and Q. But in reality A shall have more privileges than B to certain data, e.g. of patients in P under A's care, and vice versa for B's privileges in Q.

Managing a resource access model is complex where there is a large number and various types of users, resource items, and access policies, user responsibilities being dynamic and ownership being distributed. The common practice of simply defining roles that aggregate all permissions required for the collection of resources to complete tasks is not realistic due to the diversity of individual needs which literally entails each individual a distinct role. Even the burden of defining and maintaining a proper set of access control policies based on roles for automating authorisation could be considerable. A security solution must be able to cope with the complexity.

4   Overview of a Layered Security Model

It has been pointed out that healthcare systems should be designed with multilateral security rather than multilevel security [9]. Unlike some military systems prevent information flow "down" from top secret to secret then to confidential, healthcare systems usually prevent information flow "across" from one clinician to another or from one hospital to another. This is evidenced by the requirements outlined in Section 2.4 and Section 2.5 where different access needs to cases and case partitions are distinguished due to distinct job responsibilities.

However, we argue a multilevel security model is more manageable, task availability being in the top level control and resource availability to tasks in lower level control. A multilateral security model resides in the lower level and complements the multilevel security model. The assignment of tasks to users is a business decision to be made by stakeholders, possibly explicitly in rules. It is sensible to regard the accessibility to tasks the organisational privileges with which organisation seniority is related and access to business functions restricted. Since tasks already exist in organisations and are routinely performed by specific user groups, they help to functionally decompose the system and ease security management. If a user can perform a specific type of task, then there must be certain resource items available to him/her to load into the task, if not all. Without the context of accomplishing one or more tasks in different privilege levels, information access makes no sense. The rational of using a combined multilevel and multilateral model is further supported by the fact that a job responsibility is determined by the level of authority and the division of work [14]. The former prevents information flow downwards and the latter prevents information flow across, being concerned about workgroup membership and job speciality under our further refinement. This forms a layered security architecture that addresses the healthcare security requirements.

1)   Privilege of performing various types/levels of tasks and executing associated interaction models is determined by job title or grade/level. Users may upgrade their job titles occasionally and this is managed locally. Semantics of job titles and task collections must be globally defined and agreed among organisations.

2) Privilege of loading case instances for performing tasks (or enactment of interaction models) is determined by real life workgroup memberships or job boundary. This is managed by the locally named doctors, who shall be flagged as owners in case records' access control lists.

3) Privilege of accessing case record partitions (e.g. patient data, biopsy data, microarray data, MRI and MRS data, diagnosis data, therapy data, surgery data, etc.) is determined by job nature or specialist one takes on in hospitals (e.g. oncologist, pathologist, radiologist, surgeon, etc.). This is managed by system administrators when the account is setup and is maintained at a high level of stability.

Thus, a user's overall privileges will be the sum of the user's access privileges in all tasks that user is involved in (being a policy), each of which is decided by the particular cases he/she can operate as a workgroup member to deliver healthcare service (being a fact upon interaction instantiation) at the time of performing tasks, which in turn will be constrained by the accessible case partitions as determined by user professional roles (being a fact).

User Privileges = $\sum$ (Interaction Model Set as determined by job level * Interaction Model's Operational Cases as determined by job boundary * Case Subset as determined by job nature)

Alternatively, the following meta-rule determines the prerequisite a user exercises privileges: a user has a title above the one required for running an interaction model can load a case, that is under the care of a workgroup which the user is a member of, and perform operations on the case parts the user's specialists allow.

user_privilege (user, im, case, part, operation) ←
   job_title(user, title1) & executable(title2, im) & above(title1, title2) &
   member(user, workgroup) & responsible(workgroup, patient) & own(patient, case) &
   job_specialist(user, specialist) & rights(specialist, part, op)

5 Secure Interaction Models for Healthagents: A Comprehensive Case Study

In this section, we describe our HealthAgents system, the elicitation of interaction models, and their secure running in our layered security model for HealthAgents.

5.1 The HealthAgents Architecture

The HealthAgents [18] system is a distributed decision support system that facilitates diagnosis and prognosis, employs a set of distributed nodes that either store patient case data, build classifiers that are trained upon case data and capable of classifying tumour types, or use classifiers for the diagnosis and prognosis of brain tumours. The magnetic resonance spectroscopy (MRS) data used by the system is built up using anonymous information from child and adult cases. Producer nodes receive requests from clinicians and generate classifiers for particular tumours. Clinicians with cases will employ classifiers (instead of the actual cases) to assist in the diagnosis of patients for particular tumours. Knowledge extracted from cases is implicitly involved for decision making and patient privacy not compromised, private case information not being revealed in the process. The HealthAgents system consists of a variety of agents each charged with a different task. A more detailed description of the HealthAgents components and architecture can be found in [19].

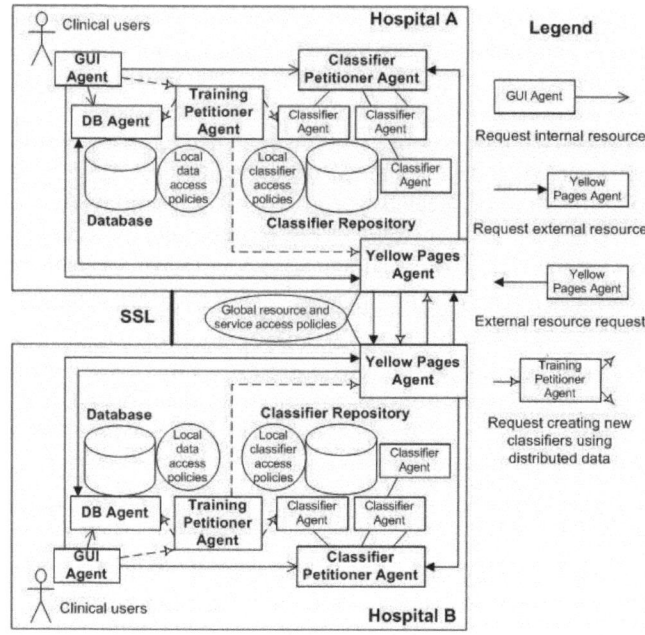

Fig. 1. The HealthAgents system architecture and resource access flow control

5.2   Building an Interaction Model Hierarchy with a Goal-Decomposition Graph

Four major interaction models, as shown in Figure 2, are identified: create classifier, execute existing classifier, update classifier reputation value, and update case profile. They are elaborated as four sub-goals under the root goal of "tumour type diagnosis" via a goal decomposition graph, useful for requirements analysis and interaction model identification. A detailed goal decomposition procedure and underpinning process

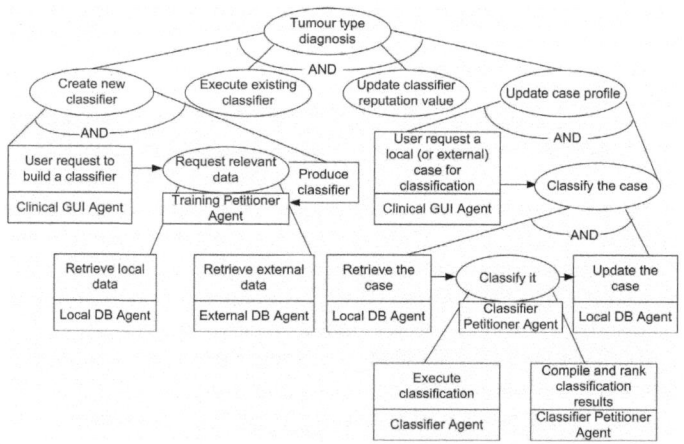

Fig. 2. The goal-decomposition graph for HealthAgents

Table 1. A high level view of selected interaction models

Goal	Sub-goals (Interaction Model)	Interaction Model privileges	Interaction Model participants	Interaction Model constraints
Tumour type diagnosis	Update case profile, etc.	N/A	All	N/A
Update case profile	Classify case	Principal clinicians or above	GUI Agent, DB Agent, Classifier Agent, and Classifier Petitioner Agent	The clinician can update the specialised data areas
Classify case	N/A	Junior clinicians or above	Classifier Agent, and Classifier Petitioner Agent	The clinician must be a workgroup member taking care of the case

elicitation can be found in [24]. Table 1 describes a specific branch of the graph ("Update case profile") for further discussion.

5.3 Secure Interaction Models and Lightweight Coordination Calculus (LCC)

Assume three job titles, senior clinician, principal clinician, and junior clinician, in that order, forms the existing clinical hierarchy, from top to bottom. Roles in a role hierarchy of RBAC have inheritance relationships. Likewise, a job title higher up in the hierarchy inherits task execution privileges from a job title further down in the hierarchy. Suppose the following rules in HealthAgents restrict task availability.

- Rule1: Senior clinicians can identify the need of new classifiers in the network and so are able to create classifiers, using all public cases and local private cases.
- Rule2: Principal clinicians have primary healthcare responsibilities and so are able to run classifiers, update case profiles and diagnosis results, as well as update classifier reputation values.
- Rule3: Junior clinicians assist in healthcare and can run classifiers and be advised of classification results.

Gaia [3] unifies responsibilities and permissions in a single role notion. It is also recognised in [21] that the coordination among agents/roles and resources must enable authorisation policy specification over interaction specification to achieve an expressive and safe interaction model. Thus, role, interaction, and constraint should be correlated. The descriptive interaction behaviour which consists of message passing and constraint solving have been defined in Lightweight Coordination Calculus (LCC) [12] that can be transmitted, interpreted, and executed by agents in the network. The LCC language has been developed in the OpenKnowledge project [13] and it uses logic expression to regulate the message exchange protocols among participant peers each of which plays a particular role. The LCC language combines role functions and constraints in a single framework and this gives us the opportunity to express permission enforcement prior to responsibility fulfilment within role playing behaviour, in the context of running interaction protocols. The following LCC clauses describe the fundamental interaction pattern for resource access control.

```
a(resource_request, RRID) ::
  request(Resource, Operation, Context) ⇒ a(resource_manager, RMID)
a(resource_manager, RMID) ::
```

request(Resource, Operation, Context) $\Leftarrow$ a(resource_request, RRID) $\leftarrow$ grantPermission(RRID, Resource, Operation, Context, Policies) then (

response(Grant_yes) $\Rightarrow$ a(resource_request, RRID) or

response(Resource_result) $\Rightarrow$ a(resource_request, RRID) $\leftarrow$ getOperationResult(Resource, Operation, Access_result))

Briefly, a(resource_request, RRID) :: Def_{RRID} and a(resource_manager, RMID) :: Def_{RMID} denotes that agents RRID and RMID play the roles of resource_request and resource_manager respectively as defined in the definitions follow. Def_{RRID} has a single and Def_{RMID} has a composite message passing behaviour constructed using the following forms: Def_a *then* Def_b (Def_a satisfied before Def_b), Def_a *or* Def_b (either Def_a or Def_b satisfied), or Def_a *par* Def_b (both Def_a and Def_b satisfied). In the Def, $M_l \Rightarrow A_m$ denotes that a message M_l is sent to agent A_m while $M_l \Leftarrow A_m$ denotes that a message M_l is received from agent A_m. In the above role definitions, a message of resource access request is sent from the agent that plays the request role to the agent that plays the manager role. Upon receipt of this message, the resource manager agent applies appropriate security policies and responds by sending back a message either saying the request has been granted (or rejected) or by providing the actual resources (or the results of their usage) being requested. In the Def, $\leftarrow Cons_n$ denotes that a constraint must be satisfied (as some running code) before the clause prior to it.

The notion *a(id, role)* defines the role a certain agent should play and its identity can be bound with executable tasks, workgroup memberships, and professional specialists at runtime. The role playing behaviour defines the common responsibilities an entitled user supposed to fulfil, being in a position with/above a given title as are in Gaia, the organisational roles in well-defined positions associated with expected behaviour. Then the memberships and professional specialists further constrain the concrete resource usage in the role's interaction model participation, being identity-specific and role-independent. This layered architecture is discussed as follows, illustrated by a principal clinician updating case profile after classification.

Level 1: Interaction Model constraints

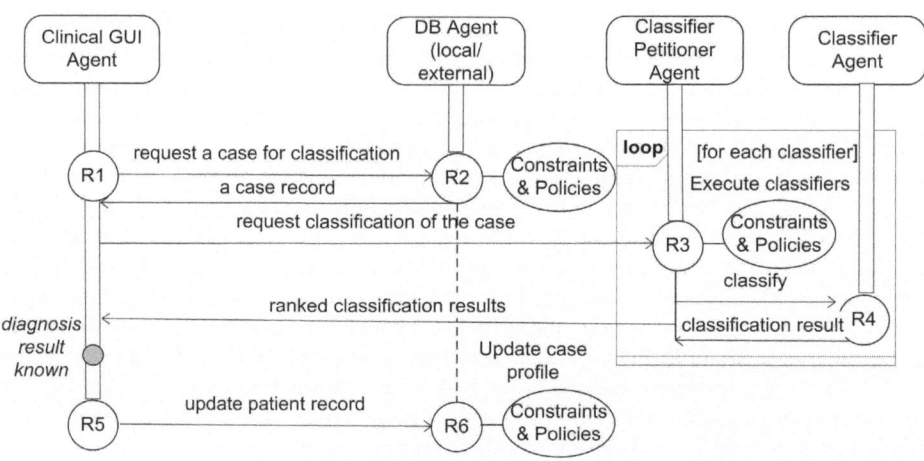

Fig. 3. Interaction Model: update case profile (including case classification)

The first layer filters interaction model availability. A principal clinician (possibly a GP) can load cases for which they have caring responsibilities and later update its profile (diagnosis result, etc.). A junior clinician can perform classification but cannot do the update. Figure 3 shows the interaction model and the following LCC clauses show its specification. The clinician plays a role of classification (R1) and updating case profile (R5). The role changes when an accurate diagnosis result is known.

```
/* R1: classify a case */
a(clinician_classify, CID) ::
  requestCaseRecordByID(I) ⇒ a(database, DBID) then
  caseRecord (R) ⇐ a(database, DBID) then
  requestClassification(R, C) ⇒ a(classifier_petitioner, CPID) then
  classificationResults(S) ⇐ a(classifier_petitioner, CPID) then
  a(clinician_followingdiagnosis, CID)
/* R5: update case record and classifier reputation following diagnosis */
a(clinician_followingdiagnosis, CID) ::
  ( updateCaseRecordByID(I) ⇒ a(database_update, DBID) then
   caseRecordUpdated(Y) ⇐ a (database_update, DBID) )
  par
  ( updateClassifier(I) ⇒ a(classifier_petitioner, CPID) then
   classifierUpdated(Y) ⇐ a (classifier_petitioner, CPID) )
```

Level 2: Case level constraints
An interaction model is uniquely defined and its running context varies, e.g. involved clinicians and cases. A resource manager must check the request (resource and operation) against the requester identity at runtime, in compliance with the access policies. Specifically, the clinician must be a member of the workgroup delivering care to the owner of the case before the case is allowed to be updated, being a meta-rule of healthcare access control. Additional local policy rule satisfaction must also be considered for extra constraints, e.g. a particular clinician can/cannot access particular resource items. A generic security policy schema for healthcare is described in [25] that can complement the meta-rule with any number of specific policies. The following shows the LCC constraints used by the database agent, being a resource manager, for permission checking before the actual role functions are carried out. The database agent issues a case record (R2) and updates the same record (R6), different levels of permissions being needed.

```
/* R2: send a case record for classification */
a(database_download, DBID) ::
  requestCaseRecordByID(I) ⇐ a(clinician_classify, CID) ← grantPermission(CID, I, Read, Normal_classify_from_local_site, Local_database_read_policy_set) then
    caseRecord(R) ⇒ a(clinician_classify, CID) ← getCaseRecordByID(I, R) then
    a(database_update, DBID)
/* R6: update a case record after classification */
a(database_update, DBID) ::
    updateCaseRecordByID(I) ⇐ a(clinician_followingdiagnosis, CID) ← grantPermission(CID, I, Update, Normal_update_from_local_site, Local_database_update_policy_set) then
    caseRecordUpdated (Y) ⇒ a(clinician_followingdiagnosis, CID)
```

It is at the point of checking the LCC constraint of "grantPermission" that user workgroup and case will be related (clinician identity of CID and case identity of I), and other locally set read or update policies applied, prior to the required operation. A clinician not in the right workgroup may be able to download a case but cannot update it. The running and execution of LCC specification is supported by the OpenKnowledge kernel.

Level 3: Case partition constraints

Similarly with level 2, a user identity is bound with professional specialists at runtime and this will constrain further permission to case partitions, e.g. only the named clinicians may update or write major diagnosis results; certain specialists may write reports in their areas; others on the case care list may only read those areas. Thus, a three dimension resource request of (user, resource, operation) will be constrained in two dimensions: user-resource must match workgroup membership and user-operation match job specialist.

6   Conclusions and Discussion

In this paper, we have analysed the general security requirements for clinical information systems and developed a layered security model, illustrated by its application to the HealthAgents system but which is also applicable to other healthcare systems.

Organisational structure and context association are key assumptions to our privilege model. Organising authorisation at user level cannot realise cooperation and inter-organisational communication in extended health networks, as stated in [17]. The authors distinguish structural roles, describing prerequisites or competencies for actions and functional roles, being bound to the realisation of actions. Such a conjunctional perspective of role is in accordance with the privilege control in business processes and then their contextual constraint. The semantic similarity of clinical user group privileges and the business processes they can perform is described in [11]. In addition to that, access decisions need to be made on the exercise of privileges in business processes depending upon contextual information. Structuring business process (or task) context related constraints, e.g. attending relation between physician and patient as well as clinician speciality, as contextual parameters to task execution that affect access control decisions is expressed in [20].

Clinical task execution privileges, therefore, should be distinguished, and represented by the privileges of running interaction models in our approach. The layered security model authorises at a higher level, the users' task accessibility based on a static organisational structure and at a lower level, within task enactment, users' case and case partition accessibility based on dynamic functional needs in order to perform tasks. This inevitably avoids the occasion that a junior clinician creates a classifier of poor quality or updates a classifier reputation value improperly.

Next, higher level business function-based constraints are coupled with lower level data-based constraints. A limited set of data, determined by user workgroup memberships, will be allowed to be populated into the limited set of task functions. Finally, data-based constrains are additionally coupled with operation-based constraints. The available operations, determined by job nature and specialists, will be allowed, e.g. write (reports) or update (diagnosis results), upon particular data sections. These constraints, as well as individually defined local policies, must be satisfied prior to interaction model running. In sum, we constrain the availability of tasks to users, case availability to tasks, and further operations availability to cases, as the overall layered security architecture. The architecture is scalable since access rights are precisely controlled by the combination of these dimensions. For example, a senior pathologist doctor who is responsible for a patient can update the pathology part of this patient profile but someone who is a senior pathologist but not involved in caring for the patient cannot, or someone who is a junior doctor, or someone who is not specialised in pathology at all.

No global user account repository is required in our system. The necessary interaction models are globally agreed. The case to workgroup assignment is locally defined and user to workgroup possibly across organisations, for enabling interaction model running. When one user invokes an interaction model and this involves resources from other sites, the permission checking is determined by this user being involved in patient care or not, e.g. a remote clinician may perform a classification on behalf of a named doctor who is on holiday and delegates the responsibility to this clinician, in emergency situations, even the local hospital has not set up a local account for the clinician.

Interaction models can be publicly accessible since the descriptive interaction logic among peers reveals no secret information itself and so no issue exists such as alternative interaction model provision to certain users under certain conditions. Rather, alternative resource peers may be selected because the access to others is restrictive or, a subset or related/alternative resource items from query returned to the requester peer with a limited set of privileges. Such an autonomic query relaxation paradigm, as part of our future work, will avoid additional user interaction and frustrating experience. Another direction of future work is via monitoring unsuccessful resource access, an interaction model adjustment is advised if an access without satisfying constraints is encountered but considered necessary. It may be useful to let such requests be recorded and routed to responsible doctors or other delegated authorisers who may or may not approve the issuing of additional privileges, either permanently or temporarily. With better understanding of the necessity of such exceptional requests possibly after real life communication, critical and timely care aimed to patients will not be compromised.

References

1. Pereira, A.L., Muppavarapu, V., Chung, S.M.: Role-based access control for grid database services using the community authorization service. In: Transactions on Dependable and Secure Computing, vol. 3(2), pp. 156–166. IEEE, Los Alamitos (2006)
2. M-Tech Information Technology, Inc.: Beyond Roles: A Practical Approach to Enterprise User Provisioning (2006)
3. Wooldridge, M., Jennings, N.R., Kinny, D.: The Gaia methodology for agent-oriented analysis and design. Journal of Autonomous Agents and Multi-Agent Systems 3(3), 285–312 (2000)
4. Zhang, L., Ahn, G., Chu, B.: A role-based delegation framework for healthcare information systems. In: 7th ACM Symposium on Access Control Models and Technologies, pp. 125–134. ACM, New York (2002)
5. Joint Computer Group of the GMSC and RCGP: GMSC and RCGP guidelines for the extraction and use of data from general practitioner computer systems by organisations external to the practice. Appendix III In: Committee on Standards of Data Extraction from General Practice Guidelines (1988)
6. Hawker, A.: Confidentiality of personal information: a patient survey. Journal of Informatics in Primary Care, 16–19 (1995)

7. Anderson, R.J.: Clinical system security: interim guidelines. British Medical Journal 312, 109–111 (1996)
8. Pitchford, R.A., Kay, S.: GP Practice computer security survey. Journal of Informatics in Primary Care, 6–12 (1995)
9. Anderson, R.J.: Patient Confidentiality - At Risk from NHS Wide Networking. Proceedings of Healthcare 96 (1996)
10. BMA - British Medical Association, http://www.bma.org.uk/
11. Chandramouli, R.: Business Process Driven Framework for defining an Access Control Service based on Roles and Rules. In: 23rd National Information Systems Security Conference (2000)
12. Robertson, D.: A lightweight coordination calculus for agent systems. In: Leite, J.A., Omicini, A., Torroni, P., Yolum, p. (eds.) DALT 2004. LNCS (LNAI), vol. 3476, pp. 183–197. Springer, Heidelberg (2005)
13. Robertson, D., et al.: Open Knowledge: Semantic Webs Through Peer-to-Peer Interaction. OpenKnowledge Manifesto (2006), http://www.openk.org/
14. Crook, R., Ince, D., Nuseibeh, B.: Modelling Access Policies Using Roles in Requirements Engineering. Information and Software Technology 45(14), 979–991 (2003)
15. Calam, D.: Information Governance - Security, Confidentiality and Patient Identifiable Information,
    http://etdevents.connectingforhealth.nhs.uk/eventmanager/upl oads/ig.ppt
16. Sandhu, R.S., Coyne, E.J., Feinstein, H.L., Youman, C.E.: Role-Based Access Control Models. Computer 29(2), 38–47 (1996)
17. Blobel, B.: Authorisation and access control for electronic health record systems. International Journal of Medical Informatics 73(3), 251–257 (2004)
18. HealthAgents, http://www.healthagents.net/
19. Xiao, L., Lewis, P., Gibb, A.: Developing a Security Protocol for a Distributed Decision Support System in a Healthcare Environment. In: 30th International Conference on Software Engineering, pp. 673–682. ACM, New York (2008)
20. Hu, J., Weaver, A.C.: Dynamic, Context-Aware Access Control for Distributed Healthcare Applications. In: 1st Workshop on Pervasive Security, Privacy and Trust (2004)
21. Omicini, A., Ricci, A., Viroli, M.: RBAC for organisation and security in an agent coordination infrastructure. Electronic Notes in Theoretical Computer Science 128(5), 65–85 (2005)
22. Anderson, R.: Undermining data privacy in health information. BMJ 322, 442–443 (2001)
23. Denley, I., Smith, S.W.: Privacy in clinical information systems in secondary care. BMJ 318, 1328–1331 (1999)
24. Xiao, L., Greer, D.: Adaptive Agent Model: Software Adaptivity using an Agent-oriented Model Driven Architecture. Information & Software Technology. Elsevier. In: Press (2008), http://dx.doi.org/10.1016/j.infsof.2008.02.002
25. Xiao, L., Peet, A., Lewis, P., Dashmapatra, S., Sáez, C., Croitoru, M., Vicente, J., Gonzalez-Velez, H., Lluchi Ariet, M.: An Adaptive Security Model for Multi-agent Systems and Application to a Clinical Trials Environment. In: 31st IEEE Annual International Computer Software and Applications Conference, pp. 261–266. IEEE, Los Alamitos (2007)

Security Challenges in Adaptive e-Health Processes

Michael Predeschly, Peter Dadam, and Hilmar Acker

Institute DBIS, University Ulm
`firstname.lastname@uni-ulm.de`

Abstract. E-health scenarios demand system-based support of process-oriented information systems. As most of the processes in this domain have to be flexibly adapted to meet exceptional or unforeseen situations, flexible process-oriented information systems (POIS) are needed which support ad-hoc deviations at the process instance level. However, e-health scenarios are also very sensitive with regard to privacy issues. Therefore, an adequate access rights management is essential as well. The paper addresses challenges which occur when flexible POIS and adequate rights management have to be put together.

1 Introduction

The personnel in clinical domains have to deal with a large number of different processes. Due to high workloads, exceptional situations, frequently changing diagnostics, treatments, and accounting procedures the risk of errors is pretty high. For that reason it is widely acknowledged that adequate process-oriented information systems (POIS) could help to improve this situation [1]. However, clinical processes are not static by nature. Therefore, POIS have to be flexible, i.e. to allow ad-hoc deviations at the process-instance level [2]. While good progress has been made in understanding how to build such powerful process management systems [3], little attention has been paid so far how an adequate access rights management system (RMS) for such systems should look like. In "classical" information systems, a few administrators are authorized to assign access rights and privileges to users. However, in flexible POIS even end-users may be authorized to perform modifications to processes at the instance level, i.e. to insert, delete, or move process steps. In this context the question arises, for example, what kind of rights should be granted to an end-user to be able to insert a new step and to assign access and execution rights to this newly inserted step without being too restrictive at the one side or to completely undermine the RMS at the other side. Some of the issues related to access rights management in conjunction with flexible POIS shall be discussed in the following.

2 Challenges and Problems

Contemporary RMS have been designed to enable or to restrict access of users to functions and data in "classical" information systems. In such information systems users may come and go but the information systems themselves, i.e. the functions

M.D. Harrison and M.-A. Sujan (Eds.): SAFECOMP 2008, LNCS 5219, pp. 181–192, 2008.

they are providing, are rather static. If these information systems are implemented in the traditional, monolithic fashion, access rights management is typically handled in a centralized way. Thus such systems have an RMS component which manages which users have which kind of permission to execute which application functions and/or which kind of access to data has been granted to them.

These RMS maintain some kind of rights matrix which associates subjects (users) and objects (functions, data) with access rights. If many subjects (s) and objects (o) have to be handled, this rights matrix becomes very large if it is directly stored as full-fledged $s \times o$ matrix. Therefore, typically only one dimension of this rights matrix is physically stored as a so called access control list (ACL). This ACL represents for every object the list of users which have access to it along with the corresponding permissions. The other dimension, namely the association between users and the objects they are allowed to access is either not supported at all or must be computed by inspecting all ACLs.

In environments with high security requirements, one wants to ensure that the security rules are always obeyed. Thus one wants to enforce *mandatory access control*. In such environments, typically the assignment of access rights to users follows the need-to-know principle. In addition, constraints like separation of duties or the enforcement of the four-eyes-principle also have to be supported. To implement such principles, one usually assigns security levels or permissions to users and respective qualifications to the objects. Only if a user's permissions matches the respective requirements of the object, the access is granted. In order to enforce mandatory access control, the RMS is typically implemented as an "active" component sometimes called a reference monitor [4] which controls the access to functions and objects.

All these aspects are well known and the alternative approaches to implement them are pretty well understood in the area of "classical" monolithically implemented information systems. This picture changes significantly, however, when flexible POIS have to be realized. Firstly, such systems are typically no longer implemented in a monolithic fashion. Instead, the information system consists of separate, individually invokable application functions ("services") which interact with each other according to a process schema which is executed by a process engine. In principle, each process schema constitutes an own schema-specific RMS which regulates which users (based on their roles, organizational units they belong to, etc.) can execute which process steps. If the processes are static, i.e. all process instances execute according to the process schema without any deviations, then there is not much difference to the "classical" monolithic information systems, at first glance. However, in many cases these application functions come from different (and often even foreign) sources and have not been designed to cooperate with a centralized RMS or with a reference monitor.

The situation becomes even worse, if we have to deal with flexible POIS, i.e. systems which may deviate from the pre-planned execution sequence by inserting new process steps, deleting process steps, or moving process steps to another place in the process schema. Furthermore it is possible to individually change the pre-planned actor assignments at the instance level. In this case every process instance now constitutes an independent instance-specific RMS (see Fig. 1. Instance rights can diverge from the schema rights.

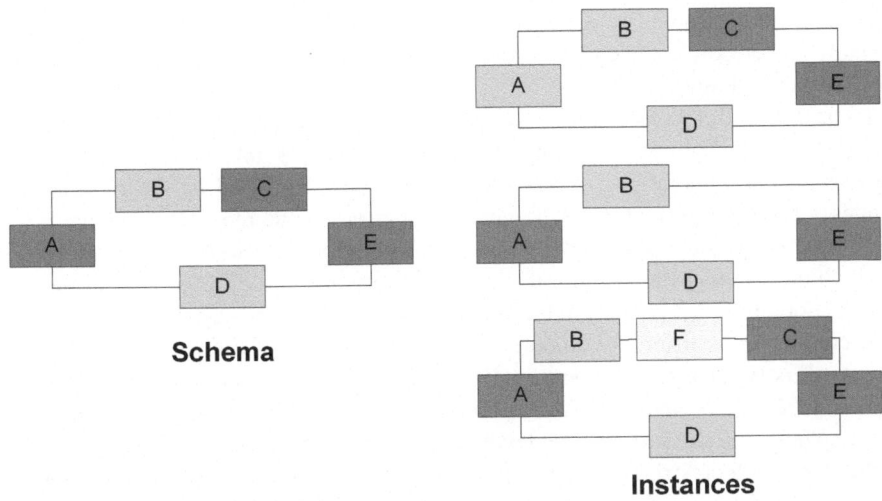

Fig. 1. Instance rights can diverge from the schema rights

Another problem is the granularity of rights used in contemporary POIS. Usually, actor assignment expressions are used to describe for a given step who is authorized to execute this step. The expression `orgunit = "radiology" AND role = "doctor"` would mean that the actor (user) has to be a doctor from the radiology department. As mentioned above, application functions are associated with these steps. Thus when executing this process step the actor is allowed to use the complete functionality this application provides. This would cause no problem if these application functions would be limited to the task to be performed. We will address this issue in the following section.

The rest of the paper is organized as follows: Section 3 discusses challenges of RMS in the context of component-oriented composition of POIS. Section 4 deals with constraints. In section 5 storage-aspects of rights information are treated and section 6 deals with the administration of RMS. Then we want to give a short introduction in the solutions we want to integrate in today's POIS. This solution is introduced in section 7. Finally, section 8 gives a short conclusion of the paper.

3 Components

As mentioned above, actor assignment expressions are used in POIS to decide which users are authorized to execute a certain process step.

Assuming the user's task is to view a certain document and, therefore, a general purpose text processor like Word for Windows, for example, is associated as application function. Once invoked, such a program would in many cases not only allow to read a document, but also allow to modify it, to store it, to print it, or to even invoke some macros which may do very strange things. This is no specific problem of such a text processor but is rather typical for most application functions or components. They are usually implemented in a generic way to support a broad variety of application domains. As a consequence the problem arises how to constrain such application functions accordingly.

To deal with this kind of problem component systems like .NET [5] offer the possibility to implement some component specific access rights management. This is done by special annotations in the source code which are used by the runtime environment of the component to check whether the user has the required privileges.

Using this approach the implementer of the component decides at the source code level which privileges are required for a user to invoke certain functions of the component. For POIS which may integrate a large variety of different application functions coming from different components, this approach is problematic for at least two reasons: Firstly, every component implementer introduces names for privileges independent from others (name inflation problem), and secondly, changes to the RMS will require changes at the source code level of the affected components in many cases.

Similar problems arise when using security frameworks like JAAS [20]. JAAS is based on PAM (*Pluggable Authentication Module* [21]) and mainly supports configurable user authentication. Thus like in .NET the method used for authenticating users must not be hard coded in the source code. Instead, the concrete authentication mechanism (e.g. by password, fingerprint or even retina scan) is implemented by individual plugins. Which concrete plugin should be used for identifying the user can be chosen in configuration files by the administrator. On the other hand JAAS also extends the Java 2 security policies so that not only the code origin but also the current user can be taken into account when the security manager needs to decide whether access from the sandbox to system resources is granted or not [19]. Although not intended to, this can be used to secure internal application functions. But therefore the developer has to encapsulate the different functions in separate classes implementing special interfaces. This leads to name inflation problems and to a quite bad system design as well.

Therefore, such approaches are not suitable for POIS. Instead a (logically) centralized approach is needed. Today, this works only in environments where all components come from the same vendor. A general vendor-independent approach is still missing.

Web services are a variant of this component oriented software development and, thus, have similar problems, in principle. So far, the main focus of web services (WS) is on establishing trustful communication and on Quality-of-Service (QoS) aspects, however. Standards such as WS-Security [6] and its extension WS-Trust [7] are addressing such issues. A centralized approach to method-based access rights management is only discussed in [8]. [8] suggests a layer model. Rights at the bottom layer are associated with the functions provided by the WS. A set of functions can be bundled into rights packages, so-called "keys", and passed on to the next higher level. There these keys can be combined to new keys and propagated to next higher level and so on. However, changes performed at lower levels have significant effects on the roles defined in higher levels. Changes to the rights at higher levels usually lead to major changes of keys at lower levels.

A problem, especially in the clinical domain, may be that the permission for a given user to invoke a certain application function may be limited to a certain type of process and perhaps even to a certain place within this process. E.g., the function to administer an examination may only be allowed within treatment and diagnostic processes but not stand-alone. To read and modify a document (using a certain application function) may be allowed during some creation phase but may be forbidden after the document has been published. In addition, the permission to execute this function may be even dependent on the data or document to be accessed. In total, this means that a rather fine-granular access rights management is required.

4 Storage Aspects

As described above, ACLs are typically used to store (parts of) the rights matrix in a compact manner. Unfortunately, this approach is not very satisfying in the POIS context, because efficient access along both dimensions is necessary. Whenever responsibilities of users are changed, users leave the organization or certain structural changes in the organization happen, one has to determine which objects (and thus their ACLs) are affected. Without adequate support of the user → object dimension this will result in an exhaustive search through all ACLs. And there may be many of them! Because fine-granular access rights management is needed (see above), it's not sufficient to store just one ACL per object (application function). One has also to discriminate in which process schema and at what position this application function is used. This, in turn, increases the number of ACLs significantly because a component may easily have hundreds of occurrences of this kind.

The situation worsens when flexibility comes into the game, i.e. new process steps are inserted at process instance level. Modifying process instance by inserting a new step, for example, means to dynamically create a new, individual process schema. This, in turn, leads to a new ACL because access rights have to associated with this step. As a consequence, one has to maintain a very dynamic and potentially large rights matrix. Solutions have to be found for both, efficient access along both dimensions and for adequate storage representations. Efficient access becomes a very important factor here, because the number of users which can modify rights increases in such flexible environments significantly (see introduction) and thus the number of accesses to the RMS will very likely significantly increase as well.

A real world example of this problem can be found in the EU project Webocrat [9], where an e-government solution is being developed. In small installations, such as single municipalities, efficient access to the rights matrix is not a big problem. However, in large application domains like a whole county with a lot of users who interact with the RMS, the realization of efficient access is a big problem. One approach under investigation here is whether some kind of intelligent caching can help to improve the situation [10].

5 Constraints

In addition to simple rights, like execute permissions, an RMS has usually also to obey several kinds of dependencies or constraints. *Separation of duties*, where the

execution of a certain step by a certain user restricts the set of actors which are allowed to execute subsequent steps, is an example for such a constraint. Another example is *binding,* where a set of process steps has to be executed by the same actor.
This means that once the first step of this set is executed by a certain actor, "binding"
for the other steps to this set is fixed.

Such constraints may also span multiple processes. This poses new challenges in
case of flexible processes. If process steps are inserted, deleted, or moved to another
position at the process instance level, it must be immediately (and efficiently) checked
that no such constraint is violated. The constraints specified at build time must be also
valid at runtime. New constraints have to be integrated and need to be checked if they
stay in conflict to the yet specified ones. Another problem occurs in the context of
process modifications: when inserting new process steps the user is only allowed to
assign a subset of the services available in the system. For example an employee adds
a new step in a process. If he is not responsible for financial transactions assigning
financial services should not be allowed in order to prevent e.g. financial transactions
from the companies account to the account of the employee.

This kind of problems is partly addressed by W-RBAC (Workflow - Role Based
Access Control) [11]. In W0-RBAC constraints like separation of duties as well as
binding across instance boundaries can be defined. The extension W1-RBAC addresses the selective replacement and adaptation of conditions to correct modeling
errors at runtime or to handle unforeseen cases. How to support these constraints at
the process instance level, especially in the context of flexible POIS is not addressed
and will be a major challenge.

Time-based restrictions of rights are considered in GTRBAC (Generalized Temporal Role Based Access control) [12][13]. The system provides a comprehensive set of
time-based conditions which cover arbitrary periodic time periods as well as single
events. Conditions can be related to fixed points in time or to events. The approach is
general and not related to processes. The application of these concepts to POIS and to
perform the necessary checking at the process instance level is a non-trivial task,
especially when flexibility comes into the game.

6   Administration

As can be seen from the discussion above, the definition of appropriate types and levels
of rights and constraints on the one side and their association with users and application
functions (and their different occurrences) on the other side is a non trivial task. Typically, some kind of structuring or leveling is used to make this task manageable. On the
users' side usually organizational units (departments, projects,...) and roles are used to
form groups of users having the same rights to perform certain tasks. On the objects'
side usually hierarchical structures are used to form appropriate groups.

Systems which address the users' aspects are ARBAC97 [14] and ARBAC02 [15]
(Administrative Role Based Access Control) and the extension RBACAM (RBAC
Administration Model) [16]. ARBAC97 distinguishes between three elements of
descriptions: users, roles, and permissions. Users can be associated with roles, roles
can be associated with other roles (in order to form hierarchies), and roles can be
associated with permissions. ARBAC02 extends this approach by introducing

organizational structures and hierarchies of permissions. RBACAM (RBAC admini-stration model) complements ARBACxx by adding verification facilities to check the modeled rights. These approaches point in the right direction. However, it remains to be elaborated how these concepts can be applied to flexible POIS and how the support of constraints can be integrated in such an RMS.

Using hierarchical structuring of objects in order to simplify access rights manage-ment is a well known principle and used, for example, in the directory structure of file systems. One single entry at the parent directory level is sufficient to set the access rights for all files in that directory. SAM Jupiter [17] simply applies this principle to data objects, with finer granularity. The grouping is done here according to responsi-bilities. It is rather unclear, however, to what extend this approach can be utilized to simplify access rights management in POIS.

Another important aspect are organizational changes and their impact on access rights management. When departments are closed, outsourced, or combined with other departments, actor assignment expressions at the process schema as well as at the process instance level have to be adjusted accordingly. A detailed discussion of this problem can be found in [18].

All approaches introduced here require a deep understanding of the RMS and knowledge from the user which privileges and permissions exists and how they have to be associated with users, roles, assignment expressions, and application functions such that they show the desired effects. With the increased number of persons in flexible POIS who interact with the RMS, user interface aspects become a very im-portant issue. Only with a proper model, with few and easy to understand basic con-cepts, one has a chance to enable end-users to deal with access rights managements in a safe way.

7   Short Introduction to ARMS

All challenges presented so far have no adequate solution in today's RMS. We propose an integration of all these topics in one approach we call it ARMS (adaptive rights man-agement system). Here, we want to introduce the concepts we have considered so far.

In RMS for process oriented information systems like WRBAC [11] or in IBMs MQ Workflow the arrangement of security policies is managed as in Figure 1. The process is created and (as already mentioned above) an actor assignment is the basis for the access rights management. The first problem about this solution is that the actor assi-gnment doesn't have to be equivalent with the rights specified at a process step.

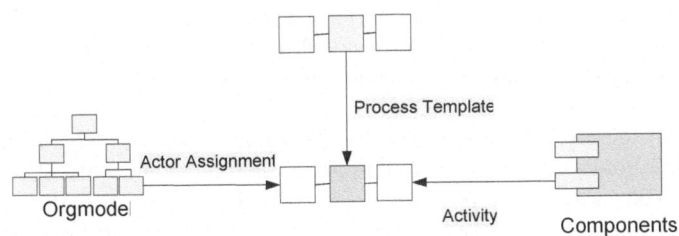

Fig. 1. Security policies in process oriented information systems

The second one is that an actor assignment is done at build time of the process and cannot be changed at runtime. This means the rights information in such systems is very static. One exception is the concept of W1-RBAC in WRBAC that makes it possible to modify and change access rules at runtime.

Therefore ARMS shall be different from other systems and divide the access rights management in two phases. Phase one is the build time phase. In this phase a lot of validation of the modeled rights are carried out to support the administrator or process modeler. This helps to deal with defined constraints or detecting actor assignments that do not fit the rights basis.

For example if there is a constraint that a service only can be used by actors out of the group of permanent employees and if a process modeler makes actor assignments that comprises a person that do not fit this constraint ARMS gives the instruction to change the assignment towards a correct one. ARMS therefore resolves the assignment and tells the user which actors mustn't be included or must be included because of other constraints. This could be the case if an application should always be supervised by a special employee, for example.

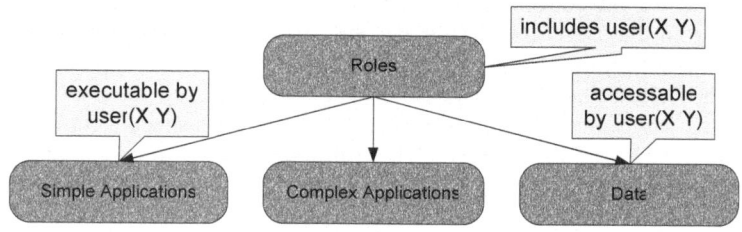

Fig. 2. Different positions for rights validation at build time

There are many other constraints on groups (users that have to be integrated in groups with a special kind of user) or data (data is accessible to users or not). In ARMS all these could be efficient checked at build time. We need these validations at build time to simplify the work of the process designer (see Figure 3) and avoid unnecessary runtime errors.

Phase two is the runtime phase. When thinking of flexible POIS supporting deviations at the process instance level, these changes are often done by normal employees. Therefore these checks are absolutely required to support the untrained user in the interaction with the RMS. This leads to a great usability and security problem which we want to avoid with these validation so the user cannot make "wrong" decisions.

So at process runtime extensive checks have to be done and although these checks need much time to be validated they must not slow down process execution. This will be solved through the calculation of special checksums on the constraints that can be checked faster.

The question therefore is which user has the right to perform which modification on the process? Solutions for this problem are a clear separation of concerns, which guarantees that no access rule is violated. In ARMS we want to realize this with a combination of roles of authorized persons and the abilities stored in an organization model. Users are assigned to roles which have different rights like in other systems.

Additional to these roles the user has an ability field assigned to him where special abilities can be stored to extend a role.

For realizing all the different security features mentioned so far just preventing the execution of a service for a certain user is not enough. For some of them the application even must have the opportunity to interact with the ARMS – therefore we have divided the rights that can be specified for a component in several layers (see Figure. 3) according to the functionalities of the service and the kind of security which should be expressed.

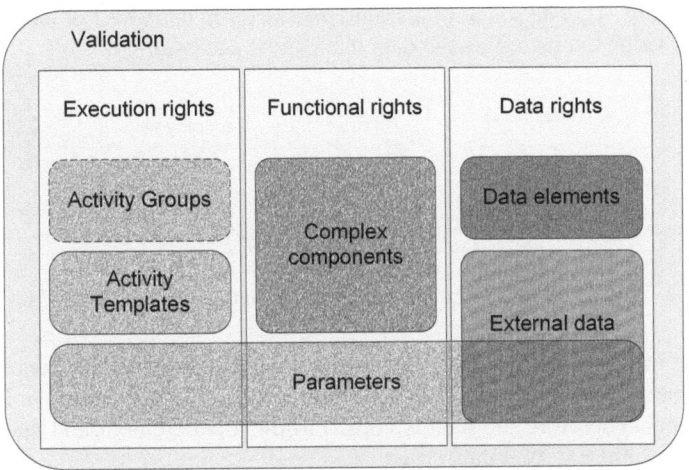

Fig. 3. Layer architecture for component integration into ARMS

The problem of the components is that they may come from different sources and need to be integrated in a complex environment with many boundary conditions. The main problem therefore is that the developer of the application should as far as possible not be engaged with the integration of his component into ARMS. So our multi layer environment makes it possible to keep the developer away from many of the access rights management activities.

At the execution rights layer simple components (i.e. ones which only provide the necessary functionality for the respective process step) can be integrated in ARMS by using Activity Templates (AT). An AT represents exactly one service function in a POIS. In such an AT the default rights information is stored and ARMS uses this information when performing its checks. Therefore the component developer doesn't have to integrate any special, security related functionality. Activity Groups are only a hierarchisation of ATs to simplify administration.

However components often not only provide the necessary functionality. In fact they often try to support different scenarios (like Word described above). Such components we call complex components, because the integration might not be done without any further work of the component developer.

One can distinguish two kinds of complex components. The first one supports parameterization so the needed functionality can be chosen before the call is actually made. In ARMS we support such components also by using ATs. Therefore the number of ATs for one component is not limited. Functional parameters can be hard coded in an AT which leads to at least one AT per component specification. With this no security related application logic by the developer is needed.

The second kind of complex components offers a lot of functionality when the component is running. But they not support to limit the available functions by parameterization of the component call. In order to integrate such components into ARMS it is necessary that the developer does some extra work. At the moment we are working on an interface to make this integration also simple for the developer.

The integration of data in the system is mainly done by parameterization of applications (see Fig. 4). The scope of a single parameter is narrowed or exclusive selected to a special value to customize the data that can be accessed.

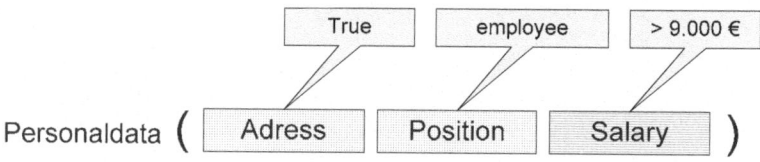

Fig. 4. Parameterization of methods to integrate data in the RMS

Through this customization the data from foreign sources can be encapsulated and integrated in the RMS. Alternatively the data can be integrated into the POIS through data elements. Therefore the data is copied to this data element and needs to be synchronized with the external data structure.

Additionally to the integration of the applications is the number of potentially used applications in a POIS. In a new system there probably are only a few applications installed and used. During the lifecycle of the system the number of components increases not as the number of stored processes schemas and instances from these schemata's increases. As mentioned above this data cannot be stored in ACLs or Capability lists so we have to split the classical matrix into instance specific matrices which makes it possible to handle the stored information. These instance matrices only store the process instance relevant data. Only components and users which are assigned to the instance are mentioned in this table.

As a result we can say ARMS shall become a prototype for adaptive POIS-RMS which needs to address all here presented challenges. We therefore have to extend our current approach and to finish the implementation of our current concepts.

8   Conclusion

The discussion above has shown that access rights management is a complex task with many facets and many open questions in the context of flexible POIS have still be to solved. We then give a short introduction in the construction of a solution we want to implement. The main purpose of this paper was to raise awareness of these issues and show that solutions for these problems must be found in the near future.

We are working on these solutions in the context of ARMS and will focus our work on the component integration and the efficient storage with included validation of the rights information. When these challenges are mastered we can go into further details of the administration aspects.

References

[1] Dadam, P., Reichert, M.: Towards a New Dimension in Clinical Information Processing. In: Proc. MIE2000/GMDS 2000, Hannover, September 2000, pp. 295–301. IOS Press, Amsterdam (2000)

[2] Dadam, P., Reichert, M., Kuhn, K.: Clinical Workflows – The Killer Application for Process-oriented Information Systems? In: Abramowicz, W., Orlowska, M. (eds.) Proc. 4th Int'l Conf. on Business Information Systems BIS 2000, Poznan, Poland, April 2000, pp. 36–59. Springer, London (2000)

[3] Reichert, M., Rinderle, S., Kreher, U., Dadam, P.: Adaptive Process Management with ADEPT2. In: Proc. Int'l Conf. on Data Engineering, ICDE 2005, Tokyo, Demo Session, April 2005, pp. 1113–1114 (2005)

[4] Anderson, J.P.: Computer security technology study ESD-TR-73-51, vol. 2.

[5] Microsoft Library: Security in the .NET Framework (2007),
http://msdn2.microsoft.com/en-us/library/fkytk30f.aspx

[6] Oasis WSS: SOAP Message Security, http://docs.oasis-open.org/wss/2004/01/oasis-200401-wss-soap-message-security-1.0.pdf

[7] Oasis WS-Trust 1.3, http://docs.oasis-open.org/ws-sx/ws-trust/200512/ws-trust-1.3-os.html

[8] Payne, C., Thomson, D., Bogle, J., O'Brien, R.: Napoleon: A Recipe for Workflow, Computer Security Application Conference, p. 134 (1999)

[9] Dridi, F., Pernul, B.M., Pernul, G.: Administration of an RBAC system. In: Proceedings of the 37th Hawaii International Conference on System Sciences – 2004 (2004)

[10] Kern, A., Kuhlmann, M., Kuropka, R., Ruthert, A.: A meta model for authorisations in application security systems and their integration into RBAC administration. In: ACM symposium on Access control models and technologies, New York, pp. 87–96 (2004)

[11] Wainer, J., Barthelmess, P., Kumar, A.: W-RBAC – A workflow security model incorporating controlled overriding of constraints. International Journal of Cooperative Information Systems, 455–485 (2003)

[12] Joshi J. B. D., Bertino E., Latif U., Ghafoor A.: Generalized Temporal Role Based Access Contol Model (GTRBAC) Part 1 - Specification and Modeling, Cerias Tech Report 2001-47

[13] Joshi J. B. D., Bertino E., Latif U., Ghafoor A.: Generalized Temporal Role Based Access Contol Model (GTRBAC) Part 2 – Expressiveness and Design Issues, Cerias Tech Report 2003-01

[14] Sandhu, R., Bhamidipati, V., Munawer, Q.: The ARBAC97 Model for Role-Based Administration of Roles. ACM Transactions on Information and System Security 2(1), 105–135 (1999)

[15] Oh, S., Sandhu, R.: A model for role administration using organization structure. In: Proceedings of the seventh ACM symposium on Access control models and technologies, Monterey California, pp. 155–162 (2002)

[16] Jiong, Q., Chen-hua, M., Jian-wei, Y., Jin-xiang, D.: Research and Implementation of Role-Based RBAC Administration Model. In: The Fifth International Conference on Computer and Information Technology (CIT 2005) (2005)

[17] Kern, A., Schaad, A., Moffett, J.: An Administration Concept for the Enterprise Role-Based Access Control Model, ACM, SACMAT 2003, Como, Italy, June 2–3, 2003, pp. 3–11 (2003)

[18] Weber, B., Reichert, M., Wild, W., Rinderle, S.: Balancing Flexibility and Security in Adaptive Process Management Systems. In: Proc. 13th Int'l Conf. on Cooperative Information Systems, Agia Napa, November 2005, pp. 59–76 (2005)

[19] Middendorf, S., Singer, R., Heid, J.: Java – Programmierhandbuch und Referenz füie Java-2-Plattform, Standard Edition, 3rd edn (2002) (last visited on January 28, 2008), http://www.dpunkt.de/java/Programmieren_mit_Java/Sicherheit/14.html

[20] Lai, C., Gong, L., Koved, L., Nadalin, A., Schemers, R.: User Authentication and Authorization in the Java Platform. In: Proc. 15th Annual Computer Security Applications Conference, Phoenix, AZ, December 1999, pp. 285–290 (1999)

[21] Morgan, A.G., Kukuk, T.: The Linux-PAM System Administrators' Guide (2008) (last visited on May 8, 2008), http://www.kernel.org/pub/linux/libs/pam/Linux-PAM-html/Linux-PAM_SAG.html

An Efficient e-Commerce Fair Exchange Protocol That Encourages Customer and Merchant to Be Honest

Abdullah Alaraj and Malcolm Munro

Department of computer science
Durham University, the UK
{a.m.alaraj,malcolm.munro}@durham.ac.uk

Abstract. A new e-Commerce fair exchange protocol is presented in this paper. The protocol is for exchanging payment with digital product (such as computer software) between customer (C) and merchant (M). It makes use of Trusted Third Party (TTP) but its use is kept to minimum when disputes arise. In this respect it is an optimistic fair exchange protocol. A new idea, in which if the parties are willing to exchange then they are encouraged to be honest, is originated in this protocol. The protocol has the following features: (1) It comprises four messages to be exchanged between C and M in the exchange phase; (2) It guarantees strong fairness for both C and M so that by the end of executing the protocol both C and M will have each other's items or no one has got anything; (3) It allows both parties (C and M) to check the correctness of the item of the other party before they send their item; (4) It resolves disputes automatically online by the help of the Trusted Third Party (TTP); and (5) The proposed protocol is efficient in that it has a low number of modular exponentiations (which is the most expensive operations) when compared to other protocols in the literature.

1 Introduction

When exchanging digital products and payments over the internet, Merchants (M) and Customers (C) want to be sure that the exchange will be fair thus ensuring that C will receive the right product and M will get the correct payment. Fair exchange also means that if either or both parties are dishonest then both parties will receive each other's items or neither will receive anything.

There are three main processes involved in fair exchange of digital products and payments:

1. C sends the payment to M;
2. M send the digital product to C; and
3. Dispute resolution if something goes wrong.

Processes 1 and 2 can be carried out in either order, so C sends the payment to M and then receives the digital product or M sends the digital product to C and then receives the payment.

This paper will firstly discuss the literature on related work, and then present the details of the ECMH (Encouraging Customer and Merchant Honesty) protocol, and then evaluates the protocol by analysing different scenarios and comparing it against existing protocols.

M.D. Harrison and M.-A. Sujan (Eds.): SAFECOMP 2008, LNCS 5219, pp. 193–206, 2008.
© Springer-Verlag Berlin Heidelberg 2008

2 Review of Literature

There is a number of fair exchange protocols described in the literature [5, 7, 10, 9, 3, 4, 1, 2, and 12].

Fair exchange protocols can be divided into two types: those that do not involve a trusted Third party (TTP); and those that do. This paper is concerned with the later type where the TTP takes all or some of the following roles: (1) ensures fairness in the exchange, (2) acts as certificate authority that is trusted by all parties, and (3) resolves disputes and/or validates items.

Protocols that involve a TTP can be of two types. The first type (such as [6]) uses a TTP for delivering the exchanged items. This involves each party sending their item to a TTP and the TTP delivering them to the parties. Involving a TTP will guarantee the fair exchange of items, but it has some drawbacks. The TTP could be the source of a bottleneck [10], it must always be available [8], and if the TTP crashes, the protocol will not deliver the items properly.

The second type is the protocols (such as [1, 2, 12, 5, 7, 10, 9, and 3]) where there is minimal use of the TTP, usually when something goes wrong. In these protocols the two parties directly exchange their items and the TTP only gets involved to resolve disputes. This type of protocol is called "Optimistic fair exchange protocols" [3].

The Ray et al [9] optimistic fair exchange protocol allows each party to verify whether the item they are going to receive from the other party is indeed the item they want, and this is done before receiving the item. A merchant (M) uploads the digital product to a TTP who encrypts it; the customer (C) downloads the encrypted digital product from the TTP to compare it with the digital product that will be received from M. The actual interaction between C and M in this protocol consists of four messages: C sends to M the first message which contains the purchase order and payment token that are encrypted; M sends a message to C which includes the encrypted digital product; C sends a message to M which includes the decryption key for the payment token; and finally M sends the decryption key of the product. If C has a dispute, then C contacts the TTP to resolve it. C needs to download the product twice (from TTP and then from M), so this will be a communication overhead.

Asokan et al [3] propose a generic optimistic fair exchange protocol that is suitable for exchanging signatures, confidential data or payments. If the items to be exchanged are payment and a digital product, the protocol can be explained as follows: a merchant (M) and a customer (C) promise to exchange their items using 2 messages; C sends the payment to M; M sends the digital product to C. If anything goes wrong then the TTP will cancel the payment. If the TTP is not able to do so then the TTP can provide an affidavit proof to be used in court to resolve the disputes. This protocol seems to be not the best solution for exchanging digital products fairly as TTP will not be able to resolve the disputes online.

The Zhang et al [12] fair exchange protocol for exchanging two valuable documents (the two documents can be a payment and digital product) comprises of four messages. The two parties involved in the protocol exchange their encrypted documents in the first two messages and then exchange the keys to decrypt them. The protocol is based on the idea of having each party verify the correctness of the key used to encrypt the document without seeing the key itself. The TTP can be contacted to recover the key if one party misbehaved.

The Alaraj and Munro [1, 15] protocol for exchanging digital products and payments consists of three messages to be exchanged between the customer (C) and the merchant (M). The messages are: M sends the encrypted digital product and its certificate to C; C verifies it and if satisfied sends to M the payment that is encrypted using a key that M already has; finally, the decryption key is sent to C by M when M is satisfied with the payment. If there is any dispute, the TTP will be contacted. This protocol enforces the customer to be honest because they cannot gain anything by being dishonest. Alaraj and Munro [2] extended the idea in this protocol to enforce the merchant to be honest. In this protocol C starts the exchange by sending an encrypted payment and its certificate to M, who verifies it and if satisfied, sends to C the digital product that is encrypted using a key that C already has. Finally, the decryption key is sent to M by C when C is satisfied. If there is any dispute, the TTP will be contacted to resolve the dispute.

In this paper, we applied the techniques used in [1, 15, 2] to encourage both C and M to be honest. This paper presents an optimistic fair exchange protocol for exchanging payment and digital product between customer and merchant. The proposed protocol allows both parties (customer and merchant) to check the correctness of the item of the other party before they send their items to them. Therefore, both parties are encouraged to be honest. The proposed protocol overcomes the drawbacks of the protocols in the literature.

3   Encouraging Customer and Merchant Honesty (ECMH) Protocol

3.1   Notations

This section defines the notation used in this paper, some of which are similar to the ones appear in [7].

- C: Customer
- M: Merchant
- TTP: Trusted Third Party which is a party neither M nor C that is trusted by all parties. TTP will not collude with any other party
- D: Digital product
- CA: Certificate Authority
- CB: the customer's Bank
- $desc.$: description of digital product
- $h(X)$: a strong-collision-resistant one-way hash function, such as SHA-1 [14]
- $pkx = (ex, nx)$: RSA Public Key [13] of the party x, where nx is a public RSA modulus and ex is a public exponent
- $skx = (dx, nx)$: RSA Private Key [13] of the party x, where nx is a public RSA modulus and dx is a private exponent
- kx: a symmetric key generated by x
- $P\text{-}Cert$: Payment Certificate that is issued by CB. $P\text{-}Cert$ contents are:
  - $amount$: the amount of payment
  - hP: hash value of payment

- o *heP:* hash value of encrypted payment with *kc*
- o *heKc:* hash value of encrypted *kc*
- o *Sig.CB:* CB's signature on *P-Cert*
- *D-Cert*: Digital-product Certificate that is issued by CA. *D-Cert* contents are:
  - o Price: price of D
  - o *d*: Description of D
  - o *hD*: hash value of D
  - o *heD*: hash value of encrypted D with *km*
  - o *heKm*: hash value of encrypted *km*
  - o *Sig.CA*: CA's signature on *D-Cert*
- *C.mt*: the certificate for the shared public key between M and TTP; *C.mt* is issued by TTP. A standard X.509 certificate is used to implement *C.mt* [11]
- *C.ct*: the certificate for the shared public key between C and TTP; *C.ct* is issued by TTP. A standard X.509 certificate is used to implement *C.ct* [11]
- *enc.pkx(Y)*: RSA encryption of Y using the public key pkx (ex, nx). That is,

  enc.pkx(Y) = Y^{ex} mod nx = Z
- enc.skx(Z): RSA decryption of Z using the private key skx (dx, nx). That is,

  enc.skx(Z) = Z^{dx} mod nx = Y
- *enc.kx(Y)* : encryption of Y using a symmetric key kx (kx can also be used for decrypting enc.kx(Y))
- *Sig.A (X)*: RSA signature of the party A on X i.e. encrypting the hash value of X using the private key skA (dA, nA) as follows:

  Sig. A (X) = $\left(h(x)\right)^{dA}$ mod nA
- A → B: X: A sends message X to B
- X + Y: concatenation of X and Y
- ECMH protocol: Encouraging Customer and Merchant Honesty protocol which is the protocol presented in this paper

3.2 Protocol Description

This protocol is for exchanging a digital product D with a payment. It is assumed that the payment in the protocol is in the form of a payment order that is issued and signed by a customer's bank and specifies the amount of payment to be paid, the payee and the payer. Double spending of the same payment is assumed to be detected and therefore will not occur. It is assumed that the communication channels between all parties (TTP, M and C) are resilient i.e. all sent messages will be received by their intended recipients [9]. C and M will agree on the TTP to be used in both the pre-exchange phase (by C) and the dispute resolution (by M) before they start the protocol.

The trustworthiness of C is governed by two things which are the payment certificate (P-Cert) issued by CB and the public key certificate (C.ct) issued by TTP. Therefore, the payment that will be sent by C is certified by CB; and the public key to be used by C to encrypt the key used to encrypt this payment is certified by TTP. The trustworthiness of M is also governed by two things which are the digital product certificate (D-Cert) issued by CA and the public key certificate (C.mt) issued by TTP. Therefore, the digital product that will be sent by M is certified by CA; and the public key to be used by M to encrypt the key used to encrypt this digital product is certified by TTP. Therefore, this protocol encourages both C and M to be honest by sending correct items as each party will be able to detect if the received item is incorrect.

The scenario of this protocol is like C and M exchanging their encrypted items (payment and digital product) and their certificates. These encrypted items and their certificates will test the trustworthy of each party. If the parties found that the other party is trustworthy then they will complete the exchange otherwise they abort it.

3.2.1 Pre-exchange Phase

In the pre-exchange phase (Fig 1), C needs to get the certificate C.ct of the shared public key from TTP to be used to encrypt the key used to encrypt the payment (PE-b-M1 of Fig1). C also needs to get the payment and its certificate P-Cert from CB (PE-b-M2 of Fig1). The P-Cert is unique for each transaction (completed exchange) because the payment can only be used once. Also in the pre-exchange phase M needs to get the certificate C.mt of the shared public key from TTP to be used to encrypt the key used to encrypt D (PE-a-M1 of Fig1). M also needs to get the digital product (D) and its certificate D-Cert from CA (PE-a-M2 of Fig1), (the CA can be thought of as the producer of the digital product).

In this protocol, there are two public keys to be shared. The first one is shared between TTP and C. The other one is shared between TTP and M. The way in which these keys are shared is as follows.

- Each party (C, M and TTP) has its own public and private keys. The TTP's public key is denoted as $pkt = (et, nt)$ and its corresponding private key is denoted as $skt = (dt, nt)$. While C's public key is denoted as $pkc = (ec, nc)$ and its corresponding private key is denoted as $skc = (dc, nc)$; and M's public key is denoted as $pkm = (em, nm)$ and its corresponding private key is denoted as $skm = (dm, nm)$.

- The shared public key between C and TTP is denoted as $pkct = (ect, nct)$ and its corresponding private key is denoted as $skct = (dct, nct)$. The nct is a product of two distinct large primes chosen by TTP.

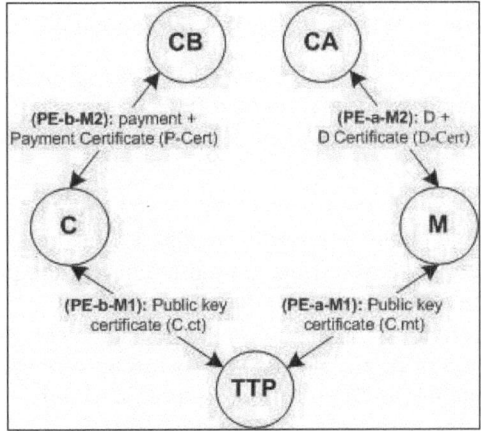

Fig. 1. Pre-exchange phase

- The shared public key between M and TTP is denoted as pkmt = (emt, nmt) and its corresponding private key is denoted as skmt = (dmt, nmt). The nmt is a product of two distinct large primes chosen by TTP

3.2.2 The Exchange Phase

It is assumed that the exchange phase will take place after C finds the wanted digital product (D) with M (either in M's website or through the search engines). It is also assumed that this phase will take place after C and M agree on the digital product and negotiated the price. Hence this phase is about the actual exchange of payment and digital product D.

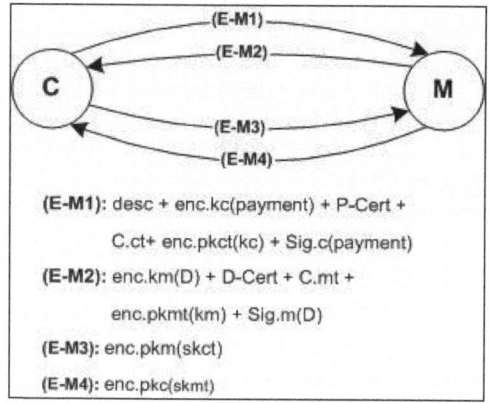

Fig. 2. The exchange phase

There are four messages to be exchanged between M and C in the exchange phase (Fig 2). These four messages are as follows.

[E-M1] C → M: desc + enc.kc(payment) + P-Cert + C.ct+ enc.pkct(kc) + Sig.c(payment)

C sends to M message E-M1 which contains the following:

- **desc:** specifies what C wants from M i.e. description of D that C wants (the description can be the digital product ID)
- **enc.kc(payment):** the payment that is encrypted with the key *kc*. *kc* is generated by C
- **P-Cert:** the payment certificate that is issued by CB
- **C.ct:** the shared public key certificate that is issued by TTP
- **enc.pkct(kc):** the key *kc* (that is used to encrypt the payment) is encrypted using the shared public key *pkct* that is certified in *C.ct*
- **Sig.c(payment):** C's signature on the payment. This signature can serve as non-repudiation of origin which allows M to be sure that the payment is sent by C. As explained in the notations section, C's signature on payment is the encryption of the hash value of payment using C's private key *skc*

[E-M2] **M → C: enc.km(D) + D-Cert + C.mt + enc.pkmt(km) + Sig.m(D)**
On receiving message E-M1 from C, M checks the correctness of *enc.kc(payment)*, *enc.pkct(kc)*, *P-Cert* and *C.ct*. The correctness of *P-Cert* can be checked by verifying CB's signature on *P-Cert*. Also the correctness of *C.ct* can be checked by verifying TTP's signature on *C.ct*.

To check that the encrypted payment is correct, M needs to check three things (1) the amount field in P-Cert against the price field in D-Cert that M has. This is to make sure that the payment meets the asked price; (2) the payment itself; and (3) the encrypted payment with *kc* i.e. *enc.kc(payment)*.

To check the correctness of payment, M needs to get the hash value of payment *(HP)* by decrypting *Sig.c(payment)* using C's public key *pkc* (the public keys of all parties are publicly available) and then compare it with hash value of payment *(hP)* that is included in *P-Cert*. That is, to check the following:

$$HP ?= hP$$

If they are the same then M can be sure that the actual payment is correct.

To check the correctness of the encrypted payment *enc.kc(payment)*, M computes the hash value of *enc.kc(payment)* *(HeP)* and then compare it with the hash value of encrypted payment with *kc* i.e. *heP* which is included in *P-Cert* (note that it is assumed that M will use the same function used by CB to compute the hash value). That is, to check the following:

$$HeP ?= heP$$

If they are the same then M can be sure that C encrypted the payment using *kc* and not another key.

M also needs to check the correctness of *kc* which is used to encrypt payment. To do so, M computes the hash value of *enc.pkct(kc)* *(HeKc)* and then compare it with *heKc* that is included in P-Cert, so M will check the flowing:

$$HeKc ?= heKc$$

If they are the same then M can be sure that the encrypted key is *kc* and not another key. The point here is to make sure that C is honest by sending the key used to encrypt the payment.

Therefore, if all comparisons are correct then, at this point, M will have the following fact. The encrypted payment is correct (i.e. it is the one described in *P-Cert*) and it is indeed encrypted with *kc*. In addition, the encrypted key in *enc.pkct(kc)* is indeed *kc* and not another key. The shared public key *pkct* used to encrypt *kc* is certified by TTP. Therefore, once M got the private key *(skct)* of the shared public key then M will be able to get the payment (by first decrypting *enc.pkct(kc)* to get *kc* and then decrypting *enc.kc(payment)* using *kc*).

Now, it is M's choice to complete the exchange or abort the protocol. If M wants to exchange D for the payment then M sends (in E-M2) the following:

- **enc.km(D):** the digital product D that is encrypted with the key *km* that is generated by M
- **D-Cert:** the digital product certificate that is issued by CA
- **C.mt:** the shared public key certificate that is issued by TTP

- **enc.pkmt(km):** the key *km*, that is used to encrypt D, encrypted using the shared public key *pkmt* that is certified in *C.mt*
- **Sig.m(D):** M's signature on D. This signature can serve as non-repudiation of origin which allows C to be sure that D is sent by M. As explained in the notations section, M's signature on D is the encryption of the hash value of D using M's private key skm

Note that if M decides to abort the transaction after receiving message E-M1 and before sending message E-M2 to C then neither M nor C lose anything.

[E-M3] C → M: enc.pkm(skct)

On receiving message E-M2 from M, C checks the correctness of *enc.km(D)*, *enc.pkmt(km)*, *D-Cert* and *C.mt*. The correctness of *D-Cert* can be checked by verifying CA's signature on *D-Cert*. Also the correctness of *C.mt* can be checked by verifying TTP's signature on *C.mt*.

To check the correctness of D, C needs to check two things which are the digital product *D* itself and the encrypted D with *km* i.e. *enc.km(D)*. Firstly, to check the correctness of D, C needs to get the hash value of D (*HD*) by decrypting *Sig.m(D)* contained in message E-M2 using M's public key *pkm* (the public keys of all parties are publicly available) and then compare it with hash value of D (*hD*) contained in *D-Cert*. That is, to check the following:

$$HD ?= hD$$

If they are the same then C can be sure that the actual D is correct. Secondly, to check the correctness of the encrypted D *enc.km(D)*, C computes the hash value of *enc.pkmt(D)* (*HeD*) and then compare it with the hash value of encrypted D with *km* i.e. *heD* which is contained in *D-Cert* (note that it is assumed that C will use the same function used by CA to compute the hash value) i.e. to check the following:

$$HeD ?= heD$$

If they are the same then C can be sure that M encrypted D using *km* and not another key.

C also needs to check the correctness of *km* which is used to encrypt *D*. To do so, C computes the hash value of *enc.pkmt(km)* (*HeKm*) and then compares it with *heKm* that is included in D-Cert, so C will check the flowing:

$$HeKm ?= heKm$$

If they are compared then C can be sure that the encrypted key is *km* and not another key. The point here is to make sure that M is honest by sending the key used to encrypt D.

Therefore, if all comparisons are correct then, at this point, C will have the following fact. The encrypted D is correct (i.e. it is the one described in *D-Cert*) and it is indeed encrypted with *km*. In addition, the encrypted key in *enc.pkmt(km)* is indeed *km* and not another key. The shared public key *pkmt* used to encrypt *km* is certified by TTP. Therefore, once C got the private key (*skmt*) of the shared public key then C will be able to get D (by first decrypting *enc.pkmt(km)* to get *km* and then decrypting *enc.km(D)* using *km*).

Now, it is C's choice to complete the exchange or abort the protocol. If C wants to exchange the payment for D then C sends to M the decryption key *skct* encrypted using M's public key *pkm* to allow M be able to decrypt the encrypted payment.

Note that C must be sure that the encrypted D matches their requirements as explained earlier, otherwise C will be at risk if they send message E-M3 to M because when C sends to M the decryption key then this means that they are satisfied with E-M2 and hence M will be able to decrypt the payment.

Note that if C decides to abort the transaction after receiving message E-M2 and before sending message E-M3 to M then neither C nor M lose anything. But once C sends message E-M3 to M then the transaction must be completed and the protocol will guarantee that the exchange of payment and D will be fair if the sent items are as they described i.e. the payment matches the price that appears in message E-M2 and also the digital product matches *desc* that appears in message E-M1.

[E-M4] M → C: enc.pkc(skmt)
On receiving message E-M3, M decrypts *enc.pkm(skct)* using M's private key *skm* to get the private key *skct*. Once M got *skct* then they decrypt *enc.pkct(kc)* to get *kc* that can be used to decrypt the encrypted payment received in E-M1.

If C encrypted the payment using different key (i.e. M was not able to decrypt the encrypted payment using *kc*) then M ignores the transaction and aborts the protocol. If however M managed to get the payment correctly then M sends to C in E-M4 the decryption key *skmt* that is encrypted using C's public key.

On receiving message E-M4, C decrypts *enc.pkc(skmt)* using C's private key *skc* to get the private key *skmt*. Once C got *skmt* then they decrypt *enc.pkmt(km)* to get *km* that can be used to decrypt the encrypted D received in E-M2.

If M encrypted D using different key (i.e. C was not able to decrypt the encrypted D using *km*) then C contacts TTP for resolution (as will be explained in the next section). If however C managed to get D correctly then the protocol finishes and the fair exchange of payment and digital product is ensured.

3.2.3 After Exchange (Dispute Resolution)
All disputes requests, if any, will come from C because M will not need to raise disputes as they get the decryption key of the encrypted payment and decrypt it before they send the decryption key of the digital product to C. Therefore, if C has a dispute, the following messages are executed (see Fig 3):

[DR-M1] C → TTP: D-Cert + C.mt + C.ct + enc.pkt(skct) + Sig.m(D)
In case C has a dispute, they need to send to TTP the following: *D-Cert, C.mt, C.ct , enc.pkt(skct)* and M's signature on D that has been received in message E-M2 of the exchange phase.

[DR-M2] TTP → M: enc.pkm(skct)
On receiving message *DR-M1* above, TTP will check the correctness of *D-Cert, C.mt, C.ct* by checking their signatures. If they are correct then TTP will decrypt the signature of M on D. That is, TTP decrypts *Sig.m(D)* to get the hash value of D included in the signature and then compares it with the hash value of D (*hD*) which is included in *D-Cert*. If TTP managed to decrypt *Sig.m(D)* correctly and the two hashes

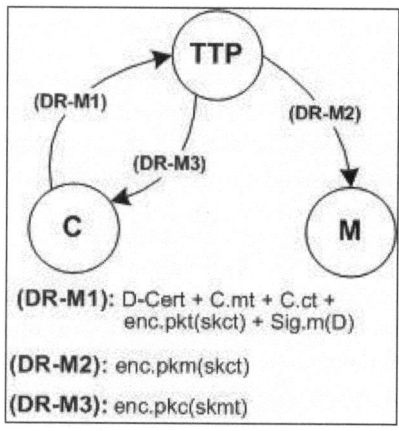

Fig. 3. Dispute resolution

are the same then TTP is sure that M was satisfied with the payment that C sent to them in message E-M1 of the exchange phase. This is because no other party can sign D as it needs M's private key which is only held by M. If M was not satisfied then they would not send *Sig.m(D)* to C in message E-M2. In other words, M will send message E-M2 (which includes *Sig.m(D)*) only if they are sure that the payment sent by C is correct. If TTP found the signature of M is correct then TTP sends to M the decryption key *skct* (encrypted with M's public key) to be used to get *kc* that decrypts the encrypted payment. The reason for sending the decryption key *skct* to M (as M is not the one who raises the dispute) is because C may have not sent the decryption key to M in message E-M3 or has sent incorrect decryption key.

Otherwise, if TTP found that the signature of M is incorrect then TTP sends an abort message to C and nothing will be sent to M.

[DR-M3] **TTP → C: enc.pkc(skmt)**
           OR
           TTP → C: aborts;

This is the same process for message *DR-M2* above, if TTP found that *Sig.m(D)* is correct then TTP sends to C the decryption key *skmt* (encrypted with C's public key) to be used to get *km* that decrypts the encrypted D. Otherwise if *Sig.m(D)* is incorrect then TTP sends an abort message to C.

It is clear that if either C has sent incorrect decryption key *skct* to M or C has not sent the decryption key at all in message E-M3 then C will not get an advantage over M because the TTP will check *DRM1* that C send to the TTP in order to check the signature of M. If the signature is correct then the TTP will send the decryption keys to both parties (C and M) to ensure fairness. Therefore, the fairness is ensured for both C and M. However, if the signature of M is incorrect then the TTP will reject C's request for the dispute.

As can be seen in the dispute resolution phase, the TTP does not need to have both C and M to be involved in order for the dispute to be resolved; rather only the dispu-tant (C in this protocol) and the TTP will be involved. That is, the TTP does not need to contact M to verify whether or not they have received the correct decryption key; rather TTP asks C to provide all evidences and finally will makes the resolution. M will only be contacted by the TTP if the dispute has a resolution. Therefore, this will reduce the number of messages needed to resolve dispute and as a result will reduce the load on the communication channels.

4 The Protocol Analysis

In ECMH protocol, the only party to raise a dispute is C. The following scenarios are presented and studied (Note that after C and M exchange their encrypted items in messages E-M1 and E-M2, they exchange the decryption keys):

- C received a correct decryption key, and M either received incorrect decryption key or has not received the decryption key at all. This case is not applicable in the ECMH protocol because C has to send a correct decryption key to M to be able to receive the correct decryption key from M
- C has either received incorrect decryption key or not received the decryption key at all, and M received the correct decryption key. In this case C will make a dispute to TTP as explained in section 3.2.3
- Both C and M have not received any decryption keys from each other. So, no dispute will be made as both of them have not revealed their items (the decryption keys). This represents the case where C received E-M2 and did not send E-M3 to M or the case where C sends E-M1 to M but M does not send E-M2 to C
- Both C and M have received incorrect items (decryption keys) from each other. That is, C received incorrect decryption key and M received incorrect decryption key. This case is not applicable in the ECMH protocol because C has to send a correct decryption key to be able to receive the correct decryption key from M. So, if M found that the decryption key is incorrect then M will not send to C neither correct the decryption key nor incorrect decryption key
- C received incorrect decryption key and M has not received the decryption key at all. This case is not applicable in the ECMH protocol because C has to send a correct decryption key to M to be able to receive the correct decryption key from M. So, if M has not received the decryption key then M will not send the decryption key at all
- C has not received the decryption key at all and M received incorrect decryption key. This case is normal to occur because if C sent incorrect decryption key then M will not send their decryption key to C. Therefore, if this case occurs then for C to raise a dispute to the TTP, C needs to send to the TTP a correct DR-M1 (see message DR-M1 in section 3.2.3). If C sends the correct DR-M1 to the TTP then the TTP will make a resolution to both C and M. However, if the TTP found that DR-M1 is incorrect then C's dispute will be rejected

It is clear how the design of the ECMH protocol reduces the possibilities for having disputes. Additionally, in the ECMH protocol only C will raise disputes as M will not send their item unless the item of C is correct. As a result, the possibilities for disputes are reduced by preventing them.

In addition to the previous cases, the following cases (scenarios) are studied:

- C disputes to the TTP that they have received incorrect digital product: this scenario is not possible because *D-Cert* guarantees that the digital product is correct; and if C found that the digital product is incorrect or not the same as they wanted then they should have not sent to M the decryption key in E-M3. So, it is C's fault to send to M the decryption key if they have a doubt about the digital product. But once C sends to M the decryption key then this means that they are satisfied with the digital product. Therefore, this scenario will not happen because C knows the rules of the protocol which allow C to check the digital product before they send the decryption key to M; and as a result C will not put themselves at risk

- It is clear that M will not raise a dispute because M will receive from C the decryption key *skct* and get the payment before they send the decryption key *skmt* to C. However, the following scenarios are studied:
  - o M claims that they have received incorrect payment from C: this will not occur because if M received incorrect payment then they will not send the encrypted digital product in message E-M2 and hence no one will get advantage over the other party. However, once M sends message E-M2 to C then this means that they are satisfied with the encrypted payment
  - o M claims that they have not received the decryption key *skct*: this is not applicable in this protocol because if the decryption key is not received then no party is hurt and the fairness is not compromised. The reason for not receiving the decryption key *skct* may be because C is not satisfied with the encrypted digital product
  - o M claims that they have received incorrect decryption key *skct* from C: this is not applicable in this protocol because if the decryption key is incorrect then no party is hurt and the fairness is not compromised as if the *skct* is incorrect then M will not their decryption key (*skmt*) to C.

5 Comparisons

The ECMH protocol presented in this paper has been compared to some of the protocols described in the literature that have the same characteristics in that they are used for exchanging digital products and payments and are based on RSA [13]. Thus ECMH protocol is compared to Ray et al [9] (denoted as Ray protocol) and Zhang et al [12] (denoted as Zha protocol).

The comparisons are made using the following criteria. (1) number of messages in both the exchange and dispute resolution phases, (2) whether or not the TTP needs to hold a copy of an item to be exchanged, (3) whether or not all parties (M and C) will be involved to allow the TTP to resolve any disputes, and (4) number of modular exponentiations in both the exchange and dispute resolution phases. The modular exponentiations are considered to be the most expensive operation [7].

The Ray et al [9] paper did not give details of the dispute resolution phase so the number of messages and the number of modular exponentiations had to be estimated manually. In addition, the number of modular exponentiations for Zha's protocol has also been estimated manually.

As can be seen in Table 1, all protocols have the same number of messages between C and M in the exchange phase. Ray's protocol lets the TTP hold M's item before the exchange between C and M takes place. This requires more storage and security assurance to be added to the TTP's jobs. Additionally, this may compromise the confidentiality of the items to be exchanged.

The Ray protocol requires both parties (C and M) to be contacted by the TTP when one party raises a dispute; whereas in the ECMH protocol and the Zha protocol only the disputant and the TTP will be involved. Involving both parties in dispute resolution would require more messages to be sent and hence more load on the communication channels.

The ECMH protocol has the lowest number of modular exponentiations needed to generate and verify messages in the exchange phase. While ECMH protocol has more modular exponentiations in the dispute resolution phase. However, most of modular exponentiations in ECMH protocol are for adding more security assurances such as encrypting the content of messages to prevent any other party (not those involved in the protocol) from gaining any useful information. This means that, 4 out of 14 modular exponentiations in the exchange phase and 6 out of 9 modular exponentiations in the dispute resolution phase are for adding such security assurances. The implication of this is that if there is an assumption that the channels are secured then the number of modular exponentiations are 10 and 3 for the exchange phase and dispute resolution phase, respectively.

Table 1. Protocols comparisons

	Ray protocol [9]	Zha protocol [12]	ECMH protocol
# messages (exchange phase)	4	4	4
# messages (dispute resolution)	3 to 5	3	3
TTP hold item	Yes	No	No
Both parties are involved in dispute resolution	Yes	No	No
# modular exponentiations (exchange phase)	27	20	14
# modular exponentiations (dispute resolution phase)	5 to 6	6	9

6 Conclusion

A new fair exchange protocol for exchanging digital products and payment has been presented in this paper. It comprises four messages to be exchanged between C and M. The protocol uses certificates that are issued by trusted parties such as a TTP, a CA and a CB. These certificates are *P-Cert* which allows M to check the correctness of *payment*, *D-Cert* which allows C to check the correctness of *D*, *C.mt* which allows C to check the origin of the key used to encrypt the key used to encrypt D and *C.ct* which allows M to check the origin of the key used to encrypt the key used to encrypt the *payment*. The only way in which M might misbehave after receiving the decryption key from C is by sending incorrect decryption key or by not sending it at all. This can be resolved automatically and online by the help of TTP. The protocol guarantees strong fairness for both C and M.

References

1. Alaraj, A., Munro, M.: An e-commerce Fair Exchange Protocol for exchanging Digital Products and Payments. In: Proceedings of IEEE ICDIM 2007, Lyon, pp. 248–253 (October 2007)
2. Alaraj, A., Munro, M.: An Efficient Fair Exchange Protocol that Enforces the Merchant to be Honest. In: Proceedings of IEEE International Conference on Collaborative Computing: Networking, Applications and Worksharing 2007, CollaborateCom 2007, New York, pp.196–202 (November 2007)
3. Asokan, N., Schunter, M., Waidner, M.: Optimistic Protocols for Fair Exchange. In: Proc. Fourth ACM Conf. Computer and Communication Security, Zurich, Switzerland, pp. 8–17 (April 1997)
4. Ben-Or, M., Goldreich, O., Micali, S., Rivest, R.: A Fair Protocol for Signing Contracts. IEEE Trans. Information Theory 36(1), 40–46 (1990)
5. Ezhilchelvan, P., Shrivastava, S.: A Family of Trusted Third Party Based Fair-Exchange Protocols. IEEE transactions on dependable and secure computing 2(4) (October-December 2005)
6. Ketchpel, S.: Transaction Protection for Information Buyers and Sellers. In: Proceedings of the Dartmouth Institute for Advanced Graduate Studies 1995: Electronic Publishing and the Information Superhighway, Boston, USA (1995)
7. Nenadic, A., Zhang, N., Cheetham, B., Goble, C.: RSA-based Certified Delivery of E-Goods Using Verifiable and Recoverable Signature Encryption. Journal of Universal Computer Science 11(1), 175–192 (2005)
8. Pagnia, H., Vogt, H., Gärtner, F.: Fair Exchange. The Computer Journal 46(1) (2003)
9. Ray, I., Ray, I., Narasimhamurthy, N.: An Anonymous and Failure Resilient Fair-Exchange E-Commerce Protocol. Decision Support Systems 39, 267–292 (2005)
10. Ray, I., Ray, I.: An Optimistic Fair Exchange E-Commerce Protocol with Automated Dispute Resolution. In: Bauknecht, K., Madria, S.K., Pernul, G. (eds.) EC-Web 2000. LNCS, vol. 1875, pp. 84–93. Springer, Heidelberg (2000)
11. Public-Key Infrastructure (X.509), The PKIX working group (accessed on 08-06-2007), http://www.ietf.org/html.charters/pkix-charter.html
12. Zhang, N., Shi, Q., Merabti, M., Askwith, R.: Practical and Efficient Fair Document Exchange over Networks. The Journal of Network and Computer Applications, the Elsevier Science Publisher 29(1), 46–61 (2006)
13. Rivest, R., Shamir, A., Adleman, L.: A method for obtaining digital signatures and public-key cryptosystems. Commun. ACM, 120–126 (1978)
14. Ferguson, N., Schneier, B.: Practical cryptography. Wiley, Indianpolis (2003)
15. Alaraj, A., Munro, M.: An e-Commerce Fair Exchange Protocol that Enforces the Customer to be Honest. International Journal of Product Lifecycle Management, IJPLM (to appear)

Creating a Secure Infrastructure for Wireless Diagnostics and Software Updates in Vehicles

Dennis K. Nilsson, Ulf E. Larson, and Erland Jonsson

Department of Computer Science and Engineering
Chalmers University of Technology
SE-412 96 Gothenburg, Sweden
{dennis.nilsson,ulf.larson,erland.jonsson}@chalmers.se

Abstract. A set of guidelines for creating a secure infrastructure for wireless diagnostics and software updates in vehicles is presented. The guidelines are derived from a risk assessment for a wireless infrastructure. From the outcome of the risk assessment, a set of security requirements to counter the identified security risks were developed. The security requirements can be viewed as guidelines to support a secure implementation of the wireless infrastructure. Moreover, we discuss the importance of defining security policies.

Keywords: Infrastructure, vehicle, wireless, security, guidelines, policies.

1 Introduction

This paper presents guidelines for creating a secure infrastructure involving wireless communication for performing diagnostics and software updates in vehicles. It is assumed that both wireless diagnostics and software updates use the same communication channel and security principles. We assume that the security requirements for the communication channel are the same for both wireless updates and wireless diagnostics.

Today, vehicles contain a number of *electronic control units* (ECU). These units are responsible for various functionality in the vehicle, ranging from small tasks such as opening a window or unlocking a door to more advanced functionality such as automatic brake systems and collision warning systems [1]. Each ECU runs its own specific and independent software. As with all software, new improved versions are created to remedy bugs and add new functionality. As new releases of software are available, the customer can update the software for the corresponding ECUs by visiting an authorized service station. The service station employee sets up a wired connection to the vehicle to update the software. The new software is downloaded and flashed to the ROM of the particular ECU, overwriting the old software. In addition to software updates, diagnostics can be performed on the ECUs to detect errors or to determine the cause of malfunctions. For example, if the head lights do not turn on, diagnostics can be performed at a service station to find the cause of the problem (e.g., a faulty fuse). Diagnostics is also performed in test environments to test functionality

M.D. Harrison and M.-A. Sujan (Eds.): SAFECOMP 2008, LNCS 5219, pp. 207–220, 2008.

(e.g., lock and unlock the passenger door) and find errors in an early phase before the software is released. These procedures which today require physical access to set up a wired connection, may be inconvenient for the customer as well as for the service station.

Thus, software updates via a wireless communication channel emerges as a promising possibility. The benefits are several, including minimal customer inconvenience, mass updates, and faster updates. In addition, it allows improved testing and reduced time from fault to action [2].

We have analyzed the wired diagnostics and software update procedures, and assessed the security risks associated with a wireless infrastructure. As a result of the risk assessment, we provide a set of security requirements that can be viewed as guidelines for creating a secure infrastructure for wireless diagnostics and software updates. This paper is a revised version of a more extensive work [3].

The paper is outlined as follows. Section 2 discusses related research in this area. In Section 3, we identify the attacker model, and define desired security properties and assumptions in the wireless infrastructure. In Section 4, we assess the risks for a wireless infrastructure. Section 5 presents the guidelines for creating an infrastructure for wireless diagnostics and software updates. In Section 6, we discuss importance of defining security policies. Section 7 provides possible future work directions, and Section 8 concludes the paper.

2   Related Work

The research in this area is often very specific and usually targets only one part of the infrastructure: the communication part. For example, Mahmud et al. present an architecture for secure software upload to vehicles using wireless communication links, where the focus is on the communication links [4]. In their solution, a set of authentication keys are installed in a vehicle at production time. A central server has the same set of authentication keys. Once authentication has been performed, the central server issues a symmetric session key to the vehicle. The symmetric session key is then used for secure communication with the vehicle during that session.

A discussion on securing vehicular communications is presented in [5]. The discussion describes challenges and vulnerabilities in vehicular communications. However, the vehicular communications only involve communications between vehicles, and not communications with third parties for updating software. A security architecture using secure hardware, vehicular public key infrastructure and new methods for certificate revocation is also presented.

A comparison of different flashing methods for software updates is described in [6]. The physical connection reflashing method is compared to software updates over the air at a service station and at the customer location. The incentives and challenges for updating software over the air are presented; however, no proposals for solving the mentioned challenges are given.

A proposal of using multicasting to update the software in a large number of vehicles is presented by Miucic et al. [7]. The security issues are carefully discussed, and the idea of a decentralized key management system, where multicast session keys are generated and distributed to group members is presented. By using an encryption key, shared by authorized members only, the security of the multicast communication, i.e., the confidentiality and integrity of the transmitted data and the authenticity of the group members, is achieved. Digital certificates are used to provide source authenticity and integrity of the multicast data. However, in this paper, we assume that vehicles use different software and software versions, which makes multicast data not useful for our scenario. We need to establish individually secure end-to-end communication.

There also exist several patents [8,9,10,11] in the area of the wireless diagnostics and software updates but the descriptions are often very high-level and do not contain any security-relevant details.

There have been substantial more work in this area; however, the work typically focuses on one aspect of the infrastructure while we take on a broader perspective of providing *guidelines for creating a secure infrastructure for wireless diagnostics and software updates in vehicles.*

3 Background

In this section, we describe an attacker model that is specific for the wireless infrastructure. We also present the desired security properties and assumptions we make about the wireless infrastructure. In addition, we discuss the hardware constraints that exist in the vehicular environment.

3.1 Attacker Model

We categorize our attacker as either an *insider* or an *outsider* [12,13]. An insider is an authorized member of a system, in this case the infrastructure. Basically, an insider can perform any action the authorized user can and, in addition, can mount attacks from inside the system.

An outsider is considered an intruder to the system and can only mount attacks from outside the system. For example, an outsider attacker can attack the wireless communication link. In order to address this problem, we adopt the Dolev-Yao attacker model [14], where an attacker can eavesdrop, intercept, modify or inject messages into the communication link. Moreover, after a successful intrusion, an attacker can gain access to the internal network of the vehicle or the portal, and thus execute attacks as an insider.

3.2 Desired Security Properties

In this section, we list the desired security properties for wireless diagnostics and software updates in vehicles.

Confidentiality
The software to be installed in the ECUs is proprietary and should be kept confidential. This includes the storage and the transmission of software binaries.

The transmitted diagnostics requests and replies as well as the stored diagnostics data should also be kept confidential.

Integrity
The software to be installed in the ECUs is used for controlling safety and security-critical features and needs to be protected against modification. The data integrity of the software must be verified such that a vehicle can assure that the correct software has been received.

Authentication
The communication between the portal and the vehicle needs to be authenticated. Mutual authentication is required to prevent impersonation of either portal or vehicle. Moreover, data authentication is needed such that the vehicle can verify that the received software comes from a trusted source.

Freshness
To protect against replay attacks, for example, replaying a diagnostics request to turn off the head lights, the protocol must ensure that the messages are fresh.

Resilience to lost packets
Since wireless communication is susceptible to packet loss, the infrastructure must be designed to handle lost packets in a graceful and secure way. The communication link must also be specifically protected against denial-of-service (DoS) attacks to preserve the availability of the link.

3.3 Assumptions about the Wireless Infrastructure

We assume that a centralized architecture is used, since the proprietary software and secret cryptographic keys are stored at a central location, which we denote the *portal*. The portal communicates with a large number of vehicles, and each vehicle is treated individually in terms of software and cryptographic keys. In other words, each vehicle has its own set of installed software and keys. Therefore, the portal must store the state (current software versions and keys) corresponding to each vehicle. We assume that the portal consists of high-computational devices with large storage areas, and therefore storing and accessing this data in the portal is *not* a problem.

Furthermore, since diagnostics and software updates are performed at an infrequent basis per vehicle, we assume that the communication, computation and memory overhead at the portal for each instance is insignificant. In other words, we assume that the portal will *not* be a bottleneck for wireless diagnostics and software updates, even for a large number of vehicles.

We further assume that necessary cryptographic keys (e.g., authentication keys) are distributed offline and installed in the vehicles during manufacturing. Therefore, key management is *not* an issue, since the portal and the vehicles already have established keys when the vehicles are deployed in the network. Moreover, since we assume that the established keys in the vehicles will be used for the rest of their lifetime, rekeying is *not* considered.

3.4 Limited CPU Processing Power and Memory Size

Most ECUs in the vehicle have very limited CPU processing power and memory size. This limits the possibility to use heavy cryptographic algorithms in the encryption and authentication procedures. Also, the downloaded software binaries might not fit in the ECU RAM[1] meaning that the binaries must be temporarily stored somewhere else. Issues that need to be resolved include the storage of encryption keys for the temporarily stored software binaries. Moreover, incorporating a firewall, IDS or logging utility in the vehicle also requires careful consideration with respect to the limited hardware resources.

4 Assessing Security Risks for a Wireless Infrastructure

A traditional wired infrastructure, containing the three regions *portal, communication link*, and *vehicle* [6], is illustrated in Fig. 1 and can be described as follows. The portal is communicating with a vehicle over a wired connection. For software updates the portal accesses data (the software to be installed in the ECU) in the internal portal network and sends the data to the vehicle over the wire. Once received in the vehicle, the data is routed through the in-vehicle network and installed in an ECU inside the vehicle. The procedure is similar for diagnostics requests.

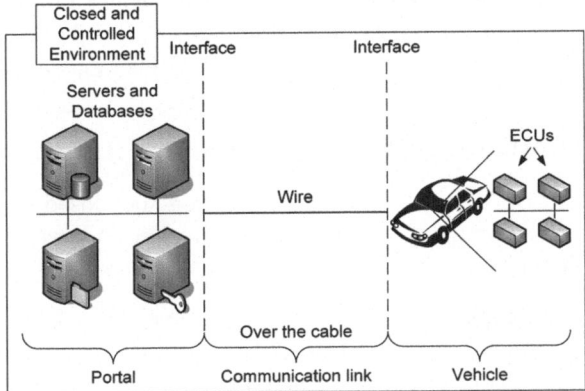

Fig. 1. Infrastructure for wired diagnostics and software updates

4.1 Risk Assessment for a Wireless Infrastructure

For the wired scenario, the procedures for diagnostics and software updates require physical access to the vehicle to connect it to the wire. Moreover, the portal, the wire, and the vehicle are in a closed and controlled environment under immediate supervision. This scenario can therefore be considered as relatively secure against attacks, especially outsider attacks, but when the same

---

[1] The software binary is downloaded to the RAM and then flashed to the ROM. The ROM could be larger than the RAM.

procedures are performed in a wireless infrastructure the scenario drastically changes. We therefore perform a risk assessment using the attacker model, desired security properties and assumptions, described in Section 3, as a basis, and list the assessed security risks of attacks for each of the three regions in the following paragraphs. We use traditional computer and network attacks [15] as a basis to develop the security risks in our scenario. In addition, we list the risks of consequences as a result of such attacks.

4.2 Portal Security Risks

The portal is still in a controlled environment but by setting up wireless communication links to the outside, the environment is no longer closed since an entry point[2] to the portal is introduced. The following risks are identified.

1. **Impersonation**
   The risk for an impersonation attack increases. For a wire it is possible to know that the vehicle is connected to the portal by physically following the wire but for wireless communication this is not possible. An attacker can impersonate the portal and establish a connection to a vehicle.
2. **Intrusion**
   The entry point to the portal also poses a security risk. A weakness in the portal could allow an intrusion, which in turn could potentially allow the outsider attacker equal access to that of an insider. An insider can access sensitive and proprietary data and execute more serious attacks.

4.3 Communication Link Security Risks

The communication link is no longer in a controlled and closed environment. The wire is replaced with communication over the Internet and over-the-air. The following risks are identified.

3. **Traffic Manipulation**
   The risk that an attacker can inject or modify packets in the communication link is increased, especially due to the added exposure caused by the wireless communication. This attack could cause diagnostics to perform actions that they were not originally intended to perform (e.g., unlock the door instead of checking if the door was locked). An attacker can also replay, for example, a diagnostics request to unlock the door.

4.4 Vehicle Security Risks

The vehicle is no longer in a controlled and closed environment, and immediate supervision may not be possible. The following risks are identified.

---

[2] We define an entry point as a communication interface that allows entry to an internal network.

4. **Impersonation**

The risk for an impersonation attack is increased. For wired communication, it is possible to physically follow a wire to know which vehicle is connected but for wireless communication this is not feasible. An attacker can impersonate a vehicle and set up communication links with the portal.

5. **Intrusion**

The wireless interface to the vehicle also introduces an entry point. A weakness in, for example, the authentication procedure in the vehicle could allow intrusions, which could potentially allow an outsider attacker equal access to that of an insider. An insider can access sensitive and proprietary data and execute more advanced attacks.

4.5   Risks of Consequences

If the attacks on the portal, communication link, and vehicle are successful, the consequences could be disastrous.

6. **Execution of Arbitrary Code**

With both wireless diagnostics and software updates, it is possible to affect the behavior of the ECUs. Thus, as a result of a successful impersonation of the portal or an intrusion attack to the portal, an attacker can issue diagnostics requests or software that execute in the vehicle which believes the requests or software originated from the real portal. Thus, an attacker can run arbitrary code on the vehicle. A rational attacker can read confidential data from the vehicle or, for example, unlock the driver door. A malicious attacker can cause damage by, for example, disabling the brakes in the vehicle. Furthermore, an attacker who has access to the internal portal network can perform attacks as an insider. In addition, a successful intrusion attack to the vehicle could allow an attacker to update the ECUs with modified versions of software, where the attacker can control the functionality of the ECUs. A rational attacker can update the software in ECUs with performance-enhanced versions of the software. A malicious attacker, on the other hand, can update the ECUs with malicious versions of the software that can cause damage to the vehicle or injury to a person (e.g., triggering the airbag remotely when a person is sitting in the seat).

7. **Disclosure of Information**

A successful intrusion to the portal may allow an attacker to learn private information about customers and access proprietary software. Moreover, a successful impersonation attack of a vehicle could lead to the attacker getting access to confidential data and proprietary software available on the portal that is meant for the impersonated vehicle. In addition, since a vehicle is susceptible to physical attacks, there is a risk that an attacker can extract, e.g., authentication and encryption keys stored in the ECUs. Using these keys, the attacker can impersonate a vehicle or eavesdrop encrypted communication. An attacker could also access private data and proprietary software stored in the vehicle or sent over the communication link.

8. **Denial of Service**

An attacker can execute a DoS attack targeting the portal, the communication link, or the vehicle causing software updates to fail or diagnostics to report incorrect values. As a consequence, legitimate users can be prevented from updating potentially vulnerable software. Furthermore, this attack could cause damage to the vehicle or injury to a person in the vehicle.

Based on these risks, we develop a set of security requirements which is presented in the next section.

5   Guidelines for a Secure Wireless Infrastructure

For the wireless infrastructure, the portal is communicating with a vehicle over the Internet and over-the-air. This infrastructure is also divided into three regions: *portal, communication link* and *vehicle*, as illustrated in Fig. 2. For each of the three regions we define a set of security requirements and discuss what protection is offered if the requirements are met. Several security requirements might seem obvious for Internet traffic and high-end Internet servers but for low-performance devices and special-purpose networks those security requirements are often lacking (cf. the complete lack of security features for wireless software updates in sensor networks [16]). Therefore, our set of security requirements can be seen as guidelines for creating a secure wireless infrastructure. In the following paragraphs a brief description of each region followed by the security requirements is presented.

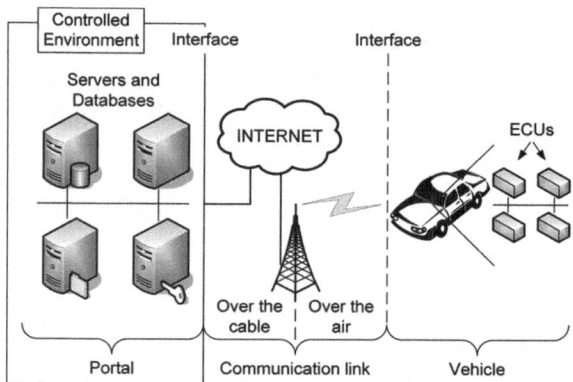

Fig. 2. Infrastructure for wireless diagnostics and software updates

5.1   Portal Security Requirements

The portal consists of servers and databases in the internal portal network and has an interface to the Internet. It has access to file servers with proprietary software that is to be installed in the vehicles and databases containing information about vehicles and what hardware and software versions they contain. Furthermore, the portal has access to databases that contain cryptographic keys for authentication with the vehicles. Thus, the portal has access to sensitive data.

1. **Preventing Impersonation of Portal**

   Security Requirement: The portal must possess something *unique* that can be used to *establish its identity.* Certificates suitable for the vehicle environment [17] should be used, and the public key of the portal should be installed in vehicles during manufacturing. A method for handling certificate revocation must also be incorporated [18]. The portal must ensure that data sent from the portal cannot be spoofed or modified. Sensitive data should, for example, be *signed* using the portal's private key.

   Achieved Security: Prevents an outsider attacker from forging the portal identity and impersonating the portal to set up communication links with the vehicle.

2. **Intrusion Protection**

   Security Requirement: The portal is a traditional environment in the sense that it consists of powerful devices, and there already exist a number of *best practices* for firewalls, logging, and Intrusion Detection Systems (IDS) [19] that should be used.

   Achieved Security: A firewall and an IDS assist in preventing and alerting on intrusion attempts on the portal.

5.2   Communication Link Security Requirements

The communication link connects the portal to the vehicle, and is divided into two parts: *over-the-cable* and *over-the-air.* The security requirements for the over-the-cable and the over-the-air communication are the same.

3. **Secure End-to-End Communication**

   Security Requirement: A *secure end-to-end channel* for diagnostics and software updates [20] must be established. Data traffic should be encrypted and integrity protected using, for example, the TLS protocol. Cryptographic algorithms must be chosen carefully to agree with the limited resources in the vehicle. A comparison of cryptographic algorithms for use in a vehicular environment is found in [13], and a performance evaluation of public-key cryptosystem operations in WTLS is found in [21]. Moreover, the software and diagnostics requests should use *timestamps* or other methods to guarantee freshness.

   Achieved Security: Secure end-to-end communication prevents packets from being read, injected, modified and replayed by both insider and outsider attackers.

5.3   Vehicle Security Requirements

The vehicle contains sensitive data, such as cryptographic keys and proprietary software, stored in the ECUs. Furthermore, received diagnostics requests and software are executed respectively installed in the ECUs.

4. **Preventing Impersonation of Vehicle**

   Security Requirement: The vehicle must possess something *unique* to *establish its identity.* An analogy is client certificates in TLS [22]. These certificates should be installed in the vehicle during manufacturing.

   Achieved Security: Prevents both insider and outsider attackers from impersonating a vehicle.

5. **Intrusion Protection**

   Security Requirement: The vehicle is a nontraditional environment in the sense that it consists of resource-constrained embedded devices. The security requirements must be adjusted accordingly. A *firewall* should be used to block incoming traffic from non-trusted parties and to allow only trusted parties to connect to the vehicle. In addition, an *IDS* should be installed in the vehicle to detect unauthorized accesses and raise alerts on intrusion attempts on the vehicle. Proper trace and network logging should be enabled.

   Achieved Security: A firewall and an IDS assist in preventing and alerting on intrusion attempts. In addition, in the event of an intrusion, log data can be analyzed and used to reconstruct the actions after the intrusion [23]. This information can be used to prevent future intrusions.

5.4   Risks of Consequences

The risks of attacks on the portal, communication link, and vehicle could be reduced by taking the proposed security requirements into consideration. If an attack is successful, the risks of consequences could be lowered by consulting the following security requirements.

6. **Preventing Execution of Arbitrary Code**

   Security Requirement: A *filter*, e.g., a blacklist, which contains a list of disallowed commands, must be used at the portal to prevent generating diagnostics requests or software that contain certain *dangerous*[3] commands that should not be allowed to be executed remotely in the ECUs. These commands should still be available when physically connecting to the vehicle. Therefore, a solution to remove these commands from the command set is not suitable. Moreover, the software and the diagnostics requests should be *signed* by the portal, and the vehicle must *verify* that the received software and requests have not been altered. The vehicle should also use a filter to prevent dangerous commands from executing remotely in the ECUs. This is comparable to server-side security enforcements for cross-site scripting attacks [24].

   Achieved Security: An insider attacker is prevented from generating and sending diagnostics requests and software that contain dangerous commands. Since these commands are not allowed in the requests and software created at the portal, they will not be executed in the ECUs. An outsider attacker is prevented from generating and modifying the software and diagnostics requests. An attacker is also prevented from executing dangerous commands in the ECUs.

---

[3] Commands such as triggering the airbag or turning off the head lights. Such commands must be well-defined before deployment.

7. **Secure Storage and Communication**

Security Requirement: The portal must *encrypt* the proprietary software binaries and the private information it stores about customers using a strong symmetric cipher, e.g., AES. Access to the data requires proper authentication and authorization. Data traffic in the internal portal network should be protected using, e.g., the transport layer security (TLS) protocol. On the other hand, the vehicle should use a *tamper-resistant storage* to store sensitive data, such as encryption and authentication keys, private information and downloaded software. For example, a trusted platform module [25] could be used. Moreover, data traffic on the in-vehicle network should be protected with respect to data authentication [26].

Achieved Security: Prevents an outsider attacker from accessing sensitive data in the portal and the vehicle. Moreover, an attacker is prevented from injecting messages as well as altering messages in the internal portal network and the in-vehicle network.

8. **Denial-of-Service Protection**

Security Requirement: The portal and the vehicle should use proper DoS protection, although such solutions exists for traditional environments [27], they are nonexistent for vehicles. In addition, the communication protocol must be resistant to packet loss caused by not only communication problems, such as bad reception, but also intentional attacks. Therefore, a *reliable* protocol which also can handle low bandwidth communication with long delays must be used. For example, the SCTP [28] protocol provides reliable message-stream communication.

Achieved Security: Proper DoS protection assists in preventing availability attacks. Furthermore, a reliable protocol helps for protecting the availability of the link.

6   Security Policies

Since wired diagnostics and software updates typically are performed in closed and controlled environments with immediate supervision, security policies have been nonexistent. However, for allowing wireless diagnostics and software updates, defining a set of security policies is imperative. If several parties are involved, e.g., portal, service station, and vehicle owner, it is especially important to define who is allowed to perform what actions. We provide a few policies as examples. Policies for various involved parties must be specified.

- Only the portal is allowed to create and sign software.
- The portal and service stations are allowed to perform software updates of signed software on the vehicle.
- The vehicle must verify the authenticity of the received software to verify that it was created by the portal.
- The portal and service stations are allowed to send diagnostics requests.

Furthermore, policies for the ECUs must be well-defined.

- A time limit, e.g., 30 minutes, for updating the software on the same ECU should be used.
- Only ECUs that do not affect the maneuverability of the vehicle are allowed to be updated over-the-air.
- Prerequisites for updates include the engine being turned off for at least one hour, a velocity of zero mph, and no driver or passengers in the vehicle.
- A time limit, e.g., 1 minute, for responding to repeated diagnostics requests should be used.
- Only non-safety critical diagnostics requests are allowed to be sent to safety-critical ECUs, and only safety-critical diagnostics requests are allowed to be sent to non-safety critical ECUs.

These examples are only a few of the policies that need to be defined. A thorough analysis of all the ECUs in the vehicle to classify them into safety-critical classes [29] and defining policies for the different ECUs and classes is required to properly specify security policies. These policies define the security of the ECUs and prevent attackers from installing malicious software and vehicle owners from boosting the performance in the ECUs. Moreover, combination of policies could prevent denial-of-service attacks on the ECUs and more advanced cyber attacks [30,31] targeting the safety of the vehicle. Thus, the policies are the vanguard of security and safety on the vehicle and the portal.

7   Future Work

The most pertinent issue for the near future is to scrutinize the in-vehicle network for possible entry points and weaknesses. A risk analysis of the ECUs is to be conducted, and measures to provide the necessary security are to be evaluated.

Another possible direction is to explore the possibilities of using an IDS in the vehicle. The IDS could trigger on reads and writes to security-critical data or on abnormal activities, and thus detect attacks on the vehicle. Finally, it would be interesting to investigate the possibility to include a firewall in the vehicle to prevent unwanted external accesses as well as an internal filtering service within the in-vehicle network to block accesses to certain ECUs with respect to safety. It would be highly interesting to investigate how an IDS, firewall or filtering service can be adapted to the typical vehicular communication, which is significantly different from Internet traffic.

8   Conclusion

This paper aims to deepen the awareness of security risks involved in creating an infrastructure for wireless software updates and diagnostics in vehicles and provides guidelines for improving the security. The security risks for a wireless infrastructure are first assessed. The result is used to develop a set of guidelines for creating a secure infrastructure. The infrastructure is subdivided into the portal, the communication link, and the vehicle, and a number of security risks in each part are identified. These risks must seriously be taken into consideration when designing the infrastructure and security must be incorporated from the very start. Consequently, we have listed a number of security requirements and discussed the importance of defining security policies.

References

1. See, W.-B.: Vehicle ECU Classification and Software Architectural Implications. Technical report, Feng Chia University, Taiwan (2006)
2. Miucic, R., Mahmud, S.M.: An In-Vehicle Distributed Technique for Remote Programming of Vehicles' Embedded Software. Technical report, Electrical and Computer Engineering Department, Wayne State University, Detroit, MI 48202 USA (2005)
3. Nilsson, D.K., Larson, U.E., Jonsson, E.: Creating a Secure Infrastructure for Wireless Diagnostics and Software Updates in Vehicles. Technical report, Chalmers University of Technology, 2008:02 (2008)
4. Mahmud, S.M., Shanker, S., Hossain, I.: Secure Software Upload in an Intelligent Vehicle via Wireless Communication Links. In: Proceedings of IEEE Intelligent Vehicles Symposium, pp. 587–592 (2005)
5. Raya, M., Papadimitratos, P., Hubaux, J.-P.: Securing Vehicular Communications. IEEE Wireless Communications 13(5), 8–15 (2006)
6. Shavit, M., Gryc, A., Miucic, R.: Firmware Update over the Air (FOTA) for Automotive Industry. Technical Report 2007-01-3523, SAE (2007)
7. Miucic, R., Mahmud, S.M.: Wireless Multicasting for Remote Software Upload in Vehicles with Realistic Vehicle Movement. Technical report, Electrical and Computer Engineering Department, Wayne State University, Detroit, MI 48202 USA (2005)
8. Parrillo, L.C.: Wireless motor vehicle diagnostic and software upgrade system. U.S. patent 5442553 (1995)
9. Lightner, B., Botrego, D., Myers, C., Lowrey, L.H.: Wireless diagnostic system and method for monitoring vehicles. U.S. patent 6636790 (2003)
10. Suman, M.J., Zeinstra, M.L.: Remote vehicle programming system. U.S. patent 5479157 (1995)
11. Chen, C.-H.: Vehicle security system having wireless function-programming capability. U.S. patent 6184779 (2001)
12. Wolf, M., Weimerskirch, A., Paar, C.: Security in Automotive Bus Systems. In: Workshop on Embedded IT-Security in Cars, Bochum, Germany (November 2004)
13. Raya, M., Hubaux, J.-P.: The Security of Vehicular Ad Hoc Networks. In: Proceedings of the 3rd ACM Workshop on Security of Ad Hoc and Sensor Networks, pp. 11–21. ACM Press, New York (2005)
14. Dolev, D., Yao, A.C.: On the Security of Public Key Protocols. IEEE Transactions on Information Theory 29(2), 198–208 (1983)
15. Howard, J.D., Longstaff, T.A.: A Common Language for Computer Security Incidents (SAND98-8667) (1998),
   http://www.cert.org/research/taxonomy_988667.pdf
16. Hui, J.: Deluge 2.0 - TinyOS Network Programming Manual (2005),
   http://www.cs.berkeley.edu/~jwhui/research/deluge/deluge-manual.pdf

17. IEEE. 1609.2. Standard for Wireless Access in Vehicular Networks (2004)
18. Raya, M., Jungels, D., Papadimitratos, P., Aad, I., Hubaux, J.-P.: Certificate Revocation in Vehicular Networks. Technical report, Laboratory for computer Communications and Applications (LCA), EPFL, Switzerland, 2006. LCA-Report-2006-006.
19. US-CERT. Current Malware Threats and Mitigation Strategies (2005),
    http://www.us-cert.gov/reading_room/malware-threats-mitigation.pdf
20. Nilsson, D.K., Larson, U.E.: Secure Firmware Updates over the Air in Intelligent Vehicles. In: Proceedings of the First IEEE Vehicular Networking & Applications Workshop (Vehi-Mobi), pp. 380–384 (2008)
21. Levi, A., Savas, E.: Performance Evaluation of Public-Key Cryptosystem Operations in WTLS Protocol. In: Proceedings of the Eighth IEEE International Symposium on Computers and Communications, pp. 1245–1250 (2003)
22. Network Working Group. The TLS Protocol Version 1.0 (1999)
23. Nilsson, D.K., Larson, U.E.: Conducting Forensic Investigations of Cyber Attacks on Automobile In-Vehicle Networks. In: Proceedings of the First ACM International Conference on Forensic Applications and Techniques in Telecommunications, Information and Multimedia (e-Forensics). ACM Press, New York (2008)
24. Jovanovic, N., Kruegel, C., Kirda, E.: Pixy: A static analysis tool for detecting web application vulnerabilities. In: Proceedings of the 2006 IEEE Symposium on Security and Privacy (S&P), pp. 258–263 (2006)
25. Trusted Computing Group. Trusted Platform Module Specification (2003),
    https://www.trustedcomputinggroup.org/specs/TPM
26. Nilsson, D.K., Larson, U.E., Jonsson, E.: Efficient In-Vehicle Delayed Data Authentication based on Compound Message Authentication Codes. In: Proceedings of the IEEE 68th Vehicular Technology Conference (VTC2008-Fall) (2008)
27. Deal, R.: Cisco Router Firewall Security. Cisco Press (2004)
28. Network Working Group. Stream Control Transmission Protocol (SCTP) Specification (2006)
29. Nilsson, D.K., Phung, P.H., Larson, U.E.: Vehicle ECU Classification Based on Safety-Security Characteristics. In: Proceedings of the 13th International Conference on Road Transport and Information Control (RTIC) (2008)
30. Hoppe, T., Dittman, J.: Sniffing/Replay Attacks on CAN Buses: A simulated attack on the electric window lift classified using an adapted CERT taxonomy. In: Proceedings of the 2nd Workshop on Embedded Systems Security (WESS), Salzburg, Austria (2007)
31. Nilsson, D.K., Larson, U.E.: Simulated Attacks on CAN Buses: Vehicle virus. In: Proceedings of the Fifth IASTED Asian Conference on Communication Systems and Networks (ASIACSN). ACTA Press (2008)

Finding Corrupted Computers Using Imperfect Intrusion Prevention System Event Data

Danielle Chrun[1], Michel Cukier[1], and Gerry Sneeringer[2]

[1] Center for Risk and Reliability, University of Maryland
College Park, Maryland 20742-7531
[2] Office of Information Technology, University of Maryland
College Park, Maryland 20742-7531
{chrun,mcukier,sneeri}@umd.edu

Abstract. With the increase of attacks on the Internet, a primary concern for organizations is how to protect their network. The objectives of a security team are 1) to prevent external attackers from launching successful attacks against organization computers that could become compromised, 2) to ensure that organization computers are not vulnerable (e.g., fully patched) so that in either case the organization computers do not start launching attacks. The security team can monitor and block malicious activity by using devices such as intrusion prevention systems. However, in large organizations, such monitoring devices could record a high number of events. The contributions of this paper are 1) to introduce a method that ranks potentially corrupted computers based on imperfect intrusion prevention system event data, and 2) to evaluate the method based on empirical data collected at a large organization of about 40,000 computers. The evaluation is based on the judgment of a security expert of which computers were indeed corrupted. On the one hand, we studied how many computers classified as of high concern or of concern were indeed corrupted (i.e., true positives). On the other hand, we analyzed how many computers classified as of lower concern were in fact corrupted (i.e., false negatives).

Keywords: Security Metrics, Empirical Study, Intrusion Prevention Systems.

1 Introduction

With the increase of attacks on the Internet, a primary concern for organizations is how to protect their network. To do so, organizations monitor their traffic using security devices such as intrusion detection systems or intrusion prevention systems. The monitored activity provides some insight into an organization's security and identifies potentially corrupted computers. While in some organizations the quantity of monitored traffic is manageable, it becomes a hassle to analyze security data for large organizations. For example, intrusion prevention systems could record thousands of alerts per day and the security team cannot investigate every alert. Moreover, although intrusion prevention systems are aimed at detecting and blocking malicious activity, they also raise false alarms. Due to 1) the potentially large quantity of data to deal with, and 2) the number of

M.D. Harrison and M.-A. Sujan (Eds.): SAFECOMP 2008, LNCS 5219, pp. 221–234, 2008.
© Springer-Verlag Berlin Heidelberg 2008

false alarms, it is of main interest to organize the generated alerts and to extract information from the collected data that would be useful to the security team.

This paper presents a method to retrieve useful information for the security team from data collected by an intrusion prevention system (IPS). The method consists in identifying potentially corrupted computers inside the organization and ranking them according to three metrics: the coefficient of consecutiveness indicating during how many consecutive weeks IPS alerts were observed, the number of weeks during which alerts were raised and the number of distinct attack types. Based on these metrics, potentially corrupted computers can be ranked. We will show that the proposed method helps the security team gaining some insight into the organization's security. The introduced method is evaluated for data collected at a large organization of about 40,000 computers. The evaluation is based on the judgment of a security expert of which computers were indeed corrupted. On the one hand, we studied how many computers classified as of high and medium concern were indeed corrupted (i.e., true positives). On the other hand, we analyzed how many computers classified as of low concern were in fact corrupted (i.e., false negatives).

The remainder of the paper is structured as follows. Section 2 describes the related work on data analysis of security logs. Section 3 introduces the concepts relative to IPSs. Section 4 defines the method. Section 5 presents the evaluation of the method. We provide conclusions in Section 6.

2   Related Work

A lot of research focuses on analyzing security logs for security assessment. To face the possibly high quantity of data to analyze, a common step is to reduce data before analyzing it. [1] describes an architecture to analyze distributed darknet traffic: first, collected data on attacks are filtered; secondly, forensics is used to analyze the reduced data. [2] focuses on analyzing data of a denial of service. In order to study the traffic volume per protocol, a categorization of the collected network traffic by protocol was made.

Analyzing large amounts of security data becomes an emerging task in the intrusion detection field. Indeed, intrusion detection systems face two main issues: 1) a high number of alarms can be raised and 2) there can be many false alarms among them. Thus, the objective is to decrease the number of false alarms. Research was conducted to retrieve normal behavior (i.e., traffic that is not malicious) from the dataset using several techniques: time series [3], data mining [4, 5, 6, 7, 8, 9, 10 and 11] and correlation [6, 12, 13, 14, 15 and 16]. A common practice is to use historical data to define normal behavior so that future alarms can be handled more efficiently. Data mining techniques can be used to achieve this goal. However, research projects differ in the data mining technique used: association rules [10], frequent episode rules [4, 9], classification [11] or clustering [5, 6, 7, 8 and 9]. A commonly used method in intrusion detection is alert correlation. [13] defines a model for intrusion detection alert correlation and presents three examples of correlation: aggregation of alerts referring to a single targeted host, aggregation of alerts referring to hosts vulnerable to an attack occurrence and aggregation according to alerts similarities (such as alerts caused by the same event or referring to the same vulnerabilities). [5 and 6] introduce a cooperative intrusion detection framework in which functions to manage, cluster, merge and correlate alerts were implemented. The objective was to correlate alerts to generate more global alerts and discard false alarms.

In [11], the authors present the Adaptive Learner for Alert Classification (ALAC) system. ALAC is a system to reduce false positives in intrusion detection systems and relies on two elements: 1) expert judgment and 2) machine learning techniques. An analyst classifies alerts as true positives or false positives. Then, ALAC autonomously processes alerts that have been classified by the analyst. The accuracy of ALAC is as good as the quality of the analyst's classification.

3 On the Use of Intrusion Prevention System Event Data

3.1 Approach

Many organizations use security devices to monitor their network activity. The quantity of data collected per day can be so substantial that every event identified by a security device cannot be investigated by the security team. Hence, retrieving meaningful information from the collected data on the malicious activity would give a more detailed insight to security administrators into the network's security. The main issue is that the data currently collected are far from being perfect. For example, the data collected by security devices, such as intrusion prevention systems (IPSs), might contain alerts for activity that is not malicious (i.e., false positives) and might not detect some malicious activity (i.e., false negatives). Moreover, they will not include new attacks in the case of signature-based IPSs. They often rely on the trust we have in the security devices and the vendors. No ground truth is provided. Details are lacking on the meaning of the data and how they are produced (the security devices are black boxes for which vendors only release few details).

Two approaches are then possible. The first one is to work on obtaining datasets clean enough so that accurate security estimations are possible. The second one is to accept that the dataset is imperfect but that useful information regarding an organization's security can be retrieved. In this paper, we adopt the second approach.

In this paper, we provide a method to extract useful information from IPS event data. The suggested method aims at extracting a list of potentially corrupted organization computers that would then be handled by the security team. Those computers manifest in the IPS dataset as the potential source of attacks. The dataset might not only contain attackers who willingly launch attacks. It might also include computers that may not have been fully patched. Once the list of suspected computers is identified, the security team can make a decision regarding these computers. For example, a decision could consist in blocking the IP address from the network until the computer is cleaned.

3.2 Intrusion Prevention Systems

An IPS is a security device that monitors malicious activity and reacts in real-time by blocking a potential attack. An IPS is considered as an extension of an intrusion detection system (IDS). An IDS is a passive device that monitors activity whereas an IPS is an active device that blocks potential malicious activity. For our study, we focus on signature-based IPSs: the IPS blocking decision relies on a set of signatures that are regularly released by the vendor as attacks are newly discovered on the Internet. When characteristics of an attack match the ones of a defined signature, the attack is blocked and an alert is recorded in the IPS logs.

We assume that the IPS is located at the edge of the organization. In other words, the IPS monitors 1) malicious activity originating inside the organization and targeting outside computers, 2) malicious activity originating outside the organization and targeting organization computers.

We define an alert in the IPS dataset as a source IP address (SIP/attacker) attacking a destination IP address (DIP/target) with a certain type of attack (signature) at a given time.

3.3 Dataset: Assumptions

As previously mentioned, the IPS dataset has several issues. We have not evaluated the IPS and thus do not know how many false positives and false negatives the IPS produces. Moreover, since the IPS is a signature-based device, new attacks will not be detected nor blocked.

Furthermore, the dataset does not include the case where a computer inside the organization attacks another computer inside the organization. The IPS is located at the edge of the organization so it cannot detect traffic within the organization. Besides, this study solely focuses on computers with static IP addresses.

Finally, we cannot prove that a blocked attack would have been harmful to the targeted computer. Indeed, for an attack to be successful, the targeted computer should have the associated vulnerability. We have scanned several computers for which an IPS alert was raised and noticed that in many cases the vulnerability associated with the alert was not present. This means that even without the IPS, the attack would not have been successful. This also indicates that the IPS identifies and detects an attack in its early stage preferring to block attacks that would not have been successful instead of not blocking a potentially successful attack.

4 Method

The next sub-sections present the method to identify potentially corrupted organization computers. First, we define three metrics to characterize the activity in the IPS dataset. Then, we present the method for ranking the potentially corrupted computers according to the three metrics values.

4.1 Metrics

A computer is of main concern to the security team if 1) it appears often in the IPS dataset as the source of an attack, and 2) it launches a wide range of different attack types. Therefore, we introduce the following metrics for a computer: 1) a coefficient of consecutiveness of the number of weeks for which at least one alert was raised, 2) the number of weeks for which at least one alert was raised, and 3) the number of different signatures (i.e. attack types) associated to the computer. We defined these metrics that we believe are appropriate for attackers. These metrics might be less relevant for targets (computers under attack).

4.1.1 Coefficient of Consecutiveness
Computers that appear in the IPS dataset for many consecutive weeks are of main concern for the organization's security team, seeming to indicate that a computer is

launching attacks during several consecutive weeks and has not been checked. We define the coefficient of consecutiveness as:

$$Cons = Week/(Max - Min + 1)$$

where *Max* is the identifier of the last week when the computer appears in the dataset, *Min* is the identifier of the first week, and *Week* is the number of distinct weeks. The consecutiveness factor is positive and the maximum value is 1. Let us consider a computer that appears in the IPS dataset at weeks 2, 3, 6, 8, 9, among 10 weeks of observation (Figure 1). In this case, *Max* = 9, *Min* = 2 and *Week* = 5. The consecutiveness factor is: 5/(9-2+1) = 0.625.

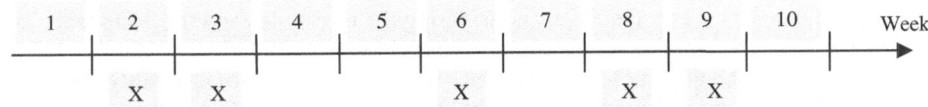

Fig. 1. Consecutiveness Factor

The closer to 1 the coefficient of consecutiveness is, the more focus the security team should put on the computer. However, if a computer only appears once in the IPS dataset, it means that *Week* = 1. Nonetheless, it does not necessarily mean that the security team should focus on that computer. This emphasizes that the number of weeks is also an important metric.

4.1.2 Number of Weeks
The number of weeks for which at least one alert was associated to the computer is the second metric.

However, the case where the number of weeks is 1 may be misleading. In this case (the computer was recorded as an attacker only for one week along the considered period of time), the coefficient of consecutiveness would be 1 and the computer would be reported to the security team. Considering the computers for which *Week* = 1 would raise a lot of alerts for computers that are in fact not corrupted. Therefore, we discard for the study all computers where week = 1.

Hence, the minimum is *Week* = 2 and the maximum is the number of weeks during which data have been collected.

The number of weeks reflects the frequency at which the computer appears in the IPS dataset. A computer with a large number of weeks reveals that the computer is potentially corrupted and has not been checked.

4.1.3 Number of Signatures
Finally, we believe that the number of distinct attack signatures associated with a given computer is important. It reflects the range of different attack types one computer seemed to have launched. Note that a great number of distinct signatures might also reveal that the computer contains several vulnerabilities.

The minimum number is 1 and the maximum is the total number of existing distinct signatures in the IPS.

4.2 Level of Criticality

We define the level of criticality of a computer as the 3-tuple {*Cons, Week, Sign*} (*Cons* stands for the coefficient of consecutiveness, *Week* for the number of weeks, *Sign* for the number of signatures). The higher the level of criticality, the more important it is for the security team to check that computer.

We identify three levels of interest: high concern, concern, and low concern. We define thresholds for each metric so that the interval is cut into three intervals: *C1* and *C2* are thresholds for the consecutiveness factor, *W1* and *W2* for the number of weeks, *S1* and *S2* for the number of distinct signatures. We decided to visualize each computer by using a Cartesian coordinate system: the coordinates are the consecutiveness factor, the number of weeks and the number of signatures. In other words, each computer is represented in a 3-D space. Hence, by considering the thresholds and the 3-D space, we can visualize a cube that is cut into 27 sub-cubes (Figure 2a).

We then introduce three colors associated with the three levels of criticality: 1) green regions (G) depict computers of low concern, 2) yellow regions (Y) group computers of concern that should be checked by the security team, and 3) red regions (R) show computers of high concern that should be addressed in priority. For each sub-cube, a security expert helped us decide on their level of criticality and thus their associated color. Figure 2b depicts the colors selected for the 27 sub-cubes.

4.3 Method for Identifying Computers of Concern

The method consists in five steps: 1) analysis of the IPS dataset to identify computers that were the source of alerts, 2) calculation of the level of criticality for each identified computer, 3) determination of thresholds for the three metrics, 4) investigation of computers in the red region and 5) investigation of computers in the yellow region.

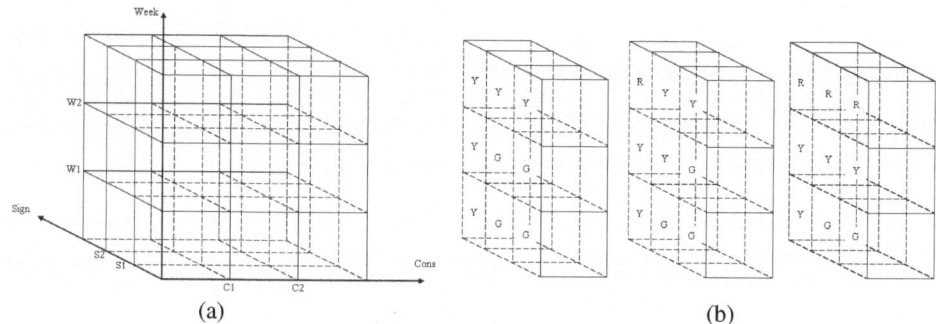

(a)   (b)

Fig. 2. Visualization of Metrics (a) and Colored Zones (b)

Step 1: Analysis of the IPS dataset
The identification of computers that were the source of alerts in the IPS dataset is done through the extraction of the internal IP addresses that appear as the source of alerts in the IPS dataset.

Step 2: Calculation of the level of criticality
To calculate a level of criticality, a period of time for which to calculate the metrics needs to be defined. We advise selecting a period long enough to allow a metric like the

coefficient of consecutiveness to be relevant (at least 5 weeks for the coefficient of consecutiveness to be meaningful).

Step 3: Determination of thresholds for the three metrics
We believe that threshold values ($C1$, $C2$, $W1$, $W2$, $S1$, $S2$) are organization dependent. Characteristics, such as the size of the organization, the type of the organization can greatly differ between organizations. In that sense, we advise each organization to choose its own thresholds.

Steps 4 and 5: Investigation of computers in the red and yellow regions
As the method consists in ranking computers in function of the level of criticality in order to focus on the computers of main concern, the security team should focus in priority on the computers in the red region.

Depending on the available sources of information, checking a potentially corrupted computer would include:

- Using the IPS dataset to look at the date and time of events,
- Using the IPS dataset to understand the attack type,
- Investigating previous incidents with that particular IP address.

The method tends to identify computers that appear frequently in the IPS dataset: those are the computers in the red and yellow regions (the frequency is reflected by the metrics *Cons* and *Week, Sign* interferes in making the distinction between the red and yellow regions). Hence, our method will not raise a flag for a computer that is involved in a single alert that could be harmful. Therefore, the method does not identify all potential corrupted computers.

Also, the method identifies computers that may be corrupted or not. For the remaining of the paper, we call:

- False negatives: corrupted computers that have not been identified by the method,
- True positives: corrupted computers that have been identified by the method,
- False positives: non-corrupted computers that have been identified by the method,
- True negatives: non-corrupted computers that have not been identified by the method.

The thresholds $C1$, $C2$, $W1$, $W2$, $S1$ and $S2$ are chosen by making a trade-off between the number of true positives and the number of false negatives.

5 Evaluation

5.1 Approach

In this section, we will evaluate the presented method. We will study IPS event data collected on a large public university (University of Maryland) composed of about 40,000 computers. The considered IPS dataset covers a period of 17 months, from

September 1st 2006 to January 31st 2008. The IPS raised an average of around 142 alerts per day during the studied period for computers inside the campus that are detected transmitting potentially malicious traffic toward computers outside the campus. Over the 17 months, 1,441 different computers inside the organization that launched at least one attack were identified.

First, we need to define a time period on which to apply the metrics. The campus is much less populated during the summer break (3 months) and the winter break (1 month). In other words, the traffic recorded by the IPS may drop during these periods due to fewer students/computers. In order not to bias the results, we should apply the metrics over a period greater than 3 months. We decided to apply the metrics over a 6-month period. In order to show how the metrics evolved over time, we calculated the metrics for increments of 2 weeks. On each period of 6 months, we extracted a list of computers and calculated the associated metrics.

We then asked the Director of Security of the Office of Information Technology at the University of Maryland, to indicate which computers were corrupted among the ones identified by our method. To do so, the Director of Security needed to investigate every computer. This step relies on expert judgment and human activity, as opposed to an automated investigation. As previously stated, we believe that computers for which $Week = 1$ are of less interest that those that appear at least two weeks over a 6-month period. By eliminating those computers, we are left with 303 computers to investigate.

We recognize that we rely on expert judgment to indicate which computers are corrupted. Another security expert might provide slightly different results. To decrease the potential bias due to expert judgment, we asked the Director of Security: 1) to use a systematic method for deciding if a computer is corrupted, and 2) to be conservative in his judgment (the Director will declare a computer corrupted (respectively non-corrupted) only if he is sure that the computer is corrupted (respectively non-corrupted)). Such requirements led to many investigated computers without clear decision. Among the 303 investigated computers, for 76 (25%) of them it was unclear whether they were corrupted. One reason is that the analyzed data went back to September 2006 making it difficult to make sure if the flagged computer was indeed corrupted.

First, in order to investigate the computers to determine if they are corrupted, the Director of Security needed the following information:

- For each computer: the number of alerts triggered in the IPS, the signature list associated to these alerts (SL), the time span for these alerts by signature, the list of computers targeted (target list TL), the list of incidents associated to the computer,
- A list of signatures known to trigger false alarms,
- A list of signatures known to be non-malicious.

Figure 3 depicts the sequential questions to answer regarding a given computer to determine if it is corrupted (C), non-corrupted (NC), or undetermined (O for other). If the answer to a question is "yes", the computer can be classified and the Director of Security investigates another computer. If the answer is "no", the Director of Security moves to the next question. These steps are the ones that were followed by the Director of Security to investigate the computers in order to evaluate the suggested method.

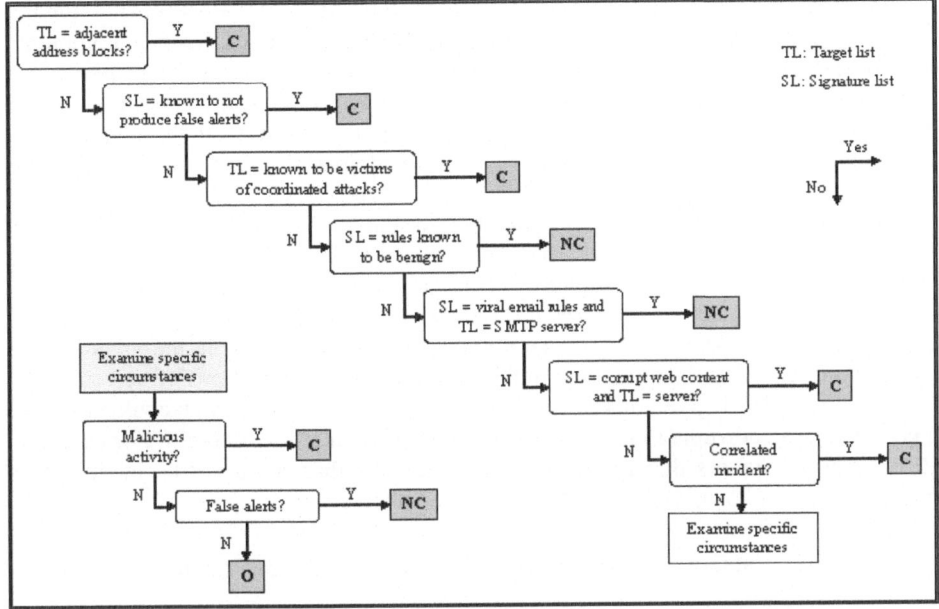

Fig. 3. Flowchart of the Steps of the Investigation

First, classifying computers as (non)-corrupted involves investigating the target list (TL): does the target list contain computers in the adjacent address blocks? If yes, it is possible that the computer is scanning the adjacent IP addresses range in order to detect computers. In that case, the computer is classified as corrupted. Otherwise, the signature list (SL) needs to be investigated: does the signature list contain signatures known to not produce false alerts? If yes, the computer is classified as corrupted. Six sequential steps consist in investigating the target list and the signature list. The seventh step aims at searching into the incident data in order to find an incident report involving the computer under investigation: if there is an incident report to support the alert associated with the computer, then the computer is classified as corrupted. If none of these steps allows classifying the computer as (non)-corrupted, the Director of Security will examine the specific circumstances of the alerts: if the investigation reveals malicious activity, the computer is corrupted; otherwise, if the investigation indicates false alerts, the computer is non-corrupted; otherwise, the computer is classified as undetermined.

Out of 303 investigated computers, 91 (30%) were identified as corrupted and 136 (45%) were identified as non-corrupted. One important measure is to find how many among the 303 identified computers led to an interesting investigation (independently of the outcome). The issue is whether the method identifies computers worth investigating or flags computers clearly of no concern leading to a waste of the time for the security team. Among the 303 flagged computers, the Director of Security found that the investigation was useful for all the identified computers. Indeed, either the computer is declared corrupted and the security team did check it or should have checked it, or the computer is not corrupted and the IPS itself needs to be retuned to reduce the number of alerts raised for non-corrupted computers. This high percentage indicates that the proposed method is already of practical use for the security team. Although the number of non-corrupted investigated computers is high, the non-corrupted computers may reveal events that could not have been identified otherwise. For example, we identified an event where 64 systems tried to access Facebook using a suspicious PHP argument and users who operated Nmap. The computers involved in these two events were identified as non-corrupted but provided an additional insight into the organization's security.

The next step is to assess our method to know if it correctly identifies the (non)-corrupted computers. Each computer will be assigned a color: red (R), yellow (Y), green (G), and an investigation result: corrupted (C) or non-corrupted (NC) computer. A computer that was in the red or yellow regions and was identified as corrupted is a true positive. On the contrary, a computer that was in the green region and was identified as corrupted is a false negative. All combinations of color and investigation result are given in Table 1. Note that when the investigation could not tell if a computer was corrupted or not, we will use O (O stands for "Other"). For example, RO groups computers that are in the red region and that could not be identified as corrupted or non-corrupted by the Director of Security.

Table 1. All Combinations of Color and Investigation Result

Color	Investigation result	Notation	Conclusion
R	C	RC	True Positive (TP)
R	NC	RNC	False Positive (FP)
Y	C	YC	True Positive (TP)
Y	NC	YNC	False Positive (FP)
G	C	GC	False Negative (FN)
G	NC	GNC	True Negative (TN)
R	O	RO	-
Y	O	YO	-
G	O	GO	-

5.2 Results

We studied 23 periods of 6 months from September 1st 2006 to January 31st 2008 with increments of 2 weeks. Period 1 is the period from September 1st 2006 to February 28th 2007. Period 2 covers September 15th 2006 to March 14th 2007, etc. Period 23 defines the period from August 1st 2007 to January 31st 2008. For each period, we extracted the address of the organization computers that raised at least one alert corresponding to an attack towards a computer outside the University of Maryland and calculated the associated metrics. We applied the following thresholds: $C1 = 0.5$ and $C2 = 0.8$ for the coefficient of consecutiveness, $W1 = 2$ and $W2 = 4$ for the number of distinct weeks, $S1 = 1$ and $S2 = 2$ for the number of distinct signatures. For each of the 23 periods of 6 months, our method automatically puts each flagged computers in a green, yellow or red region. According to the identification of the (non)-corrupted computers by the Director of Security, we can calculate 1) the number of true/false positives based on the results in the yellow and red regions, and 2) the number of true/false negatives based on the results in the green region. The results are shown in Table 2.

Note first that the number of computers for which it could not be decided whether they were corrupted or not highly depends on the region. In the red region, they represent 12% (Period 1), 0% (Period 12) and 20% (Period 23). In the yellow region, we find 26% (Period 1), 38% (Period 12), and 36% (Period 23). In the green region, we have 71% (Period 1), 54% (Period 12), and 32% (Period 23). It is interesting to note that often the red region has the lowest percentage and the green region has the highest percentage of computers that could not be clearly identified as (non)-corrupted. This increases the confidence in our method since the computers in the red region should have the highest likelihood of being corrupted and the green region should have a much lower likelihood of being corrupted. This shows that the information provided to the security team should be useful as it seems to rank the computers based on the likelihood of corruption.

Graphs of the evolution of true positives, false positives, true negatives and false negatives over the 23 periods are shown in Figure 4. The results show that the method is improving regarding the number of true negatives. At Period 1, among the computers in the green region (i.e., computer of low concern), only 10% were not corrupted. However, the trend significantly changes over time. At Period 23, among the computers in the green region, 91.7% were not corrupted.

Table 2. Results of the Evaluation

Period	RM	RNM	YM	YNM	GM	GNM	OR	OY	OG	TP	FP	TN	FN	TP/P (%)	FP/P (%)	TN/N (%)	FN/N (%)
1	19	3	24	10	18	2	3	12	49	43	13	2	18	76.8	23.2	10.0	90.0
2	19	4	30	10	17	1	3	13	46	49	14	1	17	77.8	22.2	5.6	94.4
3	21	5	29	8	17	1	3	14	44	50	13	1	17	79.4	20.6	5.6	94.4
4	17	5	34	7	19	1	2	16	39	51	12	1	19	81.0	19.0	5.0	95.0
5	14	5	35	7	17	1	2	15	38	49	12	1	17	80.3	19.7	5.6	94.4
6	13	5	33	6	20	8	2	15	35	46	11	8	20	80.7	19.3	28.6	71.4
7	13	5	30	5	20	14	2	11	40	43	10	14	20	81.1	18.9	41.2	58.8
8	10	4	35	4	18	14	1	13	34	45	8	14	18	84.9	15.1	43.8	56.2
9	10	3	33	5	17	13	2	12	36	43	8	13	17	84.3	15.7	43.3	56.7
10	11	3	29	3	17	11	2	11	39	40	6	11	17	87.0	13.0	39.3	60.7
11	10	2	24	4	17	12	2	8	38	34	6	12	17	85.0	15.0	41.4	58.6
12	12	2	13	5	16	12	0	11	33	25	7	12	16	78.1	21.9	42.9	57.1
13	9	1	13	6	13	12	0	10	34	22	7	12	13	75.9	24.1	48.0	52.0
14	6	0	11	6	13	18	0	6	30	17	6	18	13	73.9	26.1	58.1	41.9
15	3	1	10	7	8	19	0	5	30	13	8	19	8	61.9	38.1	70.4	29.6
16	3	0	6	8	6	20	0	7	30	9	8	20	6	52.9	47.1	76.9	23.1
17	3	2	4	9	5	39	0	6	30	7	11	39	5	38.9	61.1	88.6	11.4
18	3	2	3	10	4	36	0	6	27	6	12	36	4	33.3	66.7	90.0	10.0
19	2	3	3	8	5	39	0	6	28	5	11	39	5	31.2	68.8	88.6	11.4
20	2	3	1	8	6	41	0	6	28	3	11	41	6	21.4	78.6	87.2	12.8
21	2	3	0	8	6	43	0	5	28	2	11	43	6	15.4	84.6	87.8	12.2
22	2	3	0	8	6	44	0	5	26	2	11	44	6	15.4	84.6	88.0	12.0
23	1	3	1	6	4	44	1	4	23	2	9	44	4	18.2	81.8	91.7	8.3

On the other hand, the method identifies a high percentage of true positives at Period 1 but a low percentage at Period 23. At Period 1, the method identified 76.8% of the computers in the red and yellow regions as being indeed corrupted. At Period 23, the method only found 18.2% of corrupted computers in the red and yellow regions. These numbers might indicate that our method is worsening over time. More details are necessary to better understand the reasons for the obtained results. As expected, over time, the security team learned how to integrate the results provided by the IPS in their overall security solution. The number of identified corrupted computers is 61 at Period 1, 41 at Period 12, and only 6 at Period 23. This clearly indicates that the IPS is helping the security team improving the overall organization's security. These numbers help putting in perspective the only 18.2% of corrupted computers in the regions of concern. At Period 23, only 5 computers were placed in the red region and 11 in the yellow region. Among them, 3 computers were incorrectly put in the red region when they were not corrupted and 6 in the yellow region. On the other hand, at Period 23, most computers were placed in the green region (71). Among them, only 4 (5.6%) were incorrectly put in the green region, i.e., they were identified as of low concern when in fact they were corrupted. The method seems thus to be able to correctly identify the biggest volumes of events, i.e. corrupted computers at Period 1 and non-corrupted computers at Period 23.

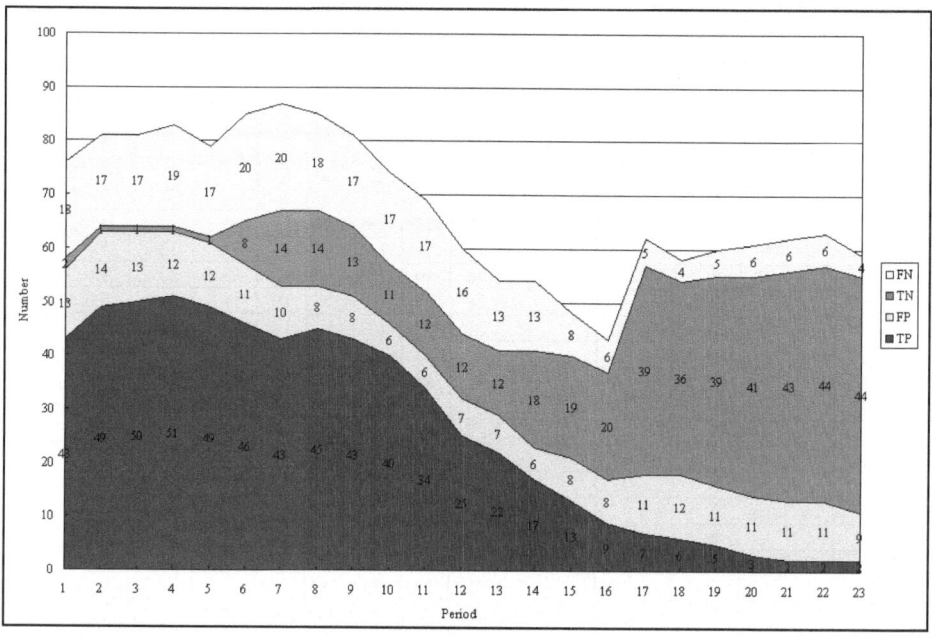

Fig. 4. Evolution of False Negatives (FN), True Negatives (TN), False Positives (FP) and True Positives (TP)

6 Conclusion

We presented a method to extract useful information from imperfect IPS event data in order to rank potentially corrupted computers in an organization. We introduced three metrics to quantify the level of criticality of a computer: the coefficient of consecutiveness, the number of distinct weeks and the number of distinct signatures. The method classifies computers into regions of main concern (red regions), concern (yellow region), or lower concern (green region). We applied the method to IPS event data collected in an organization of about 40,000 computers. We evaluated our method by comparing the results obtained by our method with the identification of (non)-corrupted computers by a security expert. We showed that: 1) the percentage of computers identified as corrupted is higher for computers in the red region than for computers in the green region, 2) the trend of the number of true negatives increases over time, 3) the security team seems to integrate the IPS in their overall organization's security as the number of computers identified as corrupted decreases over time.

Acknowledgements

The authors would like to thank Robin Berthier for the fruitful discussions during the development of this paper.

This research has been supported in part by NSF CAREER award 0237493.

References

1. Bailey, M., Cooke, E., Jahanian, F., Provos, N., Rosaen, K., Watson, D.: Data Reduction for the Scalable Automated Analysis of Distributed Darknet Traffic. In: Proceedings of the USENIX/ACM Internet Measurement Conference, New Orleans (2005)
2. Sung, M., Haas, M., Xu, J.: Analysis of DoS attack traffic data. In: 2002 FIRST Conference, Hawaii (2002)
3. Viinikka, J., Debar, H., Mé, L., Séguier, R.: Time series modeling for IDS alert management. In: Proceedings of the 2006 ACM Symposium on Information, computer and communications security, pp. 102–113. ACM Press, New York (2006)
4. Clifton, C., Gengo, G.: Developing custom intrusion detection filters using data mining. In: MILCOM 2000. 21st Century Military Communications Conference Proceedings, vol. 1 (2000)
5. Cuppens, F.: Managing alerts in a multi-intrusion detection environment. In: Proceedings of the 17th Annual Computer Security Applications Conference, vol. 32. IEEE Computer Society, Los Alamitos (2001)
6. Cuppens, F., Miege, A.: Alert correlation in a cooperative intrusion detection framework. In: IEEE Symposium on Security and Privacy, pp. 202–215 (2002)
7. Julisch, K.: Mining Alarm Clusters to Improve Alarm Handling Efficiency. In: Proceedings of the 17th Annual Computer Security Applications Conference (ACSAC), pp. 12–21 (2001)
8. Julisch, K.: Data mining for Intrusion Detection. Applications of Data Mining in Computer Security. Kluwer Academic Publishers, Dordrecht (2002)
9. Julisch, K., Dacier, M.: Mining intrusion detection alarms for actionable knowledge. In: Proceedings of the eighth ACM SIGKDD international conference on Knowledge discovery and data mining, pp. 366–375. ACM Press, New York (2002)
10. Manganaris, S., Christensen, M., Zerkle, D., Hermiz, K.: A data mining analysis of RTID alarms. Computer Networks 34(4), 571–577 (2000)
11. Pietraszek, T.: Using Adaptive Alert Classification to Reduce False Positives in Intrusion Detection. In: Recent Advances In Intrusion Detection: 7th International Symposium. Springer, Heidelberg (2004)

12. Debar, H., Wespi, A.: Aggregation and correlation of intrusion-detection alerts. Recent Advances in Intrusion Detection. Springer, Heidelberg (2001)
13. Morin, B., Me, L., Debar, H., Ducasse, M.: M2D2: A Formal Data Model for IDS Alert Correlation. In: Recent Advances in Intrusion Detection: 5th Internatonal Symposium. Springer, Heidelberg (2002)
14. Ning, P., Xu, D., Healey, C., Amant, R.S.: Building attack scenarios through integration of complementary alert correlation methods. In: Proceedings of the 11th Annual Network and Distributed System Security Symposium, pp. 97–111 (2004)
15. Valdes, A., Skinner, K.: Probabilistic Alert Correlation. In: Proceedings of the Fourth International Workshop on the Recent Advances in Intrusion Detection (2001)
16. Valeur, F., Vigna, G., Kruegel, C., Kemmerer, R.: Comprehensive approach to intrusion detection alert correlation. IEEE Transactions on Dependable and Secure Computing 1(3), 146–169 (2004)

Security Threats to Automotive CAN Networks – Practical Examples and Selected Short-Term Countermeasures

Tobias Hoppe, Stefan Kiltz, and Jana Dittmann

Otto-von-Guericke University of Magdeburg
ITI Research Group on Multimedia and Security
Universitaetsplatz 2
39106 Magdeburg, Germany
{tobias.hoppe,stefan.kiltz,
jana.dittmann}@iti.cs.uni-magdeburg.de

Abstract. The IT security of automotive systems is an evolving area of research. To analyse the current situation we performed several practical tests on recent automotive technology, focusing on automotive systems based on CAN bus technology. With respect to the results of these tests, in this paper we discuss selected countermeasures to address the basic weaknesses exploited in our tests and also give a short outlook to requirements, potential and restrictions of future, holistic approaches.

Keywords: Automotive, IT-Security, Safety, Practical tests, Exemplary threats and countermeasures.

1 Introduction / Motivation

The complexity of current automobiles is constantly increasing. Modern cars contain a variety of Electronic Control Units (ECUs) that are connected to each other via different kinds of bus systems in order to reduce the amount of cables needed.

But this growing complexity and added functionality might increasingly attract attackers to misuse these systems for their individual purposes, which has already been speculated about by IT security researchers like Eugene Kaspersky [1]. Another factor is the trend of increasing information exchange between automotive systems and the outside world: For example, [2] demonstrated a technique to inject forged traffic information into navigation systems using the wireless protocols RDS (Radio Data System and TMC (Traffic Message Channel). And future technologies like car-to-car (C2C) [3] or car-to-infrastructure (C2I) communication are already discussed to implement several new automotive applications.

Looking at these trends and the high *safety* risks of such fast-moving computing systems, automotive IT *security* is an important emerging area of research: Unlike within typical home PC systems, a successful security violation on an automotive IT system might not only cause nuisance and disclose sensitive data but also directly endanger the safety of its human users (drivers, occupants) and environment [4].

In this paper we illustrate that already today the IT security of current automotive systems has to be addressed more forceful. We demonstrate this by summarising

M.D. Harrison and M.-A. Sujan (Eds.): SAFECOMP 2008, LNCS 5219, pp. 235–248, 2008.
© Springer-Verlag Berlin Heidelberg 2008

results of several practical tests we performed on current automotive hardware based on the controller area network (CAN) bus system [5]. The basic weaknesses exploited in these tests are identified to discuss potential countermeasures for the future. Though suggestions for holistic approaches for long-term solutions are shortly introduced, we do focus on short-term countermeasures which address the basic weaknesses identified so far and might help achieve a reasonable security compromise until such a major redesign in future.

The paper is structured as follows: In the following section 2 we shortly present the state of art of automotive IT security measures, starting with existing applications. In the section 3 we describe our practical tests investigating attacks on exemplary automotive components, which have been partly extended for this publication and illustrate potential impacts to safety and comfort. They also serve as a basis for section 4 to identify what security aspects have been violated and which basic weaknesses have been exploited in these tests. In that section we also discuss potential countermeasures (some of which could already been demonstrated practically) as well as their potential and restrictions. The last section 5 concludes this paper with a summary and an outlook.

2   State of the Art

Whilst car manufacturers have improved the *safety* of their automobiles a lot during the past decades, adequate holistic concepts for IT *security* are not available yet. As state of the art, IT security mechanisms based on encryption or digital signatures can already be found in today's cars [6], but only in a very local scope protecting single components or functionalities:

Anti-theft systems like central locking or the immobiliser use cryptographic protocols. One example is the keyless entry which typically uses a cryptographic challenge-response to protect against replay-attacks: The car generates a random value (challenge) which has to be processed by the key remote using its secret key. After passing back the correct result, the car doors will be opened. Even if an attacker records the entire communication between the car and the key remote during this process, a replay of these logs does not allow him to enter the car in the absence of the authentic driver. However, such systems have to be designed carefully. Recently a successful side channel attack on the proprietary system "Keeloq" has been presented by [7]. It yields a manufacturer specific master key allowing an attacker access every car after sniffing two messages from a distance up to 300 ft.

Other potential attack targets car manufacturers are trying to protect are the contents of memory chips, especially of rewritable flash memory holding updateable programme code and configuration data. One motivation is the protection of their intellectual property represented by this data. Other threats are posed by common attacker types like car tuners who frequently modify programme code or configuration data to achieve a higher power output (or, increasingly, also less fuel consumption / eco tuning). Since such unauthorised manipulations also affect issues like safety and liability, therefore the integrity of flash updates has to be ensured, too. In the context of the HIS ("Herstellerinitiative Software") group in Germany [8] several car manufacturers joined and developed a common specification for secure flashing, which employs digital signatures as cryptographic mechanism.

Although these examples for sound IT security approaches can already be found in current cars of many manufacturers, they are only covering a very local scope. They are not conceived to provide a holistic protection for the entire system. This is demonstrated in the following section 3 by presenting results from practical tests we performed in the past months.

3 Practical Demonstration of Exemplary Automotive IT Security Threats

Several practical test setups have been created to demonstrate IT security threats of current automotive technology, to analyse potential safety implications and to define and evaluate first countermeasures. In this section we summarise the basic principles and results of these tests to give an overview on our previous work. While most of these tests have been described in more detail in previous publications, we also have extended some of them recently to offer new results for this publication.

The tests were performed on a test setup consisting of real automotive hardware. It contains a wiring harness and different electronic control units (ECUs) of a recent model (built in 2004) of a big international car producer. Cars of this series use the CAN bus for the communication between the separate devices. Supported by different bus interfaces, a PC system can be used to investigate or interact with the automotive system. Fig. 1 illustrates this test setup.

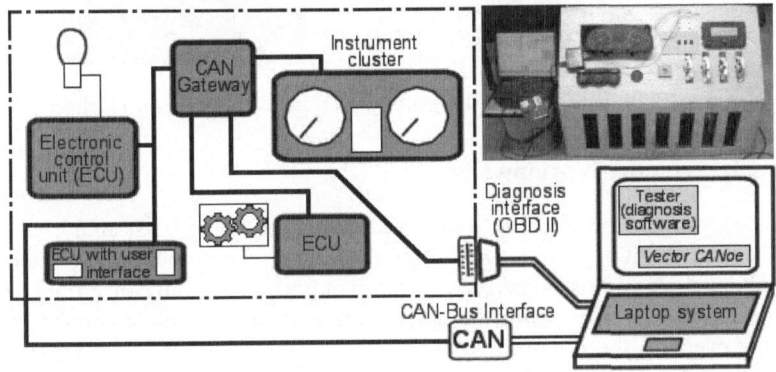

Fig. 1. Illustration of the practical test environment of automotive hardware

3.1 Analyses on the Electric Window Lift

The first potential attack target we investigated was the electric window lift. Early practical tests performed on this target were done within a simulation environment. For this purpose we used a simplified car environment which is part of CANoe, an established development and simulation software from Vector Informatik [9] widely used throughout the automotive industry.

In this test, a few lines of malicious code have been added to any ECU attached to the simulated Comfort CAN subnetwork. By waiting until some condition occurs (in this case when the car's speed exceeds 200 km/h) the code then replays the CAN message containing the flag for opening the driver window. Although the real console still sends its messages in the same frequency indicating that no button is currently pushed, the window opens and blocks until the driver reacts by pushing the "close" button. More details about this test can be found in [10] (as well as [4] and [11]).

Meanwhile, the completion of the aforementioned physical test setup allowed us to demonstrate similar results on a real window lifter (being part of the door control modules in our practical test setup, see left part of Fig. 2) during a student project.

After identifying the CAN messages relevant for triggering the window lifts, an attack strategy similar to the simulated attack has been conceived: Every time a CAN message is observed on the comfort CAN subnetwork containing a flag set to open the window, a new copy is generated onto this bus specifying an opposite (close) or cleared (no action) flag. This practical test on current real automotive hardware constitutes a Denial-of-Service (DoS) attack on the window lifts (availability aspects).

The implications of a successful attack can affect both, *comfort* (the window cannot be moved any more) and *safety* (if the shocked driver loses control).

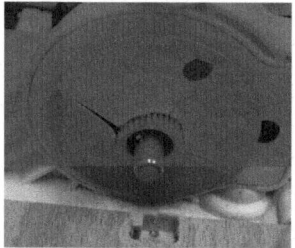 

Fig. 2. Electric window lift (section 3.1), Indicator bulb, off (section 3.2)

3.2 Analyses on the Warning Lights

As a second target, the warning lights (the indicators) have been analysed. Amongst others, the anti-theft system triggers them once an intrusion into a parked and secured car is detected. A common scenario is an unauthorised opening of a door. Triggered by a corresponding event from the door contact sensor, the door control module reports this event to the Comfort system ECU, which also contains anti theft system functionality. Now, an alarm is generated for a few minutes by sending alternating command telegrams to the vehicle electronics ECU to set or unset the warning lights.

This scenario served as another test case. In our evaluation we found that every component with access to the Comfort CAN subnetwork (this might be an original ECU after the injection of malicious code or an additionally attached device like a developer's circuit board) can heavily interfere with this process by immediately sending an "off" command once an "on" command (sent by the Comfort system ECU) is observed. Even though the "on" commands do not get removed from the Comfort CAN subnetwork, in our tests [11] this attack proved to be quite powerful: The indicator bulbs (see right part of Fig. 2) stay completely dark most of the time, while (apparently due to timing reasons) sometimes only a short, weak glowing appeared (though this is not expected to be noticeable through orange glass covers).

While for this attack target *comfort* implications are hardly relevant, it could affect the *safety* e.g. if it activates while the car broke down and hinders it to indicate a warning to other road users.

3.3 Analyses on the Airbag Control System

Another automotive component which we checked for security vulnerabilities was the airbag control system. In this attack, which is described in more detail in [12], the airbag control module was removed from the system. This might be done by an attacker to endanger the cars occupants (by the loss of a safety system), but much more common purposes are monetary interests. Unfortunately, as more and more police and press reports state, the theft of airbag systems is already quite common.

Within the attack examined, the attacker tries to suppress several signs of this removal which might sooner or later raise suspicion. One example clearly visible to the driver is the airbag warning lamp within the within the instrument cluster which indicates a failure (or absence) of the airbag control system. Another sign would be the failure of a communication with the "defective" system using the diagnostics protocol, which might be performed in the car service station by connecting to the car's diagnostics interface.

In [12] we managed to emulate the behaviour of a fully functional airbag control module within a diagnostics session by any device with access to the powertrain CAN subnetwork (where the removed system also was attached to). In practice this might be another original device after some software manipulation or an additionally attached cheap circuit board; in our tests we used a PC system attached to the powertrain network via the CAN bus interface. After recording the reactions to diagnostic queries during a regular diagnostics session, these replies could successfully replayed in the absence of the airbag control module. The diagnostics software reports the presence of the device (including its name, part no., etc.) and attests the absence of any error conditions.

Since this technique only covers the diagnostics protocol so far, it does not yet also lead to an expiration of the airbag warning light within the instrument cluster, which is triggered by the CAN gateway ECU. To monitor the presence of each other, ECUs generally do not use the diagnostics protocol, but monitor other messages usually transmitted by the respective device – in this case by the airbag control module. In [12] we preliminarily addressed this problem by removing the airbag system from the gateway's device list. To the gateway it looks as if no airbag system was installed in the first place (which is an option in some countries), therefore no error condition is generated and the airbag warning light is not triggered. However, for an attacker this approach still had a few drawbacks. One is the removed device list entry which might raise attention when listing it during a diagnostics session. An attentive driver might also discover that, directly after entering ignition state, the airbag lamp does not show up shortly during the startup checks.

In additional tests we conducted for this publication, we could also practically demonstrate a more appropriate solution: we identified the relevant CAN message the gateway ECU expects from the airbag control module. This allows emulating also the general communication of the airbag control system (beyond the diagnostics protocol already covered). By replaying this message in its original frequency onto the power-train CAN subnetwork, the malicious device can also pretend the presence of the airbag system among the other ECUs. Since this message also contains a bit flag to set and unset the airbag lamp in the instrument cluster, also a successful startup check could be emulated this way by the malicious device.

While not reducing *comfort* (the driver will not notice any lack of functionality in regular operation) potential *safety* implications in emergency cases could be severe.

4   Analysis of the Underlying Problems; Capabilities and Restrictions of Potential Countermeasures

In this section we identify basic weaknesses in today's automotive systems that made the exemplary attacks in our practical tests possible. Based on this, potential counter-measures for future systems are discussed, some of which have already been tested in our test environment.

In the practical tests described in section 3, we accessed the car's IT infrastructure from within its internal bus systems. In the scope of this paper, we do not focus on the question, what technique a potential attacker might have chosen to get into this posi-tion. As already mentioned earlier, he might simply have placed some additional circuit board onto the bus wires, like we did with the CAN bus adapter we used (on most current cars adequate, exposed positions can be found where wires of the corre-sponding buses are located). But an attacker could also reduce the required amount of physical access and equipment by injecting malicious code into an existing device, e.g. by exploiting unsecured diagnostics interfaces, manipulated update discs for me-dia systems distributed by social engineering or exploiting potential weaknesses of wireless communication systems (like future C2C/C2I systems).

Consequently, also the internal communication of a car will have to be secured more in future. The following five central security aspects and privacy concerns known from IT security help to identify weaknesses in section 4.1 and discuss poten-tial countermeasures afterwards:

- Confidentiality / Privacy
- Integrity
- Availability
- Authenticity
- Non-Repudiation

4.1   Analysis of Underlying Problems Relevant for the Exemplary Tests

The exemplary attacking strategies that we utilised in the practical tests primarily exploited drawbacks of the CAN bus protocol frequently employed in today's auto-mobiles. For this reason we concentrate on discussing exemplary requirements for a secure automotive bus communication, using the CAN bus as example.

Though the CAN bus does provide measures to ensure aspects like the *integrity* of the transmitted information from the functional safety perspective (protection against unintended transmission errors by Cyclic Redundancy Checks / CRC), the existing measures do not meet the requirements from the IT security perspective. For example, a CRC checksum is not sufficient for detecting falsified contents of a CAN message which has intentionally been generated by an attacker – just because he would also re-adjust the CRC information accordingly.

When looking at the IT security aspects listed at the beginning of section 4, for none of them sufficient measures are provided at the CAN bus level, yet:

Confidentiality / Privacy: A message sent onto a CAN bus can at least be received by all other ECUs connected to that bus system. Based on the type identifier (ID) of the message, each ECU decides if or if not to use it. If a gateway is amongst these nodes and transmits the message into another subnetwork, even more nodes are affected. So in general, each of the receiving nodes can principally read the up to 8 bytes transported with each message. However, in some applications the transmitted information might be regarded confidential; by collecting information from CAN bus systems, an attacker could for example be empowered to conclude privacy-relevant information (e.g. driving behaviour) of the current (or during diagnostic sessions even about previous) drivers. Encryption or anonymisation would reduce threats like these.

Integrity: With reference to the example given at the beginning of this subsection, a checksum is not a sufficient measure to ensure integrity from the IT security perspective. Appropriate measures known from desktop IT would be cryptographic hash functions, message authentication codes (MAC) or digital signatures, which cannot be "re-adjusted" by an attacker without knowledge of a secret (private) key.

Authenticity: The CAN bus protocol provides no authenticity measures, CAN bus messages do not even contain a sender or receiver address. If a node is not configured to be a regular receiver of the respective type of message (with respect to its ID), the message and its contents are ignored. The usual sender of each message type is implicitly known, but a node has no possibility to verify this assumption. As our practical tests showed, malicious nodes can easily spoof messages usually sent by others. Receiving devices cannot detect that these come from a non-authentic source, rely on the forged contents and consequently perform unauthorised actions. In future automotive networks this could be addressed e.g. by MACs or digital signatures.

Availability: Using techniques like repeatedly sending unauthorised error flags or high-priority messages, a malicious node can easily overload an entire CAN (sub-) network. During such a DoS-attack, none of the other devices in this network would be available. To ensure availability in the face of DoS-attacks is a difficult problem in general. The specification of the oncoming FlexRay bus system [13] considers the option of disconnecting malfunctioning devices or branches from the network by node-local or central "bus guardians". However, this also seems to be more a *safety* measure against unintended malfunctions than to address *security* viewpoints.

Non-repudiation: After an incident like the spoofing attacks in our practical tests it is hard for the attacked devices to deliver proof of their innocence (i.e. that they did really receive such a malicious command or, respectively, that they did not send such a message). In the absence of mechanisms for the four aspects above, this is even more difficult to ensure.

In the following two subsections, exemplary countermeasures are being discussed that could help to increase the IT security of future automobiles by addressing these problems like the basic weaknesses exploited in our practical tests.

As mentioned before, a holistic approach obviously would be the best choice. But ensuring a maximum number of the IT security aspects introduced before would require an expensive, major redesign. In section 4.3 some current efforts of automotive IT security researchers are described.

While such extensive solutions are expected to be inevitable in the long-term, simpler and cheaper solutions might be a way to address the most urgent weaknesses in the near future. In section 4.2 we therefore focus on discussing first concepts that might help to address basic weaknesses which made our practical tests succeed, which are mainly the missing *authenticity* measures in CAN communication.

4.2 Discussion of Short-Term Countermeasures to Address the Demonstrated Threats, Their Potential and Restrictions

To implement a minimal protection against basic attacking techniques like the ones presented in the practical tests, in this subsection we discuss two different approaches:

Approach a) Intrusion Detection techniques
Often when a given system has no effective means to prevent some kind of attacks initially, it should at least be tried to detect them. In the desktop IT domain such components are usually called Intrusion Detection Systems (IDS) [14]. Once an incident has been discovered by such a system (having discovered suspicious activity patterns in the network activity or at some end system), it might generate warnings or trigger reactions to limit the consequences of the attack (in that case such systems are often also called Intrusion Response or Intrusion Prevention Systems / IPS).

A potential application of Intrusion Detection approaches to automotive systems could be useful as well: In an emergency case where an attack is detected which has not been thwarted by other existing measures, a warning could be generated to the driver and advise him to perform an appropriate reaction (e.g. stop the car at the next safe position). Automatic, autonomous reactions of an automotive IPS could also be discussed as a further option. However, due to the high safety risks in an automotive environment and the ever-present risk of potential false positive classifications or the choice of inappropriate reactions, such an extension would have to be developed with great care.

With reference to the practical attacks investigated in section 3, we already identified several patterns which could be applied to detect such attacks. We shortly introduce these patterns below, one of which we have already tested in practice and discussed in more detail in the context of [15].

Pattern 1: Increased Message Frequency
Often CAN messages of a given ID are broadcasted by a single sending device and in a constant frequency. In our examples this applies to the state of the window switches (first part of section 3.1) as well as to the message triggering the warning lights (section 3.2). As we demonstrated in the tests, another (malicious) device with access to the respective (sub-) network can simply add contradicting messages of the same type to the bus communication to achieve unauthorised actions by the receivers. However,

since removing existing messages is a lot harder to achieve, this often results in a notably higher occurrence rate and frequently changing semantic contents of messages having the respective ID. Such features can serve as a simple detection pattern for this kind of attack, indicating authenticity and integrity violations. We could already demonstrate the effectiveness of this approach practically: in [15] we implemented this detection pattern for a prototypical IDS component and successfully tested it within our setup for the attack on the warning lights described in section 3.2.

Pattern 2: Obvious Misuse of Message-IDs

In the practical tests, unauthorised messages have been put on the bus by a device different from the original sender. Since the receiving nodes have no proof of the *authenticity* of the message (i.e. if it really has been sent by the original sender), this attack proved to be very effective. However, these injected messages will also arrive at the original sending ECU. Currently, from the perspective of an attacker, this is no serious problem, because that device is not expected to evaluate this type of message, if this is usually only sent by itself. Consequently, with little effort some IDS functionality could be added to any ECU looking for suspicious incoming messages like such ones using its exclusive message ID. This could also be applied to gateway ECUs: Given, a gateway is configured to pass messages of type m_a from a subnetwork n_a to another subnetwork n_b using the (maybe differing) ID m_b. If in this setup a malicious message with the ID m_b is injected to the target network n_b (which would not be visible to the originally sender, which is only responsible to detect forged messages of type m_a in the source network n_a), the gateway would be able to detect this incident (unauthorised use of its exclusive ID m_b within n_b) accordingly.

Pattern 3: Low-Level Communication Characteristics

In addition to the techniques chosen in the previous patterns, the last pattern discussed in this section uses a substantially different approach to detect forged messages that have been injected into a CAN network from an arbitrary bus location. While the previous patterns only regarded information available from the data link layer (OSI level 2), we assume that for this purpose also information from the physical layer (OSI layer 1) could be useful: To put a CAN message onto the bus, every ECU has to pass it to some CAN controller which generates the corresponding electrical signal at the bus wires. These controllers are available from different manufacturers (partly as CPU integrated circuitry). While all of them are supposed to fulfil the CAN specifications in the end, it might be possible to identify features characteristic for each individual chip when looking more closely at the electrical signal generated. Such features might be voltage amplitudes and their stability, the shape of the clock edges, propagation delays, signal attenuation due to wire lengths etc. While still being within valid intervals or above/below acceptable thresholds, these low-level communication characteristics could be analysed by a special detection unit to identify the authentic device which has sent the current message. Such a system could provide useful additional information allowing the verification of the *authenticity* of sending nodes within CAN networks (without the need of any change to existing bus specifications).

Discussion of restrictions

However, with respect to the three patterns mentioned above, a few restrictions can be identified: As already mentioned, pattern 1 is only applicable to messages transmitted cyclically. It cannot be applied to message types that only appear occasionally (e.g.

which are only sent once as an indication of some event). Furthermore, pattern 1 and pattern 2 can obviously only be used to detect an incident, as long as the original sender is still present and functional. Pattern 3 is principally capable of compensating these restrictions of pattern 1 and 2. However, if malicious messages are sent by the same device (i.e. the attacker managed to modify the original sending ECU directly, e.g. by injecting malicious code), their low-level characteristics do not differ. Another expected problem might be that different ECUs can use the same CAN controllers (same manufacturer, same product line). Amongst these, the differences can be expected to be much smaller. So an interesting point of research would be finding appropriate features with an adequate resolution also for these cases. Also the problem of a legitimate swap of an ECU (e.g. due to component failure) would have to be addressed.

Approach b) Proactive Forensics Support
Assuming that IT security related attacks will increase in future, also post-incident inquiries on automotive systems might get more and more common (driven by police, insurance companies etc.). As the practical attack introduced in section 3.3 shows, finding a responding and faultless airbag control system during a diagnosis session is no reliable indication against a theft suspicion. Currently on the one hand diagnostics are only designed to detect unintended failures (safety violations) and are not secured against intended attacks (security violations). On the other hand, it would be too time consuming to dismantle a huge set of potentially affected cars to look for the physical presence of the components – and a clever attacker could have even placed dummies.

To also speed up the search for suspected security incidents, the diagnosis system would have to be extended accordingly. Not only safety related events (more or less random component failures, blackouts and other malfunctions) would have to be logged but also additional information especially relevant for security related inquiries. This might contain information about flash operations (updated device, timestamp, source etc…), systems being connected from the outside, power downtimes and many more. If present, also the intrusion detection components discussed above could notify the black box about suspicious events, e.g. to increase the logging intensity. To protect this sensitive data and avoid additional costs for the regular components, it could be stored in a single protected device like a black box and additionally get configured to be privacy preserving for the drivers.

When discussing this approach, also a few downsides of this approach have to be mentioned. Although memory devices are constantly getting cheaper and more powerful, especially the physical protection requirements would make such a black box relatively expensive without an obvious benefit to the customer. In the past, such a system for safety purposes (accident recorder) was already offered as option (e.g. [16] by an international car manufacturer. Although due to concerned customers it was made possible to erase the stored information at any time, they did not accept the system and it finally did not establish at the market in great numbers. So maybe privacy concerns were not addressed well enough in the system and its marketing. Another problem would be that malicious code, once present in the system, might try to flood the data recorder by spoofing useless information. This way an attacker might try to overwrite stored evidence or to hide them in a vast number of irrelevant entries.

4.3 The Need for Long-Term Solutions for Holistic Automotive IT Security Concepts, Their Potential and Restrictions

In the long run, holistic security concepts for automotive systems are inevitable. Research about an appropriate basis for the implementation of such security measures has just started in the last few years (e.g. [17]). This subsection gives a short overview on selected approaches currently discussed, their potential and remaining restrictions.

Looking at the special requirements of automotive systems and their role in every day life yields a few important requirements individual to this domain: Unlike home or office computer systems, cars are a kind of target frequently being physically exposed to different kinds of attackers (even the owner can be interpreted as an attacker if he tries to 'tune' or unlock some features in his home garage). This means, beneath a protection against software-based attacks like prevailing in desktop IT, the design of a holistic security concept for automotive IT systems should also put special focuses on hardware-related attacks. Another important factor is economy, i.e. the high cost restrictions car manufacturers have to face. The components to establish a holistic automotive security platform have to be as cheap as possible.

Especially to guarantee aspects like authenticity or integrity, current IT security measures rely on asymmetric cryptography which is known to be computationally very expensive. To reduce computation and therefore hardware costs, alternative asymmetric algorithms like elliptic curve cryptography (ECC) are currently discussed [6], which are more efficient (compared to RSA, for example). An additional measure to address this is implementing these consuming algorithms in hardware.

To provide trustworthy computing platforms in the desktop IT domain, several international companies joined in the Trusted Computing Group (TCG) [18]. So-called Trusted Platform Modules (TPMs) developed by the TCG can already be found in many computers sold today and first security-related applications increasingly use the features of these hardware components. The potential of the underlying Trusted Computing (TC) technology for the protection of automotive IT systems is currently being researched (for example see [19]). Due to the special requirements for the automotive domain (see above) current TPMs have been identified as inappropriate for the automotive application. Since current TPMs are separate chips being connected via bus systems, they are vulnerable to hardware attacks and are not suited for the automotive application with users not being trustworthy. Instead, one-chip solutions are being discussed combining CPU and TPM in a single, secured chip. To be as cost efficient as possible, it might only contain the least subset of TC functionality necessary for the automotive application.

Once such a secure hardware basis will be available in future, the automotive applications will also need to use these newly provided functions in order to really tap the potential this new security basis offers. So we expect a major redesign of automotive components and networks to be necessary in that stage. With reference to the results of our practical tests in section 3, the following example illustrates this: A car manufacturer might decide to utilise such an automotive Trusted Computing basis only for securing the different kinds of software updates (flashing, update media etc.) and selected sensitive information like the mileage counter. Consequently, this will not cover attacks from the bus level, if the communication between single, protected ECUs will still use unsecured communication channels (like automotive bus systems established today – at least an additional security layer would be required on top that utilises the functions provided by the TC basis).

Other remaining questions are how to keep the deployed crypto algorithms up to date to face the continuous improvements in cryptanalysis. Currently, the life cycle of cryptographic algorithms is significantly lower than the typical life time of current cars (which might easily be on the road for around 20 years). Hardware implementations of cryptographic algorithms (as discussed) are performing better and are cheaper than software implementations. On the other hand they are harder to maintain. Field-Programmable Gate Array (FPGA) chips might be a compromise to address this.

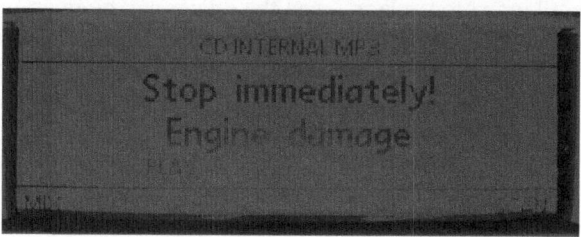

Fig. 3. Exemplary low-tech attack on multimedia system interfaces

Besides the fact that every future automotive security solution will only be a compromise between the achievable security level and the resulting costs, the following last scenario demonstrates that even a technically perfect IT security solution (if actually possible) could not be expected to provide a full protection against intended attacks without respecting the human factor, as already known from the desktop IT security domain. Users tend to ignore warning messages and click them away if they bother them too frequently (e.g. whilst surfing through the web). Others enter sensitive information into forged phishing web pages because an authentic looking email advised them to do so.

As an example for such "Social Engineering" attacks in the automotive domain we prepared a multimedia disc containing MP3 music. An attacker might give or send this disc to his victim as a 'kind' gift, knowing that the victim might listen to it at his next car ride. The multimedia system, which is part of our automotive test environment, plays the music and, for comfort reasons, always shows artist and title information about the current track (read from tag information contained) on its display using a large font. After a few regular songs, a specially prepared section might have been inserted by the attacker. In our tests we have split one song into short fragments and specified a seriously looking warning message as track information on every second fragment, while letting the entries in the other fragments (nearly) blank. When the player reaches this location during playback, it starts to display a flashing warning message (Fig. 3). This attack might even get extended by mixing a horrific warning signal into the sound material. Frightened by this situation, the driver might not realise the simplicity of this hoax and be seduced to follow such a malicious advice, and e.g. stop the car immediately – while the system still operates as designed.

Obviously, this attack does not need to break any *technical* security mechanisms in order to be effective. Beneath a secure technical platform, for a sound design of an automotive system in its entirety also *non-technical* aspects need to be addressed – like a very careful design of the user interfaces. For example, passing metadata of entertainment media (like MP3 tags) also to the instrument cluster (which seems not to be supported in our test setup) would be even more critical. Where such arbitrary information is to be displayed, the designers should take great care to always emphasise the context of information being displayed. Although it consumes a bit more valuable display area, leading "artist:" or "title:" strings in the same font size, which are displayed by default, might be an appropriate measure to address this.

5 Summary and Outlook

With the focus on CAN based attacks on automotive IT systems, in this paper we motivated the development of more efficient automotive IT security measures in the future. Based on the results from our practical tests, we identified basic weaknesses in today's automotive communication networks and discussed future countermeasures. In this publication we focused on short-term solutions addressing the most basic weaknesses that made our test results possible. We discussed a few exemplary approaches for such mechanisms (some of which we already tested in practice) with their individual advantages, potential and restrictions. In the long run, holistic long-term solutions will be inevitable. We shortly introduced some basic approaches that are currently discussed by automotive IT security researchers and also discussed exemplary advantages, potential and restrictions of these more holistic approaches.

Acknowledgments. The work described in this paper has been supported in part by the European Commission through the EFRE Programme "Competence in Mobility" (COMO) under Contract No. C(2007)5254. The information in this document is provided as is, and no guarantee or warranty is given or implied that the information is fit for any particular purpose. The user thereof uses the information at its sole risk and liability.

European Commission
European Regional Development Fund
INVESTING IN YOUR FUTURE

References

1. Kaspersky, E.: Viruses coming aboard?, Viruslist.com Weblog January 24, 2005 (June 2008), http://www.viruslist.com/en/weblog?discuss=158190454&return=1
2. Barisani,A., Daniele, B.: Unusual Car Navigation Tricks: Injecting RDS-TMC Traffic Information Signals. In: Can Sec West, Vancouver (2007)
3. Car-2-Car Communication Consortium (June 2008), http://www.car-2-car.org/
4. Lang, A., Dittmann, J., Kiltz, S., Hoppe, T.: Future Perspectives: The Car and its IP-Address - A Potential Safety and Security Risk Assessment. In: Saglietti, F., Oster, N. (eds.) SAFECOMP 2007. LNCS, vol. 4680. Springer, Heidelberg (2007)
5. BOSCH CAN, Website (June 2008), http://www.can.bosch.com/

6. Wolf, M., Weimerskirch, A., Wollinger, T.: State of the Art: Embedding Security in Vehicles. EURASIP Journal on Embedded Systems 2007, 16 (2007); Article ID 74706, 16 pages, 2007. doi:10.1155/2007/74706

7. Press release of Ruhr-Universität Bochum: Remote keyless entry system for cars and buildings is hacked, may 31st, Link (2008), http://www.crypto.rub.de/imperia/md/content/projects/keeloq/keeloq_en.pdf

8. HIS: Herstellerinitiative Software (June 2008), http://www.automotive-his.de/

9. Vector Informatik (June 2008), http://www.vector-informatik.com/

10. Hoppe, T., Dittmann, J.: Sniffing/Replay Attacks on CAN Buses: A Simulated Attack on the Electric Window Lift Classified using an Adapted CERT Taxonomy. In: 2nd Workshop on Embedded Systems Security (WESS 2007), A Workshop of the IEEE/ACM EMSOFT 2007 and the Embedded Systems Week, October 4 (2007)

11. Hoppe, T., Kiltz, S., Lang, A., Dittmann, J.: Exemplary Automotive Attack Scenarios: Trojan horses for Electronic Throttle Control System (ETC) and replay attacks on the power window system. In: Automotive Security - VDI-Berichte 2016, 23. VDI/VW Gemeinschaftstagung Automotive Security, Wolfsburg, Germany, 27-28 November 2007, pp. 165–183. VDI-Verlag (2007) ISBN 978-3-18-092016-0

12. Hoppe, T., Dittmann, J.: Vortäuschen von Komponentenfunktionalität im Automobil: Safety- und Komfort-Implikationen durch Security-Verletzungen am Beispiel des Airbags. In: Sicherheit 2008; Sicherheit - Schutz und Zuverlässigkeit, Saarbrücken, Germany, April 2008, pp. 341–353 (2008) ISBN 978-3-88579-222-2

13. FlexRay - The communication system for advanced automotive control applications (June 2008), http://www.flexray.com/

14. Stakhanova, N., Basu, S., Wong, J.: A Taxonomy of Intrusion Response Systems. nternational Journal of Information and Computer Security 1(1), 169–184 (2007)

15. Hoppe, T., Kiltz, S., Dittmann, J.: IDS als zukünftige Ergänzung automotiver IT-Sicherheit. In: DACH Security 2008, June 24-25, 2008, Technische Universität Berlin (to appear, 2008)

16. Website Kienzle-Automotive, product page of the Unfalldatenspeicher UDS system (June 2008), http://kienzle-automotive.com/index.php?108&tt_products=33

17. Jan Pelzl: Secure Hardware in Automotive Applications. In: 5th escar conference – Embedded Security in Cars, November 6./7, Munich, Germany (2007)

18. Trusted Computing Group (June 2008), https://www.trustedcomputinggroup.org/

19. Bogdanov, A., Eisenbarth, T., Wolf, M., Wollinger, T.: Trusted Computing for Automotive Systems; In: Automotive Security - VDI-Berichte 2016, 23. VDI/VW Gemeinschaftstagung Automotive Security, Wolfsburg, Germany, 27-28 November 2007. VDI-Verlag, pp. 227-237, (2007) ISBN 978-3-18-092016-0

Constructing a Safety Case for Automatically Generated Code from Formal Program Verification Information

Nurlida Basir[1], Ewen Denney[2], and Bernd Fischer[1]

[1] ECS, University of Southampton, Southampton, SO17 1BJ, UK
{nb206r,b.fischer}@ecs.soton.ac.uk
[2] USRA/RIACS, NASA Ames Research Center
Mountain View, CA 94035, USA
Ewen.W.Denney@nasa.gov

Abstract. Formal methods can in principle provide the highest levels of assurance of code safety by providing formal proofs as explicit evidence for the assurance claims. However, the proofs are often complex and difficult to relate to the code, in particular if it has been generated automatically. They may also be based on assumptions and reasoning principles that are not justified. This causes concerns about the trustworthiness of the proofs and thus the assurance claims. Here we present an approach to systematically construct safety cases from information collected during a formal verification of the code, in particular from the construction of the logical annotations necessary for a formal, Hoare-style safety certification. Our approach combines a generic argument that is instantiated with respect to the certified safety property (i.e., safety claims) with a detailed, program-specific argument that can be derived systematically because its structure directly follows the course the annotation construction takes through the code. The resulting safety cases make explicit the formal and informal reasoning principles, and reveal the top-level assumptions and external dependencies that must be taken into account. However, the evidence still comes from the formal safety proofs. Our approach is independent of the given safety property and program, and consequently also independent of the underlying code generator. Here, we illustrate it for the AutoFilter system developed at NASA Ames.

Keywords: Automated code generation, formal program verification, Hoare logic, fault tree analysis, safety case, Goal Structuring Notation.

1 Introduction

Model-based design and automated code generation have become popular, but substantial obstacles remain to their widespread adoption in safety-critical domains: since code generators are typically not qualified, there is no guarantee that their output is safe, and consequently the generated code still needs to be fully tested and certified. Here, formal methods such as formal software safety certification [6] can be used to demonstrate safety of the generated code (i.e., that the execution of the code does not violate a specified property) by providing formal proofs as explicit evidence or *certificates* for the

M.D. Harrison and M.-A. Sujan (Eds.): SAFECOMP 2008, LNCS 5219, pp. 249–262, 2008.
© Springer-Verlag Berlin Heidelberg 2008

assurance claims. However, several problems remain. For automatically generated code it is particularly difficult to relate the proofs to the code; moreover, the proofs are the final stage of a complex process and typically contain many details. This complicates an intuitive understanding of the assurance claims provided by the proofs. Hence, it is important to make explicit which claims are actually proven, and on which assumptions and reasoning principles both the claims and the proofs rest. Moreover, the complexity of the tools used can lead to unforeseen interactions and thus causes additional concerns about the trustworthiness of the assurance claims. We thus believe that traceability between the proofs on one side and the certified program and the used tools on the other side is important to gain confidence in the formal certification process.

Here, we address these problems and present an approach currently under development to systematically derive safety cases from information collected during the formal software safety certification phase, in particular the construction of the necessary logical annotations. The purpose of these safety cases is to provide a "structured reading guide" for the program and the safety proofs that will allow users to understand the safety claims without having to understand all the technical details of the formal machinery. We use a fault tree analysis to identify possible risks to the program safety and the certification process, as well as their interaction logic, and thus to derive the structure of the safety cases. We then use a generic, multi-tiered argument [3] that is instantiated with respect to a given safety property and program. Its three tiers together constitute a single safety case that justifies the safety of the program. The upper tier simply instantiates the notion of safety and the formal definitions for the given safety property while the two lower tiers argue the safety of the program as governed by the property. The lower tiers are constructed individually to reflect the program structure. This can be done systematically because their structure directly follows the course the annotation construction takes through the program. In principle, our approach is thus independent of the given safety property and program, and consequently also independent of the underlying code generator [10].

We have developed the overall structure of the generic safety case and manually instantiated it for several examples, using only information logged during annotation construction. We expect that this process can be automated easily and that it will furthermore be straightforward to integrate with existing tools to construct safety cases such as Adelard's ASCE tool [1]. The program safety case will eventually be complemented by an additional safety case that will argue the safety of the underlying safety logic (the language semantics and the safety policy) with respect to the safety property (i.e., safety claims), as well as other components such as the theorem prover. This will clearly communicate how the safety claims, key safety requirements, and evidence for the program safety are connected. We expect that this will alleviate distrust in code generators, which remains a problem for their use in safety-critical applications.

2   Background

Here, we give a brief overview of automated code generation; we focus on the certifiable code generation approach, where the assurance is not implied by the trust in the generator but follows from an explicitly and independently constructed argument for the generated code.

2.1 Assurance for Automated Code Generation

Automated code generation [5] is a technique for automatically constructing software from (high-level) problem specifications or models. Code generators typically work by adapting and instantiating pre-defined code fragments for (parts of) the problem specification, and composing these partial solutions. They have a significant potential to eliminate manual coding errors and reduce costs and development times. Obviously, to realize any benefits from code generation, the generated code needs to be shown correct or at least safe. In correct-by-construction techniques such as deductive synthesis [23] or refinement [22] this is done by a mathematical meta-argument. However, such techniques remain difficult to implement and extend and have not found widespread application. A formal verification of the generator would provide a similar level of assurance, but remains unfeasible with the existing program verification techniques. Currently, generators are thus validated primarily by testing [24], in line with software development standards for safety-critical domains such as DO-178B [21]. However, this time-consuming and expensive process slows down generator development and application, and only few generators have been qualified.

We believe that product-oriented assurance approaches are a viable alternative to the process-oriented approaches outlined above. Here, checks are performed on each and every generated program rather than on the generator itself. Hence, assurance is not implied by the trust in the generator but follows from an explicitly constructed argument for the generated code. In our approach [8,9,11], we focus on safety properties, which are generally accepted as important for quality assurance and are also often used in code reviews of high-assurance software. We then use program verification techniques based on Hoare logic to formally demonstrate that the generated code satisfies the safety properties of interest. Our approach generally follows similar lines as proof carrying code [16] but it works on the source code level instead of the object code level [6]. However, both approaches exploit formal safety proofs as explicit evidence or *certificates* for the assurance claims over the untrusted code.

2.2 Formal Software Safety Certification

The purpose of software safety certification is to demonstrate that a program meets its high-level requirements and remains safe in the presence of known hazards. *Formal software safety certification* uses formal techniques based on program logics to show that the program does not violate certain conditions during its execution. A *safety property* is an exact characterization of these conditions, based on the operational semantics of the programming language. Each safety property thus describes a class of hazards. A *safety policy* is a set of Hoare rules designed to show that safe programs satisfy the safety property of interest. In our framework, the rules are formalized using the usual Hoare triples extended with a "shadow" environment which records safety information related to the corresponding program variables, and a *safety predicate* that is added to the computed verification conditions (VCs) [6]. However, here we focus on the information provided by constructing the annotations, and leave the details of constructing (i.e., applying the Hoare rules) and proving (i.e., calling the theorem prover) the VCs to the complementary system-wide safety case.

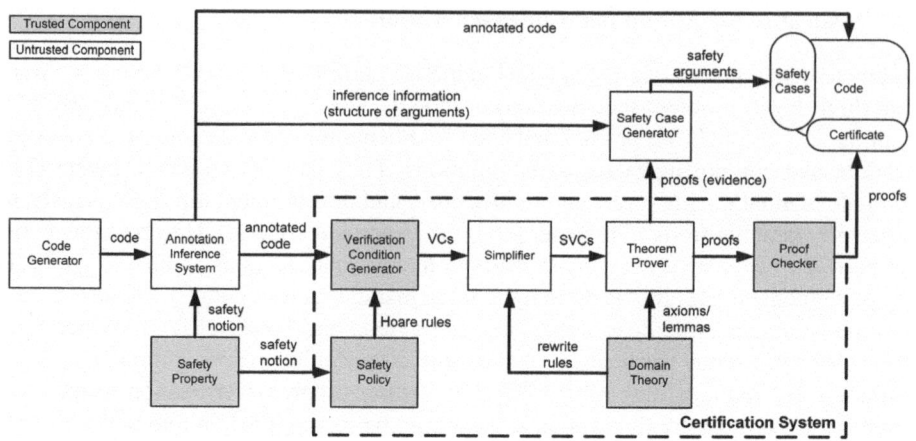

Fig. 1. System Architecture

Formal software safety certification follows the same technical approach as program verification. A VC generator (VCG) traverses the code backwards and applies the Hoare rules to produce VCs, starting with any safety requirements on output variables. If all VCs are proven by an automated theorem prover (ATP), we can conclude that the program is safe wrt. the given safety property. This approach shift the trust burden from the program to the certification system: instead of having to trust an arbitrary program to be safe, users have to trust the certifier to be correct.

Figure 1 shows the overall system architecture of our certification approach. In this, the original code generator (in this case, the AutoFilter system [28]) has been extended with the annotation inference subsystem and the standard machinery of Hoare-style verification techniques (i.e., VCG, simplifier, ATP, domain theory, and proof checker) to achieve a fully automated verification of the generated code. The architecture distinguishes between trusted (in grey) and untrusted components (in white) as shown in Figure 1. Trusted components must be correct because any errors in them can compromise the assurance provided by the overall system. Untrusted components, on the other hand, are not crucial to the assurance because their results are double-checked by at least one trusted component.These components and their interactions are described in more detail in [6,8,9].

Rather than acting as a black-box verification tool which provides a simple pass/fail result, our certification approach provides a structured safety arguments, supported by a body of evidence (i.e., safety cases) to demonstrate why the generated code can be assumed to be sufficiently safe. The safety case is generated from the analysis of the code and provides a high-level traceable argument of how the code complies with the specified safety property. The inference engine supplies information to the safety case generator, which renders this along with the code. The safety case generator identifies each part of the program that can draw attention to potential certification problems and select appropriate evidence to reason correctness of the underlie safety claims and the certification process. By elucidating the reasoning behind the certification process, there is less of a need to trust the tool.

Here, we use initialization safety (i.e., each variable or individual array element has explicitly been assigned a value before it is used) as an example, but our framework can handle a variety other safety properties including absence of out-of-bounds array accesses [6]; we expect that other properties handled by proof-carrying code such as null pointer dereferences [16] can be formalized easily. However, we are not restricted to showing exception freedom but can also encode domain-specific properties such as matrix symmetry or coordinate frame consistency (which requires significant proofs involving matrix algebra and functional correctness), whose violation will not immediately cause a run-time exception but still renders the code unsafe.

The Hoare-approach to safety certification is more flexible than special-purpose static analysis tools such as PolySpace [18] that can only handle the comparatively simple language-specific properties. It also provides explicit evidence in form of proofs, which static analysis tools typically lack.

2.3   Annotation Inference

In order to achieve a fully automated verification, a program logic requires annotations (i.e., pre- and post-conditions, and loop invariants) at key program locations. These annotations serve as lemmas that facilitate the proof of VCs, but they have to be established in their own right (i.e., they will produce VCs that show that they hold at their given location). The purpose of annotation inference [8,9] is to construct these annotations automatically, by analyzing the program structure. In our case, the annotations must formalize all pertinent information that is necessary for the ATP to prove that all *potentially* unsafe locations are in fact safe. If the program is safe, this information will be established or "defined" at some location (which we thus call a *definition*) and maintained along all control-flow paths to all the potentially unsafe locations, where it is used. The idea of the annotation inference algorithm, therefore, is to "get the information from definitions to uses", i.e., to find the endpoints of all such generalized *def*-*use*-chains, to construct the formulae used in the annotations, and to annotate the program along the paths.

The annotation inference algorithm itself is generic, and parametrized with respect to a library of coding patterns that depend on the safety policy and the code generator. The use of these patterns isolates the annotation construction from the internal details of the code generation and also allows us to a certain degree to handle code that has been modified manually. The patterns characterize the notions of definitions and uses that are specific to the given safety property. For example, for initialization safety, definitions correspond to variable initializations while uses are statements which read a variable, whereas for array bounds safety, definitions are the array declarations (where the shadow variables get their values from the declared bounds), while uses are statements which access an array variable. The inferred annotations are thus highly dependent on the actual program and the properties being proven. For example, for initialization safety, an invariant on a for-loop might express that an array has been initialized up to the loop index ($\forall j \leq i \cdot A_{\text{init}}[j] = \text{INIT}$). The VCG will turn this annotation into three VCs, corresponding to establishing the invariant on loop entry, preservation of the invariant by the loop body, and implication by the "exit form" of the invariant (i.e., over the loop bounds) of the loop post-condition. For other safety properties, the annotations can be seen as encapsulating the safety requirements directly. In the case of the symmetry policy, a postcondition $\forall i, j \cdot M[i,j] = M[j,i]$ expresses the symmetry of M. Again, this will be converted into VCs and checked by the prover. However, it is the *def-use*-dependencies, rather than the annotations or the VCs, which govern the overall structure of both the safety argument and the safety case.

3 Hazard Analysis for Formal Program Verification

While formal program verification has become a viable alternative in demonstrating program safety, doubts about the trustworthiness of the verification proofs remain. These doubts concern not only the correctness of the proofs (i.e., whether each proof step is legal in the underlying calculus) or the correctness of any of the other tools that handle the verification conditions, but also the question whether the proofs actually entail program safety. Since there are many possible ways in which the trustworthiness can be compromised, a fault tree analysis is required to identify the chain of causes and their interaction logic that initiate this undesired event.

However, our situation is complicated by the fact that the code generator is a meta-level system, and we do not know the application context of the generated program. In order to analyze the situation already at this meta-level (rather than deferring this to the final application), we need to make the simplifying but conservative assumption that every violation of the safety property is a "potential condition that can cause harm to personnel, system, property or environment", i.e., a hazard [15].

A further complication is caused by the fact that the certification system is purely observational in the sense that it cannot introduce any additional hazards as defined above, but should nonetheless be included in the hazard analysis. We thus need to look at the interaction between the code generator and the certification system to identify faults of the combined system. We consider two sets of indicators, namely the output of the code generator, or more precisely, whether the generated code is safe or unsafe, and the output of the certification system, or more precisely, its claim about the safety of the code (i.e., safe, unsafe, or unknown). We then consider all situations in which these two indicators do not agree as abnormal or faults of the combined system. The most critical fault, on which we concentrate here, occurs if the code exhibits an unsafe behavior when it is executed but the certification system claims that all safety properties were proven to hold.

The fault tree shown in Figure 2 demonstrates how the combinations of events that could lead to the top-level hazard (i.e., an undetected violation of the safety property) are linked together. It focuses on showing possible events that might invalidate the safety claim construction as it follows the structure of the generated code. A complete analysis would also need to look at other hazards, e.g., incorrect proofs or inconsistent axioms; the corresponding fault tree will lead to the system-wide safety case and is left for future work.

Figure 2 shows that there are two potential causes for the top-level hazard, either a missed potentially unsafe location in the code or the certification system erroneously concluded that all locations in the code are safe. Potentially unsafe locations in the generated code can be missed because of

- an incomplete or incorrect formalization of the safety policy corresponding to the given safety property (i.e., the failure to detect a location as potentially unsafe),
- an incomplete or incorrect representation of the safety requirements in critical annotations (e.g., a wrong global post-condition on the output variables),
- missing VCs (e.g., due to errors in the VCG), or
- incomplete coverage of the program, missing claims for any variable, occurrence or path in the program.

Here, our safety case will focus on the last cause, as it is the only cause directly related to the code generator. All other causes will be handled by the complementary system-wide safety case.

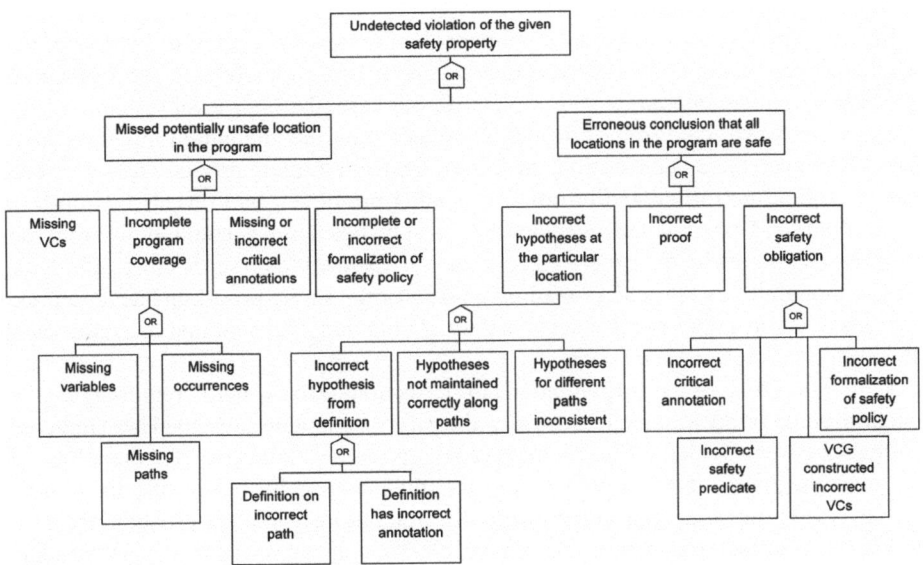

Fig. 2. Fault Tree for Program Verification

Since any location is considered safe if a proof for its corresponding safety obligation can be found, assuming the hypotheses available at that location, the conclusion that the program is safe at all locations can be wrong due to three reasons:

- the proof can be technically wrong (i.e., not conform to the inference rules of the underlying calculus), or
- the safety obligation that is proven can be wrong (i.e., does not imply the safety of the location), or
- the hypotheses used in the proof can be wrong (i.e., do not hold at the location).

Here, we concentrate on the last two reasons and rely on proof checking [29] to mitigate the hazards connected with the first cause. The safety obligation can be wrong if any of the critical annotations are wrong (similar to the case of missing a potentially unsafe location described above), or if the safety policy (including the safety predicate) or its implementation in the VCG are wrong. The hypotheses can be wrong because they have been constructed wrongly at a definition or result from a definition that is on an incorrect path, or because they are not maintained along the paths from the definition to the use, or because the different hypotheses from the different paths are inconsistent to each other.

4   Constructing Safety Cases Via Annotation Inference

In our work, we consider each violation of the given safety property by the generated code as a hazard. The purpose of the safety case described here is to construct a safety case that argue that the safety property is in fact not violated and thus that the risk associated with this hazard (as identified in section 3) is controlled or mitigated and can not lead to a system failure.

Safety cases [4] are structured arguments, supported by a body of evidence, that provide a convincing and valid case that a system is acceptably safe for a given application in a given operating environment. In our case, the high-level structure of this argument is constructed from information collected by the annotation inference algorithm. However, the evidence still comes from the formal safety proofs. The safety case makes explicit the formal and informal reasoning principles, and reveals the top-level assumptions and external dependencies that must be taken into account. It also provides information about why the generated code can be assumed to be sufficiently safe. It can thus be thought of as "structured reading guide" for the safety proofs and act as a traceable route to the safety requirements, safety claims and evidence that are required to show safety of the generated code.

We use the Goal Structuring Notation [14] as technique to explicitly represent the logical flow of the safety argument. Basically, the safety arguments presented here indicates a linkage between evidence (i.e., formal proofs) and safety claims i.e., that there is no violation to the given safety property that lead to the incorrect formal proofs, and thus the code is indeed safe with respect to the initialization before use safety property. Here, we provide a simplified overview of this safety case. We concentrate on its generic structure and describe its different tiers. We further concentrate on the program itself, leaving the remaining elements (i.e., the formal framework, the certification system and its individual components, and the safety proofs) of the combined safety case for future work.

4.1   Tier I: Explaining the Safety Notion

Figure 3 shows the the top tier of the safety case. It starts with the top-level safety goal (i.e., the safety of the generated code with respect to the safety property of interest) and shows how this is achieved by a defensible argument based on the partial correctness of the generated code. The argument stresses the meaning of the Hoare-style framework, specialized to the given safety property. However, the argument structure remains independent of the property. Here, contexts explain the informal interpretation of key notions like "safe" and "safety property". Constraints outline limitations of the approach, in particular, the fact that certification works on an intermediate representation of the source code and only shows a single property, e.g., init-before-use. Hyperlinks refer to additional evidence in the form of documents containing, for example, the model from which the source code has been generated.

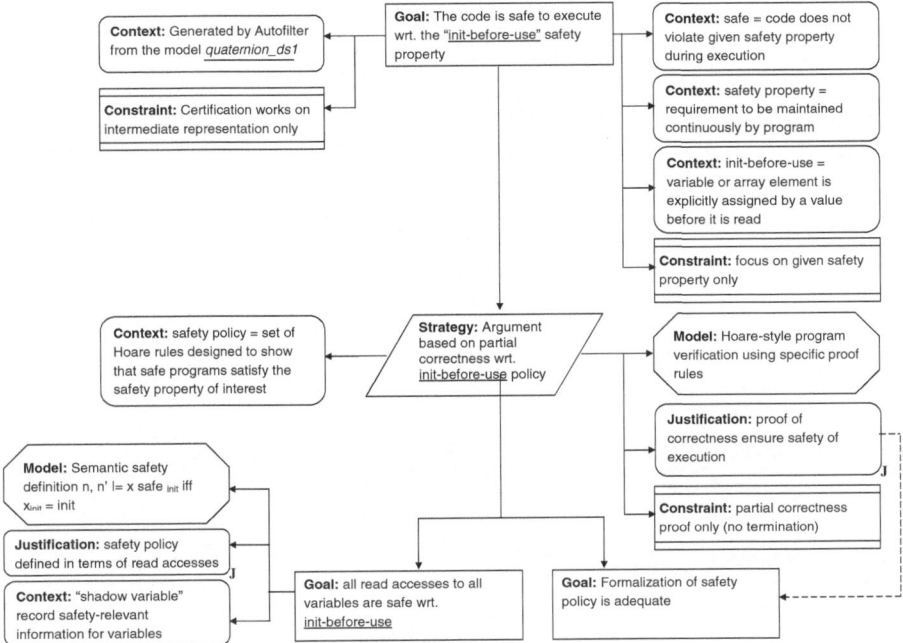

Fig. 3. Tier I of Derived Safety Case: Explaining the Safety Notion

The key strategy at this tier and its model (i.e., a Hoare-style partial correctness proof using the dedicated proof rules of the init-before-use safety policy) as well as its limitations (i.e., no termination proof) are made explicit. The strategy reduces showing the safety of the whole program to showing the safety of all read accesses, which emerges as first subgoal. This is justified by the fact that the safety property is defined in terms of variable read accesses. The subgoal is further elaborated by a model of the semantic safety definition, which exactly defines what is meant by "safe", using the notion of shadow variables given as context. The strategy's second subgoal is to show that the safety policy adequately represents the safety property, which is also the foundation of the strategy's original justification (i.e., the claim that the proofs ensure the safe execution of the program). This subgoal is not elaborated further in this safety case but leads to the complementary safety case for the safety logic.

4.2 Tier II: Arguing over the Variables

The second tier reduces the safety of all variables in two steps, first to the safety of each individual variable (justified by the fact that the safety property is defined on individual variables) and then to the safety of the individual occurrences. Note that the number of subgoals of both strategies (see Figure 4 for the goal structure) and the safety conditions are program-specific. This information is provided by the annotation inference.

Both strategies are predicated on the assumption that they iterate over the complete list of variables (resp. occurrences). Each individual occurrence then leads to a subgoal

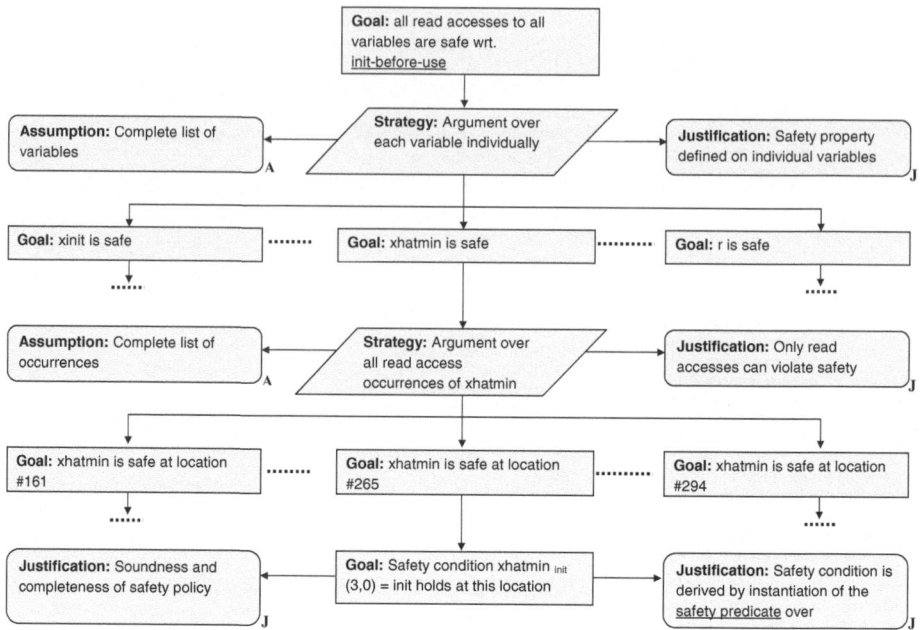

Fig. 4. Tier II of Derived Safety Case: Arguing over the Variables

to show that the computed safety condition is valid at the location of the variable's occurrence. This reduction to a formal proof obligation is justified by the soundness and completeness of the safety policy; in addition, the specific form of the safety condition is also justified. Note that some of the root cause identified in the fault tree remain as assumptions in the safety case (i.e., the list of variables and their occurrences are assumed to be complete). However, these can be checked easily, since they require no deep analysis of the generated code; in fact, the check could be automated easily.

4.3 Tier III: Arguing over the Paths

The final tier (see Figure 5 for the goal structure) argues the safety of each individual variable access, using a strategy based on establishing and maintaining appropriate invariants. This directly reflects the course the annotation inference has taken through the code. The first subgoal is thus to show that the variable safety is established on all paths leading to the current location, using an argument over all definition locations. Here, the model for the subgoal corresponds to the pattern that was applied during annotation inference to identify the definition. Each definition thus leads to a corresponding subgoal and then further to any number of VCs, although here only a single VC emerges in both cases. The proof from these VCs demonstrate that the risk identified in the hazard analysis (cf. Figure 2) does not occur for the given program.

Goals that concern properties of the program (e.g., "xhatmin is defined") are decomposed into subgoals that comprise program-independent tasks for the prover, i.e., VCs. The validity of the construction of the VCs depends on the soundness of the rules of the

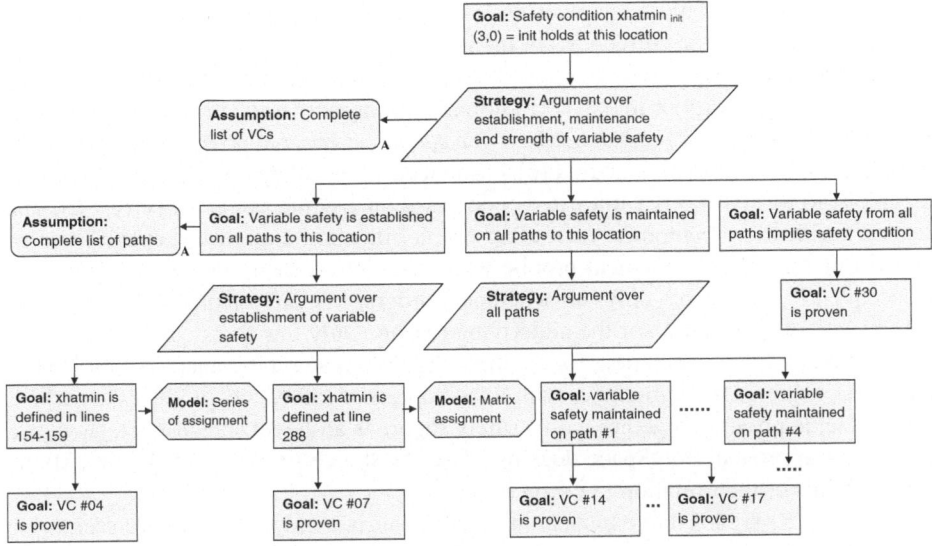

Fig. 5. Tier III of Derived Safety Case: Arguing over the Paths

VCG, the simplifier, and the definition of the safety policy, while the correspondence to program locations is based on tracing information added by the VCG and retained during the certification process. We have omitted these details from the safety case.

The second subgoal of the top-level strategy is to show that the established variable safety is maintained along all paths. This proceeds accordingly and the VCs again demonstrate that the identified risk is mitigated. The final subgoal is then to show that the variable safety implies the validity of the safety condition. This can again lead to any number of VCs. If (and only if) all VCs can be shown to hold, then the safety property holds for the entire program. The evidence for the VCs is provided by the formal proofs; we plan to convert these into safety cases as well.

5   Related Work

Most previous work on assurance for automated code generation has focused on techniques to ensure the correctness of the code generator. Whalen *et al.* [27] describe a minimum set of requirements for creating code generators that are fit for application in safety-critical systems. However, this set includes a formal correctness proof of the translation implemented by the generator (more precisely, an equivalence proof between model and generated code), which practically feasible only for generators with very similar input and output notations. Our approach, in contrast, is applicable for a much wider range of generators. Stürmer *et al.* [24,25] present a systematic testing approach and safeguarding techniques for model-based code generation tools. However, the effort easily becomes excessive and testing on its own is insufficient to provide enough assurance for safety-critical systems. Instead, some other basis is required to trust automatically generated code. Both O'Halloran [17] and Denney *et al.* [11] thus suggest that there should be explicit proofs for the correctness of the generated code rather than just trust the correctness of the generator itself.

Only program verification can prove that of program is free of certain defects or does have a certain property of interest. Traditionally, program verification concentrates on showing full functional equivalence between specifications and programs, as for example the KIV system [19]. Necula [16] introduced proof-carrying code (PCC) as new technique to formally verify untrusted code based on specific safety property. PCC and related verification techniques (including our certification approach) generate a large amount of formal mathematical proofs, which cannot be easily understood by users. Consequently, the proofs only tell whether a program is safe or unsafe, but offer no insight into or explanation of the underlying reason. Only few tools combine program verification and documentation, for example the PolySpace static analysis tool [18]. It analyzes programs for compliance with fixed notions of safety, and produces a marked-up browsable program together with a safety report as an Excel spreadsheet. However, unlike our approach, PolySpace does not describe the construction of the underlying safety claims or their relation to the program.

Hughes [13] argues that explanations are appropriate only when we are seeking to understand why something occurred while arguments are appropriate when we want to show that something is true. The argumentation (i.e., safety cases [4,14]) has been adopted across many industries especially in safety-critical systems. For example, Weaver [26] presents arguments that reflect the contribution of software to critical system safety and Reinhardt [20] presents arguments over the application of the C++ programming language in safety-critical systems. Audsley *et al.* [2] present arguments over the correctness of specification mapping from system model to code and subsequent translation into code. In [12], Galloway *et al.* present a generic argument for technology substitution i.e., argue for the safety of substitution of testing with proof-based verification in the context of certification standards like DO-178B [21]. They present an argument on how can we reasonably conclude, from the evidence available, that the replacement technology is at least as convincing as the evidence produced by testing and there is no impact on system safety. All of this work remains completely generically. While our approach uses a generic argument over safety of the generated code with respect to the safety property of interest, it then shows how this is achieved for the specific code, by constructing a specific defensible argument based on the partial correctness of the generated code. However, our approach remains independent of the given safety property and program, and consequently also independent of the underlying code generator.

6   Conclusions

We believe formal methods such as formal software safety certification can provide the highest level of assurance of the code's safety, and have described an approach whereby the inference of annotations drives both formal safety proofs and the construction of a safety case. Here, assurance is not implied by the trust in the generator but follows from an explicitly constructed argument for the generated code.

However, the proofs by themselves are no panacea, and it is important to make explicit which claims are actually proven, and on which assumptions and reasoning principles both the claim and the proof rest. We believe that purely technical solutions such as proof checking [29] fall short of the assurance provide by our safety case, since they do not take into account the reasoning that goes into the construction of the VCs. Here, we use formal proofs only as evidence and base the argumentation structure derived from the course the annotation inference has taken through the code. We consider the safety case as a first step towards a fully-fledged software certificate management system [7].

The work we have described here is still in progress. So far, we have developed the overall structure of the generic program safety case and instantiated it manually. The example shown here uses code generated by our AutoFilter system [28], but the underlying annotation inference algorithm has also been applied to code generated from Matlab models using Real-Time Workshop, and we are confident that the same derivation can be applied there as well. Future work will focus on complementary safety cases that argue the safety of the certification framework itself, in particular the safety of the underlying safety logic (the language semantics and the safety policy) with respect to the safety property (i.e., safety claims) and the safety of other certification components such as the domain theory and the theorem prover.

We believe that the result of our research will be a combined safety case (i.e., for the program being certified, as well as the safety logic and the certification system) that will clearly communicate the safety claims, key safety requirements, and evidence required to trust the generated code.

Acknowledgements. This material is based upon work supported by NASA under awards NCC2-1426 and NNA07BB97C. The first author is funded by the Malaysian Government, IPTA Academic Training Scheme.

References

1. ASCE home page (2007), http://www.adelard.com/web/hnav/ASCE
2. Audsley, N.C., Bate, I.J., Crook-Dawkins, S.K.: Automatic Code Generation for Airborne Systems. In: Proc. of the IEEE Aerospace Conference, p. 11. IEEE, Los Alamitos (2003)
3. Basir, N., Denney, E., Fischer, B.: Deriving Safety Cases for the Formal Safety Certification of Automatically Generated Code. In: Huhn, M., Hungar, H. (eds.) SafeCert 2008 Intl. Workshop on the Certification of Safety-Critical Software Controlled Systems, ENTCS. Elsevier, Amsterdam (2008)
4. Bishop, P., Bloomfield, R.: A methodology for safety case development. In: Redmill, F., Anderson, T. (eds.) Industrial Perspectives of Safety-critical Systems: Proc. 6th Safety-critical Systems Symposium, pp. 194–203. Springer, Heidelberg (1998)
5. Czarnecki, K., Eisenecker, U.W.: Generative Programming: Methods, Tools, and Applications. Addison-Wesley, Reading (2000)
6. Denney, E., Fischer, B.: Correctness of source-level safety policies. In: Araki, K., Gnesi, S., Mandrioli, D. (eds.) Proc. FM 2003: Formal Methods. LNCS, vol. 2805, pp. 894–913. Springer, Heidelberg (2003)
7. Denney, E., Fischer, B.: Software certification and software certificate management systems (Position paper). In: Proc. ASE Workshop on Software Certificate Management Systems, pp. 1–5. ACM, New York (2005)

8. Denney, E., Fischer, B.: A generic annotation inference algorithm for the safety certification of automatically generated code. In: Jarzabek, S., Schmidt, D.C., Veldhuizen, T.L. (eds.) Proc. Conf. Generative Programming and Component Engineering, pp. 121–130. ACM, New York (2006)

9. Denney, E., Fischer, B.: Annotation inference for safety certification of automatically generated code (extended abstract). In: Uchitel, S., Easterbrook, S. (eds.) Proc. 21st ASE, pp. 265–268. IEEE, Los Alamitos (2006)

10. Denney, E., Trac, S.: A Software Safety Certification Tool for Automatically Generated Guidance, Navigation and Control Code. In: Electronic Proc. IEEE Aerospace Conference. IEEE, Los Alamitos (2008)

11. Denney, E., Fischer, B.: Certifiable program generation. In: Glück, R., Lowry, M. (eds.) GPCE 2005. LNCS, vol. 3676, pp. 17–28. Springer, Heidelberg (2005)

12. Galloway, A., Paige, R.F., Tudor, N.J., Weaver, R.A., Toyn, I., McDermid, J.: Proof vs testing in the context of safety standards. In: The 24th Digital Avionics Systems Conference, vol. 2, p. 14. IEEE Press, Los Alamitos (2005)

13. Hughes, W.: Critical Thinking. Broadview Press (1992)

14. Kelly, T.P.: Arguing safety a systematic approach to managing safety cases. PhD Thesis, University of York (1998)

15. Leveson, N.G.: Safeware: System Safety and Computers. Addison-Wesley, Reading (1995)

16. Necula, G.C.: Proof-carrying code. In: Proc. 24th Conf. Principles of Programming Languages, pp. 106–119. ACM, New York (1997)

17. O'Halloran, C.: Issues for the automatic generation of safety critical software. In: Proc.15th Conf. Automated Software Engineering, pp. 277–280. IEEE, Los Alamitos (2000)

18. PolySpace Technologies, http://www.polyspace.com

19. Reif, W.: The KIV Approach to Software Verification. In: KORSO: Methods, Languages and Tools for the Construction of Correct Software. LNCS, vol. 1009, pp. 339–370. Springer, Heidelberg (1995)

20. Reinhardt, D.W.: Use of the C++ Programming Language in Safety Critical Systems. Master Thesis, University of York (2004)

21. RTCA, Software Considerations in Airborne Systems and Equipment Certification. RTCA (1992)

22. Smith, D.R.: KIDS: A semi-automatic program development system. IEEE Trans. on Software Engineering 16(9), 286–290 (1990)

23. Stickel, M., Waldinger, R., Lowry, M., Pressburger, T., Underwood, I.: Deductive composition of astronomical software from subroutine libraries. In: Proc. 12th Conf. Automated Deduction. LNCS (LNAI), vol. 814, pp. 341–355. Springer, Heidelberg (1994)

24. Stürmer, I., Conrad, M.: Test suite design for code generation tools. In: Proc. 18th Conf. Automated Software Engineering, pp. 286–290. IEEE, Los Alamitos (2003)

25. Stürmer, I., Weinberg, D., Conrad, M.: Overview of Existing Safeguarding Techniques for Automatically Generated Code. In: Proc. of 2nd Intl. ICSE Workshop on Software Engineering for Automotive Systems, pp. 1–6. ACM, New York (2006)

26. Weaver, R.A.: The Safety of Software–Constructing and Assuring Arguments. PhD Thesis, University of York (2003)

27. Whalen, M.W., Heimdahl, M.P.E.: On the requirements of High-Integrity Code Generation. In: Proc. 4th High Assurance in Systems Engineering Workshop, pp. 217–224. IEEE, Los Alamitos (1999)

28. Whittle, J., Schumann, J.: Automating the implementation of Kalman filter algorithms. ACM Transactions on Mathematical Software 30(4), 434–453 (2004)

29. Wong, W.: Validation of HOL proofs by proof checking. Formal Methods in System Design: An International Journal 14, 193–212 (1999)

Applying Safety Goals to a New Intensive Care Workstation System

Uwe Becker

Dräger Medical AG & Co KG
Moislinger Allee 53 – 55
23542 Lübeck, Germany
uwe.becker@draeger.com

Abstract. In hospitals today, there is a trend towards the integration of different devices. Clinical workflow demands are growing for the integration of formally independent devices such as ventilator systems and patient monitoring systems. On one hand, this optimizes workflow and reduces training costs. On the other hand, testing complexity and effort required to ensure safety increase. This in turn gives rise to new challenges in the design of such systems. System designers must change their mindset because they are now designing a set of distributed systems instead of a single system which is only connected to a central monitoring system. In addition, the complexity of such workstation systems is much higher than that of individual devices. This paper presents a case-study on an intensive care workstation. To cope with this complexity, different use-cases have been devised and a set of safety goals have been defined for each use-case. The influence of the environment on the use-cases is highlighted and some measures to ensure data integrity within the workstation system are shown.

Keywords: Medical devices, health care systems, systems design, resilience, reliability, safety goals, safety cases.

1 Introduction

Cost pressure and other influences are leading to scrutiny of workflows present in hospitals today. Device handling and treatment of patients shall be optimized to improve patient outcome and/or the number of patients that can be treated within a certain timeframe. Optimization of the clinical workflow calls for the integration of formally independent devices such as ventilator systems and patient monitoring systems. This trend towards the integration of different devices can be observed in hospitals all around the world. Facing this demand, a manufacturer of medical devices has to find a way to integrate different types of medical devices into one workstation system. Such an intensive care workstation system (ICWS) optimizes customer workflow and reduces their training costs. This paper will highlight some of the new challenges in the design of such integrated workstation systems. Due to the possibility of combining different medical devices to form a workstation system, the complexity of designing such an ICWS is much higher than that of designing a single, individual device. Therefore testing complexity and effort required to ensure safety are increased.

M.D. Harrison and M.-A. Sujan (Eds.): SAFECOMP 2008, LNCS 5219, pp. 263–276, 2008.
© Springer-Verlag Berlin Heidelberg 2008

This paper is organized as follows: section 2 describes the clinical environment the workstation system is to be found in and the components of the ICWS itself. This demonstrates the complexity of system design which started by defining uses-cases for the ICWS. Section 3 describes example use-cases and possible transitions from one use-case to other use-cases. For each use-case a set of safety goals has been defined. Section 4 covers aspects of safety goal realization. The following two sections (sections 5 and 6) describe how errors during data exchange and internal soft-errors are handled. The paper concludes with a summary.

2   The ICWS – A Case Study

In a critical care area various medical devices are used to treat a patient. As an example, the patient environment on a neonatal intensive care unit (NICU) typically consists of the following devices:

- ❖   An Incubator
- ❖   A Radiant Heater
- ❖   Light Therapy Means
- ❖   A Patient Monitor
- ❖   Infusion Devices
- ❖   A Ventilator
- ❖   An Hospital Information System

In a NICU, the patient is often in an incubator. This device performs more than one function. First, it serves as the patient's bed. Second, it is equipped with some means to warm the infant. There is a warming mattress in almost every case. Additionally, a radiant heater may be present. Some of the incubators are even equipped with light therapy means. The vital signs of the patient are monitored using a bedside patient monitor, [which may or may not be connected to a central monitor]. Medicine and/or nutrition are administered using fluid infusion devices. If pulmonary functions of the patient are insufficient, a ventilation system is needed. In some situations, there will even be additional devices such as imaging devices and the like.

In the past, all these systems and devices were independent of each other. One can easily imagine that space is tight in such a scenario and that clinical staff has always sought a way to silence multiple alarms. In addition, a way was sought to get the data from the different devices into the hospital information system. Manufacturers have sought to combine devices to answer the needs of the caregivers. But until recently, interfacing or exchanging data between the devices required custom design both in hardware and in software. In most cases, this was due to the differing protocols used. To solve the space problem, it is common practice to group some of the devices in tower-like assemblies. Interfacing is either not performed at all or solved using some kind of custom-made solution. The effectiveness of these solutions is more or less limited by the lack of information available to the customer. In some cases, not all of the information the devices could provide was made available externally to third party devices. To make matters even worse, each of the devices had – and still has – its own human interface. This was true even for devices of the same manufacturer. Operation was different for each device and is even more difficult in a "tower" of devices as described above. It is very confusing to the customers that the setting for one device is changed in a certain way and for the next device this is done in a totally different manner. There are diverse alarm tones and varying alarm philosophies. Even silencing an alarm differs from device to device.

If a patient reaches a critical situation, a plurality of devices begins alarming. Each device performs its own measurements and issues an alarm based on its own settings and alarm philosophy. In such a critical situation, the variety of alarms is perceived as annoying. Sometimes the alarms are perceived as disturbing during treatment in such critical circumstances. For one, the attention of the personnel is drawn towards the device instead of the patient and valuable time is lost. Further, despite the variety of devices, the information required for decision-making regarding next steps for treatment are not at hand. For these reasons, it can easily be understood that today there is a strong demand for integration, or at least standardization of medical devices. This demand is increasing. If the functions of all standalone devices could be integrated into a single device, there could be a single point where all the alarms are collected and displayed. There would be only a single source for all alarm tones. In addition, there would be only a single point of control for all the different devices. Alarms could be silenced and settings changed for all devices via this single point of control. Also, integration of the devices could lead to better treatment of patients because it would offer the chance for new or improved services. Among other things, this would be possible because more information would be available only through the fact of integrating the devices. Information would be available from different sources. This would provide a second channel required for safety reasons for some new or improved service or feature. Further, it could also be used to cross check the information from sensors or devices. Perhaps the information from a second device would lead to an improved diagnosis or decision for the treatment because it could show a different light on the value measured.

Clinical demand and possible advantages seen from a Sales and Marketing perspective are only one side of the coin. The other side is safety and resilience. Integration of different devices is much easier if done by a manufacturer offering a broad range of devices. In that case, all the knowledge is on hand to modify existing devices towards the integration needs. It turns out to be advantageous if there is a single human interface with a unique philosophy behind it. This helps to reduce human errors not only in routine but also in critical situations. In addition, it lowers the costs for staff training. But one will not get all these advantages for free. The major drawback is that adding a single point of control also increases the probability of adding a single point of common failure. Failure of any one piece of the system must be handled appropriately. Otherwise, failure of the common point of control could lead to failure of multiple devices. This is of course highly undesirable. Even though the safety of the patient may not be affected, the availability of the complete system is. The resilience and the availability of the whole system, at least as perceived by the caregiver, should be at least as high as that of a single or non-interfaced device. Otherwise customers would return to the various standalone devices they are used to have and cope with. Even if the multifunctional device is well designed, with fault tolerance and defaulting to the safest possible state for the patient, the user's perception of the device may be that the failure of a central component such as a user interface will mean loss of all functionality. A workstation system must be very flexible. Users should be able to adopt the ICWS to the actual needs of the respective patient. The ability to (re-)configure the workstation while it is in use requires special attention during the design phase. Using the case-study of the ICWS, the following paragraphs will highlight some aspects regarding the safety of such a highly flexible workstation system.

3   Use-Cases of the Workstation System

For most safety-related systems, operators have to demonstrate a systematic and thorough consideration of safety. In most countries regulatory practice differentiates between manufacturers of medical devices on the one hand and users on the other, taking into consideration the limited control of the manufacturers over the operational context of the devices. Sujan, Koornneef, and Voges [34] showed that it is essential for the safety of medical devices that the (regulatory) gap between the two is closed. They argue that the user has to know the safety goals from the manufacturer's side in order to guarantee safety of the medical device throughout its lifecycle. This knowledge is normally not accessible by the user. In addition, the current situation does not require this step neither by law nor by any directive or standard.

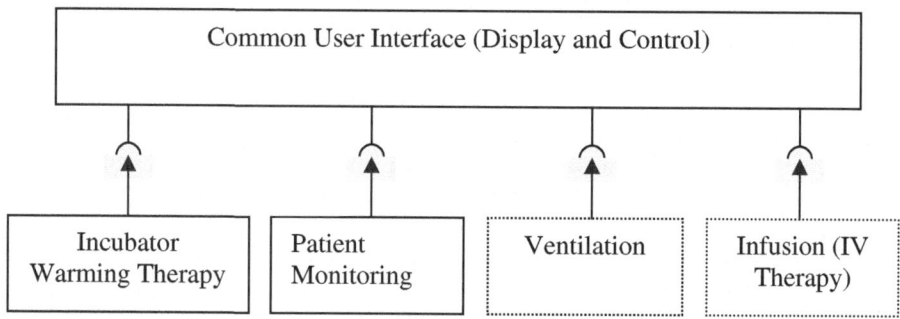

Fig. 1. Intensive care workstation system with all devices – each device independently connectable

From the manufacturer's point of view, it may not be desirable to provide users with information about the internal safety concept of medical devices. Manufacturers strive to maximize safety for users and patients. If a smart solution to provide safety has been found, a manufacturer may not be willing to share that special knowledge with the public, i.e. competing manufacturers. For the reasons mentioned, it was decided to take a new approach in developing the safety concept for a new critical care workstation system. The system itself breaks new ground in the area of critical care. The ICWS integrates many critical care devices and therefore is much more complex than the systems and devices built before. Operation of the ICWS might be different from that of the other devices. This was one of the reasons why it was decided to integrate the knowledge and the requirements of the users even during the step of defining safety goals.

Consider an intensive care workstation system (ICWS) for the neonatal intensive care unit (NICU). All low birth weight, premature patients on the NICU need warming therapy because they are not able to control their body temperature. In addition, the vital signs of the patients are monitored. The parameters monitored are ECG and oxygen saturation at a minimum. Fig. 1 shows a scenario for a critically ill patient. Such critically ill patients require an ICWS which includes infusion devices and mechanical ventilation. If the state of the patient improves, mechanical ventilation is no longer required. Thus the ICWS is reconfigured accordingly (see Fig. 1 – without the dotted part). After the situation stabilizes, the infusion of medicine may not be required, either. Therefore, the infusion devices are removed from the ICWS. The elements of Fig. 1 having solid lines show the resulting configuration.

Use-Case 1: User Stationary Reconfigures the Workstation
When first developing the safety goals, it was assumed that only the configuration of the ICWS would change (Fig. 3). It was assumed that no other changes would be performed. During the interviews with the hospital staff, it turned out that the environment of the system may change as well. Patients are grouped according to their acuity level. This grouping may even be performed locally. If the acuity level of the patients changes, they are moved to another area of the NICU. In terms of the ICWS use-cases this implied that the reconfiguration of the system had to be extended from reconfiguration during stationary use to reconfiguration during movement to another location. (First use-case in Fig. 2 – changes in the use-case underlined) In addition, even the caregivers (nurses) and the degree of supervision may change. The assessment showed that even the skill level of the nurses may vary i.e., nurses with lower skill levels treat patients at a lower acuity level.

Use-Case 2: User Changes Settings and Visually Checks Patient's Reactions
Another use-case of the ICWS was that caregivers must always be able to see the patients while changing the settings. It turns out that caregivers in certain situations turn the ICWS during the reconfiguration phase. This optimizes workflow but results in the fact that they are no longer able to see the patients while changing the settings. (Second use-case in Fig. 2 – changes underlined) This change in the use-case had an additional effect on safety. If the patient is not directly visible, safety can no longer rely on the observation of the patient's reaction upon changing a setting.

Use-Case 3: Wireless Collection of Data during Transport
Another interesting fact gathered from the assessment was that there is a difference in the mindset between doctors on one hand and nurses and parents on the other. Doctors are very positive toward wireless data transmission between sensors and/or different devices of the ICWS. Nurses and parents, however, fear negative influences from the energy of the transmitters and at least want to have a means to disable the wireless data transmission. For this reason, safety of the workstation may not rely on wireless data transmission.

Inclusion of customer input at the early stage of safety goal definition showed that some of the previous assumptions no longer hold true. The invalidation of an assumption will exclude that assumption from the current and future safety cases. For this reason some safety cases, as well as actions resulting from these safety cases, had to be re-evaluated. This re-evaluation resulted in much more complex safety cases. The term "system-wide" had to be redefined. At the starting point, "system-wide" meant the critical care workstation system by itself. At the end, the term "system-wide" did not only include the ICWS but also many parts of the environment in which the workstation system is used.

Another aspect affecting safety is the fact that caregivers tend to use every horizontal surface as a depot for diapers and other things, thus, possibly impeding device cooling. In some countries children are allowed to see their brothers and sisters on the NICU. For this reason the ICWS had to be equipped with means to disable the user

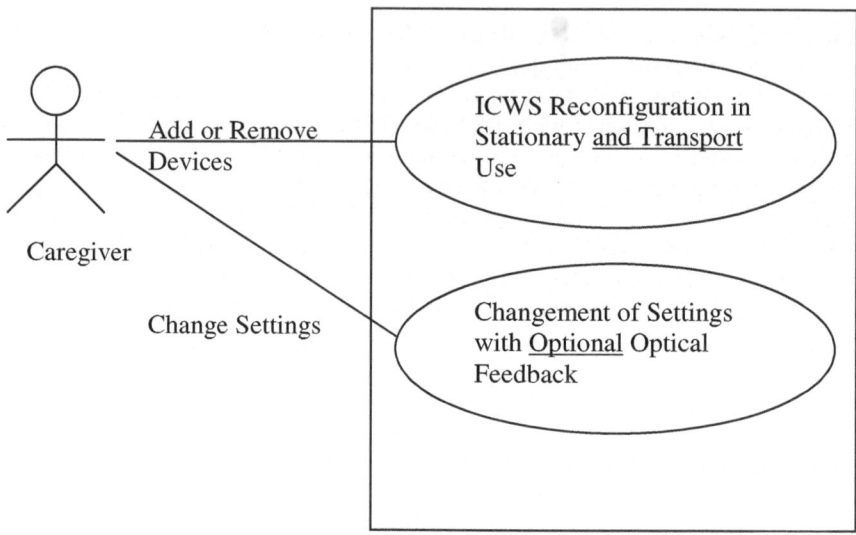

Fig. 2. Two ICWS Use-Case Examples

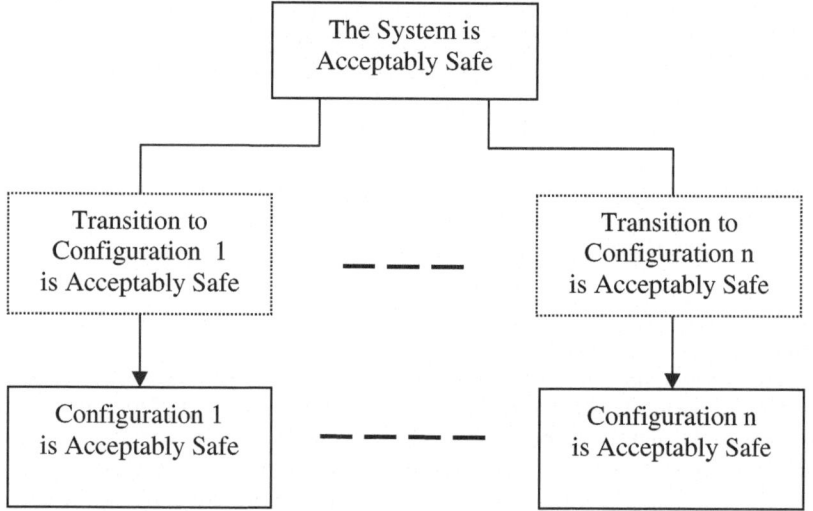

Fig. 3. Safety Goal split into Sub-Goals for each Configuration – and Extended Safety Goals including Transitions

interface until a valid password is entered. All information gathered resulted in the fact that additional safety cases had to be generated. Other safety cases got a much broader scope. Their content had to cover more aspects (Fig. 3). Evaluating and fulfilling the new safety requirements was rather complex, but eventually resulted in a safer system because daily practice is better taken into account. It additionally resulted in higher safety throughout the system's lifecycle because some special knowledge could be gathered, too. This knowledge was used to provide the user with extended instructions for the annual safety checks (STK). Besides the required measurement values, instructions for these STKs now contain some safety-goals the user should achieve when using the system. As a matter of fact, no company secrets are disclosed but it provides some help to the user to better understand the system and how to ensure the safety of the whole system throughout its lifecycle.

4 Meeting Safety Goals

Manufacturers of medical devices have to document information that states why a system can be considered sufficiently safe in a certain situation. Usually safety cases are used to document this kind of information. For each particular context it is described why the system can be considered acceptably safe [9, 10, 14]. The argumentation required has to be clear, comprehensive and defensible [21]. In addition, the argumentation has to be such that there are no rebuttals [29] and it has to show that it is safe to use the system in its intended environment [8]. This describes a goal-based approach, where the justification is constructed via a set of claims about the system, its behavior, and the process used to produce the system [34]. One might state that for the justification in most cases engineering judgment is used rather than applying formal logic. The judgment is performed carefully though, and all steps have been taken to deal with the potential hazard in an appropriate manner. The sum of all safety cases provides evidence that all hazards associated with the system have been taken into account and the risk associated with the system is below an acceptable minimum [13, 16, 17, 18, 19, 20, 21]. In addition, there are some mandatory international standards for performing hazard analysis and risk management [9, 10, 14]. Moreover, each manufacturer of medical devices has to install and maintain a quality management system. In Europe this is required by law and by the European Medical Devices Directive 93/42/EEC [7]. The manufacturing of medical devices relies on a process-based approach. It is assumed that following a well defined set of process steps during development will result in the production of a safe system.

With this in mind it is reasonably clear that the process of manufacturing medical devices, as well as the process of ensuring the safety of medical devices is acceptably safe. For this reason the way to demonstrate that the safety goals of a system are reached is to show that all safety goals and thus all safety cases result in requirements for the system. For each requirement of the system there has to be a test ensuring the fulfillment of this special requirement. At the end of the development stage, a traceability analysis is done. This traceability analysis links the requirements with the respective development documents and the respective test or test case. There may not be a single requirement which is not linked with a document or a test. It is allowed, though, that a certain requirement is fulfilled using different ways or measures. The traceability towards the tests also ensures that a developer has carefully considered the requirement and that the implementation of the requirement is effective. In addition, the quality of the implementation is improved due to peer-programming or similar techniques. Effective realization of requirements is essential for safety requirements. Testing for such requirements may to a certain degree be comparatively expensive and time consuming. It may even be required to instrument parts of the code to perform the test. Though there has to be a test for each requirement, testing can not be exhaustive because for most systems this is either impractical or simply impossible.

5 Handling Errors during Data Exchange

Safety, like resilience, is a system issue. The system is safe, not the hardware, not the electronics, not the mechanics, and not the software. Of course each of these impacts the safety, but it is ultimately the interaction of all these elements that provides the hazards as well as the risk reduction.

This, in other words, means that it is not useful to spend huge effort to make the software safe as long as the hardware is error prone, might fail at any time, and might cause potential harm. Thorough investigation has to be done if the safety level of hardware, for example, is much lower than that of the software. It may look tempting to let, at least under some circumstances, one element take over measures for risk reduction from another element. These circumstances and the measures taken have to be considered very carefully to avoid compromising the safety level of the system. For this reason, when developing a critical care medical device, it is in general desirable for a system at a certain safety level that all the elements of the system should have nearly the same safety level [10, 11].

The use-cases of the ICWS show that any device can be connected and disconnected at any time, triggering a reconfiguration of the system. There are three scenarios regarding reconfiguration. First, an additional device may be connected to the ICWS. The system will detect that a new device is connected. Messages to identify the device will be exchanged. If the device is recognized and the ICWS is able to integrate that certain device, the workstation is reconfigured and the newly connected device will be usable in the workstation environment. Second, the user may inform the ICWS that a certain device is about to be disconnected. In that case the system will set the device to be disconnected in the required state, e.g. "Standby", and reconfigure the rest of the workstation accordingly. After completion of all required steps, the ICWS will inform the user that it is safe to remove the selected device from the system. In the third scenario, a device is disconnected from the system without any previous indication. This may either be caused by a failure or by an action of the user. The ICWS will detect that it is no longer possible to exchange data with the respective device. The ICWS will try to re-establish data communication. After a defined number of trials, the device will be considered as disconnected. The ICWS reconfigures itself and the user is informed accordingly.

Users can plug into and unplug devices from the workstation system at any time. Therefore the topology of the system has to be such that if one of the therapy devices fails, no other device will fail. The devices should send out their messages and every other device that is interested in the data may take it. The data should be marked with a descriptor for identification and should contain information describing the source of the respective data. To further ensure message integrity, every message should contain a check value and a sequence number. The sequence number can be realized using a time-stamp. In most cases error correction is not required. Therefore overhead can be reduced by selecting a check value that is only capable of detecting faults. In case a consumer of the data detects an error it requests the sender to send the data again.

The central control and display device requires all connected devices to send their sensor data for display purposes. Some data will only be sent if its value changes. Other data will be sent on a regular basis. For each of the different devices there is a description of the data that is required to be received on a regular basis. This data then by itself may act as keep-alive indicator. This provides an easy way for the central control and display device to detect if data exchange with a device ceased. A broken connection can be assumed if either the keep-alive data is not received within a certain amount of time or if a certain amount of consecutive messages containing an error are received. After a certain amount of trials to re-establish the data exchange, it will stop data communication with this device and disable the respective functions. If any other device received data from the device that failed, the central control and display device will switch the respective device to a mode which does not require that particular data. The required information is present at the central display and control device because such an interaction requires the system to be in a certain mode of operation to ensure patient safety. The system then is reconfigured to ensure that it is in a consistent and defined state. Of course, an alarm is issued in such a case. In some cases even an active therapy has to be stopped and continued with different settings. This naturally requires a high priority alarm and further steps to ensure patient safety.

Reconfiguration of the system is not performed on the first occurrence of an error, though. Transient errors of short duration will be accepted by this scheme. In general, there is a tolerated time for errors depending on the effect the disturbed data exchange can have on the patient. Consider controlling an infusion pump delivering a medicine that affects the heart. If it delivers a bolus[1] and the system does not receive the correct confirmation on the "End of bolus" command, a safety reaction has to be performed relatively soon. For such safety-relevant features only a few number of trials (say two or three within a few milliseconds) to re-establish a data exchange are allowed to fail before the reconfiguration of the system is triggered. The system will try to perform as many safety measures as possible, e.g. to trigger a reboot of the device in error. In the example above, the infusion pump will stop if the data communication to the host, i.e. the ICWS, is lost. Features that are not safety-relevant allow for a higher number of trials. For instance the transmission of a patient's name to a device is not safety-relevant and thus a reaction can be triggered after more trials.

6   Coping with Internal System Errors

In safety-relevant systems like intensive care medical devices it is essential to monitor or supervise the function of their software. This is especially true for life supporting systems like the critical care workstation system described. The primary concern is the safety of the system. An error in one of the software modules or a soft error in

---

[1] A bolus is an infusion with high infusion rate, usually applied to a patient for a short time.

memory may under no circumstance result in harm to a patient. The supervision of the software modules is even more important for a workstation system. The number of software modules dramatically increases with the number of devices. The approach to ensure safety is twofold. Some parts of the software are considered to be part of the dedicated "safety software". These modules ensure that safety-relevant settings are distributed through the system in a safe and consistent manner. Some type of redundancy is used for this purpose. For instance, these modules check that if the user selects a pressure of 10.0 mbar, the pneumatic controller really will receive 10.0 mbar and not 100 mbar. The respective data is transmitted via two independent channels to the recipient. If a deviation between the data of the two channels is detected, the value will be rejected and the transmission has to be repeated. To increase the availability of the system, a certain threshold for re-transmission is introduced. The system tolerates a certain (low) amount of erroneous transfers. Once the first trials of re-transmission are successful and the data is consistent on both channels, the system will continue with that data. The error counter will be reset after some (say 10) transfers without error. If the threshold is exceeded or connection to the central display and control device is lost, some parts of the system, e.g. the ventilator system, will continue their function, i.e. ventilation, using the current settings. This increases resilience of the system and ensures continuous treatment of the patient.

The second approach to ensure safety is to use additional online test software. When introducing this online test software, the intention is to get high error coverage while only inducing minimal performance impact. One goal of this test software is to test and supervise the function of the application software. The primary focus is not on testing the hardware but on finding transient errors in RAM and on finding errors in the application software. The latter may either be caused by soft errors or by other applications that erroneously have changed memory cells they were not supposed to change. As described above, there are several possible configurations of the ICWS. Furthermore, the system can be in different operating modes which may even be independent of the current configuration. This leads to the fact that the ICWS does not always use all the modules of its software simultaneously. One way to reduce test overhead is to test only active software modules. It turns out that this reduces performance impact significantly. To obtain the required high error coverage, different algorithms have been evaluated. Zhou et al. [37] proposed a sophisticated method to test software. Unfortunately the method to check the software includes some modifications either in hardware or in the driver for that hardware. As a matter of fact, it is not always possible to do that. As far as off-the-shelf hardware is concerned, a modification is impossible in almost every case. Our experience is that only implementation in FPGAs or other (full) custom logic can be changed with reasonable overhead. With custom made chips, the possibility to perform changes depends on the costs, the minimum order quantity and in some cases on other designs the chips are used in. Modifications to operating systems may not be possible either. In aerospace or in medical applications the operating systems are certified or at least validated. A change in such an operating system would require a re-certification or re-validation. The cost – both in time and money – will be prohibitive in most cases. For the reasons mentioned above, these elements of the system are not changed.

Reconfiguring the system is not limited to reconfiguration of the display, but may also affect the applications and the tests running. The applications concerning a disconnected device are switched to an inactive state or even swapped out of the memory – depending on the application. In the case of swapping out the application, the respective memory will be marked as free and may be added to the memory checked by the memory test routine. In either case, the supervision of the task will be stopped because the task will no longer consume any processor time. As the task will not own the processor, it is not able to perform any harmful action. For this reason, system tests / self-tests are dynamically adapted to the current system configuration and requirements.

In most situations, it is known in advance which modules of the software will be in use in a certain situation. One could change the scheduler to access its process table to get the information. Nevertheless the modules using processor time can be determined easily, even if the scheduler of the operating system can not be changed. With this information, the online test instance, i.e. the test task, can be tailored to the actual needs. The tailoring results in only these tests being executed that relate to an active module of the software. This saves testing time, but does not compromise reliability. The seconds lowest process priority is assigned to test task, with only the idle task having a lower process priority. This will guarantee that the test task will not affect system performance. It is called only with a larger period, e.g. every second. In addition, it is executed only if there is no task with a higher processing priority requiring the processor.

The online test software uses an approach which combines a watchdog and test patterns. A watchdog timer is triggered at certain points of the control flow of the software. If the system hangs, the watchdog is not retriggered and will provoke a system reset. It is ensured by code reading that the watchdog is not triggered within a timer interrupt subroutine. The parts of the software that are supervised by the safety software are not tested separately. Each calculation and transmission of these modules is checked. For this reason, the test software would not increase fault coverage with these modules. For the other modules, the test software will provide input values and compare the output values obtained with pre-computed values. In addition, the time to produce the output values is checked. For this reason the latency of the modules and thus performance of the ICWS is checked as well. Faults in software modules currently inactive will not influence the function of the ICWS. Therefore, testing only active modules of the software will not reduce fault coverage of the relevant faults. It is further assumed that the application fed with test data will be affected by the fault in the same way it would be with application data.

Even the modules responsible for producing display output are checked. Some parts of the display are considered safety-relevant. For this reason, the check of the display and the respective drivers is very thorough. Users shall not notice that the display drivers are being checked. Therefore, the content of the display may not be altered. A virtual screen is introduced that is larger than the display that can be seen by the user. For testing purposes, the module which writes to the display is stimulated to produce a certain bit pattern at a region of the screen that is not visible by the user. The test software then checks the video memory i.e. the virtual screen. Some parts of the screen that are not visible by the user shall be blank. The respective memory shall therefore contain the data which was written to the memory during initialization. The other region shall contain the bit pattern produced by the driver. All the conditions mentioned have to be fulfilled. After the test is completed, the bit pattern is overwritten by the value used to initialize the memory. For the next test, a new output pattern and a new region of the virtual screen is chosen.

7 Conclusion

An intensive care workstation system has been described. Safety and system design are largely influenced by the input of the perspective users. This results in new and extended safety goals for the system. User requirements led to a design with a central display and control device. Such a central point of control is likely to introduce a single point of common failure. Certain measures have been chosen to avoid such negative effects. To increase resilience of the system some components are capable of autonomous operation. In case the central display and control device fails, the ventilation part, for instance, will continue with its last settings. The patient monitoring part will switch to its own small display and continue monitoring. In general, errors in a module only influence the respective module. All other modules continue operation. If required, graceful reduction of system functionality is performed to increase resilience. If data exchange to one of the devices of the system fails and can not be reestablished, the ICWS will reconfigure itself accordingly. Again, graceful reduction of functionality leads to a higher availability of the system as a whole.

Smart testing software is used to check system health. To limit performance impact, only active modules of the software are checked. If a module of the software becomes inactive because the respective device is removed from the system, it is not tested any longer. A soft error in the memory of that module will have no effect on the system. Thus resilience is increased. If the device is brought back to the system, the respective module of the software is loaded again into memory. As this is done during the reconfiguration phase of the system the small extra time required to re-load the software module is not noticed by the user. Re-loading the software to memory increases safety, because potential soft errors in memory are overwritten.

Further work will be done to provide patterns for safety goals that can be used with other, new devices. Some of these patterns should be generic, such that they could be used for other care areas. The assessment of the perspective users performed during development can be improved also. Nevertheless, the information gained was very valuable. This leads to the decision to let such an assessment become a mandatory process step of the development. This will ensure both that safety is enhanced during the whole lifecycle of a device and that the assessment itself will continually be improved and optimized. Additionally, some new algorithms for the testing software will be evaluated. This will include both online software-based self test (SBST) and online test of the software running.

References

1. Bishop, P., Bloomfield, R., Guerra, S.: The Future of Goal-Based Assurance Cases. In: Proc. Workshop on Assurance Cases, pp. 390–395 (2004)
2. Bloomfield, R., Bishop, P., Jones, C., Froome, P.: ASCAD – Adelard Safety Case Development Manual, Adelard (1998)
3. Bloomfield, R., Littlewood, B.: On the use of diverse arguments to increase confidence in dependability claims. In: Besnard, D., Gacek, C., Jones, C.B. (eds.) Structure for Dependability: Computer-Based Systems from an Interdisciplinary Perspective, pp. 254–268. Springer, Heidelberg (2006)

4. Bridal, O., et al.: Deliverable D3.1 Part 1 Appendix E: Safety Case, Version1.1. Technical Report, EASIS Consortium (February 2006), http://www.easis-online.org
5. CENELEC EN 50129 – Railway Applications – Safety related electronic systems for signaling, CENELEC Brussels (2003)
6. Chinneck, P., Pumfrey, D., McDermid, J.: The HEAT/ACT Preliminary Safety Case: A case study in the use of Goal Structuring Notation. In: 9th Australian Workshop on Safety Related Programmable Systems (2004)
7. European Council: Council Directive 93/42/EEC of 14 June 1993 concerning medical devices. Official Journal L 169, 12/07/1993, pp. 0001 – 0043 (1993)
8. Greenwell, W.S., Strunk, E.A., Knight, J.C.: Failure Analysis and the Safety-Case Lifecycle, Department of Computer Science, University of Virginia
9. IEC 60601-1 – Ed. 3.0 – Medical electrical equipment – Part 1: General requirements for basic safety and essential performance. IEC Geneva (2005)
10. IEC 60601-1-4 – Ed. 1.0 – Medical electrical equipment – Particular Requirement for the Safety of Programmable Medical Devices. IEC Geneva (2000)
11. IEC 62304 – Ed. 1.0 – Medical device software – Software life cycle processes. IEC Geneva (2006)
12. IEC 62366 – Ed. 1.0 – Medical devices – Application of usability engineering to medical devices. Draft. IEC Geneva (2006)
13. Intl. Electrotechnical Commission. IEC 61508: Functional Safety of Electrical/ Electronic/Programmable Electronic Safety-Related Systems. Technical Report (April 1999)
14. ISO 14971:2007 – Application of risk management to medical devices. ISO Geneva (2007)
15. Karapetian, A.V., Some, R.R., Beahan, J.J.: Radiation Fault Modeling and Fault Rate Estimation for a COTS Based Space- Borne Supercomputer. In: Proc. IEEE Aerospace Conf., Mar. 2002, vol. 5, pp. 5-2121–5-2131 (2002)
16. Kelly, T., McDermid, J., Weaver, R.: Goal-Based Safety Standards: Opportunities and Challenges. In: Proc. of the 23rd International System Safety Conference (2005)
17. Kelly, T., McDermid, J.: A Systematic Approach to Safety Case Maintenance. Reliability Engineering and System Safety 71, 271–284 (2001)
18. Kelly, T.: A Systematic Approach to Safety Case Management. In: Kelly, T. (ed.) Proc. of SAE 2004 World Congress (2004)
19. Kelly, T.: Managing Complex Safety Cases. In: Proc. 11th Safety Critical Systems Symposium. Springer, Heidelberg (2003)
20. Kelly, T.P., McDermid, J.: Safety Case Construction and Reuse using Patterns. In: Proceedings of 16th International Conference on Computer Safety, Reliability and Security (SAFECOMP 1997), September 1997. Springer, Heidelberg (1997)
21. Kelly, T.P.: Arguing Safety: A Systematic Approach to Managing Safety Cases. PhD Thesis, University of York, UK (September 1998)
22. Leveson, N.G.: Safeware: System Safety and Computers. Addison-Wesley, Boston (1995)
23. McDermid, J.: Support for safety cases and safety argument using SAM. Reliability Engineering and System Safety 43(2), 111–127 (1994)
24. Mukherjee, S.S., Emer, J., Reinhardt, S.K.: The Soft Error Problem: An Architectural Perspective. In: Proc. 11th Int'l Symp. High-Performance Computer Architecture, pp. 243–247 (Febuary 2005)
25. Nicolescu, B., Velazco, R.: Detecting Soft Errors by a Purely Software Approach: Method, Tools and Experimental Results. In: Proc. Design, Automation and Test in Europe Conf. and Exhibition, pp. 57–62 (March 2003)

26. Nordland, O.: Safety Case Categories – Which One When? In: Redmill, F., Anderson, T. (eds.) Current issues in security-critical systems, pp. 163–172. Springer, Heidelberg (2003)
27. Pradhan, D.K.: Fault-Tolerant Computer System Design. Prentice Hall, Englewood Cliffs (1996)
28. Pullum, L.L.: Software Fault Tolerance Techniques and Implementation. Artech House (2001)
29. Ridderhof, W., Gross, H.-G., Doerr, H.: Establishing Evidence for Safety Cases in Automotive Systems – A Case Study. In: Computer Safety, Reliability, and Security, 26th International Conference, SAFECOMP 2007, Nuremberg, Germany, pp. 1–13 (September 2007)
30. RVSM Pre-Implementation Safety Case, Eurocontrol (2001)
31. Shirvani, P.P., Saxena, N.R., McCluskey, E.J.: Software- Implemented EDAC Protection against SEUs. IEEE Trans. Reliability 49(3), 273–284 (2000)
32. Storey, N.: Safety Critical Computer Systems. Addison-Wesley, Reading (1996)
33. Sujan, M., Harrison, M., Pearson, P., Steven, A., Vernon, S.: Demonstration of Safety in: Healthcare Organisations. In: Górski, J. (ed.) SAFECOMP 2006. LNCS, vol. 4166. Springer, Heidelberg (2006)
34. Sujan, M.-A., Koornneef, F., Voges, U.: Goal-Based Safety Cases for Medical Devices: Opportunities and Challenges. In: Saglietti, F., Oster, N. (eds.) SAFECOMP 2007. LNCS, vol. 4680, Springer, Heidelberg (2007)
35. Weaver, R., Despotou, G., Kelly, T., McDermid, J.: Combining Software Evidence: Arguments and Assurance. In: Proceedings of the 2005 workshop on Realising evidence-based software engineering, St. Louis, Missouri, pp. 1–7 (2005)
36. Weaver, R.A.: The Safety of Software – Constructing and Assuring Arguments. DPhil Thesis, Department of Computer Science, University of York, UK (2003)
37. Zhou, Y., Lakamraju, V., Koren, I., Krishna, C.M.: Software-Based Failure Detection and Recovery in Programmable Network Interfaces. IEEE Transactions on Parallel and Distributed Systems 18(11), 1539–1550 (2007)
38. Ziegler, J.F., et al.: IBM Experiments in Soft Fails in Computer Electronics (1978-1994). IBM J. Research and Development 40(1), 3–18 (1996)

Safety Assurance Strategies for Autonomous Vehicles

Andrzej Wardziński

Gdansk University of Technology, Department of Software Engineering
Narutowicza 11/12, 80-952 Gdansk, Poland
andrzej.wardzinski@eti.pg.gda.pl

Abstract. Assuring safety of autonomous vehicles requires that the vehicle control system can perceive the situation in the environment and react to actions of other entities. One approach to vehicle safety assurance is based on the assumption that hazardous sequences of events should be identified during hazard analysis and then some means of hazard avoidance and mitigation, like barriers, should be designed and implemented. Another approach is to design a system which is able to dynamically examine the risk associated with possible actions and then select the safest action to carry it out. Dynamic risk assessment requires maintaining the situation awareness and prediction of possible future situations. We analyse how these two approaches can be applied for autonomous vehicles and what strategies can be used for safety argumentation.

1 Introduction

Nowadays we can notice a tendency towards designing systems which are more and more autonomous. In some countries we can get on autonomous buses (e.g. People Mover [1]). In DARPA Urban Grand Challenge the autonomous vehicle mission was to drive 100 km in urban environment and obey traffic rules (however vehicles were not required to interpret street signs nor traffic lights) [2, 3].

Intuitively we feel that autonomous systems are not only more complex, but autonomy introduces new problems that may require quite novel approach for design and development of such systems. This raises a question how much autonomous systems differ from non-autonomous ones and whether we can apply the same approach to safety assurance. It is not certain if the existing methods and techniques would be adequate and sufficient.

Our objective is to investigate the issues of autonomy for safety-critical systems and the ways of safety assurance. In Section 2 we introduce the problem of autonomous vehicle safety and two approaches for safety assurance. The traditional approach is presented in Section 3. We discuss the structure of safety argument, advantages and disadvantages of the approach. The second approach is more complex as it uses the situation awareness model to identify safe and risky actions. The approach is discussed in Section 4. We analyse the difficulties for constructing convincing safety argument and possible types of evidence. The results are summarised in Section 5.

2 Autonomous Vehicle Safety

Autonomy is a broad concept and there are many definitions of this term and a few levels of autonomy can be identified [4, 5]. Generally autonomy relates to freedom to

M.D. Harrison and M.-A. Sujan (Eds.): SAFECOMP 2008, LNCS 5219, pp. 277–290, 2008.
© Springer-Verlag Berlin Heidelberg 2008

determine own actions and behaviour. For the needs of the paper we will assume that autonomy of a mobile system is the ability to accomplish a given mission without human intervention. That means that the system should be able to make decisions how the mission goals could be achieved and how to cope with changes in the environment and threats.

Autonomous vehicle control system should:

- plan the mission taking into account the context of the environment (for example other vehicles),
- act according to the plan to accomplish the mission,
- adapt the plan to changes in the environment,
- ensure safety (avoid collisions),
- efficiently use energy and prevent energy loss.

In the paper we will focus on safety of a single vehicle. We make no assumptions about the communication with other vehicles. Vehicles can communicate or carry out their missions without any communication. We will not refer to vehicle communication in the paper.

When we say that the vehicle should ensure safety we primarily think of avoiding collisions with other vehicles or objects. The vehicle should keep "safe" distance from other vehicles and objects. There are probably many ways of designing a system which satisfies these requirements but generally we can say that there are two approaches. The approaches differ in the way system safety is perceived and assured.

1. The first approach is based on the analysis of possible accident scenarios and designing protection mechanisms (barriers) that prevent transitions to unsafe states. The way the hazardous situations are detected and accidents are avoided is determined during hazard analysis.
2. The second approach is based on dynamic risk assessment. The vehicle control system evaluates the risk of possible actions and then selects the one that is the least risky in the context of current situation and environment conditions.

We will discuss and compare these two approaches. The description of the a pproaches will be somewhat simplified to show the contrast how safety can be perceived and assured. In our discussion we will focus on the rationale of the approaches and ways of demonstrating safety – what strategy can be used to argue that the system is safe.

3 The Approach Based on Predetermined Vehicle Risk Assessment

The first method to achieve autonomous vehicle safety is based on the traditional approach to hazard analysis and safety assurance. The objective is to identify event sequences leading to accidents and then design mechanisms to control the risk. System safety is usually achieved by implementation of safety barriers. The concept of a barrier explains the idea of the approach.

3.1 The Concept of a Barrier

Hollnagel [6] defines a barrier as an obstacle, an obstruction or a hindrance that may either (a) prevent an action from being carried out or an event from taking place, or (b) thwart or lessen the impact of the consequences. There are many forms of barriers. Hollnagel classified barriers as [6]:

Material barriers, e.g. a fence,
- *functional barriers*, when a specific precondition is defined which has to be fulfilled before an action can be carried out,
- *symbolic barriers*, e.g. signs and signals that have to be perceived and interpreted,
- *immaterial barriers*, that is using the knowledge to follow the rules of allowed behaviour (e.g. Highway Code).

This classification shows that many different means can be used as barriers for autonomous vehicles. Usually we will combine different types of barriers to achieve more confidence in vehicle safety.

The mechanism of a barrier is intentionally simple to assure its high reliability. When a barrier is detected (we will use symbol e_1 to denote the event of barrier detection) then a specific action a_1 is to be carried out. At the moment it is not relevant whatever technology we use – barrier detection and reaction can be implemented as a mechanical, hydraulic, electric or software system. We can use safety analysis techniques like Event Trees to describe barriers (see Fig. 2).

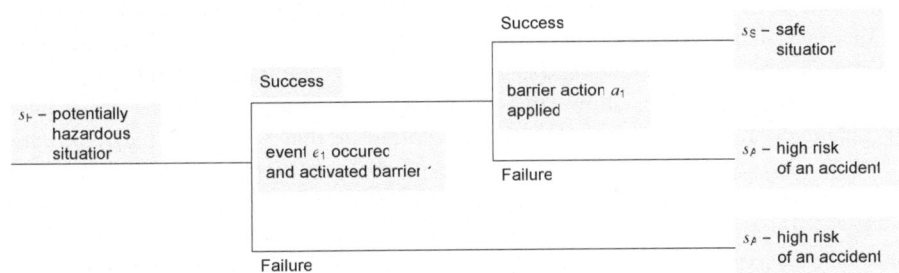

Fig. 1. Event Tree for a barrier activated when an event e_1 occurs and action a_1 defined for the barrier

In the example presented in Fig. 1 we assume that barrier action a_1 guarantees prevention of the accident (or gives some probability of a success). Usually we assign probabilities to events and the tree branches (events outcomes) to calculate probabilities of possible scenarios.

When the system barriers are implemented as software functions we can generalize the code structure to look like this:

```
if        e₁ detected   then   a₁
else if   e₂ detected   then   a₂
...
else      other actions
```

Different events e_1, e_2, ..., e_n can be defined for each barrier. The efficiency of barriers is analysed on the system level during hazard analysis and then barriers usually can be implemented and verified separately.

We call the approach the predetermined risk assessment as the risk is assessed during the hazard analysis stage of the system development. The barrier conditions and actions definitions remain unchangeable during system operation however there are possibilities for some degree of flexibility as presented in the next Section.

3.2 Examples of Barriers Use for Autonomous Vehicles

Barriers used for autonomous vehicle safety assurance can be as simple as bumper sensors or distance sensors to detect an obstacle and then stop. However barriers can be far more complex and sophisticated.

Spriggs in [7] describes an autonomous system which operates near to an airport's runaways and uses GPS coordinates to ensure operation in the allowed area only. The vehicle control system has a definition (map) of a safe designated area. GPS - coordinates serve as a barrier to ensure that the vehicle will not leave the safe area.

Robertson in [3] presents kinematic motion study for a vehicle operating in an urban environment and competing in DARPA Urban Challenge [2]. The result of the study was used to define safety regions for the vehicle motion. The safety region is an area that is required to be free from other vehicles in order to continue driving. If the safety region is occupied by any vehicle the system should stop and wait until the safe region is clear. The system uses a set of sensors to calculate its position in the terrain (on the road) and positions of other vehicles and objects. This knowledge is presented as a situation awareness model and is used to plan the vehicle movement.

These two examples show that a barrier can be a complex mechanism. The characteristic of the approach is that the condition that activates a barrier is set up during the hazard analysis. The occurrence of the condition is binary – the barrier should be activated or not. We can say that this represents a binary view on safety.

The assumption of the binary view on safety is that all the risk can be avoided (or reduced according to ALARP principle) when we define a set of conditions that activate barriers (safety functions). This often leads to vast "safety margins". For example an autonomous vehicle has to stop and wait while most of human drivers would assess that it is safe to go ahead.

3.3 Safety Argument Strategy

The approach is intended to be simple and easy for verification. We expect that the safety argument structure would be relatively simple. We will discuss a simplified generic model of safety argumentation using GSN notation [8, 9]. There are many possible ways of structuring safety arguments for barriers and we have chosen a structure which emphasizes the fact that usually each barrier can be designed, implemented and verified almost independently from other barriers. The main part of such safety argument is presented in Fig. 2.

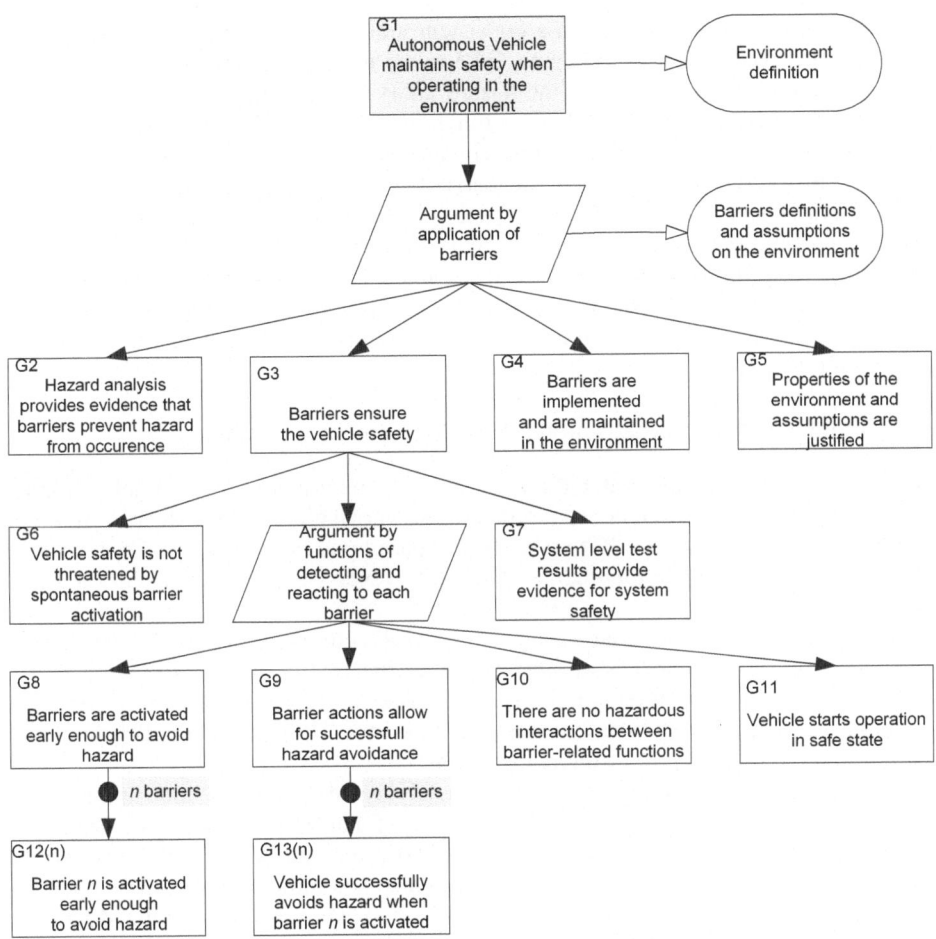

Fig. 2. General safety argument schema for barriers

There are four main claims that have to be justified in order to demonstrate system safety. The first claim (**G2**) relates to the system hazard analysis which should provide evidence that barriers prevent hazard occurrence. That means that the vehicle will not enter unsafe area (e.g. will avoid collision). This requires a thorough analysis of the vehicle and environment properties to create a conceptual model of the system states and possible transitions. The analysis should:

- identify safe states of the systems and hazardous states,
- identify possible accident scenarios,
- define means to prevent hazardous transitions leading to accidents (e.g. barriers),
- check barriers completeness and assess probability of hazard occurrence for conditions that do not activate barriers,
- assess probability of barrier failures and evaluate likelihood of the hazard,
- provide evidence that the analysis is complete and no relevant factors had been overlooked.

We should note that barriers form a consistent and comprehensive system in hazard analysis phase only. Claims **G12** and **G13** relate to each barrier separately. All the logic of system safety is built on the hazard analysis level. In the system we implement and then demonstrate each barrier function independently. The safety argument for each barrier relates to its design, implementation, tests and verification.

Only during system level tests the barrier system mechanism is validated as a whole (**G7**). There are also some conditions in the environment which are to be fulfilled or barriers may fail otherwise. Usually barriers need some kind of devices or equipment (e.g. GPS satellites) to be maintained in the system environment (**G4**). Sometimes we also need some justification for environment properties or assumptions on such properties (**G5**). An example of such assumption is maximum possible acceleration of other vehicles in the environment.

3.4 Summary of the Predetermined Risk Assessment Approach

The main objective of the predetermined risk assessment is to provide highly reliable, simple, manageable, efficient, economical and verifiable solution that assures the system safety goals. This works very well for non-autonomous systems. The approach can also be applied for autonomous vehicles however we should be aware of some limitations.

It is important to notice that binary safety model (division of possible states into safe ones and unsafe ones) is an abstraction of the reality aimed in making safety assurance techniques simpler, more reliable, easier to demonstrate and verify. That does not mean that in reality safety is a binary attribute. It is just a simplification that makes safety assurance process more effective.

The great advantage is that during hazard analysis we decompose the system safety problem and then analyse each barrier separately. Each barrier can be independently designed, implemented and tested. Additional work is required when barriers depend on each other. When barriers interrelationship becomes more complex the amount of work and difficulty of safety argumentation rises. If we had ten interrelated barriers we would have to analyse hundreds of combinations.

Another disadvantage of the approach is that it is not flexible and the system performance deteriorates when it is operating in an open environment as the vehicle cannot adapt its behaviour to the changes in the environment. For example a vehicle mentioned in Section 3.2 can wait for a long time until the safety region is clear. Most of human drivers would asses the safe region as too vast and would assess the vehicle behaviour as very protective. This approach is well suited for a situation of low traffic and low speed. The approach works well when there is only a limited number of vehicles and for most of the situations the safety region is clear.

There are some areas of applications for which such limitation could be a big disadvantage or even not acceptable. The approach is difficult to apply when the application domain requirements relate to:

- need for efficient space utilization, like congested road traffic;
- reactions for unexpected failures and events especially when stopping is not a proper way to ensure safety;
- driving in a terrain for which barriers are not implemented and maintained;

- competition between vehicles and situations like racing;
- military missions, escorting and guarding when the mission goals require taking some risk.

It seams that application of this approach alone to autonomous systems would make it difficult to achieve performance goals.

4 Dynamic Risk Assessment

In real world humans do not perceive safety in a binary way. We are not used to distinguish only two states: "this is not safe – I have to react to this" and "this is safe – no reaction is needed" (however we often react to some events). We are used to talk about the risk of an activity or a situation. We often say that something is more or less risky in some situations. That leads to a conclusion that situation safety should not be perceived as a binary condition but we should rather say that a situation can be characterised with a specific risk level depending on the attributes relevant to safety.

4.1 Dynamic Risk Assessment Approach

The concept of risk as an attribute of a situation is quite widely used in hazard analysis. When we analyse accident scenarios we identify situation attributes (events) as risk factors which contribute to hazard.

The idea of the risk assessment approach is to design a system which is able to perceive and interpret risk factors and then assess how far it is on the scale starting from an absolutely safe state and ending with an accident. The system should be able to assess the risk of the situation *before* carrying out a specific action. In that way the system would be able to select safe actions and avoid actions leading to hazards.

The concept of situation risk assessment for autonomous vehicles was described in [10, 11]. The general requirement for the system is to maintain *situation awareness* which allows for action planning taking into account risk level of particular actions. The concept of situation awareness is used in psychology and in robotics but it is quite new for safety-critical systems.

The general architecture of an autonomous vehicle control system using situation awareness model is presented in Fig. 3. In our analysis we will focus on the situation awareness model and *Task planning* process. The general algorithm of the *Task planning* process consists of following steps:

1. Select possible scenarios of actions to be analysed and assessed.
2. Assess each scenario for:
   - mission progress,
   - compliance with formal safety rules (e.g. Highway Code),
   - situation risk level.
3. Choose the optimal scenario (according to the vehicle strategy).
4. Communicate tasks of the chosen scenario to the Control layer.

One should note that the "rules" mentioned in point 2 can be barriers. Barriers can be used for dynamic risk assessment approach however it is only one of three factors of the situation assessment. An example of a rule is "vehicle to the right has the right of way" when two or more vehicles approach a crossing at the same time. Another rule can be "do not drive across the pavement". Rules often can relate to barriers.

The goal of step 3 is to select the best action to carry it out. The selection criteria depend on the vehicle strategy. The strategy can give priority to safety or mission goals depending on the mission context and current situation. For the purposes of the paper we will assume that the strategy to assure vehicle safety and safety has higher priority then mission goals.

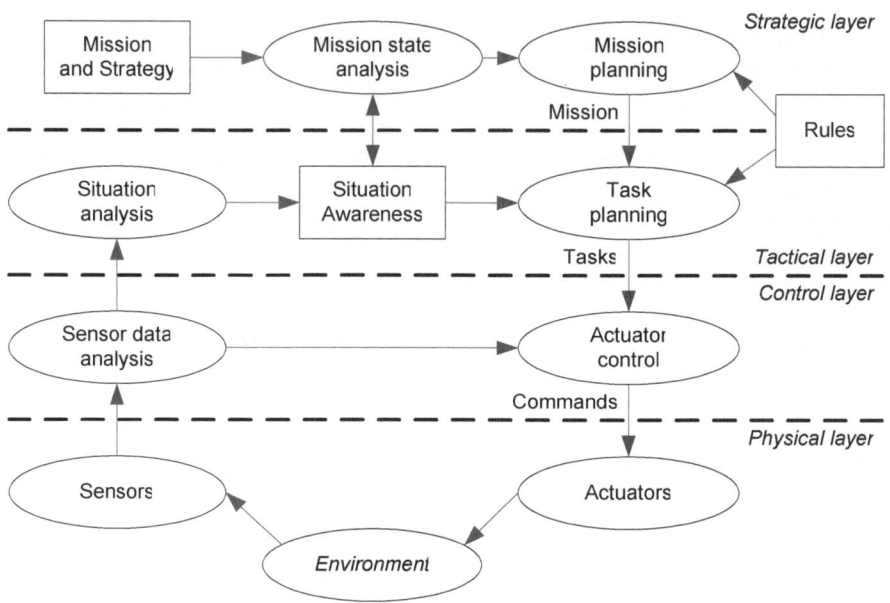

Fig. 3. Autonomous vehicle control system architecture

We assume that dynamic risk assessment would allow for:

- better system resilience and survivability in unexpected situations or in emergency;
- better performance in an open (non-controlled) environment when other entities (systems, vehicles or humans) can act independently,
- better ability to perform risky missions where some risk level has to be accepted and the system should balance between safety and mission goals,
- operation in areas where barriers are non existing or difficult to define and implement.

4.2 Safety Assurance Using Dynamic Risk Assessment

The system safety is assured by a complex mechanism of selecting the safest possible action for a given current situation. We will analyse this for the example similar to presented in Section 3.1. In Fig. 4. we have a set of hazardous situations SH and some possible scenarios of actions presented as arrows leading to other situations. Barrier actions described in Section 3.1 are presented as transitions a_1 and a_2. The main difference in comparison to predetermined risk assessment is that the task of the vehicle control system is not to activate automatically the barrier function a, but assess the risk for each possible actions and then carry out the safest one. Depending on the specific situation it can be the action a or any other action, like action b_2 for situation s_{H2}.

The main prerequisite for the approach is situation awareness model which should provide means to distinguish situations attributes that are relevant for the system safety. The model should allow for situation perception, identification of possible actions and prediction of their results. The prediction should take into account changes in the environment and actions of other entities (vehicles).

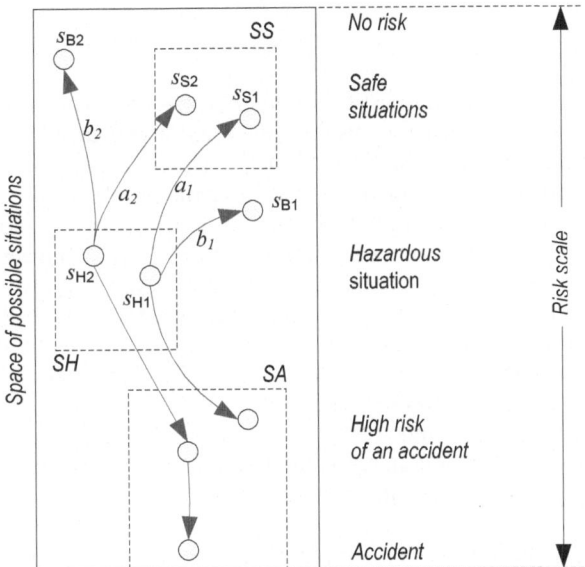

Fig. 4. Set of example situations and actions in context of the risk scale

A kinematic vehicle model forms the basis for the vehicle motion safety analysis. The model will tell us what the position of the vehicle will be when specific actions (like braking, acceleration and turning wheels) are carried out. When other vehicles operate in the same environment then the model should be extended with behaviour of other vehicles and possibly communication with them. When necessary the model should also take into account the possibility of presence of humans.

The critical function of the situation awareness model is to assess the risk of any current or predicted situation. Possible set of situations to be analysed and assessed will usually be bigger then a set of scenarios analysed for barriers (compare to Fig. 1).

For any current situation the control system will identify possible actions and assess the risk level associated with them and then select the safest one to be carried out (or other action according to the vehicle strategy). Selection of the safest action can be described as a following function using VDM-like notation:

vehicleAction(s : *Situation*) as : *ActionScenario*
post
  $as \in$ possibleScenarios(s)
  $\wedge$
  riskAssessment(s, as) =
    min{ $ra \mid ra =$ riskAssessment(s, ax) $\wedge ax \in$ possibleScenarios(s) }

It is important to note that for a given situation s_{H1} the control system does not necessary need to activate the barrier function a_1. The objective of the approach is to create a model which can be used to identify the safest action for a given situation and the selected action can be different then action a_1 (like action b_1).

4.3 Safety Argument Strategy

Our analysis led us to the main question in this paper – how can we gain confidence that this approach will really ensure system safety and acceptably low risk of autonomous vehicle operation? Safety analysis of such systems is a subject of research and probably there are no established methods that solve the problem.

The situation awareness model plays the main role in safety assurance as the vehicle uses it to interpret the situation and to assess the risk of possible actions. The main difference in comparison with the barrier approach is the way we handle cause-consequences dependencies. In the traditional approach the model of system safety and hazardous cause-consequences dependencies is created by humans in hazard analysis phase. As a result separate safety functions for all barriers are designed and implemented. This is quite different when we intend the vehicle control system to maintain situation awareness and dynamically assess the risk. The model is to be analysed as a whole and its elements cannot be analysed in separation. Providing arguments for the system safety is more complex. A general schema of argumentation structure is presented in Fig. 5.

First we have to demonstrate that situation awareness model with dynamic risk assessment function is correct in terms of consistency with real vehicle and its operating environment and is sufficient and adequate for vehicle safety assurance (claim **G2**). In our analysis we have identified the main safety requirements for the situation awareness model:

1. The model should distinguish situations that are relevant for system safety (including safe, hazardous situations and accidents).
2. Attributes that are used to identify and classify situations should be possible to measure with the use of sensors or their values should be possible to be deduced from accessible information (risk factors are examples of situations attributes).
3. The model gives information what are the possible actions for current situation.
4. The model can predict and assess safety of a situation that will be the result of a given action (or action scenario). Prediction should take into account behaviour of other vehicles and events in the environment.
5. The model should allow for representing incomplete or uncertain data, what leads to uncertain risk assessment [10]. The model should preserve safety when the risk assessment results are uncertain. The loss of situation awareness (when the control system is not able to assess the situation) is interpreted as unsafe situation.

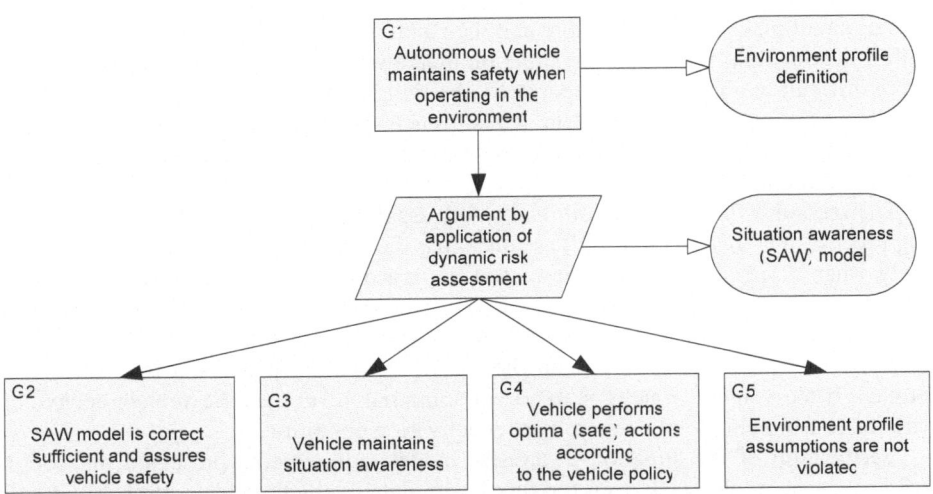

Fig. 5. Safety argument structure for dynamic risk assessment approach

6. Depth of prediction of future situations is sufficient to ensure avoidance of unsafe state what means that the vehicle after perceiving any dangerous situation has enough time to avoid accident.
7. The model implementation is effective within required time limits.

The satisfaction of the requirements should be verified and validated for a given vehicle and specific environment profile.

The next two claims relate to requirements that the systems should maintain the situation awareness (**G3**) and should use it to control the risk (**G4**). Claim **G3** relates to processes that provide situation awareness information: *sensor reading*, *sensor data analysis* and *situation analysis* processes in system architecture presented in Fig. 3. The processes should provide reliable and up-to-date information to maintain the situation awareness. The next claim (**G4**) is related to the use of situation awareness knowledge and its functions (like risk assessment function) to steer the vehicle and to ensure vehicle safety. In system architecture presented in Fig. 3 the processes *task planning* and *actuator control* perform these tasks.

The system should be demonstrated to work safely when the operating environment is consistent with the assumed environment profile. Therefore we add the fourth claim (**G5**) to provide evidence why we think the operating environment will be consistent with the intended profile and how will we assure that the vehicle will not be used in other environment what could cause errors of the situation awareness model. Part of the claim justification should be related to assumptions about the behaviour of other vehicles.

Justification for claims **G3** and **G4** can be based on the system design analysis and demonstration of the traceable process of design, development, testing, verification and validation. Justification for the situation awareness model (**G2**) is more difficult. At the moment we do not have effective methods for the situation awareness model analysis. One of the problems is that the model covers overall vehicle safety. That is quite different to traditional hazard analysis approach where we divide the analysis into as small pieces as it is possible and then analyse separately each hazard and each failure mode. Risk factors in the situation awareness model are interrelated and it is more difficult to analyse them separately. We will not be able to decompose the problem into separate and independent items. When we adjust the situation awareness model for better reaction for a single risk factor then we usually alter assessment of a broad set of situations.

The second difference to traditional safety assurance methods is the decision making process. The barrier approach assumes that the control system will react immediately when a specific condition is met. This is a direct cause-consequence relation. We can use analytical techniques like Event Tree Analysis or Fault Tree Analysis to examine effectiveness of barriers. When the system maintains situation awareness then the direct connection between the barrier activating event and the reaction is broken. If the vehicle strategy is to accept some risk level then the vehicle behaviour will differ from simple reaction to barrier activation condition.

Verification of the situation awareness model requirements presented in Section 4.2 is a very complex task even for simple simulated vehicle models. Probably it will not be possible to verify it manually and tool support will be needed.

The main problem that we had encountered during definition of a situation awareness model for a simple simulated autonomous vehicle was the correctness of the risk assessment function. The risk level for a safe situation or for an accident is easy to verify. However the risk level for intermediate states is more difficult for verification as we do not have real world values that can be measured and compared to. Humans can also differently assess the risk of the situation and our experience shows that the assessment can be subjective. The solution of the problem that we use is scenario risk profile analysis. The profile is used to present changes of the assessed risk level for a scenario. For example when a scenario starts with a safe state and ends with an accident we can observe how the risk level rises. We can also identify points in time when risk factors are detected by the vehicle or some specific conditions occurred. As a result of the analysis we can add additional risk factors or change ratings of existing risk factors to improve system ability to avoid hazard. It will be difficult for complex systems to design the correct situation awareness model and then prove its safety. The approach will be rather to analyse the system safety and then improve the model using methods like risk profile analysis.

Another problem is how to assess the system safety level as the environmental conditions has great influence on it. For example number of accidents will depend on behaviour of other vehicles, weather and road conditions and so on. The system safety requirements and safety performance can be defined and measured in the context of a specific environment profile only.

The last of the main problems is the certainty level for prediction of other vehicle actions. We have to accept some level of prediction uncertainty however there is some breaking point when the vehicle control system loses the ability to preserve safety. We discuss the problem of uncertainty in [10].

This all together gives us four types of evidence that can be used when constructing arguments for situation awareness model correctness (Fig. 6).

The first type of evidence (E1) is based on the analysis of the situation awareness model (kinematic model) and accident sequences. We analyse the model in the context of identified safety requirements.

Evidence **E2** is based on the simulation results. We define some number of scenarios (safe, near-miss and accidents) and use them for analysis and simulation. The objective of a simulation scenario is to check if the vehicle can safely operate for a given scenario. To use large number of simulated scenarios an efficient simulation tool is needed and also tool support for scenarios definition and validation. This method is especially useful when it is used for testing of modified vehicle control system using a set of already validated and documented scenarios.

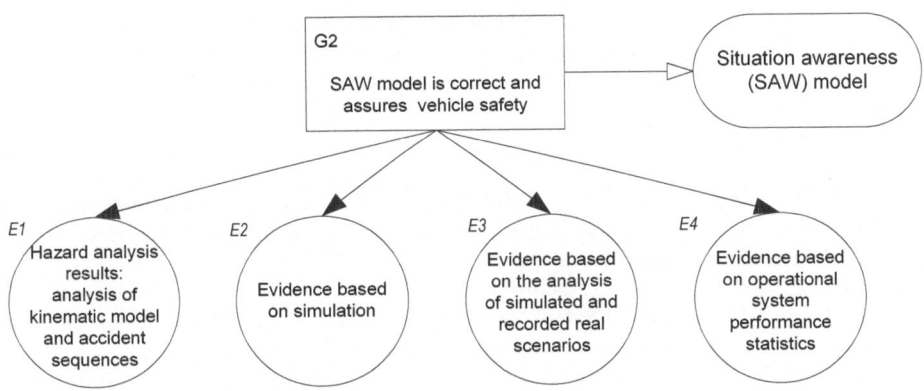

Fig. 6. Analysed types of evidence for claim of situation awareness model correctness

The third kind of evidence is the analysis of simulated and recorded real scenarios. We analyse the scenario risk profile to check how the risk level was rising before an accident, when the system became aware of the high risk level, how the risk level is assessed in absence of any threats and if we can explain observed variations in the scenario risk profile. Comparison of simulated scenarios and scenarios recorded during system operation will provide evidence for the situation awareness model consistency with real vehicle behaviour.

The last kind of evidence is based on data from the system operation. This data can be a subject to reliability growth modelling. Accidents experienced during system operation can be analysed to improve situation awareness model.

5 Summary

The use of situation awareness model and dynamic risk assessment is a novel approach for safety-critical systems. Nowadays most of safety-critical systems are not autonomous. For simple autonomous systems the traditional approach based on predetermined risk assessment and a concept of a barrier is sufficient for achieving safety and performance goals. Safety assurance methods and techniques for this approach are mature and efficient. Main safety argument strategy is to demonstrate traceable process from safety requirements analysis, through design, development, tests, verification, validation to finally operation and maintenance.

In the future we can expect a tendency to use autonomous systems operating in more complex and less controlled environments. This will raise problems presented in the paper. The presented dynamic risk assessment approach would allow for development of resilient autonomous systems. The idea of a resilient system is to sail close to the area where accidents will happen, but always stay out of the dangerous area [13]. Dynamic risk assessment approach promises such abilities however it is very difficult to provide sound and convincing safety evidence. We have discussed our experience what strategy can be used for presenting safety argument for the use of dynamic risk assessment. The main problem is that the safety argument is not so straightforward as in the case of the predetermined risk assessment approach. Appropriate safety analysis methods for situation awareness model and dynamic risk assessment are not mature and are the subject of further research.

References

1. Lohmann, R.H.C.: About Group Rapid Transit and Dual-Mode Applications. In: APM 2007, 11th International Conference on Automated People Movers, Vienna (2007)
2. DARPA: Urban Challenge Rules (2006),
   http://www.darpa.mil/grandchallenge
3. Robertson, S.W.H.: Motion Safety for an Autonomous Vehicle Race in an Urban Environment. In: 2006 Australasian Conference on Robotics & Automation (2006)
4. Clough, B.T.: Metrics, Schmetrics! How The Heck Do You Determine A UAV's Autonomy Anyway? In: PerMIS Conference Proceedings, Gaithersburg, pp. 1–7 (2002)
5. Sholes, E.: Evolution of a UAV Autonomy Classification Taxonomy. In: IEEE Aerospace Conference (2007)
6. Hollnagel, E.: Accidents and Barriers. In: Hoc, J.-M., et al. (eds.) Proceedings of Lex Valenciennes, Presses Universitaires de Valenciennes, vol. 28, pp. 175–182 (1999)
7. Springs, J.: Motion Safety for an Autonomous Vehicle Race in an Urban Environment. In: Redmill, F., Anderson, T. (eds.) Currect Issues in Safety-critical Systems – Proceeding of the Eleventh Safety-critical Systems Symposium. Springer, London (2003)
8. Bishop, P.G., Bloomfield, R., Guerra, S.: The future of goal-based assurance cases. In: Proceedings of Workshop on Assurance Cases. Supplemental Volume of the 2004 International Conference on Dependable Systems and Networks, pp. 390–395 (2004)
9. Kelly, T.P.: Arguing Safety – A Systematic Approach to Managing Safety Cases, PhD thesis, University of York (1998)
10. Wardziński, A.: The Role of Situation Awareness in Assuring Safety of Autonomous Vehicles. In: Górski, J. (ed.) SAFECOMP 2006. LNCS, vol. 4166. Springer, Heidelberg (2006)
11. Wardziński, A.: Dynamic Risk Assessment in Movement Planning for Autonomous Vehicles. In: International IEEE Conference on Information Technology, IT 2008, Gdansk (Poland), May 18-21 2008, pp. 127–130 (2008)
12. Hollnagel, E., Woods, D.D., Leveson, N.: Resilience Engineering, Ashgate (2006)

Expert Assessment of Arguments:
A Method and Its Experimental Evaluation

Lukasz Cyra and Janusz Górski

Gdansk University of Technology, Department of Software Engineering
Narutowicza 11/12, 80-952 Gdansk, Poland
lukasz.cyra@eti.pg.gda.pl, jango@pg.gda.pl

Abstract. Argument structures are commonly used to develop and present cases for safety, security and other properties. Such argument structures tend to grow excessively. To deal with this problem, appropriate methods of their assessment are required. Two objectives are of particular interest: (1) systematic and explicit assessment of the compelling power of an argument, and (2) communication of the result of such an assessment to relevant recipients. The paper gives details of a new method which deals with both problems. We explain how to issue assessments and how they can be aggregated depending on the types of inference used in arguments. The method is fully implemented in a software tool. Its application is illustrated by examples. The paper also includes the results of experiments carried out to validate and calibrate the method.

Keywords: Argument assessment, Dempster-Shafer model, Argument structures, Safety Case, Trust Case, Assurance Case.

1 Introduction

Arguments are commonly used in 'cases' (safety cases [13, 17, 18], assurance cases [2], trust cases [9, 10], conformity cases [4, 5], etc.) to justify various qualities of objects (like safety, security, privacy, conformity with standards and so on). Recently, there is a growing interest in these subjects, which leads to the development of relevant methodologies and finding new application areas for argument structures [6, 7].

The idea which lies behind the development of argument structures is to make expert judgment explicit in order to redirect the dependence on judgment to issues on which we can trust this judgment [22]. In this way, it is possible to analyze the argument structure and take a position on it.

However, argument structures tend to grow excessively, becoming too complex to be analyzed by non-experts. Therefore appropriate methods of assessment of argument structures are required. Two objectives are of particular interest: (1) assessment of the compelling power of an argument structure, and (2) communication of the result of such an assessment to relevant recipients.

The paper gives details of such an appraisal method. Although, this approach is developed in connection with the Trust-IT methodology [9, 10], which focuses on trust cases, it is general enough to be applied to other kinds of cases. The appraisal

M.D. Harrison and M.-A. Sujan (Eds.): SAFECOMP 2008, LNCS 5219, pp. 291–304, 2008.

mechanism described in the paper is based on the Dempster-Shafer model [20, 21]. It provides for issuing assessments and their aggregation depending on the types of inference used in arguments. The mechanism has been already implemented in the TCT (Trust Case Toolbox) software tool [14]. The paper includes examples of its application which have been borrowed from the trust case developed for a real system [19]. The paper also reports on the results of some experiments carried out to validate and calibrate the appraisal method.

The presented method extends and modifies the approach to arguments appraisal proposed in [12], which had problems with adapting itself to the types of arguments occurring in trust cases. It proposes linguistic appraisal scales which are then represented in terms of Dempster-Shaffer belief functions and can be mapped onto the Josang's opinion triangle [15, 16]. Some of the aggregation rules (not described in detail in the paper) are following Yager's modification of Dempster's rule of combination [20]. A general discussion of the role of confidence in dependability cases can be found in [3].

2 Representing Arguments

The proposed approaches to argument representation in 'cases' [1, 9, 17] are influenced by the Toulmin's argument model [23]. In our approach (the Trust-IT methodology) we adopt this model in a fairly straightforward way as shown in Fig. 1.

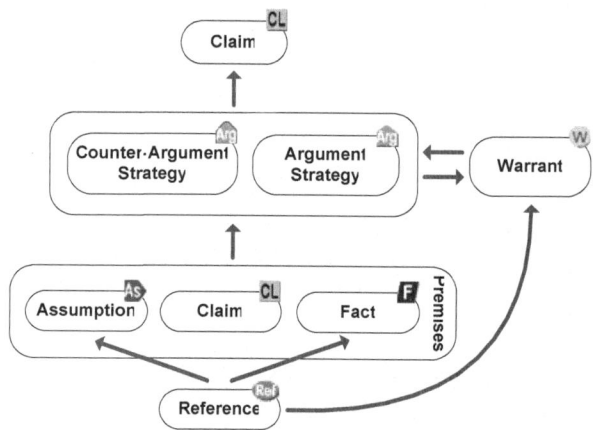

Fig. 1. Trust-IT argument model

The presented structure includes a conclusion to be justified represented as a *claim* (denoted **CL**). The claim is supported by an *argument strategy* (denoted **Arg**), which contains a basic idea how to support the conclusion. In the case of *counter-argument strategies* (denoted **Arg**) it includes the idea of rebuttal of the claim. The argument strategy is related to a *warrant* (denoted **W**) which justifies the inference from *premises* to the conclusion. This justification may require additional, more specific arguments, which is shown by the arrow leading from the argument to the warrant node.

A premise can be of three different types: it can be an *assumption* (denoted 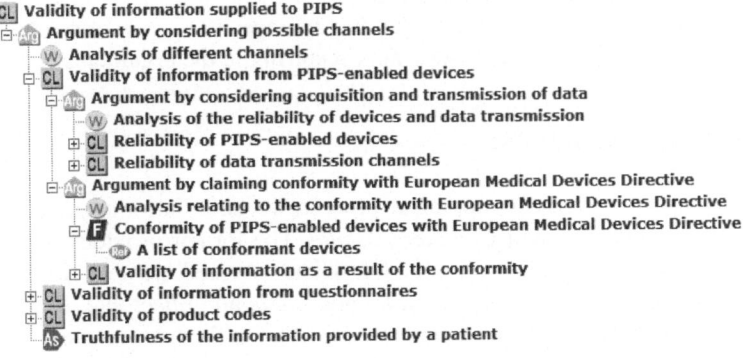), in which case the premise is accepted without further justification; it can be a more specific claim which is justified further; or it can represent a *fact* (denoted **F**) which is obviously true or, otherwise, is supported by some evidence. Evidence is provided in external (to the trust case) documents, which are pointed at by nodes of type *reference* (denoted ⊕). In the case of an assumption, the referenced document can contain explanation of the context in which the assumption is made.

As claims and warrants can be demonstrated using other (sub-)claims the argumentation structure can grow recursively. The icons labeling different nodes shown in Fig.1 and implemented in the supporting tool (TCT [14]) to denote elements of argument structure will be used in the subsequent examples presented in this paper.

An example argument following the model introduced in Fig. 1 is presented in Fig. 2. The example refers to the PIPS system, the system delivering to its users health and lifestyle related personalized services [19]. The top claim of the structure postulates validity of information supplied to PIPS. It is demonstrated by considering different channels through which information related to a patient's state is supplied to the system. This leads to four premises which are used by the argument. Three of them: *'Validity of information from PIPS-enabled devices'*, *'Validity of information from questionnaires'* and *'Validity of product codes'* are claims and are supported by more detailed arguments. The fourth one, *'Truthfulness of the information provided by a patient'*, is an assumption and is not further analyzed.

Fig. 2. An example argument coming from a trust case for the e-health system PIPS

To appraise the compelling power of such an argument we have introduced an appraisal method which is presented in the following part of the paper. The method consists of two steps:

Step 1 – appraisal of warrants and premises
    1.1. Estimate the 'strength' of warrants (these, which do not have their supporting arguments) occurring in the argument. (This assessment is based on the assessment of the evidence linked to the warrant but also the common knowledge and the logical bases for the inference.)
    1.2. Estimate the 'strength' of the facts and assumptions occurring in the argument. (This appraisal is mostly based on the assessment of the evidence linked to the premises by the reference nodes.)

Referring to the example shown in Fig. 2, the appraisal of the *'Analysis of different channels'* warrant would take into account if the validity of information received from the devices, questionnaires and by reading product codes with the additional assumption that patients are not cheating intentionally are sufficient to conclude the validity of information supplied to the system.

The appraisal of the premises would assess the acceptability of the assumption that patients are not cheating intentionally (note that this is context dependent and the result would depend on the knowledge about the system and its environment). The appraisal of the *'Conformity of the PIPS-enabled devices with European Medical Devices Directive'* fact would take into account the evidence linked to this fact by the corresponding reference node.

Implementation of **Step 1** requires that we have an appropriate scale to express the appraisals of warrants, facts and assumptions. This should be complemented by appropriate guidelines supporting the assessor.

Step 2 – automatic aggregation of the partial appraisals
    2.1. Starting from the leaves of the argumentation tree, aggregate the appraisals of the premises and warrants to obtain the appraisal of the conclusions.
    2.2. Repeat the process until the top conclusion has been reached.

Referring to the example from Fig. 2, this step would result in the appraisal of the top claim taking as an input the appraisals of warrants, facts and assumptions occurring in the argumentation and recursively applying the aggregation rules.

Implementation of **Step 2** requires that the appropriate aggregation rules were defined covering all relevant types of warrants occurring in the arguments.

3 Appraisal Mechanism

To support experts during the appraisal process we have introduced two linguistic scales, the *Decision scale* and *Confidence scale*. The former provides for expressing the attitude towards acceptance or rejection of the assessed element. It distinguishes four decision values: *acceptable, tolerable, opposable* and *rejectable*. The latter scale provides for expressing the confidence in this decision. It distinguishes six levels of confidence: *for sure, with very high confidence, with high confidence, with low confidence, with very low confidence* and *lack of confidence*.

The scales can be combined together which results in twenty four values of the *Assessment scale* as shown in Fig. 3. The elements of the scale, which are represented as small circles, have intuitively understandable linguistic values. For instance, the element represented as the white circle reads: *'with very low confidence tolerable'*.

The semantics of the scales can be formalized using Dempster-Shafer's belief and plausibility functions [20, 21]. If s is a statement, then

-   $Bel(s) \in [0,1]$ is the *belief* function representing the amount of belief that directly supports s,
-   $Pl(s) \in [0,1]$ is a *plausibility* function representing the upper bound on the belief in s that can be gained by adding new evidence.

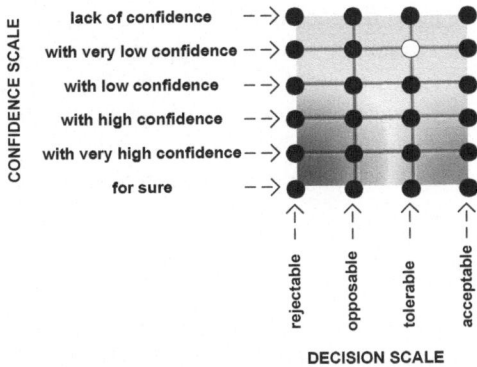

Fig. 3. The *Assessment scale* as a product of *Confidence scale* and *Decision scale*

We can formally represent *confidence* as:

$$Conf(s) = Bel(s) + 1 - Pl(s), \quad Conf(s) \in [0,1] \tag{1}$$

and map the interval [0,1] onto linguistic values from the *Confidence scale* in such a way that *lack of confidence* = 0, and *for sure*=1.

Decision scale distinguishes four levels to express the ratio between belief (acceptance of a statement) and the overall confidence in the statement (without distinguishing if we want it to be accepted or rejected).

Using Dempster-Shafer's functions we can formally represent the decision concerning *s* as:

$$Dec(s) = \begin{cases} Bel(s)/(Bel(s)+1-Pl(s)) & Bel(s)+1-Pl(s) \neq 0 \\ 1 & Bel(s)+1-Pl(s) = 0 \end{cases}, \quad Dec(s) \in [0,1] \ . \tag{2}$$

The interval [0,1] is mapped onto linguistic values of the *Decision scale* in such a way that *'rejectable'*=0 and *'acceptable'*=1.

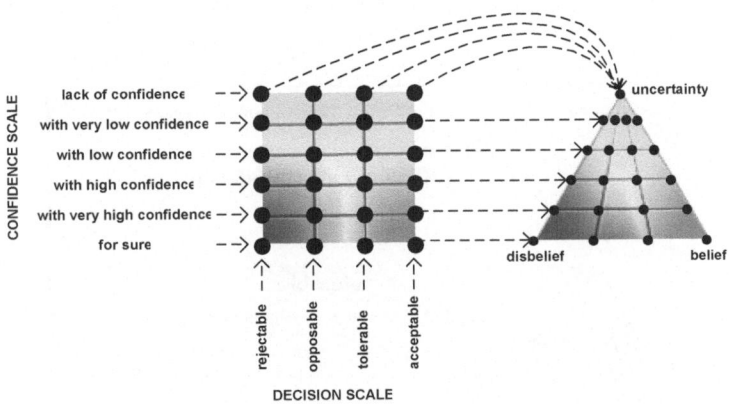

Fig. 4. Mapping *Assessment scale* on Josang's opinion triangle

We can observe that the difference between stating that something is acceptable or rejectable is significant if enough evidence supporting such assessment is available (which corresponds to e.g. *'for sure'* or *'with very high confidence'* assessments), however, there is no difference if no evidence is present (which corresponds to the *'lack of confidence'* assessment). To address this particular aspect, the *Assessment scale* can be represented as a triangle shown in Fig. 4 and called Josang's opinion triangle [15, 16].

3.1 Appraisal Procedure

The *Assessment scale* is applied to express opinions and the level of confidence in these opinions in relation to assumptions, facts and warrants which are not supported by a (counter-)argument strategy.

The assessment of a single node of the structure proceeds as follows:

1)   If no evidence for or against the statement representing the node is available the *'lack of confidence'* assessment is issued and the procedure is broken.
2)   In the other case, the ratio between the evidence supporting the acceptance and rejection of the statement is assessed and an appropriate value from the *Decision scale* is chosen.
3)   Then, it is assessed how much evidence could additionally be provided to be sure about the decision chosen in step 2. This amount of missing evidence drives the selection from the *Confidence scale*.
4)   The final assessment from the *Assessment scale* is obtained by combining the two partial assessments from steps 2 and 3.

Fig. 5 presents a fragment of the user interface for issuing assessments of the TCT tool [14]. The user can drag a small white marker over the opinion triangle shown on the left hand side. Then, the linguistic values corresponding to the current position of the marker are displayed in the Confidence level: and Decision: windows. It is also possible to directly choose an appropriate linguistic assessment. Additionally, the current levels of belief, disbelief and uncertainty are displayed as horizontal bars just above the opinion triangle.

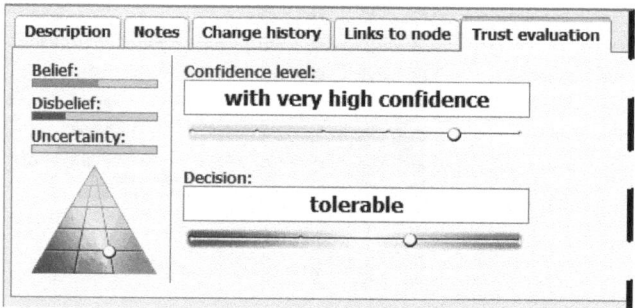

Fig. 5. User interface for issuing assessments of the TCT tool

3.2 Appraisal Examples

As an example let us consider fact F stating that *'PIPS-enabled devices are conformant with European Medical Devices Directive'* (see Fig. 2).

Let us assume that:

(1) There is some evidence relevant to F.
(2) All the available evidence supports F which results in choosing the *'acceptable'* value from the *Decision scale*.
(3) The evidence is almost complete, however, the certification process has not been performed yet – which leads to the *'with very high confidence'* assessment.

Consequently, the final appraisal of F is: *'with very high confidence acceptable'*.

Let us now consider F in a different situation:

(1) There is some evidence relevant to F.
(2) The evidence demonstrates that most of the requirements of the directive are met, however, one significant requirement is not yet fulfilled; this results in the *'opposable'* assessment.
(3) The evidence is substantial, however, not complete which gives the *'with high confidence'* assessment.

Consequently, the final appraisal of F is: *'with high confidence opposable'*.

As another example let us take warrant W (*'Analysis of different channels'*) from Fig. 2. The warrant identifies different types of channels providing information to the system and explains that if they provide valid information, the information available to the system is also valid.

Let us assume that:

(1) An inventory of types of channels exists.
(2) It identifies four major types of channels represented in the argument and in addition, some other less important ones, which were not considered in the argument; this leads to the *'tolerable'* assessment of the warrant.
(3) The inventory resulted from a formalized procedure of review of system design; this leads to the *'for sure'* assessment.

Consequently, the final appraisal of W is: *'for sure tolerable'*.

As yet another example let us take assumption A (*'Truthfulness of the information provided by a patient'*) from Fig. 2, which states that patients will not intentionally input false data into the system.

Let us assume that:

(1) There are bases to assess the assumption as there is some information about in what situations and what kind of data patients input into the system.
(2) The assessor tends to accept the assumption although sees some situations where it does not necessarily hold; the decision is to assess it as *'tolerable'*.
(3) The assessor has no doubts that she/he sees the whole scope of relevant situations; this leads to the *'for sure'* assessment.

Consequently, the final appraisal of A is: *'for sure tolerable'*.

4 Aggregation Rules

Aggregation rules define how the appraisals of the premises and the appraisal of the warrant are used to calculate the appraisal of the conclusion. We will briefly discuss four basic argument types and the corresponding aggregation rules which we have identified and illustrate their application by examples. An interested reader is referred to [8] for more detail and the formal definition of the aggregation rules. The examples show the results of application of the aggregation mechanisms as they are presently implemented in the TCT tool [14].

C-argument (*Complementary argument*) is such where the premises provide complementary support for the conclusion. In the case of *C-argument* not only the assessments of the premises and the warrant but also the weight associated with each premise is taken into account. The final assessment of the conclusion is a sort of weighed mean value of the contribution of all the premises.

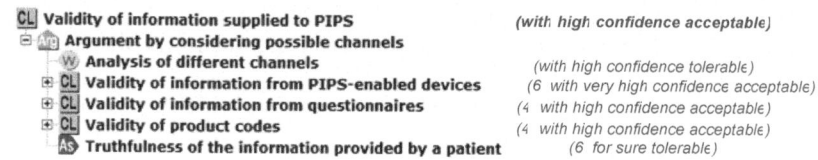

Fig. 6. Assessment of the conclusion of *C-argument*

Let us consider a *C-argument* shown in Fig. 6. The assessments of the warrant and premises (together with the associated weights) are shown on the right hand side in italic. The resulting assessment of the conclusion is '*with high confidence acceptable*' (printed in bold). Note that despite the fact that one of the premises is '*tolerable*' the conclusion is '*acceptable*'. This results form the fact that the other premises ranked '*acceptable*' outweighed in this case. Additionally, it can be seen that the confidence in the conclusion is slightly lower than it could be expected while looking at the assessments of the premises. This results from the fact that there were some doubts concerning the strength of the inference rule, reflected in the assessment of the warrant.

Let us consider the example from Fig. 6 but with the assessment of the assumption modified to *(6, lack of confidence)*. In such a case the assessment of the conclusion would be *(with low confidence acceptable)* which results form the fact that the other premises are fairly high assessed and there is relatively high assessment of the warrant. Nevertheless, the assessment of the conclusion would be lower than in the example shown in Fig. 6.

Another type of argument, called *A-argument* (*Alternative argument*) is encountered in situations where we have two or more independent justifications of the common conclusion. In *A-arguments* the confidence in the assessments coming from different argument strategies is reinforced if the assessments agree, or it is decreased if they contradict each other.

Let us illustrate the situation with the example shown in Fig. 7. Both arguments support the conclusion providing high confidence. In this case the resultant assessment of the conclusion is '*with very high confidence acceptable*'.

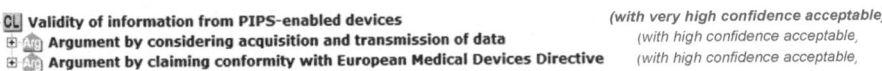

Fig. 7. Assessment of the conclusion of *A-argument*

In the case the arguments contradict each other, the effect is opposite. If one of the arguments in Fig. 7 would support rejection of the conclusion and another recommend acceptance, there would be no confidence in the conclusion at all and the *'lack of confidence'* assessment would result.

In *NSC-arguments* (*Necessary and Sufficient Condition list argument*), negative assessments are strongly reinforced. In such arguments the acceptance of all premises leads to the acceptance of the conclusion, whereas rejection of a single premise leads to the rebuttal of the conclusion. An example of such an argument is shown in Fig. 8. Each premise is a necessary condition for the conclusion. Therefore, if even one of them is rejected, the conclusion cannot be accepted. Consequently, low assessments of the premises leads to a rapid drop in assessment of the conclusion.

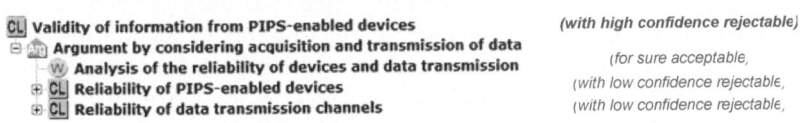

Fig. 8. Assessment of the conclusion of *NSC-argument*

In *SC-arguments* (*Sufficient Condition list argument*) acceptance of the premises leads to the acceptance of the conclusion similar to *NSC-arguments*. An example of such an argument is presented in Fig. 9. The difference to *NSC-argument* is that in this case rejection of a single premise leads to the rejection of the whole inference. For instance, the lack of conformity with EMD Directive does not result in invalid information received from the devices. The only reasonable conclusion is that in such a case we do not know anything new concerning the validity of this information.

Fig. 9. Assessment of the conclusion of *SC-argument*

Detailed analysis of the inference types encountered in argument structures is presented in [8] and the aggregation rules for different types of algorithms have been proposed. Each warrant occurring in the argumentation has its type explicitly identified and the corresponding aggregation rule assigned, which is done by the developer of the argument structure. Additionally, for *C-arguments* it is necessary to assign weights to the premises, which indicate the influence of a given premise on the conclusion.

5 Assessment Scenarios

Arguments can be used in different contexts and their appraisal serves two basic purposes:

1. to assess the compelling power of the argumentation,
2. to construct a simple and understandable message communicating the strength of the argument to the receivers who do not have capacity or resources to study and assess the argument themselves.

In the 'standard' scenario of (safety, assurance, trust) cases the argument aims to justify some distinguished property of a considered object in its application context. Such justification is often complex and difficult to understand without sufficient expertise and resources. In such situations the appraisal mechanism can be used by experts to record and accumulate their opinions about the argument. The mechanism provides full traceability to the elementary assessments and the way they were combined into the final one. Different experts' opinions can be compared and, if necessary, the resultant assessment can be easily computed. Such opinions can then be communicated to the managers and other decision-makers to support their decisions concerning suitability of the considered object for the expected purpose (for instance, granting a license for using the object in its target context). The results of the appraisal can be also communicated to broader public which can promote trust and the feeling of safety/security.

In [11] a collaborative development process for trust cases was proposed. Its extension, taking into account the proposed appraisal mechanism which supports the assessment phase in the process is illustrated in Fig. 10. The assessment phase provides, as part of its feedback, the report on structural errors in the trust case which is a positive side effect of the application of the appraisal mechanism. It provides also an assessed trust case which can be presented to viewers and used as a base for the next development cycle.

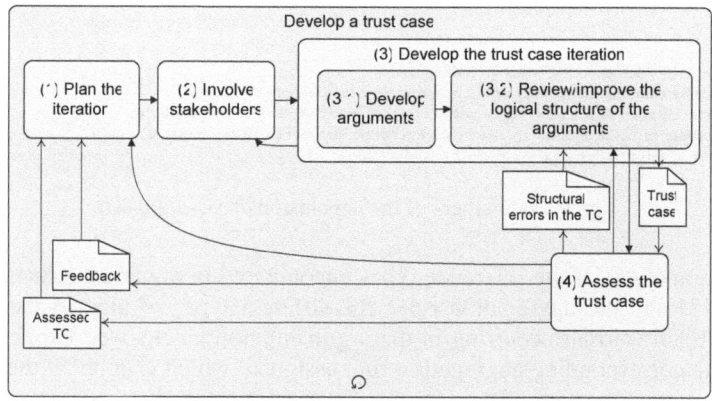

Fig. 10. Trust Case development process

Another application scenario, to which the appraisal mechanism can bring a significant value, is application of standards. Conformity assessment is a process which ends with a binary assessment (acceptance or rejection of the claim of conformity). In this process, fulfillment of numerous requirements is checked applying the same binary decision scale. This is mainly because of the lack of practical and usable mechanisms supporting more fine differentiation of the fulfillment level of various requirements of the standard. In every assessment project, some of the requirements are fulfilled better than others and are supported by stronger evidence. To exploit this fact, more sophisticated methods of evidence and justification appraisal are needed, which would make it possible to perform appraisal in parallel to the conformity achievement process to get feedback on how the project is approaching full satisfaction of the conformity criteria. This aspect is of particular interest if self-declaration is taken into account (first party conformity assessment). Application of the appraisal mechanisms could help to make self-assessment more objective and provide useful feedback during the conformity achievement process.

To this end *Standards Conformity Framework (SCF)* [4, 5] has been proposed which supports demonstration of conformity with standards. The aim of SCF is to develop and maintain a document which justifies the claim of conformity. Its component is a conformity case template which is a skeleton of argumentation about conformity, derived from a given standard. The template has gaps to be filled in during the process of conformity demonstration. The gaps are supplemented with project specific evidence which proves that an examined object fulfils the requirements of the standard. It results in a complete argument which is called a *Conformity Case*. Such a conformity case can be easily assessed using the appraisal mechanism proposed in this paper. The whole process of development of templates of conformity cases, their conversion to complete arguments and later appraisal of those arguments is fully supported by the TCT tool.

6 Experimental Evaluation

The linguistic scales for the appraisal of arguments have been chosen deliberately to support assessors by offering them a (not too large) set of intuitively understood values. However, in order to perform the calculations defined in the aggregation rules, it was necessary to represent the linguistic values as numbers from the [0,1] interval and this mapping could have a significant impact on the computed results. There was no evidence, that the most obvious, even distribution of the linguistic values over the [0,1] interval is the most proper one. And in fact, the experiments showed the opposite.

We decided to find the mapping between linguistic scales and the [0,1] interval experimentally by calibrating the aggregation rules relating them to the expert assessments of conclusions of the selected set of arguments.

A group of 31 students of the last year of a computer science university course took part in the experiment. The students were divided into three groups, each of which was supposed to apply one of the aggregation rules: *A-rule, NSC-rule* and *C-rule*. *SC-argument* type was dropped because of its similarity to *NSC-argument* type. Each student was provided with five simple trust cases composed of a claim, an argument strategy, a warrant and premises (in the case of C-rule and NSC-rule) or a claim with a few argument strategies (in the case of A-rule).

The experiment participants were asked to assess the warrant and, in the case of C-rule to assign weights to the premises. Then, assuming the pre-defined assessments of each premise (in the case of C-rule and NSC-rule) or the assessments assigned to each of the argument strategies (in the case of A-rule) the participants were asked to give their assessment of the conclusion using the Assessment Scale. They were supposed to repeat this step for 10 different sets of initial assessments of the premises (chosen randomly) for each trust case. That makes the total of 50 assessments of the conclusions issued by each participant. To check for consistency, each participant was additionally asked to repeat his/her assessments for 10 randomly selected situations.

Some students were excluded from the experiment for formal reasons or because their assessments apparently were not reasonable (for instance, they declared high confidence in acceptance of a conclusion in a situation where the premises were with high confidence rejectable). Finally, 8 questionnaires related to A-argument type, 6 questionnaires related to NSC-argument type and 10 questionnaires related to C-argument type were used in the following analysis.

In the experiment the consistency of the students' answers was measured (i.e. the average change in a repeated assessment of the same item) and the accuracy of assessments obtained by application of the aggregation rules (measured as the average distance between the students' answers and the results obtained by application of the aggregation rules). The results are presented in table 1. The numbers in the table are normalized, which means that 1 represents the distance between two adjacent positions on the linguistic scale.

The data shows that the accuracy of the results obtained by application of the aggregation rules is similar to the consistency of the participants' answers (i.e. the corresponding metrics for each aggregation rule do not differ significantly). This is the maximum of what could have been achieved regarding the data set used to calibrate the aggregation rules. The data show that using the (calibrated) aggregation rules we can expect to obtain the results which are fairly close to the results which would be obtained while engaging humans in the aggregation process.

Table 1. Results of experiments

		Aggregation rule		
		A-rule	*NSC-rule*	*C-rule*
Consistency of students' assessments	*Confidence scale*	1,03	0,94	0,84
	Decision scale	0,64	0,62	0,87
Accuracy of assessments obtained by application of aggregation rules	*Confidence scale*	1,06	1,10	0,90
	Decision scale	0,80	0,78	0,66

Further calibration requires more data which we plan to collect in the subsequent experiments.

7 Conclusion

This article introduced an innovative method of argument structures appraisal. The method provides for gathering expert opinions about the inferences used in the argumentation and the value of the supporting evidence. It can be applied to assess the compelling power of arguments used in different contexts. In particular, it can be used with respect to the arguments contained in different cases, like safety cases, security cases, assurance cases or trust cases. It can also be used to support standards conformity processes. The method has been fully implemented in the TCT tool which supports full-scale application of our Trust-IT methodology. Some experimental validation of the method has been already performed and further experiments are under preparation.

Acknowledgement

This work was partially supported by the project ANGEL - 'Advanced Networked embedded platform as a Gateway to Enhance quality of Life' (IST project 2005-IST-5-033506-STP) and by the project PIPS – 'Personalized Information Platform for health and life Services' (Contract No. 507019 IST2.3.1.11 e-Health) within the European Commission 6th Framework Programme.

Contributions by Michal Nawrot and Michal Witkowicz in the development of the appraisal mechanism in TCT tool are to be acknowledged.

References

1. Bishop, P., Bloomfield, R.: A Methodology for Safety Case Development, Industrial Perspectives of Safety-critical Systems. In: Proceedings of the Sixth Safety-critical Systems Symposium, Birmingham (1998)
2. Bloomfield, R., Guerra, S., Masera, M., Miller, A., Sami Saydjari, O.: Assurance Cases for Security, A report from a Workshop on Assurance Cases for Security, Washington, USA (2005)
3. Bloomfield, R.E., Littlewood, B., Wright, D.: Confidence: Its Role in Dependability Cases for Risk Assessment. In: 37th Annual IEEE/IFIP International Conference Dependable Systems and Networks, pp. 338–346 (2007)
4. Cyra, L., Gorski, J.: Supporting Compliance with Safety Standards by Trust Case Templates. In: Proc. of ESREL 2007, Norway, vol. 2, pp. 1367–1374 (2007)
5. Cyra, L., Gorski, J.: Standard Compliance Framework for Effective Requirements Communication. Polish Journal of Environmental Studies 16(5B), 312–316 (2007)
6. Cyra, L., Gorski, J.: Extending GQM by Argument Structures. In: 2nd IFIP Central and East European Conference on Software Engineering Techniques CEE-SET, pp. 1–16 (2007)
7. Cyra, L., Gorski, J.: Using Argument Structures to Create a Measurement Plan. Polish Journal of Environmental Studies 16(5B), 230–234 (2007)
8. Cyra, L., Gorski, J.: Supporting Expert Assessment of Argument Structures in Trust Cases. In: Ninth International Probabilistic Safety Assessment and Management Conference PSAM 9, Hong Kong, China, pp. 1–9 (2008)
9. Gorski, J., Jarzebowicz, A., Leszczyna, R., Miler, J., Olszewski, M.: Trust Case: Justifying Trust in IT Solution, Elsevier, Reliability Engineering and System Safety, vol. 89, pp. 33–47 (2005)
10. Gorski, J.: Trust Case – a Case for Trustworthiness of IT Infrastructures, Cyberspace Security and Defence: Research Issues. NATO ARW, pp. 125–142. Springer, Heidelberg (2005)

11. Gorski, J.: Collaborative Approach to Trustworthiness of IT Infrastructures. In: Proc. of IEEE International Conference on Technologies for Homeland Security and Safety TE-HOSS 2005, pp. 137–142 (2005)
12. Gorski, J., Zagorski, M.: Reasoning about Trust in IT Infrastructures. In: Proc. of ESREL 2005, pp. 689–695 (2005)
13. Greenwell, W., Strunk, E., Knight, J.: Failure Analysis and the Safety-Case Lifecycle, Human Error, Safety and Systems Development 2004, pp. 163–176 (2004)
14. Information Assurance Group: TCT User Manual, Gdansk University of Technology (2007),
    http://kio.eti.pg.gda.pl/trust_case/download/TCTEditor_Users_Manual.pdf
15. Josang, A., Grandison, T.: Conditional Inference in Subjective Logic. In: Proc. of the 6th International Conference on Information Fusion, Cairns, pp. 471–478 (2003)
16. Josang, A., Pope, S., Daniel, M.: Conditional Deduction Under Uncertainty. In: Godo, L. (ed.) ECSQARU 2005. LNCS (LNAI), vol. 3571, pp. 824–835. Springer, Heidelberg (2005)
17. Kelly, T.: Arguing Safety – A Systematic Approach to Managing Safety Cases. PhD Thesis, University of York, UK (1998)
18. Kelly, T., McDermid, J.: A Systematic Approach to Safety Case Maintenance. In: Felici, M., Kanoun, K., Pasquini, A. (eds.) SAFECOMP 1999. LNCS, vol. 1698, pp. 271–284. Springer, Heidelberg (1999)
19. PIPS Project website, http://www.pips.eu.org
20. Sentez, K., Ferson, S.: Combination of Evidence in Dempster-Shafer Theory, SANDIA National Laboratories (2002)
21. Shafer, G.: Mathematical Theory of Evidence. Princetown University Press (1976)
22. Strigini, L.: Formalism and Judgement in Assurance Cases, Workshop on Assurance Cases: Best Practices, Possible Obstacles, and Future Opportunities. In: Proc. of DSN 2004, Florence, Italy (2004)
23. Toulmin, S.: The Uses of Argument. Cambridge University Press, Cambridge (1969)

Formal Verification by Reverse Synthesis

Xiang Yin[1], John C. Knight[1], Elisabeth A. Nguyen[2], and Westley Weimer[1]

[1] University of Virginia
Department of Computer Science, Charlottesville, Virginia, U.S.A.
{xyin,knight,weimer}@cs.virginia.edu
[2] The Aerospace Corporation
Software Systems Engineering Department, Chantilly, Virginia, U.S.A.
elisabeth.a.nguyen@aero.org

Abstract. In this paper we describe a novel yet practical approach to the formal verification of implementations. Our approach splits verification into two major parts. The first part verifies an implementation against a low-level specification written using source-code annotations. The second extracts a high-level specification from the implementation with the low-level specification, and proves that it implies the original system specification from which the system was built. Semantics-preserving refactorings are applied to the implementation in both parts to reduce the complexity of the verification. Much of the approach is automated. It reduces the verification burden by distributing it over separate tools and techniques, and it addresses both functional correctness and high-level properties at separate levels. As an illustration, we give a detailed example by verifying an optimized implementation of the Advanced Encryption Standard (AES) against its official specification.

Keywords: Formal verification, formal methods, software dependability.

1 Introduction

In previous work, we introduced a novel approach to software verification called *Echo* [22]. In this paper we present details of a critical component of Echo, *reverse synthesis*, and we show how it is used in the overall verification process. We also present an evaluation in which we applied it to a non-trivial system.

In many cases, verification is undertaken by testing the developed software artifact against its specification. Testing, however, is not adequate for high levels of assurance [5]. Formal verification is an attractive alternative under such circumstances for systems in which safety and security are critical concerns. It provides confidence with mathematical rigor that many classes of errors in software development have been avoided or eliminated. In some cases—such as at Evaluation Assurance Level 7 of the Common Criteria [19]—it is required. Verification of functional correctness helps to avoid defects introduced in software development that manifest themselves as security vulnerabilities or safety hazards. We note that this complements the notion of proving that a system possesses certain specific safety or security properties.

Our approach is aimed at making formal verification of functional correctness more practical. It uses existing notations, tools and techniques, distributing the verification

M.D. Harrison and M.-A. Sujan (Eds.): SAFECOMP 2008, LNCS 5219, pp. 305–319, 2008.

burden over separate levels. At its core, a high-level specification is extracted from a low-level, detailed specification of a system. We refer to this activity as reverse synthesis. This low-level specification is shown to both describe the program and also adhere to the high-level specification. Thus, formal verification by reverse synthesis involves two proofs each of which is either generated automatically or mechanically checked. These proofs are: (1) a proof that the source code implements the low-level specification correctly; and (2) a proof that a high-level specification which is extracted from the low-level specification implies the original system specification. The two proofs can be tackled with separate specialized techniques.

In order to facilitate both proofs, a variety of *semantics-preserving transformations* are used to refactor the implementation. These refactorings reduce the complexity of verification caused by program refinements and optimizations that occur in practice. They are either effected or checked mechanically, and they are a crucial element of our verification approach because they can be used to simplify both of the proofs, in some cases making proofs feasible that otherwise would not be.

The introduction of a low-level specification as an intermediate point and the application of semantics-preserving refactorings allow our approach to dovetail with standard development processes more easily than existing approaches to formal verification. As a result, relatively few limitations are imposed on developers and many existing software engineering development methods can continue to be used, yet formal verification and all of its benefits can be applied.

In this paper, we begin by summarizing our approach to formal verification by reverse synthesis and then discuss the process and elements involved in detail. Next we present a detailed example of the use of reverse synthesis: verifying an implementation of the Advanced Encryption Standard (AES) against the official AES specification. Finally, we compare our approach to formal verification to other approaches.

2 Formal Verification by Reverse Synthesis

A crucial element of our overall approach is the use of a low-level specification since it is the intermediate representation of the software upon which our proofs are based. The level that we define for this is an *annotated implementation*, i.e., an implementation supplemented with declarative property annotations such as preconditions, postconditions, and invariants. These annotations can be defined and inserted into the source code by the developers or partially generated directly from the code, to describe the desired behaviour of subprograms in the code. Existing annotation-and-proof systems [3, 16] can verify source code against such annotations mechanically, and in our prototype system we use SPARK Ada [3]. Although we have not done so, our approach could be used with languages other than our choice of SPARK Ada, and so this choice is not a fundamental limitation. Annotations and proofs of the kind we require have also been adopted by Microsoft in both Vista and Office [8].

As part of our Echo approach [22], we assume that the original specification from which the software was developed is complete and its semantics have been restricted to those that can be implemented, and we assume a reasonable development practice has been followed to create an executable implementation together with proper annotations. Then our verification approach, shown in Fig. 1, consists of the following steps:

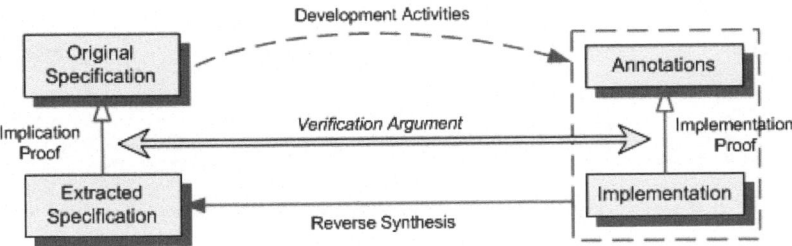

Fig. 1. Formal Verification by Reverse Synthesis

(1) Implementation Proof: A proof that the implementation implements the annotations correctly. Our prototype uses the SPARK Ada system [3] for this proof.

(2) Reverse Synthesis: A mechanical extraction (with human guidance) of a high-level abstract specification from the annotated implementation. Tools we have built perform this extraction, and the abstract specification is written in PVS.

(3) Implication Proof: A proof that the properties of the extracted specification imply the properties of the original specification. Our prototype uses the PVS system for this proof.

The implementation proof, reverse synthesis, and the implication proof are partly automated and partly mechanically checked. Thus, with this process we have a complete formal argument that the implementation behaves according to the specification.

This approach makes verification more practical. It does this in part by combining existing powerful techniques, in part by introducing reverse synthesis, and in part by allowing an engineer to work with an existing implementation rather than requiring that an implementation be designed to show compliance. Showing compliance of an implementation with a specification should not necessitate a specific method for constructing the implementation: development decisions should be minimally restricted by the goal of verification. This is not the case currently with refinement-based approaches such as the B method [1].

By exploiting existing notations and tools, the approach offers the opportunity to make progress more quickly since existing tools both solve part of the problem and point in a positive technical direction. Annotations are tightly coupled with the source code, thus are suitable to prove low-level functional correctness. High-level specification languages and are more expressive and are better at reasoning about high-level properties. Reverse synthesis provides a mechanical link between annotations and high-level specification proofs thereby filling in the gaps left by tools already available.

3   The Reverse Synthesis Process

Reverse synthesis, shown in Fig. 2, is composed of three phases: (1) *implementation* refactoring; (2) *implication* refactoring; and (3) *specification extraction*. The refactoring phases each transform the program being verified so as to preserve its semantics but to make the associated proof easier. Implementation refactoring assists the user in enhancing and completing the annotations of the source program and thereby

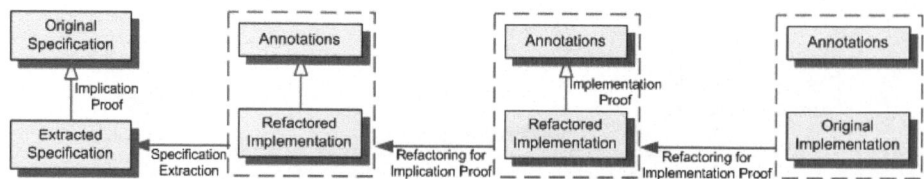

Fig. 2. Detailed Reverse Synthesis Process

facilitates the proof that the source code matches the annotations. Implication refactoring aids the specification extraction phase and reduces the effort in the proof that the extracted specification implies the original one. The specification extraction phase mechanically extracts a high-level abstract specification from the refactored annotated implementation.

We examine these reverse synthesis steps and proofs in turn in the remainder of this section. Implementation proof uses code-level tools such as static code analyzers, proof obligation generators, and proof checkers. This technology is well established, and we do not discuss it further.

3.1 Refactoring for Verification

Software implementations are often influenced by the need for efficiency in time or space. More complex algorithms are used to reduce executions times and data structures are sometimes chosen to reduce computation (and vice versa). Such implementation decisions tend to add considerably to a program's overall complexity. It is often easy to show that refactoring a program and reducing its efficiency does not change its computed function. Reducing efficiency can, however, reduce complexity and thereby facilitate verification. Hence instead of directly extracting a high-level specification from the annotated implementation and performing proofs on them, our approach first tries to refactor the implementation and reduce the complexity of proofs to the extent possible.

Refactoring for verification is the application of semantics-preserving transformations to the annotated implementation. The transformations modify the implementation in some way, and this usually simplifies the implementation, decreasing the implementation's efficiency. This is in sharp contrast to the usual role of semantics-preserving transformations where some form of improvement in efficiency is the goal. The standard approach is exemplified by the use of optimizing transformations in compilation.

We hypothesize that semantics-preserving transformations are easier to carry out, understand and prove correct at the level of the program than at the level of the proof system. That is, given complex proof obligations for a program, it is easier to simplify the program than to simplify the logical terms directly. A loop and its unrolled form yield proof obligations that are equi-satisfiable, but those obligations have different structures and are not equally easy to verify. Refactoring in reverse synthesis, therefore, reduces complexity while leaving the program semantics unaltered, thereby assisting the proofs involved in the verification.

Refactoring for verification involves both computation and storage. Programs can be made more amenable to verification by adding redundant computation or storage, by adding intermediate computation or storage, or by restructuring the program. Examples of adding redundant computation include moving computations out of conditionals, changing a loop that computes several things into a sequence of single-purpose loops, increasing loop bounds to a convenient limit, and replacing iteration with recursion. Retaining values after their initial computation so that they can be used in other (possibly redundant or intermediate) computations is an example of adding redundant storage.

Refactoring is based on the following four stages: (1) *identify candidate refactoring transformations*—since refactoring might address certain optimizations and refinements introduced during development, this usually needs guidance from developers to identify the occurrences of optimizations, although some can be found mechanically; (2) *determine the order to apply the transformations*—the order matters if there are dependencies among the transformations; (3) *prove the transformations are semantics-preserving*—all transformations should be proved to preserve the semantics and should not require the user to discharge complex proof obligations. In order to make the proofs reusable, we identify common refactoring transformations, characterize them into templates, and prove that they are semantics-preserving; and (4) *apply the transformations to the code*—all of the transformations should be applied mechanically to avoid introducing errors. In our prototype toolset, we adopt the Stratego [4] program transformation language and associated XT tools to achieve this.

Presently, refactoring for verification in our reverse synthesis approach has two phases, namely refactoring to facilitate the implementation proof and refactoring to facilitate the implication proof:

(1) Implementation refactoring: These transformations are intended to simplify the proof between the code and the annotations. The transformations are usually applied within subprograms and do not change the existing pre- and post-condition annotations for the subprograms. However, corresponding proof obligations for these annotations are likely to become much simpler to discharge. After the refactoring, the user also has the chance to enhance and complete the annotations for those elements that were otherwise obscured by the optimizations done in the original development process.

(2) Implication refactoring: These transformations are intended to aid the later specification extraction and to simplify the proof between the extracted specification and the original one. The transformations usually involve changes to the structure of the entire program with the goal of aligning the extracted specification and the original specification. This alignment simplifies the implication proof. Each transformation might involve several subprograms and the annotations usually need to be modified, although the modification can in many cases be done mechanically.

The two refactoring phases can be overlapped since some transformations may help both proofs. Neither one of them is strictly required. However, if they are applied, the resulting proof obligations are likely to be much simpler to discharge than in most traditional verification circumstances because the proof involves a transformation from a more-complex to a less-complex program. Refactoring for verification plays an important role in the whole process, and we detail an example of its application in Section 5.

3.2 Specification Extraction

The specification extraction step extracts an abstract specification from the refactored annotated implementation to be used in the proof of implication with the original specification. Presently, specification extraction exploits three basic techniques: (1) architectural and direct mapping; (2) component reuse; and (3) model synthesis, which are discussed in detail below. For any particular program, combinations of techniques will be used, each contributing to the goal of successful specification extraction for that program. We have developed a prototype toolset for specification extraction that handles architectural and direct mapping from SPARK Ada implementations to PVS specifications completely, along with minor elements of the other two techniques.

Specification extraction is automated or mechanically checked, which ensures the extracted high-level specification is a correct representation of the annotated implementation. However, to make the verification sound, we must also make sure that the extracted specification is complete, suitably abstract but not too abstract, so that we can construct and complete the implication proof. Since we extract the high-level specification mostly from the low-level annotations, it means we have to make sure the annotations in the source code describe the entire semantics. Presently we have no completely automated way to check this property, and we rely on human review and cross-check with the derivation relations between input/output variables to do this.

Architectural and Direct Mapping. We hypothesize that it is often the case that the architectural or high-level design information in a specification is retained in the implementation. While an implementation need not mimic the specification architecture, in practice it will often be similar in structure because repeating the architectural design effort is a waste of resources.

As an example, consider a model-based specification written in a language like Z that specifies the desired operations using pre- and post-conditions on a defined state. The operations reflect what the customer wants, and the implementation architecture would almost certainly retain those operations explicitly.

The above hypothesis is implicitly assumed in the well-known Floyd-Hoare approach, which requires a stepwise proof that a function implementation complies with its specification. This implicitly requires a mapping from functions and variables in the specification to those in the implementation. Thus, we have not added assumptions, only evaluated existing ones in more detail.

In a case where the implementation retains the architectural information from the original restricted specification, a simple way to begin the process of specification extraction is to directly translate elements of the annotated implementation language, such as packages, data types, state/operation representations, preconditions, postconditions, and invariants, into corresponding elements in the specification language. The extracted specification will be structurally similar to the restricted specification. Such a strategy is straightforward, but it does have considerable potential in our approach.

Component Reuse. Software reuse of both specification and code components is a common and growing practice. If a source-code component from a library is reused in a system to be verified and that component has a suitable formal specification, then that specification can be included easily in the extracted specification [24].

Model Synthesis. In some cases, specification extraction may fail for part of a system because the difference in abstraction used there between the high-level specification and the implementation is too large. In such circumstances, we use a process called model synthesis in which the human creates a high-level model of the portion of the implementation causing the difficulty. The model is verified by conventional means and then included in the extracted specification.

At present, our implementation of model synthesis relies on human insight. In future work, we plan to mechanize model synthesis by exploring ideas such as hypothesizing invariants in extended static checking [11] and obtaining partial models and invariants from iterative abstraction refinement and software model checking.

3.3   Implication Proof

The extracted specification needs to be matched to the original specification to complete the verification argument. The property that needs to be shown here is implication, not equivalence; by showing that the extracted specification implies the original specification, but not the converse, we allow the original specification to be non-deterministic, and allow more behaviours in the original specification than the implementation.

The implication argument is shown by matching the structures and components of these two specifications and setting up and proving an implication theorem using the prover associated with the specification language. The formal definition of implication we use for this is that set out by Liskov and Wing known as behavioral subtyping [18]. Behavioral subtyping was studied in the context of languages that permit inheritance, in order to define what it meant for a subtype to comply with the type constraints of a supertype. Intuitively, the requirement is similar in verification: we want to ensure that the function implementation complies with the constraints defined in its specification. While our instantiation is more general, not making assumptions on what is or is not required of a type system, the principles are the same.

Then, by implication, we mean that the types and functions in the extracted specification are subtypes of the matching types and functions in the original specification. More specifically, the extracted function specification (which represents the implementation) should have a weaker precondition and a stronger postcondition than the original function specification:

$$Pre_{original} \Rightarrow Pre_{extracted} \quad \land \quad Post_{extracted} \Rightarrow Post_{original}$$

To set up the theorem, we need human guidance to match elements such as variables and functions between the two specifications, but in many cases they can be suggested automatically. The resulting proof obligations need to be discharged automatically or interactively in a mechanical proof system. When the extracted specification shows structure similarity to the original one, the proof usually does not require considerable human efforts as will be illustrated in Section 5. Also, by setting up the implication proof theorem function by function, not property by property, we can easily locate the error if the implication theorem fails to be proved, since it must be inside the structure or component that cannot be proved.

4 An Example Application

In this section we present an example of applying formal verification by reverse synthesis to a small but important application. This example illustrates the various aspects of the approach and provides some preliminary evaluation. A comprehensive evaluation and development of industrial-strength support tools is relegated to future work.

Recall that one of our goals was to allow developers the maximum freedom possible in building a system. We sought a way to assess our success in meeting this goal as well as the utility of the overall technique. The approach we followed was to apply the technique to an important yet publicly-available system written entirely by others. Clearly, the system's development was not constrained by our verification requirements.

For this assessment, we used an implementation of the Advanced Encryption Standard (AES) [10]. We employed the following two artifacts: (1) the Federal Information Processing Standard (FIPS) specification of the AES [10] that specifies the AES algorithm, a symmetric iterated block cipher, mostly in natural language, with mathematical descriptions of some algorithmic elements; (2) a publicly available implementation written in ANSI C that contains various optimizations such as loop unrolling and function inlining. We assume that these artifacts were created by a traditional software development process, and that the developers took no actions that would make formal verification infeasible or very difficult.

We supplemented these artifacts as necessary to apply the reverse synthesis process. We translated the official FIPS specification into a formal specification in PVS. We formalized all the behaviors and constraints described in the FIPS specification in PVS and included them in the formal specification (as the original specification from Fig. 1). In practice, a formal specification might be produced by developers, making this type of translation unnecessary. We translated the ANSI C implementation into SPARK Ada and added annotations for pre- and post-conditions of functions (the annotated implementation from Fig. 1). Again, in practice an annotated implementation might be produced by developers, making this type of translation also unnecessary.

With these artifacts developed, we applied our reverse synthesis approach to formally verify the functional correctness of the SPARK Ada implementation with respect to the PVS specification. The details of the verification are described in the next section.

5 Verification of the AES Implementation

To verify the AES implementation, we applied refactoring and performed a series of complexity-reducing, semantics-preserving transformations using Stratego/XT tools. A proof that the code—with applied refactoring—adheres to its annotations was completed using the SPARK toolset with some straightforward human intervention. A PVS specification was derived from the refactored annotated implementation using our automatic specification extraction tool. The implication proof between the extracted specification and the original one was then established using the PVS theorem prover with some straightforward human intervention. Fig. 3 shows the detailed tool configuration we set up and the process we followed to conduct this case study. In all cases we included and verified only functions related to encryption and decryption; we

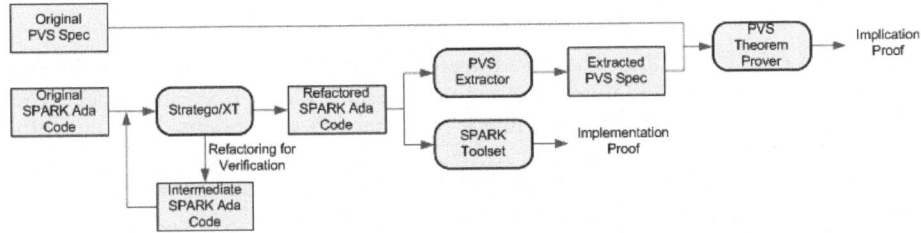

Fig. 3. Tool Configuration for AES Verification

did not describe or verify functions related to key expansion, or any of the NIST APIs. The relevant PVS specification contains 335 lines of functional specification, excluding lemmas and theorems that are required to prove its correctness. The relevant SPARK Ada code we are verifying has 733 lines of function declarations (including lookup tables), and 584 lines of function definitions excluding comments and annotations.

5.1 The Refactoring Process

According to the original AES documentation [7], the following four major optimizations had been applied to create the implementation: (1) loop unrolling; (2) word packing; (3) table lookup; and (4) function inlining. Table lookup and function inlining were dependent since the table entries encoded part of the defined functions. For each of the optimizations we identified, we developed a template defining the refactoring transformation so that they could be reused in other programs. We then characterized them and proved them to be semantics-preserving using PVS. Finally, we applied the transformations mechanically using Stratego. Besides the four major transformations, we also effected several minor transformations including adjusting intermediate variables, removing redundant statements, and aggregating data assignments. These transformations helped match the code to the transformation templates and clean up the code after the transformations. Each was proved to be semantics-preserving.

Table 1 lists details of the versions of the AES code used in verification. AES1 is the original, optimized code and each subsequent version is the result of applying a refactoring transformation. The rightmost two columns in Table 1 present the sizes of SPARK Ada code associated with function definitions and declarations (including lookup tables) respectively. We used bytes instead of lines of code to more precisely denote the size of the code since our tool does not generate proper line breaks for intermediate refactored code.

Table 1. AES versions transformed via refactoring for verification

	Transformation	Definitions (bytes)	Declarations & Tables (bytes)
AES1	Original	25,415	41,924
AES2	undo loop unrolling	8,561	41,924
AES3	undo word packing	7,180	103,389
AES4	undo table lookups	8,036	7,545
AES5	undo func inlining	8,620	8,128

5.2 The Refactoring Transformations

Reversing Loop Unrolling. The first transformation we applied was to undo loop unrolling in AES1. Undoing loop unrolling involved locating the repeated code, redefining it as a for-loop, and changing literal references to use the new loop induction variable. This transformation introduced two new loop induction variables and dramatically shrank the code size as shown in Table 1 since vast amount of repeated code were removed. After the transformation, loop invariants could be annotated to facilitate the verification. This transformation assisted the implementation proof, because by introducing new loop invariants and removing replicated loop bodies, it substantially reduced the states involved in the proof.

Loop unrolling is a well-known compiler transformation, and it might seem unusual for it to have been applied explicitly at the source code level in AES. However, it is not specific to AES, because not all compilers unroll loops and because manual unrolling is still a widespread practice (e.g. to expose concurrency). With further tool support, both identifying unrolled loops and verifying the reversing transformation can be done automatically (e.g., [17]). Here we manually identified two unrolled loops, but selecting the transformation spots, performing the transformation, and proving the preservation of the semantics were all machine checked using Stratego and PVS.

Reversing Word Packing. The second transformation involved undoing a word-packing representation optimization. The AES standard describes encryption in terms of bytes, but the original implementation packs the bytes into 32-bit words to utilize efficient word-level operations. AES1 and AES2 include utility functions to split and combine 32-bit words; the bytes inside a word are referenced by bit shifting. In AES3, we replaced references to 32-bit words by arrays of four bytes. Thus splitting, combining, and references to bytes used native array operations. Specialized procedures for manipulating packed data were removed, but every line of code that referenced packed data had to be updated to use the new representation. As a result, the function definitions shrank slightly while the lookup tables expanded considerably. This is because the tables were originally composed with 32-bit words but were composed of four-byte arrays after undoing word packing. This transformation assisted the implication proof since the code and the specification used the same basic type to refer to data after it and were thus easier to verify.

Data structure transformations and efficient representations are also not specific to AES. While there has been some work toward automatically locating likely spots for such transformations (e.g., [15]), we assume that this step is manually guided. We let the user indicate the links between the old and new representations or provide a type transformer. Once the types and the operations on the types have been selected, the behavioral equivalences of the representations are checked mechanically using PVS. Then transformation spots are selected and the code is transformed mechanically by Stratego.

Reversing Table Lookup. The third transformation replaced table lookups with explicit computations. A major optimization in the AES implementation was combining different cryptographic transformations into a single set of table lookups. The tables contain pre-computed outputs and thus reduce the run-time computation. The properties of those tables have been documented [7], and AES4 replaced references to these tables with inlined instances of the appropriate computations using Stratego. As a result, all tables were removed causing a dramatic code-size reduction as shown in Table 1. This transformation supported the next one (reversing function inlining), because some inlined functions were encoded in the tables. It also made the implication proof easier since the specification was phrased in terms of the computations, not the tables.

This transformation can be viewed as a general form of property substitution. The original implementation maintains the invariant `Table[i] = computation(i)`; the transformation replaces reads of `Table[i]` with instances of `computation(i)`. Reasonable sites for such a transformation cannot, in general, be selected automatically, but the number of computations so described in the specification is limited, and the conventional software development artifacts may well record why and where such pre-computed tables were applied. In general, once a human has identified a table and the computation, the transformation can be checked mechanically by going through all the table entries and comparing them with corresponding computations. Selecting sites and performing the transformation can be done automatically and was in our example.

Reversing Function Inlining. The final transformation we applied was to undo function inlining. After the above transformations, inlined functions continued to obscure events that are explicitly required by the specification. Reversing such inlining aided both the implementation proof and the implication proof. By finding cloned code fragments, it removed replicated or similar proof obligations in the implementation proof. By reversing the inlining, it aligned the code structure with the specification structure so that the implication proof was easier to be constructed. In this example, we identified and factored nine specified functions, each of which was quite small. After undoing inlining, the verbose function-definition syntax actually increased the source code size shown in Table 1, but the conceptual complexity was reduced.

Inlining functions is certainly not specific to AES. Finding places to undo function inlining is known in the compiler literature as procedural abstraction [20] and is used when optimizing for code size. Finding appropriate sites for this transformation can thus be done automatically, or it can be guided based on the specification structure. We prove it is semantics-preserving and perform the transforming mechanically using PVS and Stratego respectively.

5.3 Specification Extraction and Proofs

The final program version, AES5, contained 262 lines of function declarations and table, and 214 lines of function definitions, including 126 lines of annotations. Proof functions and rules are also provided in additional files to facilitate the proof of the annotations. Most of the annotations were simple postconditions that could be straightforwardly derived, while others were loop invariants. The compliance of the code to the annotations was proved using the SPARK toolset. It automatically discharged 93% of the verification conditions, and the remaining ones needed very little human guidance to be discharged.

Using our prototype tool, a PVS specification was then automatically extracted using architectural and direct mapping. The result contained 606 lines of PVS and showed great similarity in structure to the original specification. Thus an implication

proof relating that extracted specification to the original specification was easily constructed, and all resulting obligations were discharged in seconds using the PVS theorem prover. More than half of the implication proof obligations could be discharged
by a simple (grind) command. Others could be discharged by applying a sequence
of proof commands and lemmas that demanded little human insight. These proofs,
combined with the proofs that the transformations were correct, provide a formal assurance guarantee that the AES implementation adheres to the specification.

To get an idea of how refactoring helped verification, we tried to verify the original
implementation as it was before refactoring. However, the off-the-shelf SPARK toolset could not even generate verification conditions. Instead it quickly exhausted heap
space and stopped, presumably because the generated proof obligations were too
large. We then tried annotating and verifying AES1, the version with loops rerolled.
The SPARK toolset generated more than 15M bytes of verification conditions which
is around 30 times larger than the refactored version. It took approximately 2 hours on
a dual 1.0 GHz UltraSparc IIIi with 2GB RAM for the tools to analyze the verification conditions, while on the same machine it only took minutes for the refactored
version. Moreover, unlike the refactored version, the verification conditions that could
not be automatically discharged here were mostly major postconditions, whose proof
simulated traditional formal verification, and required significant human insight and
efforts.

6 Related Work

Light-weight program analyses [9] are often used to find bugs in or gain confidence
about programs. Compared to more complete formal verification, their expressive
power is limited and no formal proof of compliance is produced. Heavier-weight
techniques like the B method [1] are more suited to full formal verification, but they
intertwine code production and verification. Using the B method requires a B specification and then enforces a lock-step code production approach on developers.

A more general technique is traditional Floyd-Hoare verification [12]. Unfortunately, it requires generation and proof of many detailed lemmas and theorems. It is
very hard to automate and requires significant time and skill to complete. Annotations
and verification condition generation, such as that employed by the SPARK Ada toolset, is used in practice. However, the annotations used by SPARK Ada (and other
similar techniques) are generally too close to the abstraction level of the program to
encode higher-level specification properties. Thus, we use verification condition generation as an intermediate step in our approach.

Automated code generation from a formal specification to an implementation, using tools such as the SCADE Suite [21], provides an alternative to verification. This
approach constructs an implementation automatically from the specification using
formal translation rules. If the translation rules are correct, it offers the possibility of
assuring that the behaviour of the implementation is consistent with the formal specification. However, for most safety-critical systems, it is very difficult to automatically
generate a well-structured or efficient implementation from a formal specification. If
the developer changes the generated code to refine its structure or increase efficiency,
the verification argument is invalidated.

Other techniques are available for the properties that we do not address. Model checking techniques [14], for example, have been quite successful at verifying hardware, protocols and temporal properties; they complement our approach in such areas. While model checking can generate proofs that the software model adheres to the specification, it does not prove that the software model is faithful to the original program. More recent model extraction [14], aims to address this problem and mechanically extracts a system model from the source code so that model checking can be applied. However, model extraction does not produce a full assurance argument since model checking is not targeted at full functional correctness.

Related work in the reverse engineering domain retrieves high-level specifications from the source code by semantics-preserving transformations and abstractions [6, 23]. These approaches are similar to reverse synthesis, but the goal is to make poorly-engineered code amenable to further analyses and not to aid verification. Our approach, which incorporates intermediate annotations, can more easily capture the properties relevant to verification while still abstracting implementation details. These techniques, however, show the feasibility of approaches similar to reverse synthesis.

Andronick et al. developed an approach to verification of a smart card embedded operating system [2]. Similar to reverse synthesis, they proved a C source program against supplementary annotations and generated a high-level formal model of the annotated C program that was used to verify certain global security properties. Our approach incorporates refactoring and allows us to show broad compliance with the original specification from which the system was built.

Heitmeyer et al. developed a similar approach to ours for verifying a system's high-level security properties [13]. Their approach is focused on verifying security properties, whereas ours is aimed at general functionality.

7 Conclusion

We have defined a verification technique based upon the use of an intermediate point of abstraction between a high-level formal specification and its concrete implementation. This intermediate point is a low-level specification documented by annotated source code. Our verification approach shows that the source code correctly implements the annotations and that the annotated source code implies the high-level specification.

We have introduced the new technique of reverse synthesis that mechanically creates a high-level specification from the low-level specification. A crucial component of reverse synthesis is the application of complexity-reducing but semantics-preserving refactoring transformations. In general, it is easier to transform the program than to transform the proof. Thus, transformations facilitate verification by reducing the complexity of the source program and thereby the proof obligation.

Human insight guides much of the process, but the analysis and thus the verification is either automatic or machine-checkable. It dovetails directly with traditional development processes and artifacts. We evaluated our approach by verifying an AES implementation against its formal specification.

Although our approach provides certain benefits over existing techniques, it is in no way a verification "silver bullet". As with any formal verification technique, it requires the use of formal languages, various analytic tools including a theorem-proving system, and considerable skill on the part of the developer. One specific additional responsibility placed on the developer is to annotate the source code with pre- and post-condition documentation. Although the various elements we have incorporated are not often part of current practice, our approach can be conducted in a production setting with comparable resources to those used now but with substantially higher assurance.

Acknowledgments. We thank Praxis High Integrity Systems for their technical support. This work was sponsored, in part, by NASA under grant number NAG1-02103.

References

1. Abrial, J.R.: The B-Book: Assigning Programs to Meanings. Cambridge University Press, Cambridge (1996)
2. Andronick, J., Chetali, B., Paulin-Mohring, C.: Formal Verification of Security Properties of Smart Card Embedded Source Code. In: Fitzgerald, J.S., Hayes, I.J., Tarlecki, A. (eds.) FM 2005. LNCS, vol. 3582, pp. 302–317. Springer, Heidelberg (2005)
3. Barnes, J.: High Integrity Software: The SPARK Approach to Safety and Security. Addison-Wesley, Reading (2003)
4. Bravenboer, M., Kalleberg, K.T., Vermaas, R., Visser, E.: Stratego/XT 0.16. A Language and Toolset for Program Transformation. Science of Computer Programming (2007)
5. Butler, R., Finnelli, G.: The Infeasibility of Quantifying the Reliability of Life-Critical Real-Time Software. IEEE Trans. on Software Engineering 19(1) (1993)
6. Chung, B., Gannod, G.C.: Abstraction of Formal Specifications from Program Code. In: IEEE 3rd Int. Conference on Tools for Artificial Intelligience, pp. 125–128 (1991)
7. Daemen, J., Rijmen, V.: AES Proposal: Rijndael. AES Algorithm Submission (1999)
8. Das, M.: Formal Specifications on Industrial Strength Code: From Myth to Reality. In: Ball, T., Jones, R.B. (eds.) CAV 2006. LNCS, vol. 4144. Springer, Heidelberg (2006)
9. Das, M., Lerner, S., Seigle, M.: ESP: path-sensitive program verification in polynomial time. Programming Languages, Design and Implementation, pp. 57-68 (2002)
10. FIPS PUB 197, Advanced Encryption Standard. National Inst. of Standards & Tech. (2001)
11. Flanagan, C., Lieno, K.: Houdini, an annotation assistant for ESC/Java. Formal Methods Europe, Berlin, Germany (2001)
12. Floyd, R.W.: Assigning meanings to programs. In: Schwartz, J.T. (ed.) Mathematical Aspects of Computer Science, Proceedings of Symposia in Applied Mathematics 19 (American Mathematical Society), Providence, pp. 19–32 (1967)
13. Heitmeyer, C.L., Archer, M.M., Leonard, E.I., McLean, J.D.: Applying Formal Methods to a Certifiably Secure Software System. IEEE Trans. on Soft. Eng. 34(1) (2008)
14. Holzmann, G.: The SPIN Model Checker: Primer and Reference Manual. Addison-Wesley, Reading (2004)
15. Kataoka, Y., Ernst, M., Griswold, W., Notkin, D.: Automated support for program refactoring using invariants. In: Int. Conference on Software Maintenance, pp. 736–743 (2001)
16. Burdy, L., Cheon, Y., Cok, D.R., Ernst, M.D., Kiniry, J.R., Leavens, G.T., Leino, K.R.M., Poll, E.: An overview of JML tools and applications. International Journal on Software Tools for Technology Transfer 7(3), 212–232 (2005)
17. Lerner, S.T., Millstein, E.R., Chambers, C.: Automated soundness proofs for dataflow analyses and transformations via local rules. Princ. of Prog. Lang., 364–377 (2005)

18. Liskov, B., Wing, J.: A Behavioral Notion of Subtyping. ACM Transactions on Programming Languages and Systems 16(6), 1811–1841 (1994)
19. National Institute of Standards and Technology, The Common Criteria Evaluation and Validation Scheme, http://niap.nist.gov/cc-scheme/index.html
20. Runeson, J., Nystrom, S., Sjodin, J.: Optimizing code size through procedural abstraction. Languages, Compilers and Tools for Embedded Systems, pp. 204–215 (2000)
21. SCADE Suite, Esterel Technologies, http://www.esterel-technologies.com/
22. Strunk, E.A., Yin, X., Knight, J.C.: Echo: A Practical Approach to Formal Verification. In: FMICS 2005, Lisbon, Portugal (2005)
23. Ward, M.: Reverse Engineering through Formal Transformation. The Computer Journal 37(9), 795–813 (1994)
24. Weide, B.W.: Component-Based Systems. In: Marciniak, J.J. (ed.) Encyclopedia of Software Engineering. John Wiley and Sons, Chichester (2001)

Deriving Safety Software Requirements from an AltaRica System Model

Sophie Humbert[1], Christel Seguin[2], Charles Castel[2], and Jean-Marc Bosc[1]

[1] Turbomeca, 64511 Bordes Cedex
{sophie.humbert,jean-marc.bosc}@turbomeca.fr
[2] ONERA Centre de Toulouse, 2 avenue E. Belin, B.P. 31055 Toulouse Cedex
{charles.castel,christel.seguin}@onera.fr

Abstract. This paper presents a methodology to derive software functional requirements from Preliminary System Safety Assessment analysis (PSSA) of helicopter turboshaft engines. The proposed process starts by extracting functional failure paths from system failure propagation models, using AltaRica models and AltaRica tools. Then the paper shows how to analyse these paths to generate minimal combinations of functional software requirements. This approach is applied to a part of the control system of a helicopter turboshaft engine.

Keywords: System safety requirements, software functional requirement, failure propagation models, AltaRica languages and tools, system control of helicopter turboshaft engines.

1 Introduction

This paper presents a methodology to derive software functional requirements from Preliminary System Safety Assessment analysis (PSSA) of helicopter turboshaft engines. The PSSA aims at assessing whether the coarse grain system architecture meets the system safety objectives. It provides not only evidences of the architecture safety but also new requirements applicable to any kind of system components (software, mechanical and electrical components, etc...). This analysis is performed before entering into the detailed software design. Thus the derived software requirements drive the traditional software development and verification/validation process. The approach relies on process, models and tools that were defined to support the safety analysis of aeronautic complex systems. Nevertheless, it remains generic enough to be applied to any kind of safety critical systems that integrate some software parts.

The safety analysis of complex systems usually generates at least three classes of software requirements:

- additional functional requirements to ensure that some specific feared events will not be induced by software subparts,
- segregation requirements to ensure the appropriate level of fault independence,
- level assurance of tasks performed in the software development process.

M.D. Harrison and M.-A. Sujan (Eds.): SAFECOMP 2008, LNCS 5219, pp. 320–331, 2008.

This paper focuses on the derivation of the new functional requirements applicable to software subparts.

It is worth noting that all the requirement classes are derived from the analysis of functional failure paths i.e. sets of components, resources, items (not necessarily limited to one system) whose anomalous behaviour (random failure or systematic error) could lead to a top level feared event. A first issue is the identification of these paths when one deals with complex systems at early design stage. The use of AltaRica models and tools was proven quite efficient to support such a kind of analysis for various complex systems [1], [2].

So we propose to start our requirement derivation process from an AltaRica system model that clarifies the assumptions about the hypothetical faults of any kind of system components (including hardware and software components). The AltaRica language and the used models are briefly presented in section 3. The section 4 presents how to extract functional failure paths from AltaRica models and how to select a minimal subset of software-related feared events from the paths. Complementary steps are requested to transform each selected software feared events into functional requirements. The reader concerned may find an overview of the global process in [3].

Finally our proposal was applied to a part of the control system of a helicopter turboshaft engine. This case study is partly introduced in section 2 and it is used as leading example for sections 3 and 4. Section 5 gives details about the whole case study and the lessons learnt from the methodology application. Finally, Section 6 discusses related and further works.

2 Case Study Introduction

The helicopter turboshaft engine role is to maintain a constant rotation speed of the helicopter rotor. To reach this goal, the power delivered by the engine has to be permanently adapted to the rotor load imposed by the helicopter flight conditions. Thus, the fuel quantity injected inside the engine must be accurately controlled. In this case study, only the hardware and software functions allowing the fuel metering unit control have been considered. The elements taken into account in this example are the following:

- An actuator which shifts the fuel metering valve (FMV).
- A sensor which recopies the FMV position after each displacement.
- An electrical harness which transmits the recopied position to an Engine Electronic Control Unit (EECU).
- An EECU which ensures the control and the monitoring of the FMV.

The EECU embedded software computes on one hand a shifting order to the actuator according to the measurements coming from the engine sensors and on the other hand, two failure detection tests. The first one is made to check the consistency between the required and the observed FMV displacement. In case of a significant difference between the two data, the EECU order a system reconfiguration in a fail-safe mode which corresponds to the FMV freezing in its last position. The second one is made to detect a transmission failure of the recopying sensor. In this case, the previous consistency check is inhibited.

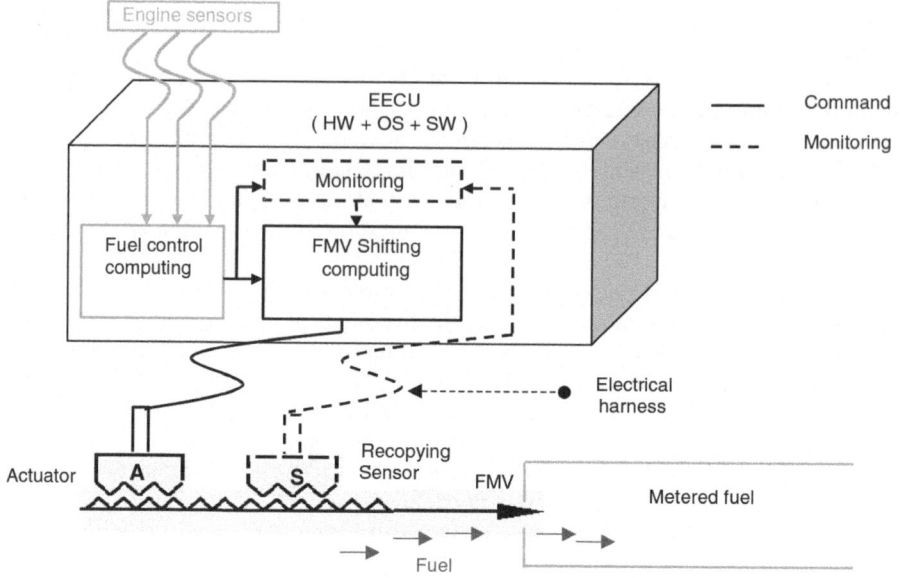

Fig. 1. Fuel-metering control system overview

Two feared events have been considered for this system:

- Loss of Power Control (LOPC) which corresponds to an undetected errone-ous shifting of the FMV. Its severity is taken as hazardous.
- Spurious Loss of Automatic Control (SLAC) which corresponds to an unjus-tified reconfiguration in a fail-safe state. Its severity is taken as major.

Safety qualitative and quantitative requirements come from the severity of the feared events. According to the aeronautic practice, only qualitative ones are used to derive software requirements. So, in this example, we will focus on the two following qualitative system requirements:

- At least two hardware failures must lead to the feared event LOPC,
- No single hardware failure leads to the feared event SLAC.

3   AltaRica Modelling

3.1   The AltaRica Language and Tools

AltaRica language is a formal high-level description language. It was especially de-signed to describe the functional and dysfunctional behaviour of industrial systems and to analyse the potential fault propagation [4], [5]. An AltaRica model can be analysed by different ways: simulation (interactive or Monte-Carlo simulations), generation of minimal cuts or sequences (ordered cuts) that lead to a feared event, model-checking... [1]. The work presented in this paper was carried out using Ce-cilia™ OCAS workshop (Dassault Aviation).

An AltaRica model consists in a set of interconnected "nodes". A node stands for a component (or function), it may be atomic or composed of subnodes. From a formal point of view, a node is a mode automaton [6] defined by three well identified parts.

The first part is the declaration of the different kinds of node parameters: state, flow and event. States are internal variables which memorise current functioning modes (failure modes or normal ones). Flows are node inputs or outputs. Possible types of states and flows are integer interval, enumeration and Boolean. Events are phenomena, which trigger transitions from an internal state to another. They can model normal actions or the occurrence of failures, or reactions to input conditions.

The second part describes the automaton transitions. A transition is a triple:

g |- evt -> e where g is the guard of the transition, evt is an event name and e is the effect of the transition. The guard is a Boolean formula over state or flow variables. It defines the configuration in which the transition can be triggered if the event evt occurs. The effect e is a list of assignations of value to state variables. So the transition part describes how functioning or failure states can evolve.

The third part is a set of assertions. Assertions are basic equalities or more structured equations using for instance case constructions. They establish relations between the states and the flows of the component and thus, describe how component outputs are determined by component inputs and current functioning mode.

These concepts are illustrated by the node of the electrical harness in the next part.

3.2 Case Study's AltaRica Model

AltaRica models that are used for preliminary safety assessment focus on potential fault propagation inside a system. Consequently, they shall depict the system components, their interconnection and normal behaviour in accordance with the system specification. Moreover, they shall highlight potential faulty behaviours of each component and each failure propagation paths.

In order to clarify the dysfunctional behaviours of the hardware components (i.e. actuator, recopying sensor, FMV and electrical harness), we use the results of the related Failure Mode and Effects Analysis (FMEA) (cf. table 1). On the same principle, we use the results of the Software Failure Mode and Effects Analysis for the studied software functions (i.e. FMV shifting computing and the two monitoring functions: the consistency check, the recopying sensor test) (cf. table 2).

Table 1. Example of a FMEA for a hardware component

Component	Failure mode	Local effect	System effect
Electrical harness between the EECU and the actuator	Erroneous data transmission	Fault detected by the EECU consistency check	The FMV is frozen (fail-safe)

Table 2. Example of a FMEA for a software function

Function	Failure mode	Local effect	System effect
Consistency check	Spurious detection	Spurious detection of an erroneous FMV shifting	Spurious fail safe : the FMV is frozen
	Loss of detection	No possible detection of an erroneous FMV shifting	None (dormant failure)

These data are used to determine the relevant error states, events and the abstraction level of each basic component of the AltaRica model. It is worth noting that the model shall focus on the fault propagation. So we depict the quality of the flows through each component rather than their concrete values.

For example, the data received from the EECU is represented by an enumerate variable whose values are {correct, erroneous}, as it appears in the AltaRica code of the electrical harness below.

```
node harness
flow
    h_in : {correct, erroneous} : in;
    h_out : {correct, erroneous} : out;
state
    h_state : {nominal, erroneous_transmission};
event
    fault;
trans
    h_state = nominal |- fault -> h_state := erroneous_transmission;
assert
    h_out = case{  h_state=erroneous_transmission : erroneous
                   else h_in };
init
    h_state = nominal;
edon
```

Let us comment the AltaRica code corresponding to this component. The harness receives one input h_in and transmits one output h_out. These flows are either correct or erroneous. Initially, the harness is in a nominal state. It may fail in the erroneous_transmission state as identified by the FMEA. This change can be observed by the event fault as written in the transition. Finally, it is asserted that in the erroneous_transmission state, the output is erroneous, whereas in the nominal mode, the output is equal to the input.

Thus, as illustrated by this example, the AltaRica nodes of the components capture a significant part of the FMEA information\n.

The whole system model is built by interconnecting all the component nodes.

Finally, the system effects of the faults or the feared events are explicitly stated in other kind of AltaRica nodes: the observers. Generally, an observer node is connected to the system outputs. It produces a Boolean output that indicates whether or not the current system state corresponds to a feared event. For example, in our case study, an observer monitors the output of the FMV node. If the FMV output is erroneous then it means that the feared event LOPC has occurred.

More details about the techniques to build AltaRica models for fault propagation analysis can be found in [7] or [1].

4 Preliminary System Safety Assessment (PSSA) Based on an AltaRica Model and Software Requirements Derivation

Let us now show how such a kind of AltaRica models can be used to analyse functional failure paths leading to a system feared event.

4.1 Analysis of Functional Failure Paths with the AltaRica Model

Using available AltaRica tools such as the minimal sequence generation tool of the Cecilia™ OCAS workshop, we can quite easily extract from an AltaRica model the functional failure paths leading to a feared event. The sequence generation tool computes the set of event sequences of bounded length that lead to a selected feared event.

Some examples of sequences leading to the feared SLAC and LOPC events are given below:

For the SLAC

```
- SW.Consistency_check.spurious_detection
```
For the LOPC
```
- SW.Consistency_check.loss & electri-
cal_harness_between_EECU_and_actuator.fault
```

It is worth noting that each event name is prefixed by a component name. So the sequence generation output highlights directly the components that contribute to the feared event.

The second step consists in extracting the sequences that infringe at least one qualitative system requirement (unacceptable sequences). The system requirements that we consider for deriving software requirements are of the following type:

Req_x = "a feared event, whose severity is s, must be caused by at least x hardware failures". Among other things, they imply that no single software failure leads to a feared event.

If all sequences are acceptable, then it means that all the qualitative requirements are met. If the quantitative requirements are also met (because it is also possible to perform quantitative analysis from AltaRica models) then the preliminary architecture is validated from a safety point of view without deriving further requirements.

In the other case, a third step is needed to eliminate the unacceptable sequences. They can be separated into two categories: those containing at least one software failure, and the others.

To eliminate the unacceptable sequences of the second category, i.e. those containing no software event, the preliminary architecture has to be modified. For example, redundancies may be added.

To eliminate a sequence S_{sw} of the first category, one may require the elimination of all the hypothetical software errors present in S_{sw}. This can be done by adding software requirements of the following type: "the software event e must not occur". Nevertheless, all these derived requirements may have different importance. We consider that a set of most unacceptable events is a smallest set of events that ensure the elimination of all unacceptable sequences. Let us now clarify how such a set can be computed.

4.2 Selection of Most Unacceptable Software Feared Events

In order to eliminate one sequence, it is sufficient to eliminate one event of the sequence. Several solutions can be adopted. Depending on the cases, it may be advantageous to eliminate an event instead of another. For example, let A, B, C and D be four software events, and "A; B", "C; A" and "A; D" three unacceptable sequences. To prevent the occurrence of the first sequence, it is sufficient to eliminate A or B. To eliminate the second sequence, it is sufficient to eliminate A or C, and for the third, A or D. In order to prevent the three sequences, it is not necessary to eliminate the four events. Two sufficient (minimal) solutions are conceivable: the first one is the elimination of the three events B, C and D; the second one is the elimination of the event A. We propose a systematic method to identify all the minimal solutions.

Notations:

Let cut_i be the cut of the sequence i i.e. set of events that occur in the ith sequence of a set of sequences.

Let E^i_j be the jth event of cut_i.

Let e^i_j be a Boolean variable associated to E^i_j, where e^i_j equals 1 if the event occurs, and 0 otherwise.

Then cut_i can be characterised by the product $c_i = \prod_{j=1}^{N_i} e^j_i$ where $N_i = \| cut_i \|$.

Algorithm principles:

Since each cut can be expressed by a Boolean product, the union of several cuts becomes a Boolean sum of Boolean products. This sum formalises a necessary condition to reach a feared event by sequences of a given bounded length. As a result, a condition sufficient to eliminate all these sequences is the negation of this sum.

Existing fault tree analysis tools can compute minimal cut sets i.e. minimal conditions that entail an input Boolean formula. Consequently, we propose to apply minimal cut computation to the Boolean formula representing the non-occurrence of the feared events. So the proposed algorithm provides all the minimal solutions to eliminate all the unacceptable sequences of a given bounded length.

Algorithm:

- 1st step: List all the minimal sequences that are unacceptable and which contain at least one software event i.e. S_{sw}. Consider the cuts of these sequences and remove all the hardware failures from these cuts. Note the list of the obtained cuts $C_{sw} = [cut_i]$.

- 2nd step: Calculate C_{sw_min} corresponding to the minimal cuts of C_{sw}.

Note $C = \| C_{sw\_min} \|$.

- 3rd step: Let F be the Boolean formula defined by $F = \sum_{i=1}^{C} c_i = \sum_{i=1}^{C} \left(\prod_{j=1}^{N_i} e^i_j \right)$. Calculate $MC = [mc_i]$, the list of the minimal cuts of $\overline{F} = \prod_{i=1}^{C} \left(\sum_{j=1}^{N_i} \overline{e}^i_j \right)$.

- 4th step: Each cut of MC represents a minimal solution, which is sufficient to eliminate all the sequences of S_{sw}. The optimal solution that minimises the number of software requirements is: $mc_{MIN} = ArgMin_{i\in[1,\|MC\|]}\{\|mc_i\|\}$.

Remark:
The 2nd and 3rd steps can be automatically made by calculating the dual of the fault tree corresponding to C_{sw}, for example using the Aralia tool [8].

Example:
Let S_{sw} the following set of unacceptable sequences where the l events represent software events and the m events represent hardware failures:

$$S_{sw} = \{"l_1 ; m_1" , "l_1 ; l_2 ; m_2" , "l_2 ; l_1 ; m_2" , "l_2 ; m_3" , "l_3 ; l_4" , "m_3 ; l_3 ; l_5" ,$$
$$"m_3 ; l_6 ; l_3" , "l_6 ; l_7" , "l_7 ; l_4 ; m_1" , "l_7 ; l_8 ; m_2" , "l_9 ; l_{10}" , "m_3 ; l_9 ; l_{11}" , "l_{11} ; l_3 ; l_9"$$
$$, "l_{10} ; l_9 ; m_4" , "l_1 ; l_8" , "l_8 ; l_2" , "m_5 ; l_4 ; l_3"\}$$

1st step:
$$C_{sw} = [\{l_1\}, \{l_1, l_2\}, \{l_1, l_2\}, \{l_2\}, \{l_3, l_4\}, \{l_3, l_5\}, \{l_3, l_6\}, \{l_6, l_7\}, \{l_4, l_7\}, \{l_9, l_{11}\}$$
$$\{l_7, l_8\}, \{l_9, l_{10}\}, \{l_3, l_9, l_{11}\}, \{l_9, l_{10}\}, \{l_1, l_8\}, \{l_2, l_8\}, \{l_3, l_4\}]$$

2nd step:
$$C_{sw\_min} = [\{l_1\}, \{l_2\}, \{l_3, l_4\}, \{l_3, l_5\}, \{l_3, l_6\}, \{l_6, l_7\}, \{l_4, l_7\}, \{l_7, l_8\}, \{l_9, l_{10}\}, \{l_9, l_{11}\}]$$

3rd step:
$$F = l_1 + l_2 + l_3 l_4 + l_3 l_5 + l_3 l_6 + l_6 l_7 + l_4 l_7 + l_7 l_8 + l_9 l_{10} + l_9 l_{11}$$
$$\bar{F} = \bar{l}_1 \bar{l}_2 (\bar{l}_3 + \bar{l}_4)(\bar{l}_3 + \bar{l}_5)(\bar{l}_3 + \bar{l}_6)(\bar{l}_6 + \bar{l}_7)(\bar{l}_4 + \bar{l}_7)(\bar{l}_7 + \bar{l}_8)(\bar{l}_9 + \bar{l}_{10})(\bar{l}_9 + \bar{l}_{11})$$

Using the Aralia tool to reduce the formula, we obtain:

$$\bar{F} = \bar{l}_1 \bar{l}_2\bar{l}_3\bar{l}_7\bar{l}_9 + \bar{l}_1 \bar{l}_2\bar{l}_3\bar{l}_7\bar{l}_{10}\bar{l}_{11} + \bar{l}_1 \bar{l}_2\bar{l}_3\bar{l}_4\bar{l}_6\bar{l}_8\bar{l}_9 + \bar{l}_1 \bar{l}_2\bar{l}_3\bar{l}_4\bar{l}_6\bar{l}_8\bar{l}_{10}\bar{l}_{11} + \bar{l}_1 \bar{l}_2\bar{l}_4\bar{l}_5\bar{l}_6\bar{l}_7\bar{l}_9$$
$$+ \bar{l}_1 \bar{l}_2\bar{l}_4\bar{l}_5\bar{l}_6\bar{l}_7\bar{l}_{10}\bar{l}_{11} + \bar{l}_1 \bar{l}_2\bar{l}_4\bar{l}_5\bar{l}_6\bar{l}_8\bar{l}_9 + \bar{l}_1 \bar{l}_2\bar{l}_4\bar{l}_5\bar{l}_6\bar{l}_8\bar{l}_{10}\bar{l}_{11}$$

All the possible solutions for eliminating all the unacceptable sequences of S_{sw} are:

$$MC = [\{\bar{l}_1, \bar{l}_2, \bar{l}_3, \bar{l}_7, \bar{l}_9\}; \{\bar{l}_1, \bar{l}_2, \bar{l}_3, \bar{l}_7, \bar{l}_{10}, \bar{l}_{11}\}; \{\bar{l}_1, \bar{l}_2, \bar{l}_3, \bar{l}_4, \bar{l}_6, \bar{l}_8, \bar{l}_9\}; \{\bar{l}_1, \bar{l}_2, \bar{l}_3, \bar{l}_4, \bar{l}_6, \bar{l}_8, \bar{l}_{10}, \bar{l}_{11}\};$$
$$\{\bar{l}_1, \bar{l}_2, \bar{l}_4, \bar{l}_5, \bar{l}_6, \bar{l}_7, \bar{l}_9\}; \{\bar{l}_1, \bar{l}_2, \bar{l}_4, \bar{l}_5, \bar{l}_6, \bar{l}_7, \bar{l}_{10}, \bar{l}_{11}\}; \{\bar{l}_1, \bar{l}_2, \bar{l}_4, \bar{l}_5, \bar{l}_6, \bar{l}_8, \bar{l}_9\}; \{\bar{l}_1, \bar{l}_2, \bar{l}_4, \bar{l}_5, \bar{l}_6, \bar{l}_8, \bar{l}_{10}, \bar{l}_{11}\}]$$

4th step:
$$\|mc_1\| = 5, \|mc_2\| = 6, \|mc_3\| = 7, \|mc_4\| = 8, \|mc_5\| = 7, \|mc_6\| = 8, \|mc_7\| = 7,$$
$$\|mc_8\| = 8. \text{ So, } mc_{MIN} = mc_1 = \{\bar{l}_1, \bar{l}_2, \bar{l}_3, \bar{l}_7, \bar{l}_9\}.$$

Conclusion: The smallest set of software events which is sufficient to eliminate in order to prevent the sequences of S_{sw} is l_1, l_2, l_3, l_7 and l_9. To be sure that those events will not occur, we formulate a non-occurrence requirement for each of them in the software specification.

5   Lessons Learnt from the Case Study

5.1   Scope of the Case Study

The lessons learnt come from the work made on an extension of the above example in order to be closer to a real industrial case study. The system studied includes an EECU composed of two independent channels (channel A and B) which are able to control the fuel flow on their own. They are redundant, however at any moment, only one channel at a time is allowed to control the fuel flow. When a channel is declared in a fail stage, the other one can take the relay, according to channel change logic. Except for very few differences, both channels work with the same embedded software.

The scope of our study includes:

- Components providing information to the EECU: engine sensors, helicopter commands, discreet inputs...
- Main functions ensured by the EECU:
    - input acquisition,
    - output driving, (including the example presented part 2)
    - communication management between the two channels,
    - monitoring,
    - actuation.

The studied safety requirements are relative to the feared events taken into account in part 2 (SLAC and LOPC).

5.2   Feedback on the Modelling Approach

The AltaRica language expressiveness is adequate for system potential propagation faults modelling. Except maybe for the model abstraction level choice, this modelling approach does not present any particular difficulty. The identification of the pertinent variables to be considered (linked to the model abstraction) can require some iteration, especially when the whole system is built by interconnecting all the component nodes. An interactive simulation tool helps evaluating if some information has been missed out or must be refined in order to obtain a correct abstraction.

AltaRica language use is quite easy. However, it requires little experience and AltaRica graphical workshops (like Cecilia™ OCAS from Dassault Aviation) make the language training easier. Additional tools make the building of a correct model easier: syntax verification, model consistency check, and interactive simulation tools.

5.3   Feedback about the PSSA and the Derivation Approach

The analysis technique proposed reaches its goals in term of identification of software events to be eliminated.

The sequence generation tool of Cecilia™ OCAS is very convenient to use. However, the time to get sequences increases proportionally with the size of the model, and exponentially in relation to the depth of the course to be achieved. The factor linked to the size of the system is the number of events of the model, and the depth of the course is linked to the length of sequences wished.

For the example presented in part 2, the time to generate sequences is very fast: about 1 second to obtain the sequences of two events, and the same time for the sequences of three events (the model only contains 18 events).

For the more complex model studied which includes 357 events, the time to obtain the list of minimal sequences of maximum two events is less than one minute and about 2h30 for the minimal sequences list that contain maximum three events (the times have been measured by using the version 4 of the Cecilia™ OCAS workshop and a PC equipped with an Intel Pentium 1.73 GHz processor, 1 Go of Ram and Windows 2000).

For the particular feared event SLAC, we have obtained the following results:

· 1168 minimal sequences which contain two events at the maximum,
· 18528 minimal sequences which contain three events at the maximum.

The sequence classification into acceptable/unacceptable classes was performed using Microsoft® Excel functions of sorting and filtering.

In order to make this data treatment easier, it is necessary to carefully choose the function names used in the AltaRica model. Here, all software functions of each channel are included in the node called SW. Thus, the software events are easily marked since they contain the SW word in their name. Another solution consists in systematically adding, in the name of events representing a software event, a reference mark, as for example _SW at the end of the name of the event, in the AltaRica models.

The identification of unacceptable sequences containing maximum two events with Excel is fast enough whereas the identification of sequences of three events deserves a more optimised implementation.

After treating sequences containing maximum two events, we obtained 428 unacceptable sequences, which include at least one software event.

To eliminate the 428 unacceptable sequences, we applied the algorithm of optimisation presented above. The search of minimal sets of software events to be forbidden was performed by computing minimal cut with Aralia tool. As a result, we found it was sufficient to forbid the occurrence of 22 software events. The first half concerns the channel A of the EECU, and the other half the channel B. However, these two channels have the same software. Thus, if two events represent the same failure mode of the same function implanted in channel A and channel B, only one requirement is necessary. Finally, we identified a set of 11 additional software requirements.

Then, we treated the 18528 minimal sequences that contain maximum three events. We first started by suppressing all sequences containing one software event among the 11 events previously identified. There remained 14188 sequences. Then we extracted the unacceptable sequences containing at least one software event. We got only 1176 new sequences.

We applied the algorithm of optimisation to get the minimal sets of software events to be proscribed. Finally, we identified 4 additional software events to be eliminated (there are 8 possible minimal solutions of 4 events).

At the end, for this case study, in order to satisfy safety goals, 15 software requirements have been added to the software specification.

6 Conclusion

The problem addressed in this paper is not new. In [9], the author advocates the use of system safety analysis techniques to derive system safety constraints which must be satisfied by software requirement. According to [10], one obstacle was the fact that safety analyses are often supported by fault trees that are quite far from the models used to design the software. Several proposals (see for instance [11], [12], [13], [14]) have been made to overcome this difficulty. In this work, we have selected the AltaRica language and additional tools because they appear to be well adapted for Preliminary System Safety Assessment (PSSA) and because they are mature enough to be used for industrial applications.

In this paper, we first focused on the technique to derive safety requirements on the software functions by extracting functional failure paths from AltaRica model. We highlighted here a key point: the AltaRica language has enabled to efficiently take into account both hardware and hypothetical software faults in a same model.

Then we showed how to extract minimal sets of most unacceptable software events from functional failure paths. We proposed a definition of these sets of events and a way to compute them thanks to a specific use of existing tools. Thus our proposal is complementary of works such as [10] that start with a prohibited feared event associated to a software function and then refine the definition of the feared event.

Finally, the proposed approach has been positively tested within a significant case study.

Future works are addressing three complementary topics. First of all, experiments are carried out to validate more extensively the proposed methodology. The sets of most unacceptable software events are transformed into Scade observers according to principles that are defined in [15]. These observers formalise requirements that can be easily tested or proved on the Scade specification of the engine control system.

Secondly, the AltaRica model of the case study is quite complex. A new PhD has started in order to study the validation of such complex models.

Finally, our approach can be extended so as to automatically allocate development assurance levels to software parts. In the aeronautic world, the definition of allocation rules is under way in the revision of the standard ARP4754 [16] and these rules take as a starting point the functional failure paths leading to feared events. The application of these news rules to the sequences extracted from AltaRica models should be fruitful.

References

1. Bieber, P., Bougnol, C., Castel, C., Heckmann, J.-P., Kehren, C., Metge, S., Seguin, C.: Safety Assessment with AltaRica - Lessons learnt based on two aircraft system studies. In: Jacquart, R. (ed.) 18th IFIP World Computer Congress, Topical Day on New Methods for Avionics Certification, Toulouse. Kluwer Academic Publishers, Dordrecht (2004)
2. Bieber, P., Blanquart, J.-P., Durrieu, G., Lesens, D., Lucotte, J., Tardy, F., Turin, M., Seguin, C., Conquet, E.: Integration of formal fault analysis in ASSERT: Case studies and lessons learnt. In: European Congress on Embedded Real-Time Software ERTS 2008, SIA, AAAF, SEE, Toulouse (2008) (electronic paper), http://www.erts2008.org/

3. Humbert, S., Bosc, J.-M., Castel, C., Darfeuil, P., Dutuit, Y., Focone, E., Seguin, C.: Déclinaison d'exigences de sécurité du système vers le logiciel assistée par des modèles formels. In: Potet, M.-L., Schobbens, P.-Y., Toussaint, Y., Saval, G. (eds.) AFADL 2007, Presses universitaires de Namur, pp. 57–73 (2007)
4. Arnold, A., Griffault, A., Point, G., Rauzy, A.: The AltaRica Formalism for Describing Concurrent Systems. Fundamenta Informaticae 40(2-3), 109–124 (2000)
5. The AltaRica project, http://altarica.labri.fr/wiki/
6. Rauzy, A.: Mode automata and their compilation into fault trees. Reliability Engineering and System Safety 78(1), 1–12 (2002)
7. Humbert, S., Bosc, J.-M., Castel, C., Darfeuil, P., Dutuit, Y., Seguin, C.: Méthodologie de modélisation AltaRica pour la sûreté de fonctionnement d'un système de propulsion hélicoptère incluant une partie logicielle. In: proceedings of Lambda Mu 15, communication 113, Lille, IMdR (2006)
8. Dutuit, Y., Rauzy, A.: Exact and Truncated Computation of Prime Implicants of Coherent and Non-Coherent Fault Trees within Aralia. Reliability Engineering and System Safety 58(2), 127–144 (1997)
9. Leveson, N.G.: Software Safety in Embedded Computer Systems. Communications of ACM 34(2), 34–46 (1991)
10. Hansen, K.M., Ravn, A.P., Stavridou, V.: From Safety Analysis to Software Requirements. IEEE Transaction on Software Engineering 24(7), 573–584 (1998)
11. Bouissou, M., Bouhadana, H., Bannelier, M., Villatte, N.: Knowledge modelling and reliability processing: presentation of the FIGARO language and associated tools. In: Lindeberg, J.F. (ed.) SAFECOMP 1991, IFAC Symposia, Trondheim, series #8, pp. 69–75. Pergamon Press, Oxford (1991)
12. Fenelon, P., McDermid, J.A., Nicholson, M., Pumfrey, D.J.: Towards Integrated Safety Analysis and Design. ACM Computing Reviews 2(1), 21–32 (1994)
13. Papadopoulos, Y., Maruhn, M.: Model-based automated synthesis of fault trees from Matlab-Simulink models. In: DSN 2001, International Conference on Dependable Systems and Networks (former FTCS), Gothenburg, pp.77–82 (2001) ISBN 0-7695-1101-5
14. Bozzano, M., Villafiorita, A., et al.: ESACS: an integrated methodology for design and safety analysis of complex systems. In: proceedings of ESREL 2003, European Safety and Reliability Conference, Maastricht, pp. 237–245. Balkema Publishers (2003)
15. Humbert, S.: Déclinaison d'exigences de sécurité du niveau système vers le niveau logiciel assistée par des modèles formels. PhD thesis of University of Bordeaux (2008)
16. Society of Automotive Engineers: ARP4754: Certification Considerations for Highly-Integrated or Complex Aircraft Systems. SAE International, Warrendale, PA (1996)

Model-Based Implementation of Real-Time Systems

Krzysztof Sacha

Warsaw University of Technology, Nowowiejska 15/19, 00-665 Warszawa, Poland
k.sacha@ia.pw.edu.pl

Abstract. A method is presented for modeling, verification and automatic programming of PLC controllers. The method offers a formal model of requirements, the means for defining and verifying safe behavior, and a technique for generating program code. The modeling language is UML state machine, which provides a widely accepted means for writing a specification at a suitable high level of abstraction. Such an abstract specification can be validated by the user, verified by means of a model-checker and translated automatically into a program code, which preserves the correctness and safety of the specification. The program code is written in one of the standardized IEC 61131 languages.

1 Introduction

This paper describes a method for modeling, verification and automatic programming of PLC controllers, which are used in industry for solving time- and safety-critical problems, like traffic or process control. A PLC controller is a computer-based device that has several inputs and outputs where two-state sensors and actuators can be plugged in. The controller executes cyclically: Polling the inputs, executing the program and updating the outputs. The duration of each cycle introduces an explicit granularity of time, which is measured and guaranteed by the operating system.

The modeling language is UML state machine [1], which provides a widely accepted means for writing a specification at a suitable high level of abstraction. Such an abstract specification can be validated by the user, verified against safety requirements and translated automatically into a program code. To do this, a method for defining the semantics of the specification is required, followed by a method of safety verification, and the rules for automatic code generation. The problem is not new and many methods have been developed for specifying safety critical real-time systems in a formal manner. Those methods are based on various mathematical theories, such as algebra [2], temporal logic [3], finite state machines [4-6] and Petri nets [7].

A UML state machine diagram describes a finite state machine augmented with hierarchical structure of nested states and time sensitive behavior. Unfortunately, whereas the syntax and the static semantics of a state machine diagram are precisely defined, the execution semantics is only given in natural language.

The method described in this paper is based on an original model of translatable finite state time machines (FSTM), which extend the classical Moore automata with a hierarchy of states and time. The model itself and a method for automatic code generation were described in detail in [8,9]. What was missing in those papers was a sound method for a formal verification of such properties as safety, liveness and

M.D. Harrison and M.-A. Sujan (Eds.): SAFECOMP 2008, LNCS 5219, pp. 332–345, 2008.

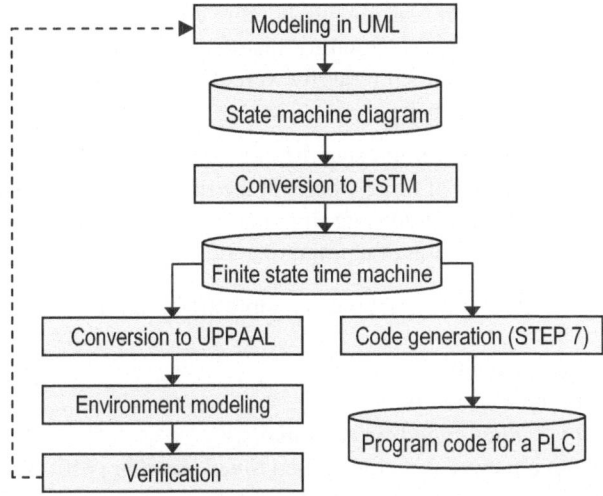

Fig. 1. Modeling, verification and implementation of the program code

reachability. This paper describes a concept of an integrated development environment with the potential for modeling of a controller, simulation of the environment (a controlled plant), verification of the compound model using UPPAAL model-checker [10] and automatic generation of IEC 61131 program code for the controller [11].

A schematic drawing of the development cycle is shown in Fig. 1. The tasks of modeling the controller in UML, modeling the environment in UPPAAL, and formulating safety requirements in a formal language of CTL formulae are done manually. The tasks of converting the model from UML to FSTM and from FSTM to UPPAAL, verifying the model, and generating the program code are done automatically.

The paper is organized as follows. Section 2 provides the reader with an overview of finite state time machine and the conversion of UML models. Section 3 describes a conversion from FSTM to UPPAAL, and Section 4 explains the rules of safety verification in UPPAAL. The process of converting finite state time machine into a program code is described in Section 5. An illustrative case study is provided in Section 6. A discussion of the results and plans for future work are given in Conclusions.

2   Conversion of UML State Machine to FSTM

UML state machine is a graph that shows the states an object can have, and the transitions between states that can be time or event triggered, and accompanied with actions. Relating this model to the execution of a PLC, one can note that an event is a combination of all the input signals of the PLC, and an actions is a combination of all the output signals of the PLC. States, transitions between states and time behavior are defined by a program code. Basic elements of a state machine can be seen in Fig. 2.

In order to provide a means for managing complexity, UML allows for a hierarchical nesting of states. Hierarchy of states does not add any new semantics to the model, in that a hierarchical diagram can always be converted into a "flat" one. A formal model of the hierarchy of states, including history indicator, entry and exit actions, and an algorithm for flattening the hierarchy were described in detail in [9] an will not be discussed in the rest of this paper.

Finite state time machine is a finite state Moore automaton extended with timer variables. It uses a discrete model of time, where all the timers progress synchronously with the same granularity of time. Finite state time machine is translatable, and can be used as a basis for automatic generation of a program code for PLC controllers [9].

Definition 1. A finite state time machine is a tuple $A = (S, \Sigma, \Gamma, \tau, \delta, s_0, \varepsilon, \Omega, \omega)$ where
S is a finite set of *states*,
Σ is a finite set of *input symbols*,
Ω is a finite set of *output symbols*,
Γ is a finite set of variables called *timer symbols*,
$\tau: \Gamma \to 2^S \times N^+$ is an injective function, called *timer function* (with two projections τ_S: $\Gamma \to 2^S$ and $\tau_N: \Gamma \to N^+$, respectively),
$\delta: S \times \Sigma \times 2^\Gamma \to S$ is a partial function, called *transition function*, such that:
$[(s, a, T) \in Dom(\delta)] \Leftrightarrow (\forall t \in T)[s \in \tau_S(t)]$
$s_0 \in S$ is the initial state,
$\varepsilon \in N^+$ is the granularity of time,
$\omega: S \to \Omega$ is an output function.

Notation: N^+ is the set of positive integers, R^+ is the set of positive reals, $Dom(\delta)$ is the domain of function δ, $card(X)$ is the cardinality of a set X, and ϕ is an empty set.

It can be noted that a finite state time machine looks much like a Moore automaton with three additional elements: $\Gamma, \tau, \varepsilon$, which add to the model the dimension of time. A timer symbol $t \in \Gamma$ is a variable, which takes values from the set R^+. The current value of a variable t is interpreted as the duration of a period of time. Timer function τ assigns to each timer a group of states $\tau_S(t)$ and a constant value $\tau_N(t)$. The meaning of those elements is such that timer t is enabled, i.e. counts time, as long as the automaton resides in one of the states from $\tau_S(t)$ and it expires when the current value of t exceeds $\tau_N(t)$.

Timer symbols in Γ can be set in an arbitrary order: $t^1 \dots t^n$. The current valuation of timer symbols $t: \Gamma \to R^+$ can now be described as a vector of values: $t^1 \dots t^n$.

The execution of a finite state time machine starts in state s_0 with the values of all timers equal to 0. For a given state s_k and a valuation of timers t_k there exists a set of expired timers, defined as:

$$\Theta(s_k, t_k) = \{ t^i \in \Gamma : s_k \in \tau_S(t^i) \text{ and } t^i_k \geq \tau_N(t^i) \}$$

The machine executes in state s_k with the valuation of timers t_k, $k = 0, 1 \dots$, by taking an input symbol a_k and moving to the next state s_{k+1} defined by the transition function:

$$s_{k+1} = \delta(s_k, a_k, \Theta(s_k, t_k))$$

When the machine enters a state s_{k+1} time advances and the values of enabled timers change reflecting the elapsed time interval ε.

$$t^i_{k+1} = \begin{cases} t^i_k + \varepsilon & \text{if } s_{k+1} \in \tau_S(t^i) \text{ and } s_k \in \tau_S(t^i) \\ 0 & \text{otherwise} \end{cases}$$

When the valuation of timers t changes, the set Θ of expired timers may change as well. This way a finite state time machine can respond to the flow of time, even if $s_{k+1} = s_k$ and $a_{k+1} = a_k$. Because the last argument of δ is a set of all timers expired in a given state and time, no conflict exists if several timers expire at the same time instant.

The finite state time machine models a time-sensitive device, which advances time with a fixed increment of ε time units. After each such increment the values of timers and the machine state are updated as described by the transition function. The device responds to a timed sequence of input symbols $a_1...a_j...$ that occur at time $\vartheta_1...\vartheta_j...$[5]. The flow of time within the input sequence is not synchronized to ε-increments of the machine. This means that a finite state time machine may or may not capture a symbol a_j of a timed input sequence, if $\vartheta_{j+1} - \vartheta_j < \varepsilon$.

A conversion algorithm of an UML state machine diagram into a finite state time machine, which defines the semantics of the diagram, can be described as follows.

S equals to the set of all states of the UML state machine.

Σ equals to the set of all events of the UML state machine; each event is a particular combination of all the input signals of the PLC.

Γ is a set of timer symbols $t^1,...,t^n$; the cardinality of Γ equals to the number of timed transitions in the diagram (i.e. transitions triggered by an *after* clause) and there is one timer symbol t^i for each timed transition in the UML state machine.

τ is the timer function, which assigns to each timer symbol t^i created for a timed transition T a pair composed of a source state of this transition and the value of the *after* clause of this transition.

δ is the transition function $\delta: S \times \Sigma \times 2^{\Gamma} \to S$, such that $\delta(s_1, a, T) = s_2$ if and only if there exists a transition in the UML state machine such that s_1 is the source and s_2 the destination state of this transition, and either a is the event that triggers this transition (in this case $T = \phi$), or $T = \{t^i\}$ and t^i is the timer symbol of this timed transition (in this case $\delta(s_1, a, T) = s_2$ for all $a \in \Sigma$).

s_o is the initial state of the UML state machine diagram.

ε is a characteristic of the PLC controller.

Ω equals to the set of combinations of all the output signals of the PLC that are set by the actions of the UML state machine.

ω is the output function, which assigns to each state $s \in S$ the output symbol $q \in \Omega$, which is set by all transitions to s.

3 Conversion of FSTM into UPPAAL

UPPAAL [10] is a toolbox for modeling and verification of real time systems, based on the theory of timed automata. The core part of the toolbox is a model-checking engine, which enables for verification of properties defined as CTL path formulae.

A timed automaton [4], as used in UPPAAL, is a finite state machine extended with clock variables that evaluate to positive real numbers and state variables that evaluate to discrete values. State variables are part of the state. All the clock variables progress simultaneously. An automaton may fire a transition between two states in response to an action, which can be thought of as an input symbol, or to a time action related to the expiration of a clock condition. Clock variables can be reset to zero at a transition.

Definition 2. A timed automaton is a tuple $TA = (S, s_0, C, A, E, I)$, where
S is a finite set of states,
C is a finite set of clock variables (called also *clocks*),
A is a finite set of actions,
$E \subseteq S \times A \times B(C) \times 2^C \times S$ is a set of transitions between states; each transition has an action, a guard and a set of clocks to be reset (a transition relation),
$s_0 \in S$ is the initial state,
$I: S \rightarrow B(C)$ is a function, which assigns invariants to states.

Notation: $B(C)$ is a set of conjunctions over simple clock conditions, e.g. $t < c$ or $t \geq c$. A valuation of clocks is a function $t: C \rightarrow R^+$. An expression $g \in B(C)$ defines a set of clock valuations that satisfy expression g; we will write $t \in g$ to mean that t satisfies g.

The execution of an automaton TA starts in state s_0 with the valuation t_0, such that all clock variables equal to 0. The machine executes in state s with the valuation of clocks t by performing an action:

$(s, t) \rightarrow (s', t')$ if there exists $e=(s, a, g, r, s') \in E$ such that $t \in g$ and $t \in I(s)$; the new valuation of clocks $t'=t$ over $C - r$ and $t'(t)=0$ for $t \in r$;

or a time action:

$(s, t) \rightarrow (s, t+d)$ if $\forall d':(0 \leq d' \leq d) \Rightarrow (t+d') \in I(s)$

The semantics of a timed automaton is a labeled graph consisting of nodes and edges. Each node defines a compound state of the automaton and is a pair $z=(s, t)$ composed of a state s and a valuation of the clock variables t. The set of all nodes $Z \subseteq S \times R^C$, and the initial state $(s_0, t_0) \in Z$. The edges in the graph are transitions, which fulfill the conditions defined above.

A set of timed automata can be composed into a network over a common set of actions. This way a model of a controller and a controlled plant can be established, such that an action of one automaton can trigger a transition in another one.

The cooperation between two automata is described in UPPAAL using synchronization channels, in which an action labeled c! (c is the channel name) in one automaton, triggers an action labeled c? in another automaton. A pair of matching actions in two component automata are performed simultaneously.

The actions are considered atomic with respect to the flow of time, which means that time can flow when the automata reside in their states. However, there are also special states, called committed states, in which delay is not allowed – such a state must be left immediately. Committed states are routinely used to separate a ?-action and !-action, in order to express causality relation between the two.

A compound state of a network of timed automata is a pair composed of a vector of states of the component automata and a valuation of all the clock variables. The semantics of a network is a graph composed of nodes, which are compound states, and edges, which correspond to transitions in component automata. The set of all nodes $Z \subseteq S^1 \times \dots \times S^n \times R^C$, and the initial state $(s_0^1, \dots, s_0^n, t_0) \in Z$.

A conversion of a finite state time machine into a timed automaton can be described as follows.

Let $A = (S, \Sigma, \Gamma, \tau, \delta, s_0, \varepsilon, \Omega, \omega)$ be a finite state time machine. The transition function $\delta: S \times \Sigma \times 2^\Gamma \to S$ is equivalent to a relation $\underline{\delta} \subseteq S \times \Sigma \times 2^\Gamma \times S$ such that:

$$\underline{\delta} = \{ (s, a, T, s'): s' = \delta(s, a, T) \}$$

For a given state $s \in S$ there exists a set of timers $T(s) = \{ t \in \Gamma: s \in \tau_S(t) \}$ that are enabled in s. Any subset $T = \{ t^1, \dots t^k \} \subseteq T(s)$ defines an expression g^T over simple time conditions:

$$t^1 \geq \tau_N(t^1) \dots t^k \geq \tau_N(t^k) \ \& \ t^{k+1} < \tau_N(t^{k+1}) \dots t^n < \tau_N(t^n)$$

which must be satisfied by a valuation t in order to enable the transition $(s, a, T, s') \in \underline{\delta}$.

Timed automaton $TA = (\underline{S}, \underline{s_0}, \underline{C}, \underline{A}, \underline{E}, \underline{I})$, which is equivalent to the given finite state time machine A can be constructed in the following way:

$\underline{S} = S \cup S_C$ (S_C is a set of committed-states)

$\underline{s_0} = s_0$

$\underline{C} = \Gamma$

$\underline{A} = \Sigma \cup \Omega$ (?-actions in Σ and !-actions in Ω)

$\underline{I} = \phi$

The set of committed states S_C and the transition relation E are created in the following way:

1. $S_C = \phi$ and $E = \phi$
2. For each $(s, a, T, s') \in \underline{\delta}$:
   - if $\omega(s) = \omega(s')$ than a transition $(s, a, g^T, \Gamma \setminus T(s), s') \in E$.
   - if $\omega(s) \neq \omega(s')$ than a new committed state s_C is added to S_C and a pair of transitions: $(s, a?, g^T, \phi, s_C)$, $(s_C, \omega(s')!, \phi, \Gamma \setminus T(s), s')$ is added to E.

Finite state time machine uses a discrete time model with an explicit granularity ε. UPPAAL uses continuous time model, in which transitions can fire at arbitrary points in time, within the boundaries defined explicitly by transition guards and state invariants. This means that the properties verified for a compound UPPAAL system does not depend on the relative speed of the component automata. Hence, they are true also for a synchronous finite state time machine.

4 Verification

The main purpose of UPPAAL is to verify the model with respect to safety requirements, which must be expressed in a formal language. UPPAAL uses a version of computational tree logic (CTL) and provides a query language consisting of state formulae and path formulae.

A state formula is an expression that can be evaluated for a particular state in order to check a property (e.g. a deadlock). Path formulae quantify over paths of execution and ask whether a given state formula φ can be satisfied in any or all the states along any or all the paths.

Path formulae can be classified into three types:

- Reachability properties (will φ be satisfied in a state of a path?) – $E<>\varphi$.
- Safety properties (will φ be satisfied in all the states along a single or along all paths?) – $E[]\varphi$ and $A[]\varphi$.
- Liveness properties (will φ eventually be satisfied? will φ respond to ψ?) – $A<>\varphi$ and $\psi \mathbin{-\!\!\!>} \varphi$.

UPPAAL model-checker enables verification of the model by evaluating path formulae over the reachability graph of a network of timed automata.

5 Code Generation

PLC controller is a technical implementation of a state machine, which yields output signals in response to input signals and to the flow of time. The controller maintains the state of the machine using flip-flops in the program code, counts time using timer blocks, and executes cyclically, firing a transition in each execution cycle.

Cyclic execution of a controller can be described in a pseudo-code, which creates a reference model for PLC execution:

```
state = initial_state ( );
loop_forever {
    input  = poll_the_input ( );
    timers = set_timers (active_timers ( state ) );
    state  = next_state ( state, timers, input );
    output = count_output ( state );
    set_the_output ( output );
}
```

The operating system of a PLC executes the following actions:

- sets the initial state (initial_state),
- executes the loop (loop_forever),
- polls the input (poll_the_input),
- counts time and sets the expired timers (set_timers),
- sets the output signals (set_the_output).

What the programmer must do is to write a code for:

- selecting the active timers, which count time in the (active_timers),
- calculating the next state of the controller (next_state),
- calculating the output (count_output).

The semantics of a PLC program, i.e. the meaning within its application domain, is a relation between a sequence of input signals and a sequence of output signals. If we establish a mapping between the input signals of a PLC and the input symbols of a finite state time machine, and a mapping between the output signals of a PLC and the output symbols of a machine, we can think about a finite state time machine as of a model of a program for a PLC controller.

The behavior of a PLC program is defined formally within the reference model by the semantics of its programming language, which may be one of the IEC 61131 languages [11], e.g. ladder diagram or structured text. The behavior of a finite state time machine has been defined in Section 2. By that means a method for translating a high level abstract model of a finite state time machine (S, Σ, Γ, τ, δ, s_0, ε, Ω, ω) into a PLC program can formally be defined. The method consists of the following steps.

1. Mapping of sets Σ, Ω into the input and output signals of PLC. The sets of input and output signals of a controller are usually defined in the requirements specification. Each combination of input (output) signals defines an event in the controlled plant, which is perceived by the controller as an input (output) symbol. This way, those two mappings are defined at the start of the modeling process.

2. Mapping of set S into the values of flip-flops. At least $\log_2(n)$, $n=card(S)$, flip-flops are needed to store all the states of set S. An arbitrary one-to-one mapping from set S to the set of n flip-flops (coding of states) can be used.

3. Mapping of set Γ into the set of timers. A separate timer with the expiration time equal to $\tau_N(t)$ is allocated for each timer symbol $t \in \Gamma$.

4. Defining function **active_timers** consistently with function τ. A timer block is a conceptual device, which has one input and one output. As long as the input equals **0**, the timer block is reset with the output equal to **0**. When the input changes to **1**, the timer block is enabled and starts counting time. The output changes to **1** as soon as the input has continued to be **1** for a predefined period of time. Function **active_timers** defines the input signals of all the timer blocks. The input signal of a timer block allocated for a timer $t \in \Gamma$, is a Boolean function over the set of flip-flops used for coding of states, such that it is true in state s if and only if $s \in \tau_S(t)$.

5. Defining function **next_state** consistently with function δ. This function defines the set and reset signals of flip-flops, which have been used for coding of states. The signal to set (reset) a flip-flop is a Boolean function over the set of flip-flops, input signals of PLC and output signal of timer blocks, such that it is true if and only if this flip-flop is set (reset) in the next state of FSTM.

6. Defining function **count_output** consistently with function ω. This function defines the values of output signals of PLC. The value of an output signal is a Boolean function over the set of flip-flops, such that it is true if and only if this output signal is set in the current state of FSTM.

6 Case Study

Consider a railroad crossing controlled by a computer system. There are two railway tracks within the crossing, and two trains can approach the crossing simultaneously (a single train on a track is allowed). The movement of trains is controlled by a set of semaphores that can prevent trains from entering the crossing. The road traffic is controlled by a gate that can be *open* or *closed*. A semaphore can be operated by a controller to display *green* light, when a train approaches, but not earlier than after the gate has been closed. Opening and closing states of the gate are confirmed to the controller by the appropriate input signals: *up* and *down*, respectively. The semaphore is *red* and the gate is *up* in the initial state of the crossing.

A train cannot be stopped instantly. When it is detected by a train position sensor, a controller has 30 seconds to *close* the gate and display *green* to allow the train to continue its course. After these 30 seconds, it takes further 20 seconds to reach the crossing. Otherwise, if the *green* signal is not displayed within these 30 seconds, the train must break in order to stop safely before the crossing. Closing the gate must last less than 20 seconds, or else an alarm must sound. The gate can be opened when the position sensor has sent a *leave* signal after the last train has left the crossing.

6.1 Modeling of the Controller

An algorithm for the railroad crossing controller, which can be a part of a broader control system, is shown in Fig. 2 in the form of a UML state machine diagram. The states within the graph correspond to states of the two trains that can appear within the crossing area. The transitions between states are labeled *event / action*, where *event* corresponds to an input symbol or the expiration of a time period, and *action* corresponds to setting an output symbol. The graph has eleven states only, but is quite complex due to the number of combinations of the input and output symbols.

The initial state, called *Outside*, corresponds to such a state of the crossing, in which no train approaches. The gate is *open* in this state, and the semaphores display *red* in order to prevent trains from entering the crossing. Such a state is safe in the application domain, because no collision between cars and trains is possible.

One problem with this model relates to a time event *after(20)*, which causes a transition from *EnteringBoth* to *Alarm3*. The requirement is such that this 20s delay

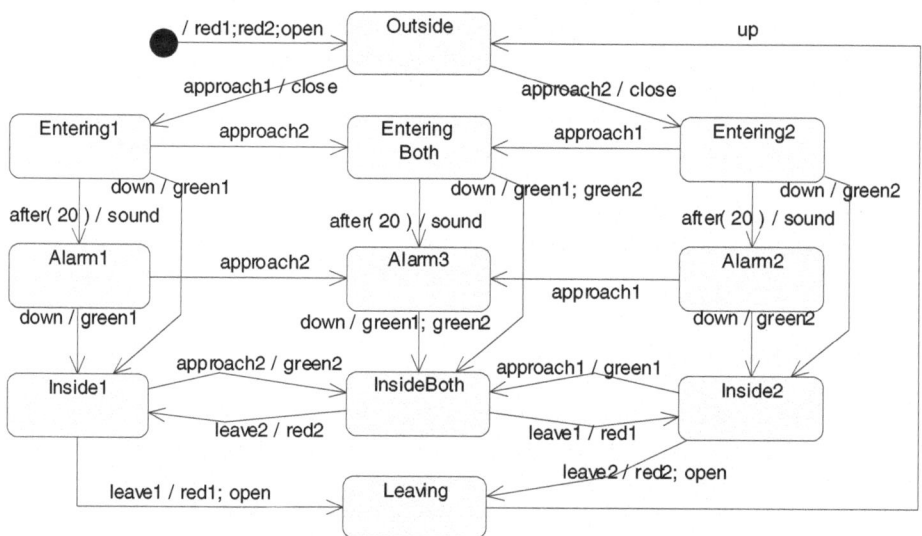

Fig. 2. UML model of the railroad crossing controller

should be measured from the moment of entering state *Entering1* or *Entering2* and the measurement should be continued through the period of being in state *EnteringBoth*. UML does not provide any simple means for expressing such a multi-state time requirement. It can only be expressed as an informal note in natural language.

FSTM model of the controller has the same set of states, input symbols and output symbols. It has a single timer symbol t, and the timer function $\tau_S(t) = \{Entering1, Entering2, EnteringBoth\}$ and $\tau_N(t) = 20$. The transition function is defined by the set of all the transitions of the UML state machine. No timing problem exists in FSTM.

6.2 Verification

UPPAAL model of the controller (Fig. 3) has the same states as the finite state time machine, plus a set of committed states. Basically, the transitions between states are in both models the same, with exception to transitions between states that differ in the finite state time machine on output symbols. Those transitions are split in UPPAAL model into two consecutive transitions separated by an added committed state.

Actions, which names bear the suffix '?', act like input symbols that enable the associated transitions. Actions, which names bear the suffix '!', act like output symbols that are passed to other automata in order to trigger the respective input symbols. This way the execution of one automaton can control the execution of a other automata.

The environment of the controller consists of two trains and a gate. Each of those elements can be modeled in UPPAAL and synchronized with the controller within a network of timed automata.

A model of a train is shown in Fig. 4. Time invariant $t \leq 30$ of state *Approaches* enforces a transition after 30 seconds have passed since the train has entered the state. This models the necessity of breaking the train if *green* has not been displayed in time. Time condition $t > 20$ assigned to the transition from *On crossing* to *Faraway* reflects the

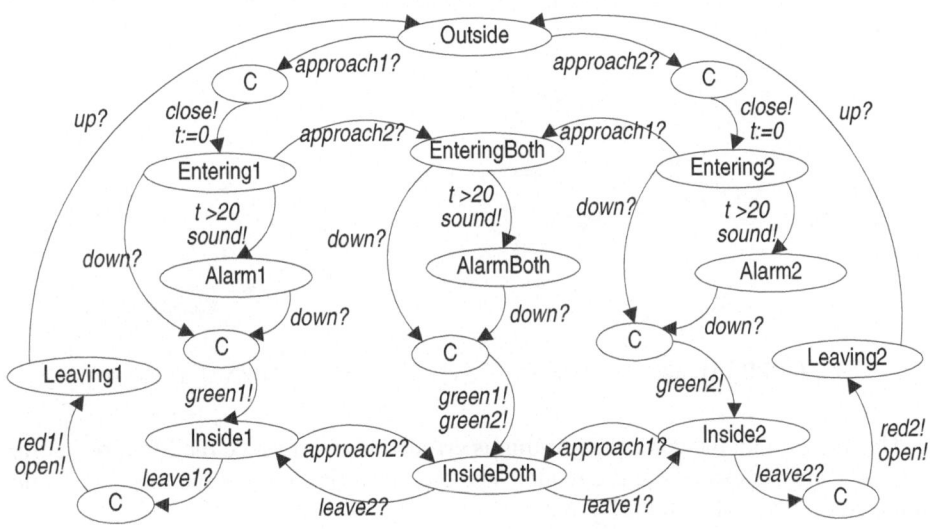

Fig. 3. UPPAAL model of the railroad crossing controller

minimum time of passing the crossing by a fast train. Time invariant $t \leqslant 40$ of the state *On crossing* reflects the maximum time of passing the crossing by a slow train.

A model of the second train is identical, with exception to the names of actions, which are: *approach2!*, *leave2!*, and *green2?*, respectively.

A model of the gate is shown in Fig. 5. Time invariants $t \leq 20$ assigned to states *Closing* and *Opening* reflect time that it takes to close or to open the gate.

The simple reachability properties can check if a given state is reachable:

- *E<> train1.On crossing*: This checks if train 1 can pass the crossing (a similar property can be checked for train 2).
- *E<> (train1.On crossing && train2.On crossing)*: This checks if both trains can move through the crossing simultaneously.

The safety properties can check that unsafe states will never happen:

- *A [] (train1.On crossing or train2.On crossing) imply gate.Closed*: This ensures that each time a train is passing the crossing, the gate is closed.
- *A [] (gate.open imply (¬ train1.On crossing && ¬ train1.On crossing)*: This ensures that each time the gate is open, a train is not on the crossing.

The liveness properties can check consequences of an event, e.g.:

- *train1.Approaches --> train1.On crossing*: This ensures that if train 1 approaches the crossing, it will eventually pass it (similar property can be checked for train 2).

All those properties can be verified by UPPAAL model-checker. In our example the liveness condition is not satisfied. A counterexample is the following: Assume that the train 2 approaches when train 1 is just leaving. The controller does not react to *approach2* in state *Leaving1*, hence, the transition to *Outside* appears without displaying *green2* for train 2. The train will stop and can never reach the crossing.

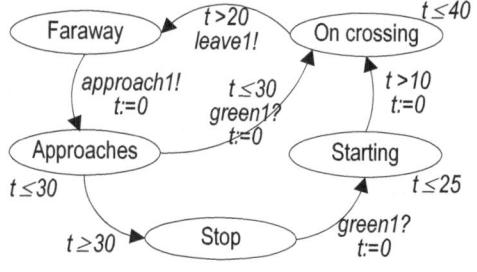

Fig. 4. UPPAAL model of a train

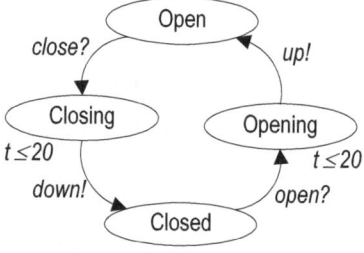

Fig. 5. UPPAAL model of the gate

This proves that the control algorithm is erroneous and must be modified by adding two additional transitions to the model. The corrected finite state time machine model of the controller is shown in Fig. 6.

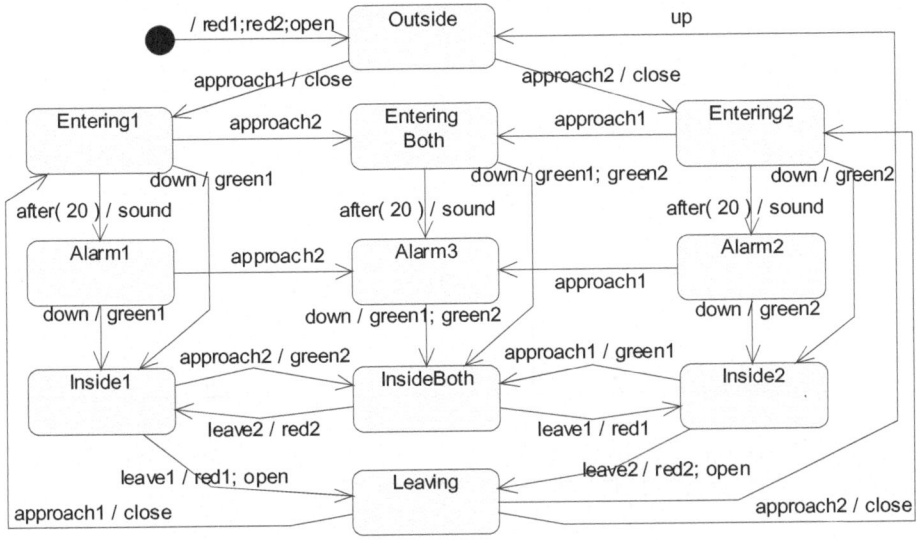

Fig. 6. The corrected model of the railroad crossing controller

6.3 Implementation

There are six input signals and seven output signals at the diagram in Fig. 6. Each combination of the input (output) signals corresponds to an input (output) symbol. This way, there are 11 states, 64 input symbols, 128 output symbols, and 1 timer in the finite state time machine, which defines the semantics of the diagram in Fig. 6.

PLC controller stores the states of the machine as values of its internal flip-flops. The coding of eleven states requires at least four such flip-flops. A selected coding for states and output signals of the railroad crossing controller is shown in Table 1.

A program for a PLC consists of a sequence of Boolean expressions to set or reset flip-flops, timers and output signals, according to the values of input signals, flip-flops and timers. These expressions implement the functions active_timers, next_state and count_output described in Sect. 5. For example, timer t must be enabled in each of the *Entering*-states, and flip-flop M1 must be set at a transition from any of the *Entering*-states to *Alarm*-states or *Inside*-states, i.e.:

(1) Set $t1 = \overline{M1} \cdot M2 \cdot (M3 + M4)$
(2) Set $M11 = \overline{M1} \cdot M2 \cdot (M3 + M4) \cdot (down + t)$
(3) Res $M11 = M1 \cdot \overline{M2} \cdot \left(\overline{M3} \cdot M4 \cdot leave1 + M3 \cdot \overline{M4} \cdot leave2 \right)$

....

( ) $M1 = M11$

To ensure atomicity of transitions, a two-phase implementation of next_state function is used. In the first phase, the next state is computed and stored using a set of auxiliary flip-flops (*M11* above), which mirror the primary flip-flops that are used to encode the model states. In the second phase, the current state is changed to the next state by copying the values of auxiliary flip-flops to the primary flip-flops [8].

Boolean expressions can be converted into the ladder diagram as shown in Fig. 7.

Table 1. The coding of states and output signals

M1	M2	M3	M4	State	red1	red2	green1	green2	close	open	sound
0	0	0	0	*Outside*	1	1	0	0	0	0	0
0	1	0	1	*Entering1*	1	1	0	0	1	0	0
0	1	1	0	*Entering2*	1	1	0	0	1	0	0
0	1	1	1	*EnteringBoth*	1	1	0	0	1	0	0
1	1	0	1	*Alarm1*	1	1	0	0	1	0	1
1	1	1	0	*Alarm2*	1	1	0	0	1	0	1
1	1	1	1	*Alarm3*	1	1	0	0	1	0	1
1	0	0	1	*Inside1*	0	1	1	0	0	0	0
1	0	1	0	*Inside2*	1	0	0	1	0	0	0
1	0	1	1	*InsideBoth*	0	0	1	1	0	0	0
0	0	1	1	*Leaving*	1	1	0	0	0	1	0

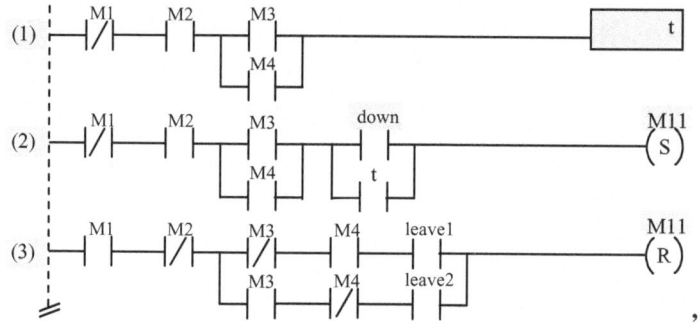

Fig. 7. A fragment of the ladder diagram program for the railroad crossing controller

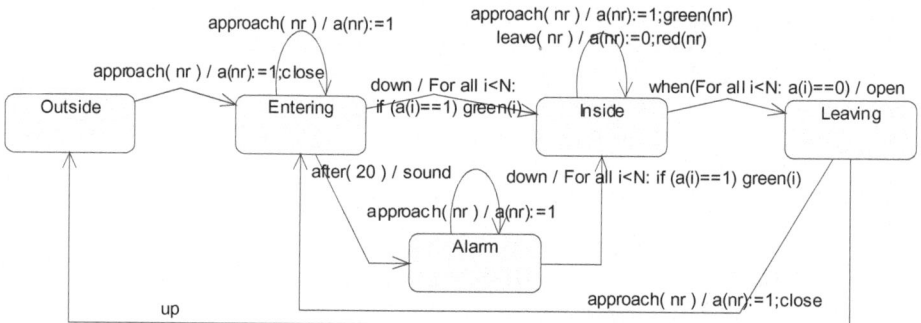

Fig. 8. A scalable model of the railroad crossing controller

7 Conclusions and Future Work

The paper describes a method for the specification, verification and automatic generation of code for PLC controllers. The method relies on a mathematical formalism based on finite state time machine model. The advantages of the method are intuitive modeling and a potential for automatic verification and implementation of the model.

A disadvantage is low scalability of the model with respect to the number of the modeled objects (trains). The problem is twofold. First, the model in Fig. 6 describes a crossing with exactly two tracks for trains. A completely new model must be built to describe, e.g., a four track crossing. Second, the number of states of the model raises exponentially. A way we want to follow to improve scalability is the introduction of variables for representing a number of similar states (vector $a[N]$ in Fig. 8). Those variables are part of the state and do not prevent the state space explosion. However, the model itself is parameterized with the number of tracks, and can be used to describe a crossing with an arbitrary number of tracks N.

References

1. OMG, Unified Modeling Language: Superstructure, version 2.0, August (2005)
2. Milner, R.: Operational and algebraic semantics of concurrent processes. In: van Leeuwen, J. (ed.) Handbook of Theoretical Computer Science, pp. 1201–1242. Elsevier, Amsterdam (1990)
3. Manna, Z., Pnueli, A.: Temporal Verification of Reactive Systems. Springer, Berlin (1995)
4. Alur, R., Dill, D.: Automata-theoretic verification of real-time systems. In: Formal Methods for Real-Time Computing, Trends in Software Series, pp. 55–82. John Wiley & Sons, Chichester (1996)
5. Kaynar, D.K., Lynch, N., Segala, R., Vaandrager, F.: The Theory of Timed I/O Automata, Technical Report MIT-LCS-TR-917a, MIT Lab. for Computer Science (2004)
6. Dierks, H.: PLC-Automata, A New Class of Implementable Real-Time Automata. In: Bertran, M., Rus, T. (eds.) Transformation-Based Reactive Systems Development. LNCS, vol. 1231, pp. 111–125. Springer, Berlin (1997)
7. Jensen, K.: Coloured Petri Nets. Basic Concepts, Analysis Methods and Practical Use. Springer, Berlin (1997)
8. Sacha, K.: Automatic Code Generation for PLC Controllers. In: Winther, R., Gran, B.A., Dahll, G. (eds.) SAFECOMP 2005. LNCS, vol. 3688, pp. 303–316. Springer, Heidelberg (2005)
9. Sacha, K.: Translatable Finite State Time Machine. In: Gaudin, E., Najm, E., Reed, R. (eds.) SDL 2007. LNCS, vol. 4745, pp. 117–132. Springer, Heidelberg (2007)
10. Behrmann, G., David, A., Larsen, K.G.: A Tutorial on Uppaal, Department of Computer Science. Aalborg University (2004)
11. IEC 61131-3, Programmable controllers – part 3: Programming languages, IEC (1993)

Early Prototyping of Wireless Sensor Network Algorithms in PVS⋆

Cinzia Bernardeschi[1], Paolo Masci[1], and Holger Pfeifer[2]

[1] Department of Information Engineering, University of Pisa, Italy
{cinzia.bernardeschi,paolo.masci}@iet.unipi.it
[2] Institute of Artificial Intelligence, Ulm University, Germany
holger.pfeifer@uni-ulm.de

Abstract. We describe an approach of using the evaluation mechanism of the specification and verification system *PVS* to support formal design exploration of WSN algorithms at the early stages of their development. The specification of the algorithm is expressed with an extensible set of programming primitives, and properties of interest are evaluated with ad hoc network simulators automatically generated from the formal specification. In particular, we build on the *PVSio* package as the core base for the network simulator. According to requirements, properties of interest can be simulated at different levels of abstraction. We illustrate our approach by specifying and simulating a standard routing algorithm for wireless sensor networks.

Keywords: WSN algorithms, simulation, PVS.

1 Introduction and Motivation

Wireless Sensor Networks (WSNs) are distributed systems consisting of a large number of spatially distributed, autonomous and cooperating nodes. The nodes of the network, referred to as *sensor nodes*, are battery-operated devices which provide limited computation capabilities, low-rate and low-range wireless communication, and are equipped with a number of sensors and actuators to monitor physical or environmental conditions. The most characterising aspect of WSNs is that they are deeply embedded in the real world, and provide unattended operation for long periods of time without infrastructure support. Due to their small size, sensor nodes can be placed in close proximity to the subject to be monitored, thus enabling *in situ* monitoring of physical phenomena. A sensor network normally constitutes a wireless *ad hoc* network, in which communication is multi-hop: due to the limited operating distance of the wireless radio compared to the physical extension of WSNs, sensor nodes must coordinate communication to forward data to a distant receiver. WSNs are highly dynamic networks: even if nodes are placed in fixed positions, node failures (e.g., due to software

---

⋆ This work was partially supported by the European Commission through the Network of Excellence ReSIST (IST-026764).

M.D. Harrison and M.-A. Sujan (Eds.): SAFECOMP 2008, LNCS 5219, pp. 346–359, 2008.

bugs or battery exhaustion), or environmental factors that are difficult to predict or avoid (e.g., physical obstacles, or humidity) may unexpectedly alter the connectivity of the network. For some application scenarios, mobile nodes may be involved in communication as well. Initially developed for military purposes such as battlefield surveillance, applications of WSNs today cover a wide spectrum of scenarios, including many safety-critical domains. For instance, WSNs have been deployed in critical infrastructures monitoring [1] to assess structural health of buildings, such as a pedestrian footbridge [2], or roads. In the area of traffic monitoring and control, a distributed application built on top of a WSN has been developed [3] to monitor a railway network for accidental or malicious system failures so as to prevent derailment of trains or even collisions. Wireless sensor networks are also deployed in health-care applications, for example to monitor vital signs of patients through tiny wearable sensor nodes [4], or as support and emergency systems for elderly people [5].

Large-scale networks are difficult to test, and the characteristics of wireless sensor networks only compound the problem. Hence, simulation plays a central role in current development processes of WSN applications. Software-based simulators are used to provide controlled environments in which experiments are to yield reproducible results. During the early stages of development, applications are commonly analysed with ad hoc simulators built as extensions of existing network simulators, such as *ns2*, or distributed system simulators, such as *ptolemy*. Once the application logic becomes consolidated, the software is evaluated in dedicated network simulators that provide emulation of real WSN hardware, such as *TOSSim* and *Avrora*. As of yet there is, however, no established standard simulation framework for WSN applications, and extensions to network simulators can usually only be accomplished by users who are familiar with the tool [6].

Inherently, simulations can only approximate real-world computation, and the challenge is to develop models that capture the behaviour of the environment in which WSNs are going to be deployed as accurately as possible. Recent studies have evidenced that for wireless networks simulations there is high risk of misleading or incorrect results because of assumptions hidden in the underlying network simulator [7]. Even in the presence of positive simulation results, failures may still occur when the system is deployed. As with any other system, remedying defects at this stage is costly at best, for WSNs it can be even impossible. Consequently, support for analysing WSN algorithms at early stages of development is essential. With a view to the reliability and safety requirements of applications of WSNs as the ones mentioned above, more rigorous analytical techniques are also desirable.

In this paper we report on our work towards developing a simulation and analysis framework for WSN algorithms within a theorem prover. Specifically, we use the *Prototype Verification System (PVS)* [8, 9] to specify and simulate WSN communication protocols in the very early stages of their design. The distinguishing characteristics of PVS are its expressive specification language and its powerful theorem prover. A less often used component is its *ground evaluator* [10] that can be used to animate functional specifications. Although PVS's specification language is based on higher-order logic and features a rich type system, a surprisingly large subset of it is executable. The ground evaluator translates the executable constructs of PVS into efficient *Lisp* code which can then be executed. The evaluation environment consists of a *read-eval-print-loop* that reads PVS expressions from the user and returns the result of their evaluation. The additional package *PVSio* [11] enhances PVS's specification language with built-in constructs for string manipulation, floating-point arithmetic and input/output operations. Thus, for certain types of applications, PVS can effectively serve as a functional programming language.

We employ the combination of PVS's rich specification language and the ground evaluator for early prototyping of a class of WSN algorithms. To this end, we introduce a series of general formal PVS models that can be refined to describe various WSN communication protocols. Specifically, we provide an extensible set of executable communication primitives to enable rapid and easy specification of protocols at different levels of detail. From these formal algorithm specifications, efficient Lisp code can automatically be generated using the ground evaluator and the PVSio extension. This implementation is suitable for simulation and allows to test and evaluate the algorithm from different perspectives. Finally, once the simulation experiments give sufficient confidence in the correctness of the algorithm, the PVS models can serve as the basis for the formal verification of the desired properties using the PVS theorem prover.

To demonstrate the effectiveness of our approach, we describe its application to the *Surge* routing protocol, which is used in various WSN systems, including prototypes in a safety-critical domain [4]. We analysed the *Surge* protocol under different aspects. During our tests of robustness of the protocol with respect to topology changes, we were able to detect a potential problem of routing loops that has gone unnoticed so far and can indeed be reproduced with one of the implementations of *Surge* provided in the library of the widely-used WSN operating system *TinyOS*.

2   PVS and PVSio

PVS is a specification and verification system which combines an expressive specification language with an interactive proof checker. It has been used for formal reasoning in several application domains (see [8] for an overview).

The PVS specification language builds on classical typed higher-order logic with the usual base types, `bool`, `nat`, `integer`, `real`, among others, and the function type constructor `[A -> B]`. Predicates are simply functions with range type `bool`. The type system of PVS also includes record types, dependent types, and abstract data types. The most powerful concept are *predicate subtypes*; e.g., the type `below(n:nat) : TYPE = {s: nat | s < n}` denotes the type of natural numbers less than a given bound n. Usage of predicate subtypes ranges from checking for violations such as division by zero, to expressing complex consistency requirements.

PVS specifications are packaged as *theories* that can be parametric in types and constants. A built-in prelude and loadable libraries provide standard specifications and proved facts for a large number of theories. A theory can use the definitions and theorems of another theory by importing it. For instance, consider the following theory `execution`:

```
execution [State : TYPE] : THEORY
BEGIN
   trans : VAR [State -> State]
   execute(trans)(n:nat) : RECURSIVE [State -> State] =
     LAMBDA (s:State): IF n = 0 THEN s
                       ELSE LET s_new = trans(s) IN
                                     execute(trans)(n-1)(s_new)
                  ENDIF
          MEASURE n
END execution
```

The theory takes one type parameter, `State`, and defines a (higher-order) function `execute` that recursively applies n steps of a state-transition function `trans`, which is provided as a parameter. As all functions in PVS must be total, the termination of the recursion has to be demonstrated; the `MEASURE` part provides the information to the typechecker and prover to ensure this. By instantiating the theory parameter with a concrete value, the `execute` function can be imported into the context of a given algorithm specification:

```
simulation : THEORY
BEGIN
   ...   % -- concrete def'n of a state type omitted

   % -- definition of a single step of the algorithm
   algorithm_step(s:State): State = ... % -- omitted

   IMPORTING execution[State]

   % -- execution of 'steps' number of steps of the algorithm
   algorithm(steps:nat) : [State -> State] =
      execute(algorithm_step)(steps)

END simulation
```

Thus, the `execution` theory provides a generic mechanism to describe the execution of an algorithm, which can subsequently be used for simulation.

Using the PVS ground evaluator one can compile the executable constructs of a specification, such as the `execute` function above, into efficient *Lisp* code. In order to still be able to simulate theories that also involve declarative specifications, the ground evaluator is augmented by so-called *semantic attachments*, through which the user can supply pieces of Lisp code and attach them to the declarative parts. Using this mechanism the *PVSio* package [11] extends the ground evaluator with a predefined library of imperative programming language features such as side effects, unbounded loops, and input/output operations, and also provides a high-level interface for writing user-defined semantic attachments. Thus, PVS specifications can conveniently be animated within the *read-eval-print*-loop of the ground evaluator.

3   Prototyping WSN Algorithms

In this section we present the basic aspects of the proposed approach. We show that prototyping of WSN algorithms can be performed with a collection of PVS models (theories), each of which represents a service installed on a sensor node (e.g., packet logger, clock), or structural properties of the network (e.g., the network graph), or communication functionalities (e.g., packet forwarding). For each theory, a number of different versions can be provided in order to specify and analyse WSN algorithms under several perspectives and at desired level of detail. The most abstract theory provides *i)* the declaration of types for a minimum set of mandatory attributes, *ii)* the declaration of interface functions. More detailed theories can be derived from the abstract definition by specifying the behaviour of interface functions, and by extending types. Abstractions enable users to create a model comprising only the parameters of interest at the desired levels of detail. Hence, lightweight models can be generated, and efficient code for simulations can be obtained.

Network Connectivity. Network connectivity is modelled with a directed graph without self-edges. We build type definition on top of directed graphs of the NASA library [12], in order to benefit from several useful lemmas and properties already proved in PVS. Custom network graphs can be generated. To simplify graph specification, we use an auxiliary topology function that identifies, for each node, the set of neighbouring nodes. Ideal and lossy links can be modelled, and topology changes can be used to model node mobility. Once the topology is given, the network graph can be instantiated with a specific interface function. Sensor nodes are identified by a unique natural number less that a given N. We developed a theory named node_th that provides a type definition for the node identifier. In the following, the theory which describes network connectivity is shown (digraph[node_id] is the NASA theory on directed graphs). An example of topology and the corresponding network graph are also included.

```
network_graph_th: THEORY
 BEGIN
IMPORTING node_th, digraphs[node_id]

   %-- network_graph: a directed graph without self-edges
   network_graph?(g: digraph[node_id]): bool =
       (FORALL (i: node_id): vert(g)(i)) AND
        (FORALL (i,j: node_id): edges(g)((i,j)) IMPLIES (i /= j))

   network_graph: TYPE = {g: digraph[node_id] | network_graph?(g)}
   topology: TYPE = [node_id -> finite_set[node_id]]
   new_network_graph(tp: topology): network_graph

   %-- instance of network graph
   fully_connected_network_graph: network_graph
      = new_network_graph(LAMBDA (i: node_id): {n: node_id | n /= i})
 END network_graph_th
```

Services. A service is identified by a unique name. A services S is associated to nodes by means of a function [finite_set[node_id] -> S]. Depending on the algorithm specification and on the property of interest, services can be installed on a single node, on a group of nodes, or on the entire network.

Currently, we have implemented the following services: *packet logger*, which stores statistics about sent and received packets, *receive buffer*, which models the buffer where packets sent by other nodes are stored, *energy consumption*, which evaluates the energy spent by nodes, *routing*, which provides the basic definitions for building routing tables, spanning trees and paths between nodes, and *node scheduler*, which gives the sequence of nodes that execute the algorithm (e.g., round robin, or random). As example of service, in the following we show the definition of energy consumption, where energy is the energy consumption of a sensor node, network_consumption is the function which associates the energy consumption to every node. Three interface functions are declared in the theory: two of them are used to compute energy consumption of senders and receivers, the other one to update consumption of the sender neighbours.

```
energy_th: THEORY
 BEGIN
    %-- type definition
    energy: TYPE = real
    network_consumption: TYPE = [node_id -> energy]
    %-- interface functions
    sender_consumption: energy
    receiver_consumption(g: network_graph, snd, rcv: node_id): energy
    update_network_consumption(ne: network_consumption,
                               g: network_graph,
                               snd: node_id): network_consumption
 END energy_th
```

Services are wrapped together into an extensible structure called *network state*. The network state is described by the set of functions that specify the allocation of services to nodes. For instance, in the following theory, a network state is defined as the collection of two services (receive buffer and log):

```
network_th: THEORY
 BEGIN
    network_state: TYPE =
         [# net_receive_buffer: [node_id -> receive_buffer],
            net_log: [node_id -> log] #]
 END network_th
```

The network state maintains the state of all nodes. The state of a node consists of the state of the services installed on a node. The state of a node can be obtained by indexing the network state: given a node x and a network state ns, the state of x is ns(x), and the state of the logging service of x is net_log(ns)(x).

Communication Primitives. Nodes can exchange packets. A packet is a structure with two mandatory fields (the sender and the destination), and a number of optional fields. The sender is a single node, while destination is specified with a finite set of nodes. Broadcast address is represented with a special constant bcast_addr, which is the full set of nodes. In the following example, a packet consists of five fields (timestamp, source, sender, destination and payload):

```
packet_th: THEORY
 BEGIN
IMPORTING node_th, time_th
   bcast_addr: finite_set[node_id] = LAMBDA (i: node_id): node_id?(i)
   packet: TYPE =
         [# timestamp: time,
            source_addr: node_id,
            sender_addr: node_id,
            destination_addr: finite_set[node_id],
            payload: finite_sequence[int] #]
 END packet_th
```

We modelled three low level single-hop primitives in order to easy the specification of communication algorithms: inject, forward and drop. Additionally, nodes are also allowed to perform an idle transition.

Inject can be used to send out packets generated by nodes (e.g., packet generated by the application executed on nodes, or control packets generated by the routing service): the function takes a packet as parameter, and sends out such packet.

Forward is suitable to relay packets previously received by nodes (e.g., when multi-hop communication is needed to reach the destination): the function takes a packet as parameter, removes the packet from the receive buffer of the node, and sends out a packet with a sender address automatically updated with the identifier of the sending node.

Drop is used to discard received packets: the function takes a packet as parameter, and removes such packet from the receive buffer of the node.

Idle is useful to update state variables of nodes, such as energy consumption, when no operation on incoming/outgoing packets is performed.

The implemented primitives are suitable for unicast, multicast and broadcast communication. The side effect of sending out the packet is that neighbouring nodes of the sender receive the packet. The graph connectivity affects reception of packets: if node x sends out a broadcast packet, it is received only by neighbours of x. A basic version of the forward primitive with the essential functionalities is the following:

```
network_th_A: THEORY
 BEGIN
   IMPORTING receive_buffer_th
   network_state: TYPE = [# net_receive_buffer: network_receive_buffer #]
```

```
% packet pk is sent out by the forwarder node
forward(pk: packet)(forwarder: node_id)
      (net: network_state, g: network_graph): network_state =
    LET fw_pk = pk WITH [sender_addr := forwarder],
        nrb0 = update_receivers_buffer(net_receive_buffer(net), g, fw_pk),
        nrb1 = update_sender_buffer(nrb0, pk)
      IN net WITH [net_receive_buffer := nrb1]
END network_th_A
```

The sender address of the packet is updated, and update_receivers_buffer and update_sender_buffer functions are invoked to update the network state. A more detailed version of the above primitive could be obtained by adding, for example, energy consumption and packet logger services to nodes. In this case, energy_th and log_th theories must be imported and the functions to update energy and log must be invoked.

Algorithms. An algorithm is specified as a cyclic procedure executed on a generic node. For instance, let us consider the flooding algorithm [13], which is designed to deliver packets to all nodes in the network. Flooding is typically used for dynamic route discovery, reconfiguration/reprogramming and to request specific data from sensors. A simple variant of flooding behaves as follows: whenever a node receives a packet, the packet is forwarded to neighbouring nodes if it is received for the first time, otherwise it is dropped. The algorithm can be specified as follows:

```
flooding_th: THEORY
 BEGIN
IMPORTING network_th, log_th
    flooding(x: node_id)(net: network_state, g: network_graph):
          network_state =
        IF empty?(net_receive_buffer(net)(x)) THEN idle(x)(net, g)
        ELSE LET pk = getpacket(net_receive_buffer(net)(x)) IN
                 IF empty?(net_log(net)(x)(fw)) THEN forward(pk)(x)(net, g)
                 ELSE drop(pk)(x)(net, g)
                 ENDIF
        ENDIF
END flooding_th
```

Flooding can be analysed at different level of abstractions by importing specific theories and leaving the specification of the flooding function unmodified. This way, different properties of the algorithm can be analysed and different implementations can be evaluated. For instance, energy consumption can be analysed in the above theory by importing a different theory for the network state. In order to discover problems of the algorithm, the underlying services can be assumed to behave correctly. Conversely, the algorithm can be also analysed by modelling malfunctions of the underlying layers in order to evaluate, for instance, service degradation.

4   A Case Study: Surge

In this section we analyse *Surge*, a popular routing protocol for WSNs. Surge is currently part of the TinyOS distribution, and it has been used as routing service during the evaluation of several WSN-based systems, including prototypes for safety-critical systems [4]. We introduce Surge with an excerpt from [14]:

> The Surge protocol forms a dynamic spanning tree, rooted at a single node (*the base station*). Nodes route packets to the root. Nodes select a new parent when the link quality falls below a certain threshold. Surge suppresses cycles in the routing by dropping packets that revisit their origin.

Modelling Surge. The Surge algorithm can be decomposed into a forwarding service built upon a dynamic spanning tree service which, in turn, relies on lower level services for single-hop communication. In order to discover bugs in the specification, such services can be analysed separately, assuming that the underlying layers behave properly. Suppose that we are interested in analysing the forwarding service. The spanning tree service can be assumed correct, i.e., it provides a correct routing table rt to the forwarding service. Hence, the forwarding algorithm of Surge applied by a single node to a received packet can be formally specified in PVS as follows:

```
surge_th: THEORY
 BEGIN
IMPORTING network_th, routing_th

surge(x: node_id)(net: network_state, g: network_graph)
     (base_station: node_id, rt: routing_table): network_state =
   IF empty?(net_receive_buffer(net)(x)) THEN idle(x)(net, g)
   ELSE LET received_pk = getpacket(net_receive_buffer(net)(x)),
           source_addr = source_addr(received_pk),
           sender_addr = sender_addr(received_pk),
           next_hop = next_hop(x,base_station)(g,rt)
        IN IF source_addr /= x
           THEN forward(received_pk
                        WITH [destination_addr := next_hop])(x)(net, g)
           ELSE drop(received_pk)(x)(net, g)
           ENDIF
   ENDIF
 END surge_th
```

Analysing Surge. To analyse Surge, a WSN application must be specified in PVS. In the following, we will explore examples of analyses.

Receive queue size. We report the results of a simulation to evaluate receive queue size of sensor nodes in a monitoring scenario, in which sensor nodes periodically send packets to report data to the base station. The routing table is

assumed to be correct, but it may change. A scheduler that selects the nodes
that execute the algorithm must be specified. Every time a node is selected by
the scheduler, such node sends out a new packet and relays all packets of other
nodes. The scheduler guarantees fairness of execution between nodes. The PVS
theory for the monitoring application is the following, where node 0 is the base
station. Two recursive functions are defined: surge_rec, which relays all received
packets, and surge_app, which invokes the scheduler (round_robin) to select a
node. The selected node, if different from the base station, sends out a packet
and relays packets of the other nodes.

```
surge_app_th: THEORY
 BEGIN
IMPORTING surge_th
base_station: node_id = 0

surge_rec(x: node_id)(net: network_state, g: network_graph)
          (base_station: node_id, rt: routing_table): RECURSIVE
        network_state =
  IF empty?(net_receive_buffer(net)(x)) THEN idle(x)(net, g)
  ELSE LET net_prime = surge(x)(net, g)(base_station, rt)
         IN surge_rec(x)(net_prime, g)(base_station, rt)
  ENDIF
    MEASURE size(net_receive_buffer(net)(x))

surge_app(ti, tf: nat)
          (net: network_state, g: network_graph,
          rt: routing_table): RECURSIVE network_state =
  IF ti >= tf THEN net
  ELSE LET sender = round_robin(ti),
           dst = next_hop(sender, base_station)(g, rt)
      IN IF sender = base_station
          THEN surge_app(ti + 1, tf)(net, g, rt)
          ELSE
            LET net_inj = inject(new_packet(sender, dst))(sender)(net, g),
                net_prime = surge_rec(sender)(net_inj, g)(base_station, rt)
             IN surge_app(ti + 1, tf)(net_prime, g, rt)
          ENDIF
      ENDIF
  MEASURE tf - ti
 END surge_app_th
```

A simulation has been performed with grid networks of different size. Networks
of hundreds of nodes can be simulated. Results for a 25 node network are shown
in Figure 1. For each node (except for the base station) we have evaluated the
maximum number of packets in the receive buffer. The application has been
simulated several times and for different number of steps. For high number of
steps, the queue size almost stabilised. As expected, the maximum number of
packets in the receive queue is bigger for nodes closer to the base station, because
they have to relay packets for a larger number of nodes.

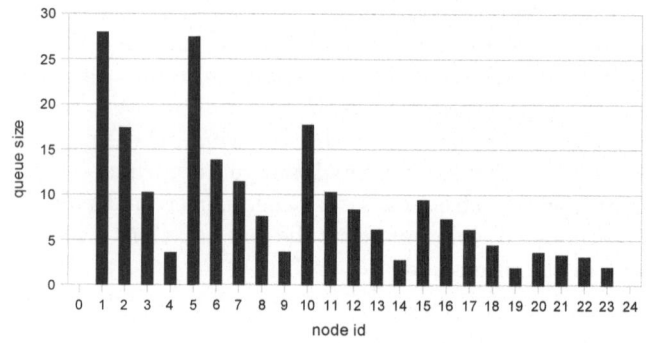

Fig. 1. Receive queue size for each node

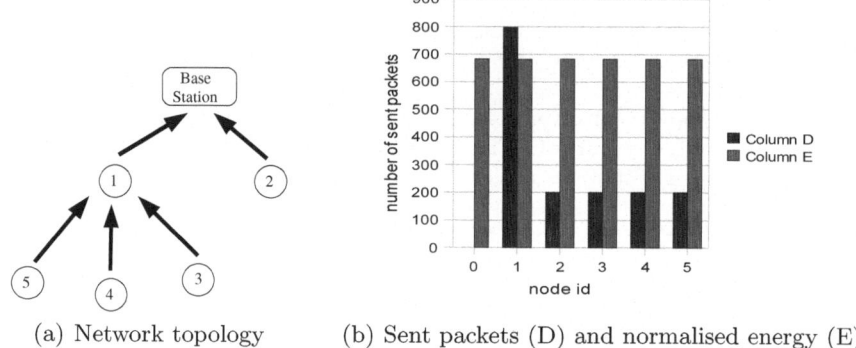

(a) Network topology   (b) Sent packets (D) and normalised energy (E)

Fig. 2. Topology and results of the simulation for energy consumption

Energy consumption. We report the results of a simulation to evaluate energy consumption for the previous monitoring application. To evaluate energy consumption, we used a theory providing an analytical model for idle consumption[1]. The PVS specification of both Surge and the monitoring application are left unchanged. The simulation can be executed by simply importing appropriate PVS theories. We compared our analysis with that described in [15], for a network of six nodes topology shown in Figure 2(a). In Figure 2(b), for each node, the total amount of sent packets and the energy consumption is shown. The application has been simulated for about one thousand simulation steps. Because we modelled idle energy, the evaluated energy consumption reflects real energy consumption. We obtained coherent results with respect to those of [15].

Robustness to topology changes. In order to test robustness of the protocol to topology changes, we consider a monitoring application in which only one node (node 5) periodically sends packets to the base station. The other nodes relay packets according to the Surge algorithm. We were able to detect a potential

---

[1] Idle consumption is the consumption of a sensor node during idle behaviour.

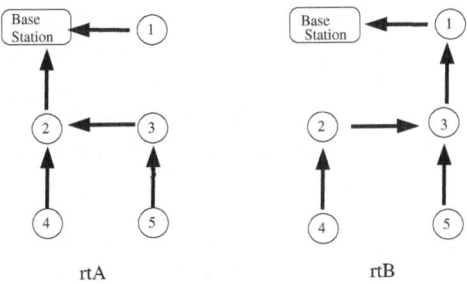

Fig. 3. Routing tables used for the robustness analysis of Surge

problem of infinite loops of routed packets in the algorithm specification. There are situations in which a packet may travel indefinitely in the network, because the routing table may change in response to topology changes. We evidenced such issue in a simulated grid network of 6 nodes by using routing tables rtA and rtB (see Figure 3). The critical situation is the following: assume that the routing table is rtA and that node 3 forwards a packet to node 2. Suppose that the routing table changes to rtB just before node 2 forwards the packet. Hence, the packet returns to node 3, which does not drop the packet because it is not the source. Just before node 3 forwards the packet, the routing table may change to rtB again, and so on. Such a pathological case is a real problem, because routing tables may actually change during Surge operations. The specification of Surge gives no constraints on routing table changes: the only assumption is that there is an underlying service which provides a correct spanning tree rooted at the base station. The design consideration discussed in [16] allows to conclude that such a bug is indeed possible. With such an assumption, loop detection based on the source address of the packet is not sufficient to avoid the problem. Infinite loops of packets may overload the network and in safety critical applications the service provided by the network could be downgraded to an unacceptable level.

5   Related Work and Conclusions

In this paper we propose a simulation and analysis framework for WSN algorithms within a theorem prover. The need for formal modelling and analysis of WSN algorithms has been pointed out in many papers. In [17], basic properties of the Reverse Path Forwarding algorithm have been analysed with FDR and Alloy Analyser. Scalability is the main problem of such approach: only very simple and small network configurations were analysed, and a proof by hand was used to prove correctness of the algorithm under specific hypotheses. In [18], TinyOS is modelled as a hybrid automaton and a sensor network is specified as a network of hybrid automata. The proposed analysis is only oriented to evaluate energy consumption of sensor nodes. Moreover, in [19] model checking is applied to a TinyOS application. In [20], Lamport's Temporal Logic of Actions is used to model and simulate diffusion protocols for discovering routing trees for gathering and disseminating data. The analysis focuses on performance variation of push and pull phases of the diffusion protocol for routing trees with different shapes, however without the objective of algorithm design evaluation. In [21] a formal model, called *Space Time Petri Nets*, has been presented to model WSNs. Time Petri Nets are augmented by adding location information to every place, and modelling broadcast transmission with a special transition. The formalism lacks both flexibility, as nodes cannot be modelled at different levels of abstraction, and scalability with respect to the generation of the reachability graph. In [22], Real-Time Maude has been applied to the OGDC density control algorithm and networks of several hundred nodes can be analysed. The specification models a node as an object, and the communication primitives are broadcast and unicast. The approach is capable of modelling the algorithm at high levels of detail, and results can be more accurate compared to other network simulators, such as *ns2*.

The framework proposed in this paper allows developers to formalise the WSN at different levels of abstraction, and it can be applied at the early stage of the development process to consolidate the algorithm design. We have used this approach to specify and simulate the Surge routing protocol for a number of networks with different topologies and number of nodes. During our analyses of Surge, we were able to detect a potential problem of routing loops due to topology changes, which has gone unnoticed so far. Further work includes the use of the theorem prover of *PVS* to verify correctness properties of WSN algorithms.

References

1. Xu, N., Rangwala, S., Chintalapudi, K., Ganesan, D., Broad, A., Govindan, R., Estrin, D.: A wireless sensor network for structural monitoring. In: Proc. Intl. Conf. on Embedded Networked Sensor Systems, pp. 13–24. ACM, New York (2004)
2. Kim, S., Pakzad, S., Culler, D., Demmel, J., Fenves, G., Glaser, S., Turon, M.: Wireless sensor networks for structural health monitoring. In: Proc. Intl. Conf. on Embedded Networked Sensor Systems, pp. 427–428. ACM, New York (2006)
3. Aboelela, E., Edberg, W., Papakonstantinou, C., Vokkarane, V.: Wireless sensor network based model for secure railway operations. In: Intl. Workshop on eSafety and Convergence of Heterogeneous Wireless Networks, pp. 623–628 (2006)
4. Lorincz, K., Malan, D.J., Fulford-Jones, T.R.F., Nawoj, A., Clavel, A., Shnayder, V., Mainland, G., Welsh, M., Moulton, S.: Sensor networks for emergency response: Challenges and opportunities. IEEE Pervasive Computing 3(4), 16–23 (2004)
5. Stanford, V.: Using pervasive computing to deliver elder care. IEEE Pervasive Computing 1(1), 10–13 (2002)
6. Chen, G., Branch, J., Pflug, J., Zhu, L., Szymanski, B.: Sense: A sensor network simulator. Advances in Pervasive Computing and Networking, 249–267 (2004)
7. Pawlikowski, K., Jeong, H., Lee, J.: On credibility of simulation studies of telecommunication networks. IEEE Communications Magazine 40(1), 132–139 (2002)
8. Owre, S., Rushby, J., Shankar, N., von Henke, F.: Formal Verification for Fault-Tolerant Architectures: Prolegomena to the Design of PVS. IEEE Trans. on Software Engineering 21(2), 107–125 (1995)
9. Owre, S., Rushby, J., Shankar, N., Stringer-Calvert, D.: PVS: an experience report. In: Applied Formal Methods. LNCS, vol. 1641, pp. 338–345. Springer, Heidelberg (1998)

10. Crow, J., Owre, S., Rushby, J., Shankar, N., Stringer-Calvert, D.: Evaluating, testing, and animating PVS specifications. Technical report, Computer Science Laboratory, SRI International, Menlo Park, CA (2001)
11. Muñoz, C.: Rapid prototyping in PVS. Technical Report NIA Report No. 2003-03, NASA/CR-2003-212418, National Institute of Aerospace, Hampton, VA (2003)
12. Butler, R., Sjogren, J.: A pvs graph theory library. Nasa technical memorandum 1998-206923, NASA Langley Research Center, Hampton, Virginia (1998)
13. Heinzelman, W., Kulik, J., Balakrishnan, H.: Adaptive protocols for information dissemination in wireless sensor networks. In: Proc. Intl. Conf. on Mobile Computing and Networking, pp. 174–185. ACM, New York (1999)
14. Levis, P., Lee, N., Welsh, M., Culler, D.: TOSSim: accurate and scalable simulation of entire TinyOS applications. In: Proc. Intl. Conf. on Embedded Networked Sensor Systems, pp. 126–137. ACM Press, New York (2003)
15. Shnayder, V., Hempstead, M., Chen, B., Allen, G., Welsh, M.: Simulating the power consumption of large-scale sensor network applications. In: Proc. Intl. Conf. on Embedded Networked Sensor Systems, pp. 188–200. ACM, New York (2004)
16. Woo, A., Tong, T., Culler, D.: Taming the underlying challenges of reliable multi-hop routing in sensor networks. In: SenSys 2003, pp. 14–27. ACM Press, New York (2003)
17. Bolton, C., Lowe, G.: Analyses of the reverse path forwarding routing algorithm. In: Proc. Intl. Conf. on Dependable Systems and Networks, pp. 485–494. IEEE Computer Society, Los Alamitos (2004)
18. Coleri, S., Ergen, M., Koo, T.J.: Lifetime analysis of a sensor network with hybrid automata modelling. In: Proc. Intl. Workshop on Wireless Sensor Networks and Applications, pp. 98–104. ACM, New York (2002)
19. Xie, F., Browne, J.C.: Verified systems by composition from verified components. SIGSOFT Softw. Eng. Notes 28(5), 277–286 (2003)
20. Nair, S., Cardell-Oliver, R.: Formal specification and analysis of performance variation in sensor network diffusion protocols. In: Proc. Symp. on Modeling, Analysis and Simulation of Wireless and Mobile Systems, pp. 170–173. ACM, New York (2004)
21. Luo, Y., Tsai, J.J.P.: A graphical simulation system for modeling and analysis of sensor networks. In: ISM 2005: Proceedings of the Seventh IEEE International Symposium on Multimedia, pp. 474–482. IEEE Computer Society, Washington (2005)
22. Ölveczky, P., Thorvaldsen, S.: Formal modeling and analysis of the ogdc wireless sensor network algorithm in real-time maude. In: Bonsangue, M.M., Johnsen, E.B. (eds.) FMOODS 2007. LNCS, vol. 4468, pp. 122–140. Springer, Heidelberg (2007)

Analyzing Fault Susceptibility of ABS Microcontroller

Dawid Trawczynski, Janusz Sosnowski, and Piotr Gawkowski

Institute of Computer Science, Warsaw University of Technology,
Nowowiejska 15/19, 00-665 Warsaw, Poland
{d.trawczynski,jss,gawkowsk}@ii.pw.edu.pl

Abstract. In real-time safety-critical systems, it is important to predict the impact of faults on their operation. For this purpose we have developed a test bed based on software implemented fault injection (SWIFI). Faults are simulated by disturbing the states of registers and memory cells. Analyzing reactive and embedded systems with SWIFI tools is a new challenge related to the simulation of an external environment for the system, designing test scenarios and result qualification. The paper presents our original approach to these problems verified for an ABS microcontroller. We show fault susceptibility of the ABS microcontroller and outline software techniques to increase fault robustness.

Keywords: Fault injection, fault tolerance, safety evaluation, real-time embedded systems, automotive systems.

1 Introduction

Recently, in automotive industry, electronic embedded systems are gaining much interest resulting in steady increase of devices controlling various car functions involved in airbags, active brakes, engine control and x-by-wire operations ([1-6] and references). These applications result in quite complex microcontrollers for which fault occurrence cannot be neglected. In particular, soft errors are becoming real problem. Hence, dependable operation of electronic control devices is a crucial point and appropriate safety norms have to be assured e.g. IEC 61508 [7] or AUTOSAR [8]. This can be achieved with various redundancy techniques as well as specific error recovery procedures ([3, 9-10] and references). An important issue is to analyze the effectiveness of the proposed solutions for various classes of faults. In particular we have to deal with permanent, intermittent and transient faults. In practice, transient faults (due to electromagnetic interference, power brownouts, and environmental disturbances) are dominating, so we are mostly interested in this class of faults.

Many approaches to analyzing fault susceptibility were proposed ([11-15] and references therein). They base on formal methods (functional analysis, fault tree and failure mode effect analysis) and various simulation experiments covering specific fault models at different abstraction levels. Most simulation techniques related to automotive systems rely on a simulation model for the entire considered system (usually in Matlab/Simulink and TrueTime [14]) enriched by some fault injection capabilities [1,15]. Typically, faults are injected at some abstract level e.g. selected state variables, abstracting from the implementation, or they are targeted at some specific

M.D. Harrison and M.-A. Sujan (Eds.): SAFECOMP 2008, LNCS 5219, pp. 360–372, 2008.

problems. In [4] the authors analyzed the impact of CAN network bandwidth (effects of the delays and jitter resulting from the use of a shared bus) on car suspension control performance, packet suppression faults, and sensor faults. The simulation experiments in [1] were targeted at high level mathematical model of a car suspension control developed in Matlab/Simulink with embedded fault injection functions.

In our approach we are closer to the real implementation and faults, which are well emulated by the software implemented fault injection (SWIFI) tool at the level of binary code of the evaluated system. SWIFI simulates faults in the real system by disturbing (e.g. performing bit-flips) the states of processor registers and memory cells (used for storing the program code, data and stack). The fault injection moments and locations can be specified explicitly or in a pseudorandom way. The fault effects are analyzed by comparing the behavior of the disturbed program with the reference execution (golden run) with no faults. The SWIFI approach is a popular and widely used dependability evaluation method for classical computational programs [11,13,16,17]. Adapting this technique to reactive control subsystem is a kind of challenge, due to the problem of taking into account the interactions of various electronic and mechanical subsystems as well as the impact of the environment e.g. driver reactions, road and weather conditions.

To qualify the controlled object's response to faults we have to define some performance parameters that describe the quality of the performed task by the controller in request of the driver or other car sub-circuits. The measured deviation of the analyzed parameter from the nominal values gives an indication on performance loss or even critical and unacceptable situations. A large class of embedded systems used in automotive or other industrial applications relates to feedback control of physical systems. Such systems usually operate in a cyclic way. Typically they get signals from sensors, process them and deliver output signals to the actuators. The control algorithm may take into account system deviation from the correct behavior (due to external disturbances or even microcontroller faults) and compensate detected error by adjusting newly calculated outputs. In this process an important issue is to meet time requirements while producing output signals.

To meet dependability expectations various techniques can be used that are based on fail-silence property. Duplex systems built by pairing two subsystems with continuously compared output signals, triplicated systems with voting or cheaper designs with limited or no hardware redundancy all can, to a certain degree, exhibit the fail-silence property ([17] and references). In any case, the analysis of fault effects and error propagation for simplex systems is important since often duplex implementation is not cost effective for an application. It is worth noting that for many applications a single temporary malfunction of the controller is not critical due to the natural inertia of the controlled system. Moreover, a simple error recovery performed after fault detection using the available idle time of the microcontroller (so as to not exceed real time requirements) maybe also effective. These features have to be validated.

In the paper we present an original methodology of analyzing fault susceptibility using SWIFI fault injector FITS [9] with appropriate adaptation to reactive systems. This methodology has been verified for the anti-lock microcontroller, and the gained experience can be easily extended for other reactive systems. We have also performed similar experiments with robot and alcohol rectification microcontrollers [16]. Section 2 describes the ABS microcontroller operation in relevance to the behavior of the braking system and the car. Section 3 outlines the fault injection platform and its adaptation to the required test scenarios. Experimental results are discussed in section 4. They show the impact of the faults on the performance of the ABS controller. The last section presents the conclusion and suggestions for the future research.

2   ABS Model

To study the fault susceptibility of the ABS controller we have developed its program and the model of the external environment covering the behavior of the car in relevance to the road conditions. The modeling of ABS is based on mathematical equations describing Newton's 2nd law of motion [18] defined separately for x, y, and z Cartesian coordinate axis. The ABS controller model has been defined using Matlab scripts for Simulink and is based on the mathematical model given in [18,19]. The developed Simulink model has later been transformed into C++ language for software fault injection based on SWIFI technology. We are mainly interested in the vehicle motion in the x direction. Whenever a vehicle brakes or accelerates the resultant instantaneous net force is lower or greater than zero. The objective of the anti-lock braking systems is to minimize the braking distance under the constraint of the tire slip. The tire slip occurs in situations where excessive braking force pressure is applied to braking pads while the friction force provided at the surface contact point of the tire and the road is insufficient. Exceeding the optimal braking force results in tire slippage, and in extreme may lead to tire locking. For the tire lock, the angular velocity of the wheel is zero, and the only friction force acting on the tire is the slippage friction. The slip friction is usually much lower than non-slip friction and may result in excessively long braking distance. Therefore, the anti-lock braking system has been developed by Bosch [18] in the 1970s, so that vehicle "slip" may be prevented in situations requiring sudden braking maneuvers. Analyzing ABS dependability we study the motion of only one wheel relative to the quarter vehicle body mass and the road surface.

The developed ABS controller unit (MCU) is composed of two blocks: the control logic block module (CLB) and the signal processing block (SPB). The input signals to the controller are *brake* signal (from the brake pedal), wheel angular velocity *Omega*, and some constants specifying various mechanical parameters. These signals are obtained from the controller environment. There are only two controller output signals namely the *inlet* and *outlet* valve control signals. These signals force appropriate brake torque within the hydraulic mechanism shown in Fig. 1. The controller (MCU) monitors sensors, calculates critical parameters and delivers control signals to actuators according to the algorithm outlined in section 2.3. While computing its outputs, the control algorithm exchanges information with a specific car behavioral model. The car behavioral model simulates the dynamics of the vehicle. The developed software is time triggered, a fixed sequence of tasks is activated periodically. There is no hardware replication. The behavior of the car is modeled as the controller environment. In the sequel we present a more detailed description.

2.1 ABS Controller

The control logic block generates two output binary signals: the *inlet* and *outlet* valve control signals. They are connected directly to the brake torque modulator module (BMM). This module is directly responsible for modulating the brake fluid pressure in individual brake lines. The binary TRUE (1) signal at the *inlet* output port commands the brake line valve to remain closed, whereas the binary *outlet* valve, depending on the situation, can be either closed - FALSE (0) or open – TRUE (1) at this time. Closing the *outlet* valve maintains current brake line pressure resulting in a constant torque, whereas opening the valve decreases the line pressure thus reducing the brake torque. Similarly, the binary FALSE (0) signal at the *inlet* valve commands this valve to be open thus increasing brake line pressure under the constraint that the outlet valve is in the closed position. Properly functioning control block module, under the condition of excessive tire slip, will generate a modulated sequence of pulses that increase, decrease or maintain brake line pressure. The concept of the used control algorithm is described in section 2.3.

The control logic block (CLB) accepts four binary control signals and its inputs are coupled indirectly through two OR gates to the signal processing block (SPB). The four input ports of the block are *decrease, hold, increase, stop decrease* and they either cause opening or closing of the associated *inlet* and *outlet* valves. Although the control action seems straight forward, the "tricky" part of the ABS control algorithm lies in the amount of time an inlet or outlet valve is either in the open or close position. This time effectively generates the necessary brake torque. The torque is a result of the brake line fluid pressure which is transmitted to the brake pads. The pads act on the wheel rotor (brake disc) surface to generate necessary friction torque. Therefore, a constant input to the logic control block produces a sequence of time varying output values forming a duty cycle varied pulse. This duty cycle modulated pulse is directly responsible for the pumping action of brakes in ABS. Its operation is described in [18]. The CLB block co-operates with the signal processing block(SPB).

The SPB block generates various signals needed to identify the state of the wheel. This block outputs two groups of binary signals. The first group is related to monitoring wheel acceleration and signaling crossing three thresholds: $-a$, $+a$ and A. The second group comprises 3 pairs of binary signals specifying the direction of crossing the above mentioned thresholds: in increasing (*pos.slope*) or decreasing (*neg slope*) direction. Moreover, SPB delivers the wheel slip coefficient *slip*, and the ratio of wheel angular acceleration to velocity *Om_dot/Om*. All these signals enable the logic controller (CLB) to determine if the tire has entered or is near the non-optimal friction region. If it happens, the wheel brake line is modulated as to bring the tire back to its optimal "friction" state. The computation of these signals and associated $A/+a/-a$ thresholds are given in [18]. The input signals to SPB block are the *vhvel, r_eff, omega,* and *brake* signals. The *vhvel* is the vehicle horizontal velocity (calculated by car dynamics module, described in the sequel), and *r_eff* is the effective rolling radius of the tire. This radius is a function of the tire stiffness and the normal force acting at the tire contact point. The *omega* is the angular velocity of the wheel and *brake* is a binary signal that is TRUE (1) whenever a driver presses the brake pedal.

2.2 ABS Environment Model

The ABS environment relates to three modules: brake modulator, tire and wheel dynamics module, and car dynamics. The brake modulator module (BTM) generates a real value of the simulated brake torque applied at the wheel rotor (disc brake). This module models a physical device and is represented in Fig. 1 as hydraulic pressure modulator. As the inlet valve remains open, the brake torque modulator integrates a constant rate of torque increase. When the integrator output exceeds the maximum brake torque that a brake pad may generate, the output of the integrator is saturated and kept constant. If the brake pedal is depressed by the driver, the torque modulator generates zero torque. This is accomplished by resetting the output of the integrator. In summary, BTM generates an appropriate brake torque as a function of inlet, outlet valve signals and the brake pedal state.

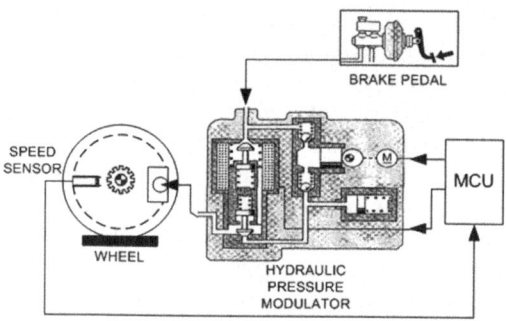

Fig. 1. Block diagram of the ABS brake system with the wheel speed sensor, hydraulic pressure modulator and electronic control unit (*MCU*)

The tire and wheel dynamics module (TWDM) is responsible for simulation of the wheel angular velocity *omega*. This value is generated based on two inputs – the *slip* (delivered by CLB) and applied *brake torque* (delivered by BTM). Additionally, the wheel angular velocity is computed based on an initial wheel velocity, polar moment of inertia of the wheel and tire, unloaded tire radius, vertical tire stiffness, effective tire radius, and normal force due to vehicle mass. These parameters are defined in [18,19]. Generally, as the slip value and brake torque increase, the wheel lock condition can be reached (the angular velocity of the wheel is zero). The controller therefore must adjust the brake torque to avoid the "wheel lock" state.

The car dynamics module (CDM) calculates the vehicle horizontal acceleration (*hac*) and velocity (*hvel*), and the *vehicle stopping distance* based on only two inputs: the wheel angular velocity *omega* and *brake status* signal. CDM calculates these signals and simulates the motion of the vehicle in the x direction taking into account the following parameters: vehicle mass, axle and rim mass, initial body translational velocity, initial axle translational velocity, tire belt translational stiffness, tire belt translational damping, vehicle translation dumping suspension, vehicle translation stiffness suspension, stop velocity, damping of translation, and normal force at the tire contact point. These parameters are defined in [18,19].

2.3 ABS Control Algorithm

The concept of the control algorithm is based on mathematical models from [18]. This concept has been implemented in software for x86 platform. The correctness of the implemented algorithm has been verified in many simulation experiments involving the developed ABS controller and the environment model. The objective of the algorithm is to decide what corrective action needs to be taken during excessive braking maneuver. The key variable of the algorithm is the wheel slip. This variable tells the algorithm if the wheel-tire system is in the normal (optimum friction) or abnormal (less than optimal friction) operating region. If the threshold slip is exceeded the tire is in the abnormal region and some corrective action needs to be taken.

The algorithm operates in an iterative way with specified time slot (0.1 ms) for each iteration. This time slot assures sufficient accuracy while controlling the brake mechanism. If an excessive wheel slip is detected during braking, the algorithm closes *inlet* and *outlet* valves of the brake torque modulator module (BTM) to maintain a constant brake pressure. The algorithm then measures the slip criterion during the subsequent iteration (time slot) and if the slip is still exceeding the required threshold, the algorithm commands the *outlet* BTM valve to open, thus reducing the torque rate at a constant rate measured in units of Nm/s. In the next phase, as the wheel speeds-up, the algorithm measures the wheel peripheral acceleration $+a$ and compares this measurement with a predetermined acceleration threshold. If the threshold is not exceeded the algorithm keeps the outlet valve open. In another case the outlet valve is closed and brake pressure is maintained constant.

In the last phase, the algorithm checks again the peripheral acceleration of the wheel to see if it exceeds the A threshold. This threshold determines when the braking torque should be increased again to maintain the safe braking action. As soon as this threshold is exceeded or a negative slope is detected in the peripheral acceleration of the wheel (here the A threshold can not be exceeded), the brake torque is increased by closing the outlet valve and opening the inlet valve of the BTM. The pressure is increased at a constant rate, either 2533 Nm/s or 19000 Nm/s, if ABS was activated for the first time in the braking interval. The first phase of ABS braking relates to the interval immediately after the driver has engaged the brake pedal. During this interval, the maximum brake pressure is applied until the slip value does not exceed the allowed maximum threshold. After the threshold has been exceeded, further brake torque increase occurs at a reduced or modulated rate. In this second control phase, the algorithm ensures that the possible wheel lock condition is avoided. The general structure of the algorithm is given below for typical values of some parameter thresholds (they can be adapted to other test scenarios):

```
WHILE (brake == TRUE AND vehicle_velocity > 1.5){
    inlet = OPEN
    outlet = CLOSE
    IF (first_braking_phase == TRUE) {
        brake_torque_increase_rate = 19000 Nm/s
    } ELSE {
    /*set second braking phase torque increase rate*/
    brake_torque_increase_rate = 2533 Nm/s}
    IF (current_vehicle_slip > 0.2) {
        /*close the BTM inlet valve*/
```

```
        inlet = CLOSE
        DO {brake_torque_decrease_rate = 19000 Nm/s}
        WHILE (wheel_acceleration < 0)
        outlet = CLOSE
        DO {outlet = CLOSE}
        WHILE              (wheel_acceleration<3           AND
    neg_slope_wheel_acceleration == FALSE)
        }
    }
```

3 Fault Simulation Platform

To analyze fault effects in the ABS controller we use software implemented fault injector FITS [9], which has been adapted to deal with real-time and reactive systems. The fault injector treats the ABS controller and car environment as an integrated application. Faults are injected by disturbing the states of processor registers or RAM locations (storing program code and data).

Each fault injection called *test* needs the execution of the application and simulating a fault at an appropriate fault triggering moment. The fault triggering moment is correlated with the program instruction i.e. its location address and execution iteration (appearance number). The fault injection (fault triggering moment and its location) can be either specified directly by the user or generated in an automatic way e.g. according to pseudorandom strategy. In the pseudorandom strategy we specify only the number of injected faults and some indications on fault location, fault type (bit flip, bit setting, resetting and bridging), fault duration, etc. This process has to be done for each test scenario (i.e. input data). The fault triggering moments may be restricted to specified program modules or even code address ranges. Fault location can be defined explicitly (e.g. specific register such as EAX or EBX, RAM memory address) or pseudorandomly within a selected group of registers, memory code or data area. Similarly, the fault type can be defined explicitly (e.g. bit flip in a specified position) or pseudorandomly within specified bit areas and related to a fixed number of faults (e.g. m-bit flips). Depending upon the goal of the analysis we can either generate the most stressing fault injection scenarios (to find critical points e.g. in specified code areas) or assure the pseudorandom selection of fault triggering moments with equal distribution within the tested code space (static strategy) or within the time of the application execution (dynamic strategy).

In the performed experiments we specify fault-triggering moments and fault locations related only to the analyzed ABS controller. The system environment is not disturbed. For each injected fault (a test) we check the system behavior. Test results are identified in relevance to the reference execution of the analyzed application - so called the golden run. The golden run delivers GR log with the registered information on the dynamic image of the application execution. In addition, it comprises some statistical data related to the number of writes, reads, state changes for each CPU register, register activity [9], etc. (they are not encountered in other SWIFI injectors). This is helpful while profiling the experiment or analyzing experimental results.

In general, test results can be qualified as: C – correct result, INC – not correct result, T – time-out, S - system exception (e.g. access violation, invalid opcode, memory misalignment or parity errors, overflow), U – user message (generated by the application). We have to define some procedure qualifying C and INC results. It can be done in a general way or targeted at the considered application. In calculation oriented applications (mostly considered in the literature [13]) the result analysis is simple and coarse-grained e.g. binary qualification based on the comparison with the final correct result. In real time applications the result qualification usually is more complex due to the fact that we have to analyze the output signals trajectories in time. Moreover, different tolerance margins and incorrect behavior severity levels can be attributed. This may lead to fined-grained result qualification with more detailed information e.g. the file comprising the generated output results of the application. We resolved this problem by defining a special result qualification module coupled to the fault injector and the model of the controlled object (environment).

In the case of the ABS controller the test result analysis can be performed by selecting some output control variables (e.g. control signals of a brake) and comparing their trajectories with the non-faulty run. The comparison can be based on calculating the mean square error. Another approach is to analyze the brake effect using two main safety parameters – the vehicle stopping distance and its final translational velocity. The vehicle stopping distance determines the total distance (in meters) traveled by a car during the braking time interval. The correct state can be defined if the final vehicle velocity (FV) is less than a specified value fv (m/sec) and the stopping distance (SD) is less than sd (meters). The relative fault severity levels can also be introduced using some knowledge on car behavior e.g. a car with a greater final velocity has a greater final momentum and thus will likely cause more damage during a head-on collision between two vehicles. The acceptable values for SD and FV can be based on the analysis given in [18,19] where the authors specify nominal values of stopping distance for a car traveling at a given initial velocity. This approach is illustrated in the next section.

The performed experiments were targeted at transient faults (bit flips) injected into registers (specified CPU or FPU registers, or all of them), the code or data area of the memory used by the ABS controller. By concentrating on specified system resources (or code segments) we can perform a deeper analysis and tune appropriate fault handling mechanisms. For each experiment, we choose a representative set of input data to assure high coverage of the code, decisions etc. The number of injected faults is sufficiently large to assure statistical significance of the obtained results.

In many applications fail-silent hypothesis is assumed i.e. the system produces correct outputs and stops producing outputs after detecting an error. For systems with some inertia as well with control loops fail-bounded hypothesis can also be assumed [17]. In this case the controller produces correct outputs, does not produce outputs after detecting some errors, produces wrong outputs within an acceptable deviation margin from the correct ones. This margin can be defined by some application dependant assertions. This approach allows postponing or even eliminating recovery. Here the assertion should allow predicting critical situations. Within the idle time of a controller we have to check if at the present control trajectory the worst case error may produce a critical situation. If so, we have to perform recovery. In the opposite case, the system follows its normal operation. Recovery can be limited to software (re-execution of some procedures, etc.) or, if needed, instantiating spare or backup hardware resources can do it.

4 Experimental Results

All simulations were performed with FITS injector within IBM PC platform (XP Windows Professional) and the model of the environment. The ABS controller and its environment model constitute an integrated program CE written in C++ language. This program was implemented based on mathematical models from [18] (compare section 2). Integrating these models we assure that the used variables and codes are disjoint. Hence, disturbing the ABS controller we do not interfere with environment model. The initial conditions of fault injection experiments are defined by setting some parameters in CE e.g. initial car speed, mechanical characteristics, road conditions (compare section 2). Each experiment is composed of a specified number of tests (a single fault injection) which are performed according to the predefined scenario (e.g. pseudorandom injections into code with static or dynamic strategy). For each test the results (system behavior) are stored in a file for the purpose of detailed analysis. We have also developed a special result qualification module (RQ) which analyzes the system behavior and identifies correct or incorrect system status. This status is based on the predefined criteria described in section 3. The qualification decision is sent to the fault injector FITS which accumulates statistics from all the tests. Moreover, it identifies timeouts and system exceptions.

The ABS control program was based on the model described in section 2. We have considered two implementations: the basic version (BV) and fault hardened versions (VH1, VH2). The basic version is a direct implementation of the mathematical model from [18,19]. In the fault hardened version we use the built in hardware and software fault detection mechanisms, which generate system exceptions. Typically these exceptions are signaled by the operating systems. It is possible to take over most of them at the application level and perform some error recovery. For this purpose we can use the *try* and *catch* construct provided by object oriented languages.

It assures taking over exceptions (specified by the filter or all of them – *catch(...)*) generated during the execution of the code within the *try* brackets (*try {segment of the application code}*). For any specified exception (in the exception filter) we define an appropriate handling procedure. For example, it may initiate reexecution of the program code starting from some specified checkpoint (previously established – backward recovery), suspending further execution of the thread, etc. The error correction can be made, if the error is uniquely correlated with the disturbed program segment. In version VH1 each captured exception initiates floating point unit (FPU) reset using the *_fpreset* function. In version VH2 additional code and state recovery is added by using a redundant DLL library and loading its static copy of the microcontroller code whenever a system exception was raised.

The size of the generated MSVC 2005 compiler binary image of the ABS controller program is approximately 100 KB for the basic version, and only 10-15 KB larger for versions with fault tolerance mechanisms VH1 and VH2. The ABS microcontroller without fault tolerance mechanisms consists of about 640 static code assembly instructions and 6 million dynamic instructions executed within the 15,000 simulation iterations. The source code was about 2000 lines in C++. The tire, wheel and suspension environment model consists of about 1000 static code instructions and 12 million dynamic instructions (for the analyzed test scenario).

In the discussed experiments, the operation of ABS is simulated for a fast braking scenario (fast pressing of the brake pedal). The braking process is performed in response to this input signal, taking into account the wheel speed, acceleration, etc. Observing the system behavior we can monitor internal variables or output control signals. The most interesting signals relate to the brake torque, car horizontal velocity and stop distance. The presented results relate to the car initial speed 60km/hr and a dry road. The golden run trajectories of these signals for the considered test scenario are given in Fig. 2a, 2b and 2c (they cover real ABS operation time of 1.5 s). The brake signal (pressing brake pedal by the driver) activates the ABS controller till the moment of achieving velocity 1.5 m/s. So the braking distance is 14.5m at this moment which corresponds to 1.5 sec time. For comparison, we give in fig. 2d-f the plots of the same variables in the case of an injected fault at time moment t=0.5438s (random bit-flip in an instruction code of the ABS controller). The plots differ

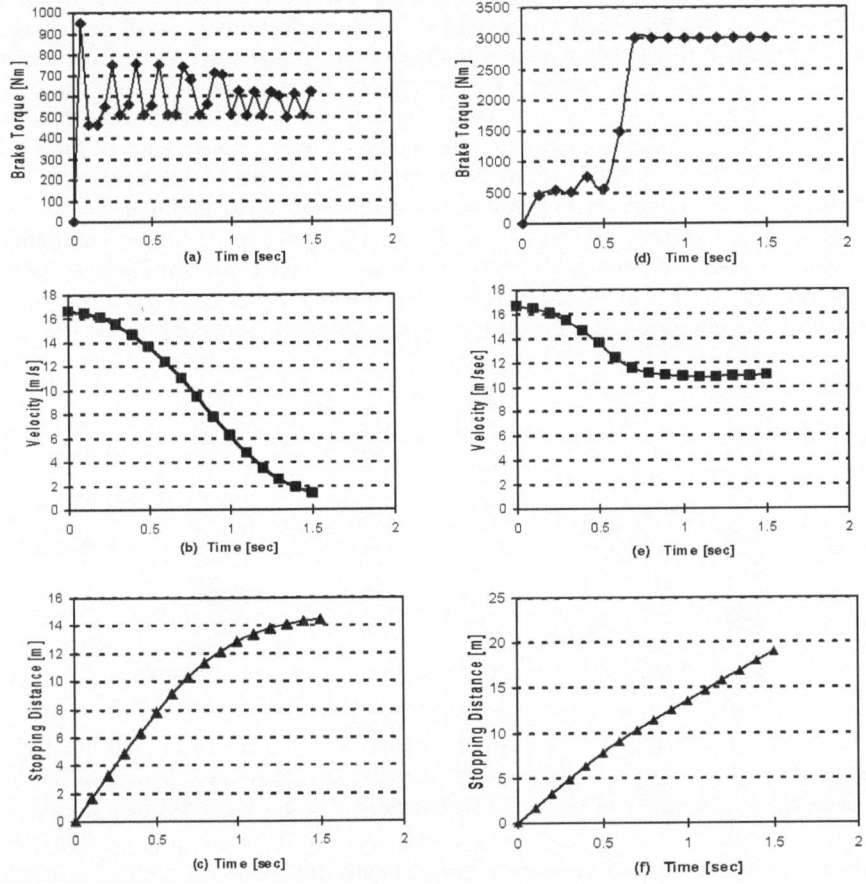

Fig. 2. Sample plots showing ABS braking: *a*), *b*) and *c*) golden run plots for the brake torque, vehicle speed and traveled distance in time, respectively; *d*), *e*) and *f*) corresponding plots related to an injected bit-flip fault in ABS code at t=0.5438s

significantly from the golden run; in particular, the braking distance at time 1.5 sec is 19 m and the final speed is 11m/s. This corresponds to a dangerous situation. Such analysis can be done manually for some selected faults to get knowledge of their impact.

More interesting are statistical results over many faults. We have injected many faults into the ABS controller code, CPU or FPU registers and data memory. We have assumed that the correct behavior corresponds to the final speed FV < 1.5 m/s at t=1.5 sec and stopping distance SD < 16 m. This criteria is easier to calculate than checking the correctness of the brake torque trajectory.

Test results for the basic controller version are shown in Fig. 3. Fault locations REG, MEM, FPU and INSTR correspond to CPU registers, data memory area, FPU registers and memory code area, respectively. The fault triggering moments are generated pseudo-randomly and distributed equally in time (dynamic). For the location CODE faults are injected with equal distribution in the memory code area space (static). A large percentage of faults resulted in system exceptions, which were not handled, and in fact they lead to dangerous situations. Relatively small percentage of incorrect results is due to some natural fault tolerance of the used control algorithm. A simple recovery based on taking over exceptions increase significantly the correct result percentage (Fig. 4). The presented results relate to latched transient faults (bit flips in registers or code/data memory). For comparison we give results of fault injections into the code for non-latched transient faults in the ABS controller memory.

This mimics the controller implementation with code stored in a non-volatile memory e.g. flash. For version VH1 we obtained INC=1%, C=94% and S=5%. This confirms significantly lower fault susceptibility for non-latched faults than for latched (compare Fig. 3). Similar results can be achieved for latched faults if more efficient error recovery mechanisms are employed (e.g. those discussed in our previous paper [9]).

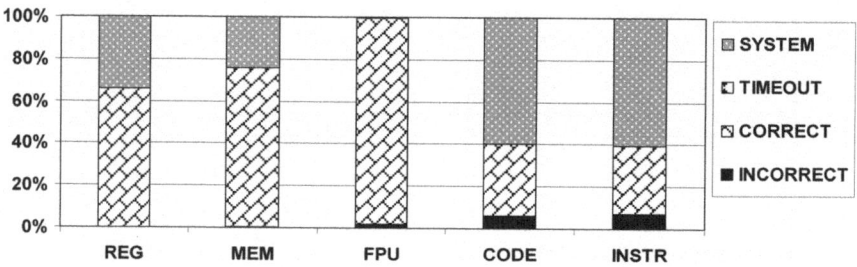

Fig. 3. Test results for the ABS micro-controller basic version *(BV)*

We have also developed the controller version VH2* adapted to platform x86 but without floating point unit (FPU). In this case all floating point calculations are done in software. The number of executed instructions for the analyzed test scenario increased from about 14 million in version VH2 to over 100 million in version VH2*. Fault susceptibility of both versions was practically the same for faults injected into registers and data memory. Faults injected into the code memory gave more correct results for version VH2* (84.5%) than for VH2 (74%).

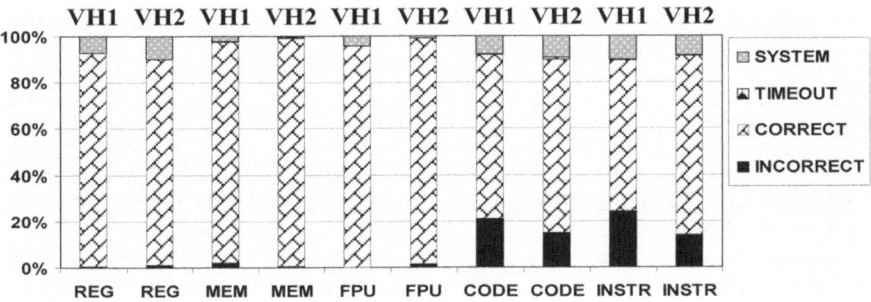

Fig. 4. Test results for the ABS with improved fault tolerance: version *VH1* and *VH2*

5 Conclusion

The main goal of this paper was to verify the developed methodology for evaluating the impact of faults on reactive systems. We have adapted our fault injector tool (FITS) for such systems by integrating it with the analyzed application (microcontroller) and its environment. Moreover, we added an interface to deal with specialized result qualification module and test scenario configuration. This approach has been successfully verified for the real ABS microcontroller and practical test scenarios. The proposed approach allows detailed behavioral analysis of the system in the presence of faults and gives statistics on susceptibility to faults injected in specified circuit areas. This approach is very useful in finding fault leakage sources, optimization of fault handling procedures as well as evaluation of the final projects. It was also verified for other applications [16]. We can identify program modules and data which are the most sensitive to faults, analyze critical behavior of the system in the presence of faults, etc. We can also evaluate the effectiveness of embedded fault handling procedures.

As opposed to classical calculation oriented applications, real-time and reactive systems require more complex result qualification methods. They can be based on observing output control signals or selected parameters describing the quality of performed tasks. This approach seems to be more effective and this was proved for the ABS controller. The experimental results showed that the basic version of the ABS controller comprises some natural fault tolerance capabilities (due to the used algorithm). This can be improved with simple exception handling procedures as well as by using non-volatile memory for the code. In the performed experiments we use a model of the control object and the system environment. Hence, the results depend upon the accuracy of the used models. This drawback can be eliminated in experiments with real objects, but such experiments are usually too expensive. Moreover there is a danger of causing critical situations or some damages in the case of injected faults.

Further research is targeted at developing and analyzing more effective fault tolerance mechanisms. Here we plan to use our experience gained with calculation oriented applications [9] and apply it to reactive systems. Result qualification will be extended by introducing more categories, e.g. loss of braking (the wheel speed is zero for some specified minimal time), lockcd wheel (the wheel speed does not decrease for more than some specified time). Moreover, we will consider distributed systems e.g. around a CAN network [10,15].

Acknowledgment. This work was supported by Ministry of Science and Higher Education grant 4297B/T02/2007/33.

References

1. Corno, F., Esposito, E., Reorda, M., Tosato, S.: Evaluating the effects of transient faults on vehicle dynamic performance in automotive systems. In: ITC 2004, pp. 1332–1339. IEEE Press, Los Alamitos (2004)
2. Dilger, E., Karrelmeyer, R., Straube, B.: Fault tolerant mechatronics. In: IOLTS 2004, pp. 214–218. IEEE Press, Los Alamitos (2004)
3. Mariani, R., Fuhrmann, P., Vittorelli, B.: Fault Robust Microcontrollers for Automotive Applications. In: IEEE On-line Test Symposium, pp. 213–218. IEEE Press, Los Alamitos (2006)
4. Gaid, M., Cela, A., Diallo, S.: Performance Evaluation of the Distributed Implementation of a Car Suspension System. In: PDS 2006. IFAC Press (2006)
5. Nouillant, F., Aisadian, X., Moreau, A., Oustaloup, et al.: Cooperative Control for Car Suspension and Brake Systems. J. of Auto. Tech. 4(4), 147–155 (2002)
6. Zalewski, J., Trawczynski, D., Sosnowski, J., Kornecki, A., Sniezek, M.: Safety Issues in Avionics and Automotive Databuses. In: IFAC World Congress. IFAC Press (2005)
7. CEI International standard IEC 61508 (1998-2000)
8. AUTOSAR partnership, http://www.autosar.org
9. Gawkowski, P., Sosnowski, J.: Experimental Evaluation of Fault Handling Mechanisms. In: Voges, U. (ed.) SAFECOMP 2001. LNCS, vol. 2187, pp. 109–118. Springer, Heidelberg (2001)
10. Short, M., Pont, M.J.: Fault tolerant time-triggered communication using CAN. IEEE Transactions on Industrial Informatics 3(2), 131–142 (2007)
11. Adermaj, A.: Slightly-of-specification failures in the time triggered architecture. In: 7th IEEE Int. Workshop on High Level Design and Validation and Test, pp. 7–12. IEEE Press, Los Alamitos (2002)
12. Anghel, L., Leveugle, R., Vanhauwaert, P.: Evaluation of SET and SEU effects at multiple abstraction levels. In: 11-th IEEE IOLTS Symposium, pp. 309–314. IEEE Press, Los Alamitos (2005)
13. Arlat, J., Crouzet, Y., Karlsson, J., Folkesson, P., Fuchs, E., Leber, G.H.: Comparison of physical and software implemented fault injection techniques. IEEE Transactions on Computers 52(9), 1115–1133 (2003)
14. Cervin, A., Henriksson, D., Lincoln, D., Eker, J., Årzén, K.: How Does Control Timing Affect Performance? IEEE Control Systems Magazine 23(3), 16–30 (2003)
15. Trawczynski, D., Sosnowski, J., Zalewski, J.: A Tool for Databus Safety Analysis Using Fault Injection. In: Górski, J. (ed.) SAFECOMP 2006. LNCS, vol. 4166, pp. 261–275. Springer, Heidelberg (2006)
16. Gawkowski, P., et al.: Software Implementation of Explicit DMC Algorithm with Improved Dependability. In: Int. Joint Conf. on Computer, Information, and Systems Sciences, and Engineering (CISSE 2007), December 3 - 12 (2007)
17. Cunha, J., Rela, M., Silva, J.: On the Use of Disaster Prediction for Failure Tolerance in Feedback Control Systems. In: Dependable Systems and Networks 2002, pp. 123–134. IEEE Press, Los Alamitos (2002)
18. Rangelov, K.: Simulink Model of a Quarter-Vehicle with an Anti-Lock Braking System. Research Report, Eindhoven University of Technology (2004)
19. MSC Software: Using ADAMS/Tire. ADAMS Software Manual (2005)

A Formal Approach for User Interaction Reconfiguration of Safety Critical Interactive Systems

David Navarre, Philippe Palanque, and Sandra Basnyat

Institute of Research in Informatics of Toulouse (IRIT)
University Paul Sabatier, 118, route de Narbonne, 31062 Toulouse Cedex 9, France
{navarre,palanque,basnyat}@irit.fr

Abstract. The paper proposes a formal description technique and a supporting tool that provide a means to handle both static and dynamic aspects of input and output device configurations and reconfigurations. More precisely, in addition to the notation, the paper proposes an architecture for the management of failure on input and output devices by means of reconfiguration of in/output device configuration and interaction techniques. Such reconfiguration aims at allowing operators to continue interacting with the interactive system even though part of the hardware side of the user interface is failing. These types of problems arise in domains such as command and control systems where the operator is confronted with several display units. The contribution presented in the paper thus addresses usability issues (improving the ways in which operators can reach their goals while interacting with the system) by increasing the reliability of the system using diverse configuration both for input and output devices.

Keywords: Model-Based approaches, ARINC 661 specification, formal description techniques, interactive software engineering, interactive cockpits.

1 Introduction

Command and control systems have to handle large amounts of increasingly complex information. Current research work in the field of Human-Computer Interaction promotes the development of new interaction and visualization techniques in order to increase the bandwidth between the users and the systems. Such an increase in bandwidth can have a significant impact on efficiency (for instance the number of commands triggered by the users within a given amount of time) and also on error-rate [21] (the number of slips or mistakes made by the users).

Post-WIMP user interfaces [24] provide users with several interaction techniques that they can choose from and provide the possibility to exploit different output devices according to different criteria such as, work load, cognitive load, or availability (of the system devices). This includes, for instance, keyboard and mouse as hardware input devices and double click, drag and drop, CTRL+click, ... as interaction techniques. Exploiting such possibilities calls for methods, techniques and tools to support various configurations at the specification level (specify in a complete and unambiguous way the configurations i.e. the set of desired interaction techniques and output configurations), at the validation level (ensure that the configurations meet the

M.D. Harrison and M.-A. Sujan (Eds.): SAFECOMP 2008, LNCS 5219, pp. 373–386, 2008.
© Springer-Verlag Berlin Heidelberg 2008

requirements in terms of usability, reliability, human-error-tolerance, fault-tolerance and possibly security), at the implementation level (support the process of going from the specification to the implementation of the configurations in a given system) and for testing (how to test the efficiency of the configurations and of the re-configured system).

A recent trend in Human-Computer Interaction addresses the issue of dynamic re-configuration of interfaces under the concept of plasticity coined by J. Coutaz [11]. However, research work on plasticity mainly addresses reconfiguration at the output level i.e. adapting the presentation part of the user interface to the display context (shrinking or expanding presentation objects according to the space available on the display). In addition, reliability issues and specification aspects of plastic interfaces are not considered. Work recently done with web site personalization/configuration [12] and [22] struggle with the same concepts and constraints even though, here again, personalization remains at a cosmetic level and does not deal with how the users interact with the web application. Our work differs significantly as users are pilots following long and intensive training programme (including on-the-fly training) and thus being trained to authorized reconfigurations while web users passively undergo the (most of the time unexpected) reconfigurations.

These issues go beyond current state of the art in the field of interactive systems engineering where usually each interactive system is designed with a predefined set of input and output devices that are to be used according to a static set of interaction techniques and are identified at design time. This set can sometimes gather many different interaction techniques and input/output devices as, for instance, in military cockpits [7]. Current safety critical systems, for example, the cockpit of the Airbus A380, presents 8 display units of which 4 of them offer interaction via a mouse and a keyboard by means of an integrated input device called KCCU (Keyboard Cursor Control Unit). Applications are allocated to the various display units. If one of the display units fails, then, according to predefined criteria (like the importance of the application according to the flight phase) the applications (displayed on that faulty unit) are migrated to other available display units. This paper proposes a formal description technique and a supporting tool that provide a means to handle both static and dynamic aspects of input and output devices configuration and reconfiguration. The justification of using formal description techniques is three fold:

- The possibility to define in a complete and unambiguous way the behaviour of the input and output devices, the interaction techniques, the authorised configurations and the reconfiguration mechanism
- The possibility to reason about that models in order to be able to assess the behaviour of the configurations (e.g. for all the possible configuration a given application is always presented to the operator)
- The possibility via the tool PetShop [6] supporting the formal notation to interactively prototype the behaviours and to modify and adjust them according to operator's requirements and global performance.

The paper is structured as follows: the next section rapidly introduces the ARINC 661 specification while section 3 briefly presents the formal description technique called ICO. Section 4 focuses on an architecture that is able to handle reconfiguration of both input and output devices, compliant with the ARINC 661 Specification, based on the ICO notation. Section 5 presents the configuration management and proposes a set of configuration manager models.

2 ARINC 661 Specification

The Airlines Electronic Engineering Committee (AEEC) (an international body of airline representatives leading the development of avionics architectures) formed the ARINC 661 Working Group to define the software interfaces to the Cockpit Display System (CDS) used in all types of aircraft installations. The standard is called ARINC 661 - Cockpit Display System Interfaces to User Systems [2, 3].

In ARINC 661, a user application is defined as a system that has two-way communication with the CDS (Cockpit Display System):

- Transmission of data to the CDS, which can be displayed to the flight deck crew.
- Reception of input from interactive items managed by the CDS.

According to the classical decomposition of interactive systems into three parts (presentation, dialogue and functional core) defined in [10], the CDS part (in Fig. 1) may be seen as the presentation part of the whole system, provided to the crew members, and the set of UAs may be seen as the merge of both the dialogue and the functional core of this system. ARINC 661 then puts on one side input and output devices (provided by avionics equipment manufacturers) and on the other side the user applications (designed by aircraft manufacturers). Indeed, the consistency between these two parts is maintained through the communication protocol defined by ARINC 661.

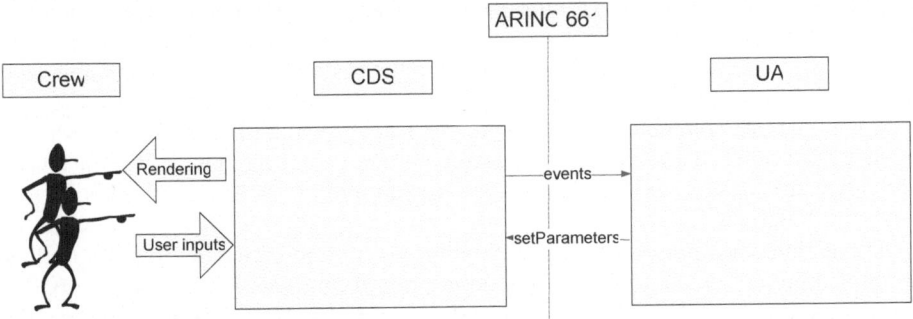

Fig. 1. Abstract architecture and communication protocol between Cockpit Display System and a User Application

3 ICOs a Formal Description Technique for Interactive Systems

The ICO formalism is a formal description technique dedicated to the specification of interactive systems [17]. It uses concepts borrowed from the object-oriented approach (dynamic instantiation, classification, encapsulation, inheritance, client/server relationship) to describe the structural or static aspects of systems, and uses high-level Petri nets [14] to describe their dynamic or behavioural aspects.

ICOs are dedicated to the modelling and the implementation of event-driven interfaces, using several communicating objects to model the system, where both behaviour of objects and communication protocol between objects are described by the Petri net dialect called Cooperative Objects (CO). The ICO formalism has been applied to other domains than user interfaces as, for instance, CORBA services specification [4] and [5].

In the ICO formalism, an object is an entity featuring four components: a **cooperative object** which describes the behaviour of the object, a **presentation part** (i.e. the graphical interface), and two functions (the **activation function** and the **rendering function**) which make the link between the cooperative object and the presentation part.

An ICO specification fully describes the potential interactions that users may have with the application. The specification encompasses both the "input" aspects of the interaction (i.e. how user actions impact on the inner state of the application, and which actions are enabled at any given time) and its "output" aspects (i.e. when and how the application displays information relevant to the user).

This formal specification technique has already been applied in the field of Air Traffic Control interactive applications [17], space command and control ground systems [19], or interactive military [7] or civil cockpits [4]. The example of civil aircraft is used in the next section to illustrate the specification of embedded systems.

4 An Architecture for Reliable and Reconfigurable User Interfaces

One of the aims of the work presented in this paper is to define an architecture that supports usability aspects of safety-critical systems by taking into account potential malfunctions in the input (output respectively) devices that allow the operators to provide (perceive respectively) information or trigger commands (perceive command results respectively) to the system. Indeed, any malfunction related to such input devices might prevent operators to intervene in the systems functioning and thus jeopardize the mission and potentially put human life at stake. In systems offering standard input device combination such as mouse + keyboard, it is possible to handle one input device failure by providing redundancy in the use of the device. For instance a soft keyboard such as the ones defined in [16] can provide an efficient palliative for a keyboard failure[1].

The architecture presented in Fig. 2 proposes a structured view on the findings from of a project dealing with formal description techniques for interactive applications compliant with the ARINC 661 specification [9, 4]. Applications are executed in a Cockpit Display System (CDS) that aim to provide flight crew with all the necessary information to try to ensure a safe flight.

We are dealing with applications that exclude primary cockpit applications such as PFD (Primary Flight Display) and ND (Navigation Displays) and only deal with secondary applications such as the ones allocated to the MCDU (Multiple Control Display Unit). For previous CDSs (such as the glass cockpit of the A320) these applications were not interactive (they only displayed information to the crew) and inputs

---

[1] This kind of management of input device failure could and should prevent the typical error message on PCs when booting with a missing keyboard "Keyboard Failure strike F1 key to continue".

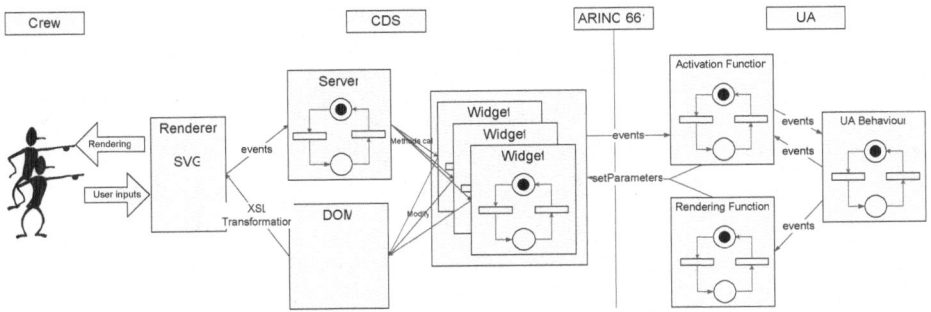

Fig. 2. Detailed architecture compliant with ARINC 661 specification not supporting interaction failures

were made available through independent physical buttons located next to the display unit. The location in the cockpit in-between the pilot and the first officer make it possible for both of them to use such application.

Within that project, we proposed a unique notation (ICOs) to model the behaviour of all the components of an interactive application compliant with ARINC 661 specification. This includes each interactive component (called widgets) the user application (UA) per se and the entire window manager (responsible for the handling of input and output devices, and the dispatching of events (both those triggered by the UAs and by the pilots) to the recipients (the widgets or the UAs).

The two main advantages of the architecture presented in Fig. 2 are:

- Every component that has an inner behaviour (server, widgets, UA, and the connection between UA and widgets, e.g. the rendering and activation functions) is fully modelled using the ICO formal description technique thus making it possible to analyse and verify the correct functioning of the entire system,
- The rendering part is delegated to a dedicated language and tool (such as SVG , Scalable Vector Graphics), thus making the external look of the user interface independent from the rest of the application, providing a framework for easy adaptation of the graphical aspect of cockpit applications.

However, this architecture does not support reconfiguration of input or output devices in the cockpit, neither in case of redesign or in case of failure while in operation. However, requirements specification for a display unit (DU) like the one of the Airbus A380 explicitly requires the possibility for the co-pilot to read information on the DU of the pilot (in case of failure on his/her side for instance).

The new architecture we propose has been extended to explicitly manage the reconfiguration of applications on the display units. In that architecture (presented in Fig. 3), the input and output devices are formally described using the ICO notation in order to be handled by a configuration manager which is also responsible for reconfiguring devices and interaction technique according to failures.

In Fig. 3 the dashed-line section highlights the improvements made with respect to the previous architecture:

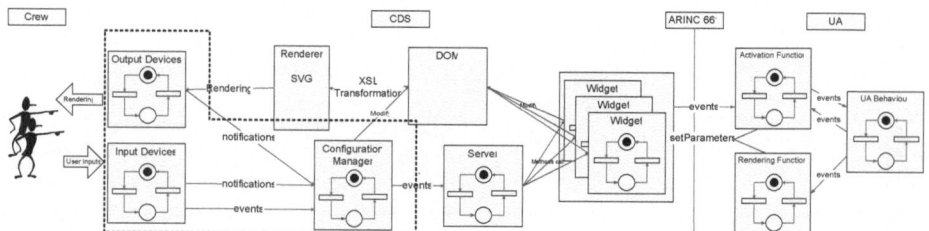

Fig. 3. Global architecture compliant with ARINC 661 specification and supporting interaction failures

- The left-hand part of the frame highlights the addition of ICO models dedicated to both input and output devices,
- The right-hand part presents the introduction of a new component named configuration manager responsible for managing the configuration of input and output devices.

Even though modelling of input devices and interaction techniques has already been presented in the context of multimodal interfaces for military cockpits [7] it was not integrated with the previous architecture developed for interactive applications compliant with ARINC 661 specification. The rest of the paper thus focuses on the configuration manager that is dedicated to the dynamic reconfiguration of user interaction (both input devices and interaction techniques).

5   Configuration Manager Policy and Modelling

This section presents the modelling of different policies to manage both input and output device configuration. We first present two policies and then present a possible modelling of such policies using the ICO formalism.

5.1   Input and Output Management Policies

Configuration management activities may occur at either runtime (while a user interacts with the application) or "pre-runtime" (e.g. just before starting an application or during a switchover of users). To illustrate the different kinds of policy, we present a pre-runtime policy where input devices are involved and a runtime policy for managing output devices.

5.2   Input Device Configuration Manager Policy

A possible use of reconfiguration is to allow customizing the interaction technique to make the application easier to manipulate. Even if it is out of the scope of the current version of the ARINC 661 Specification, customization of interaction techniques may become necessary bringing a better user experience [10] in the same way as with personal computers.

We focus only on a very simple scenario of input device configuration policy which is based on the difference between a pilot and the associated co-pilot. The first may be familiar with using the double-click interaction technique, but the second one may be more familiar with using the combination of keyboard and mouse to do the same interaction (lets say the combination of the key ALT and the mouse left click event). This reconfiguration may be possible at both pre-runtime or at runtime.

5.3 Output Device Configuration Manager Policy

A policy has to be defined on what kind of changes have to be performed when a display unit fails. This policy is highly based on the windowing system adopted by the standard ARINC 661 specification.

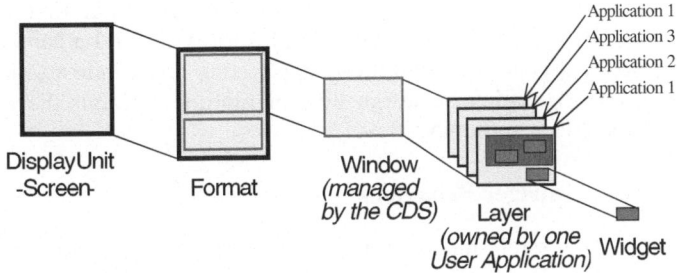

Fig. 4. ARINC 661 Specification windowing architecture

The ARINC 661 Specification uses a windowing concept which can be compared to a desktop computer windowing system, but with many restrictions due to the aircraft environment constraints (see Fig. 4). The windowing system is split into 4 components:

- the display unit (DU) which corresponds to the hardware part,
- the format on a Display Unit (DU), consists of a set of windows and is defined by the current configuration of the CDS,
- the window is divided into a set of layers (with the restriction of only one layer activated and visible at a time) in a given window,
- the widgets are the smallest component on which interaction occurs (they corresponds to classical interactors on Microsoft Windows system such as command buttons, radio buttons, check buttons, …).

When a display unit fails, the associated windows **may** have to be reallocated to another display unit. This conditional assertion is related to the fact that:

- There might be not enough space remaining on the other display units (DU),
- The other applications displaying information on the other DU might have a higher priority.

The ARINC 661 Specification does not yet propose any solution to this particular problem but it is known as being critical and future supplements of the ARINC 661 specification may address this issue[2]. However at the application level, the UADF

---

[2] ARINC 661 specification is continuously evolving since the first proposal. The draft 2 of supplement 3 containing 374 pages has been released on August 15th 2007.

(User Application Definition File) defines a priority ordering among the various layers included in the user application. At any given time only one layer can be active. At runtime, the activation of a new layer must be preceded by the deactivation of the current layer.

The policy that we have defined lays in the definition of a set of compatible windows i.e. windows offering a greater or equal display size. This is related to a strong limitation imposed by ARINC 661 which states that some methods and properties are only accessible at design time i.e. (according to ARINC 661 specification vocabulary) when the application is initialized. Methods and properties related to widget size are not available at runtime and thus any reorganisation of widgets within a window is not possible.

The only policy that can thus be implemented is a policy where first a compatible window has to be found and then the question of priority has to be handled. As only layers have a priority it is not possible for an application or a window to have a priority. This cannot be done either at design time or runtime and thus the management policy can only take place at the layer level.

5.4   Configuration Manager Behaviour

This section presents possible models for the configuration management according to the policies described above. We first present how input device configurations are managed and then deal with output devices managements.

Input devices Management
The user interface server manages the set of widgets and the hierarchy of widgets used in the User Applications. More precisely, the user interface server is responsible in handling:

- The creation of widgets
- The graphical cursors of both the pilot and his co-pilot
- The edition mode
- The mouse and keyboard events and dispatching it to the corresponding widgets
- The highlight and the focus mechanisms
- …

As it handles many functionalities, the complete model of the sub-server (dedicated in handling widgets involved in the MPIA User Application) is complex and difficult to manipulate without an appropriate tool.

Events received by the interaction server are in some way high level events as they are not the raw events produced by the input devices drivers. In our architecture, the main role of an input configuration is the role of a transducer [1]; it receives raw events and produces higher level events. The events used by the interaction server, and so produced by an input configuration are (*normalKey*, *abortKey*, *validationKey*, *pickup*, *unPickup*, *mouseDoubleClicked*, *mouseClicked*). These events are produced from the following set of raw events: *mouseMoved*, *mouseDragged*, *mousePressed*, *mouseReleased*, *mouseClicked* and *mouseDoubleClicked* from the mouse driver, and *pickup* and *unPickup* from the picking manager.

Fig. 5 shows how the reception of raw events (left side of the figure) leads to the production of higher events (right side of the figure). For instance, when a key is pressed, an event occurs from the keyboard driver that leads to firing the transition called *keyPressed_T1*. A token that holds the key value is then produced and put into the place *p1*. Depending on the value held by this token, and using the precondition mechanism in transitions, one of the three transitions *sendAbortKey, sendValidation-Key* or *sendNormalKey* is fired.

The rectangle at the bottom of Fig. 5 represents the subpart responsible for producing both click and double-click events. As this configuration is very simple, this production is more or less the forward of the raw events as new events. The conversion of the object that holds the event values is beyond the scope of this paper.

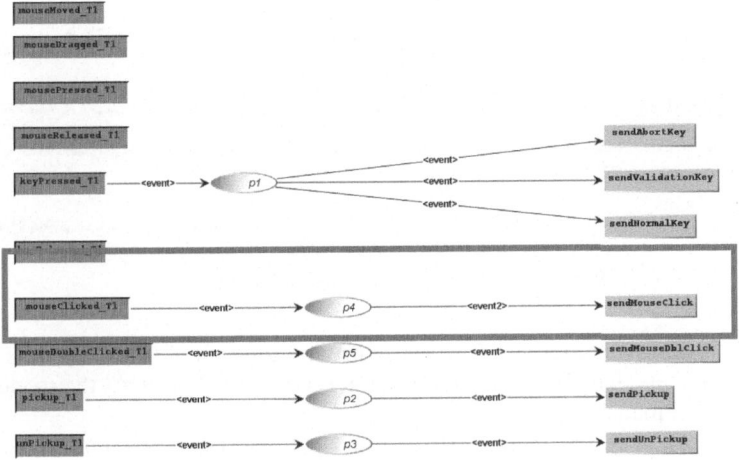

Fig. 5. Model of the raw Double Click configuration

The left and right hand sides of Fig. 6 are exactly the same as those in Fig. 5 as the input configuration represented here receives the same set of raw events and produces the same set of higher level events.

The difference between these two configurations is the part highlighted by the rectangle in the middle of Fig. 6. The production of both a single click and double click event now leads to the use of the Alt key from the keyboard. When a key event occurs, a token is put in place *p1*. As in the previous case, depending on its value one transition out of *sendAbortKey, sendValidationKey, sendNormalKey* or *altPressed* is fired. In the first three cases, this configuration behaves the same way as the previous configuration, but when the key code corresponds to the Alt key, a token is put in *p6*. If the Alt key is released, a token is put in *p4* and fires the transition *altReleased*, discarding the token in place *p6*. If the Alt key is not released, and the mouse is clicked, then a token is put in place *p7* by the firing of transition *altAndClick*, and then the transition *sendMouseDblClick* is fired, producing a double-click event using the combination of a single mouse click and the keyboard. The management of the different configurations can be seen as the management of the connections between the input device drivers, the configuration itself and the interaction server.The input

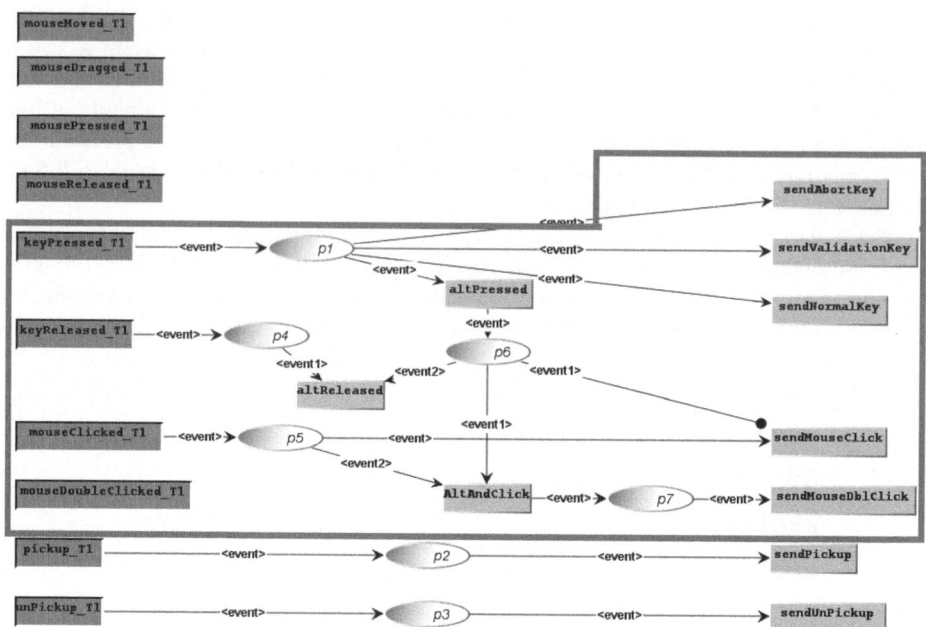

Fig. 6. Model of the raw Alt+Click configuration

devices produce events received by the configuration, which translate them into another set of events, handled by the interaction server. Fig 7 presents the model of such a configuration manager.

The four places in the central part of Fig 7 (*MouseDriver, KeyboardDriver, PickingManager* and *InteractionServer*) contain a reference to the set of models corresponding to the input devices and to the interaction server. When a new configuration is requested to be set, a token with a reference to the new configuration is put in place *NewConfiguration*. Following this, the four transitions highlighted on the left hand side are fired in sequence (could be modelled as parallel behaviour as well) in order to register the new configuration as a listener of the events produced by the mouse driver, the keyboard driver and the picking manager. The fourth transition registers the interaction server as a listener of the events produced by the new configuration.

If a configuration is already set, when the new configuration is requested, a token is put in place *UnregisterCurrent* in order to fire the four transitions highlighted on the right hand side, which corresponds to the unregistering from the different models, in parallel with registering the new configuration.

Output devices Management

In Fig. 8, we present an implementation of the previously defined policy for handling output devices using the ICO formalism. This model is a subpart of the complete configuration manager that can be added to the previous modelling we have done and thus be integrated in the behaviour of our (Cockpit Display System) CDS model [9].

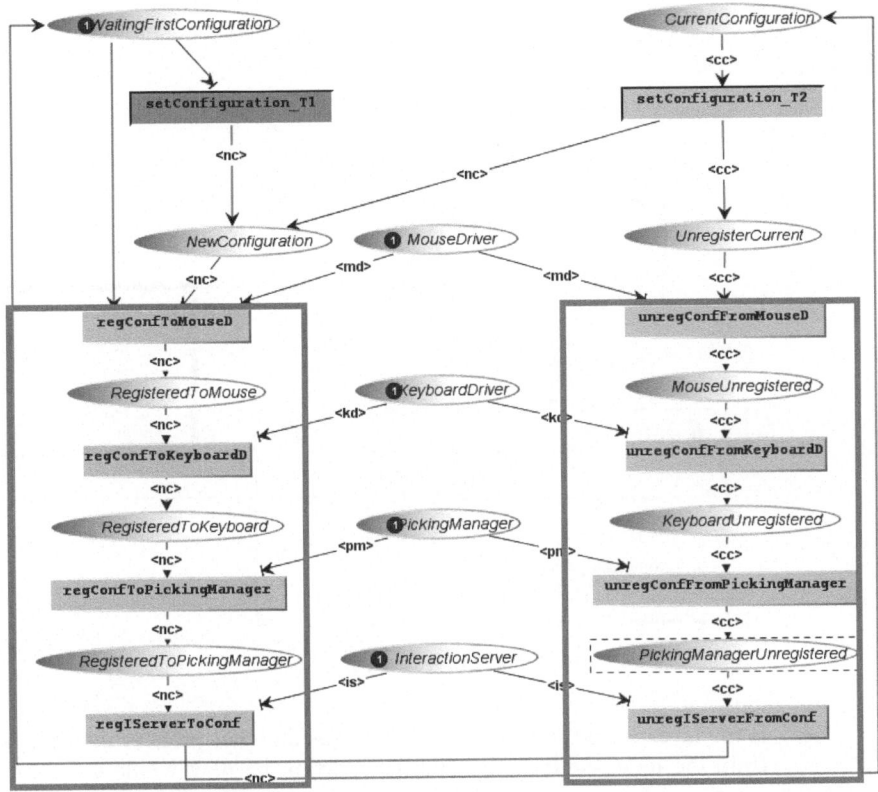

Fig. 7. ICO model of the configuration manager part dedicated to the input devices

The model presented here is based on a very simple case (1 layer per window and 1 window per display unit). This information flow and the operation to be performed remain the same, but it is possible to build models for a much more complex case as ICO proposes means to handle such complexities:

1. The display unit (DU) notifies its failure (the event may be triggered by a sensor), and then the configuration manager located the window currently displayed in that DU.
2. The configuration manager finds a compatible window for a reallocation of the contained layers (here all compatible windows are listed at creation time) and the layers are transferred to the new window.
3. As in a given window only one layer can be activated, when layers are reallocated, the configuration manager must identify the layer to be activated (among the new set of layers related to the window presented on the non functioning DU).
4. That part of the model determines which layer must be activated according to the layer priority defined at creation time:

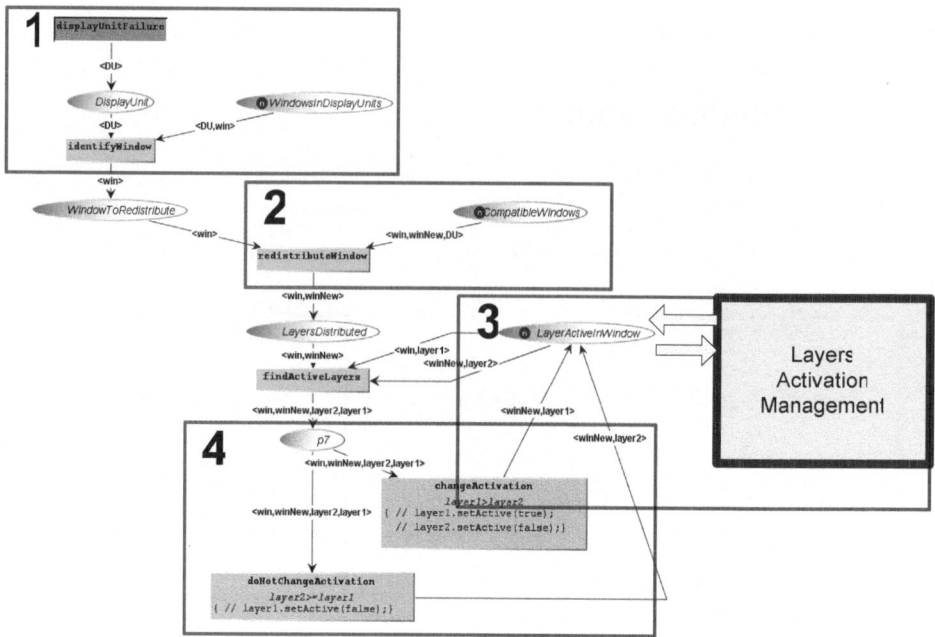

Fig. 8. An ICO model of a configuration manager

- If the layer from the previous window has a higher priority than the one from the new window, then the layer from the new window is deactivated, sending a notification to the corresponding user application according to the ARINC 661 Specification protocol (the UA may (or may not) request to reactivate the layer depending on its defined behaviour).
- Otherwise, the layer from the previous window is deactivated (leading to the same effects).
- In both case, the list of activated layers is updated.

6 Conclusion and Perspectives

This paper has addressed the issue of user interface reconfiguration in the field of safety critical command and control systems. The application domain is civil aircraft cockpit systems compliant with the ARINC 661 specification (which defines communication protocols and window management policy for cockpit displays systems). This work complements previous work we have done on this topic [9] by extending the behavioural model of cockpit display system with fault-tolerant behaviour and with a generic architecture allowing static configuration as well as dynamic reconfiguration of interaction techniques.

It is important to note that such fault-tolerance is only related to the user interface part of the cockpit display system even though it takes into consideration input and output devices as well as the behaviour of the window manager.

While the safety aspects have not been at the centre of the paper the entire work presented here serves as a basis for supporting the design and construction of safer interactive embedded applications and to improve operations.

Acknowledgements. This work is supported by the EU funded Network of Excellence ResIST http://www.resist-noe.eu under contract n°026764.

References

1. Accot, J., Chatty, S., Maury, S., Palanque, P.: Formal Transducers: Models of Devices and Building Bricks for Highly Interactive Systems. In: 4th EUROGRAPHICS workshop on design, specification and verification of Interactive systems, Spain, 5-7 june 1997, pp. 143–159. Springer, Heidelberg (1997)
2. ARINC 661, Prepared by Airlines Electronic Engineering Committee. Cockpit Display System Interfaces to User Systems. ARINC Specification 661 (2002)
3. ARINC 661-2, Prepared by Airlines Electronic Engineering Committee. Cockpit Display System Interfaces to User Systems. ARINC Specification 661-2 (2005)
4. Bastide, R., Palanque, P., Sy, O., Navarre, D.: Formal specification of CORBA services: experience and lessons learned. In: Proceedings of the 15th ACM SIGPLAN Conference on Object-Oriented Programming, Systems, Languages, and Applications (Minneapolis, Minnesota, United States). OOPSLA 2000, pp. 105–117. ACM, New York (2000)
5. Bastide, R., Sy, O., Navarre, D., Palanque, P.: A formal specification of the CORBA event service. IFIP TC6/WG6.1. In: 4th international conference on formal methods for open object-based distributed systems (FMOODS), Stanford univ., California, USA, pp. 371–395. Kluwer, Dordrecht (2000)
6. Bastide, R., Palanque, P., Sy, O., Le, D.-H., Navarre, D.: PetShop a case tool for Petri net based specification and prototyping of Corba Systems. In: Tool demonstration with Application and Theory of Petri nets ATPN 1999, Williamsburg (USA). LNCS, pp. 66–83. Springer, Heidelberg (1999)
7. Bastide, R., Navarre, D., Palanque, P., Schyn, A., Dragicevic, P.: A Model-Based Approach for Real-Time Embedded Multimodal Systems in Military Aircrafts. In: Sixth International Conference on Multimodal Interfaces (ICMI 2004), Pennsylvania State University, USA, October 14-15, 2004, pp. 243–250. ACM Press, New York (2004)
8. Barboni, E., Navarre, D., Palanque, P., Basnyat, S.: Exploitation of Formal Specification Techniques for ARINC 661 Interactive Cockpit Applications. In: Proceedings of HCI aero conference (HCI Aero 2006), Seattle, USA, pp. 81–89 (September 2006)
9. Barboni, E., Conversy, S., Navarre, D., Palanque, P.: Model-Based Engineering of Widgets, User Applications and Servers Compliant with ARINC 661 Specification. In: Doherty, G., Blandford, A. (eds.) DSVIS 2006. LNCS, vol. 4323, pp. 25–38. Springer, Heidelberg (2007)
10. Csíkszentmihályi, M.: Flow: The Psychology of Optimal Experience. Harper and Row, New York (1990) ISBN 0-06-092043-2
11. Thevenin, D., Coutaz, J.: Plasticity of User Interfaces: Framework and Research Agenda. In: Proceedings of Interact 1999, Edinburgh: IFIP TC 13, vol. 1, pp. 110–117. IOS Press, Amsterdam (1999)
12. Eirinaki, M., Lampos, C., Paulakis, S., Vazirgiannis, M.: Web personalization integrating content semantics and navigational patterns. In: WIDM 2004: Proceedings of the 6th annual ACM international workshop on Web information and data management, pp. 72–79. ACM Press, New York (2004)

13. Feiler, P., Li, J.: Consistency in dynamic reconfiguration. In: International Conference on Configurable Distributed Systems, Annapolis, MD, pp. 189–196. IEEE, Los Alamitos (1998)
14. Genrich, H.J.: Predicate/Transitions Nets. High-Levels Petri Nets: Theory and Application. In: Jensen, K., Rozenberg, G. (eds.), pp. 3–43. Springer, Heidelberg (1991)
15. Lakos, C.: Language for Object-Oriented Petri Nets. #91-1. Department of Computer Science, University of Tasmania (1991)
16. MacKenzie, S., Zhang, S.X., Soukoreff, R.W.: Text entry using soft keyboards. Behaviour & Information Technology 18, 235–244 (1999)
17. Navarre, D., Palanque, P., Bastide, R.: A Tool-Supported Design Framework for Safety Critical Interactive Systems in Interacting with computers, vol. 15/3, pp. 309–328. Elsevier, Amsterdam (2003)
18. Navarre, D., Palanque, P., Bastide, R.: A Formal Description Technique for the Behavioural Description of Interactive Applications Compliant with ARINC 661 Specifications. In: HCI-Aero 2004, Toulouse, France, 29 September-1st October 2004. CD-ROM proceedings (2004)
19. Palanque, P., Bernhaupt, R., Navarre, D., Ould, M., Winckler, M.: Supporting Usability Evaluation of Multimodal Man-Machine Interfaces for Space Ground Segment Applications Using Petri net Based Formal Specification. In: Ninth International Conference on Space Operations, Rome, Italy, June 18-22, CD-ROM proceedings (2006)
20. Petri, C.A.: Kommunikation mit Automaten. Technical Univ. Darmstadt (1962)
21. Reason, J.: Human Error. Cambridge University Press, Cambridge (1990)
22. Ríos, S.A., Velásquez, J.D., Yasuda, H., Aoki, T.: Web Site Off-Line Structure Reconfiguration: A Web User Browsing Analysis, in Knowledge-Based Intelligent Information and Engineering Systems. In: Gabrys, B., Howlett, R.J., Jain, L.C. (eds.) KES 2006. LNCS (LNAI), vol. 4252, pp. 371–378. Springer, Heidelberg (2006)
23. User Interface Management Systems, Eurographics Seminar, Seeheim, 1983. In: Pfaff, G. (ed.). Springer, Heildberg (1983)
24. van Dam, A.: Post-WIMP user interfaces. Commun. ACM 40(2), 63–67 (1997)

The Wrong Question to the Right People.
A Critical View of Severity Classification Methods in ATM Experimental Projects

Alberto Pasquini[1], Simone Pozzi[1,2], and Luca Save[1,3]

[1] Deep Blue srl, Rome, Italy
[2] Sapienza University of Rome, Department of Psychology of Social and
Developmental Processes, Rome, Italy
[3] University of Siena, Media and Communication Department, Siena, Italy
{alberto.pasquini,simone.pozzi,luca.save}@dblue.it

Abstract. The knowledge of operational experts plays a fundamental role in performing safety assessments in safety critical organizations. The complexity and socio-technical nature of such systems produce hazardous situations which require a thorough understanding of concrete operational scenarios and cannot be anticipated by simply analyzing single failures of specific functions. This paper addresses some limitations regarding state-of-the-art safety assessment techniques, with special reference to the use of severity classes associated to specific outcomes (e.g. accident, incident, no safety effect, etc.). Such classes tend to assume a linear link between single hazards considered in isolation and specified consequences for safety, thus neglecting the intrinsic complexity of the systems under analysis and reducing the opportunities for an effective involvement of operational experts. An alternative approach is proposed to overcome these limitations, by allowing operational people to prioritize the severity of hazards observed in concrete operational scenarios and by involving them in the definition of the possible means of mitigation.

1 Introduction

Every time a new system is introduced or an existing system is significantly modified, a safety assessment must be performed to identify if potential new risks are introduced as a result of the innovation. Safety assessments, especially in complex socio-technical domains such as Air Traffic Management (ATM), always require some kind of involvement of people with operational experience (e.g. controllers and pilots) whose knowledge is deemed essential for an adequate understanding and evaluation of risk. However most of the standard safety assessment techniques adopt as a central strategy the use of a safety matrix, aimed at classifying each hazard in terms of its expected severity and acceptable frequency. This method is generally intended to rely on *expert judgment* for an appropriate evaluation of the severity of hazards, but operational experts tend to experience difficulties when working at such a task. The assessment of severity is normally based on so-called *severity classification schemes* which identify a set of *severity classes*. Each class is associated to a different severity level and to a specific outcome (e.g. accident, incident, no safety effect, etc.).

M.D. Harrison and M.-A. Sujan (Eds.): SAFECOMP 2008, LNCS 5219, pp. 387–400, 2008.

In this paper we elaborate on our experience with severity classification schemes in the ATM domain to discuss the reasons why operational experts cannot easily use these schemes: (i) Hazards are typically identified as failures to a single function of the system, without considering the potential interactions of such function with other parts of the system, (ii) The severity of each hazard is assessed by considering its potential "final" effect, assuming only a linear chain of events and infringed barriers, thus neglecting the non linear dynamics of incidental scenarios.

2   Accident Models and Limits of Probabilistic Risk Assessment

In recent years a number of theoretical contributions have investigated the complex nature of accidents in socio-technical and safety critical systems like nuclear power plants, chemical industry and transportation systems. These contributions pointed out the limits of accident models based on linear sequences of events and cause-consequence configurations. The seminal studies of Charles Perrow [1] revealed that accidents can be seen as due too unexpected combinations or aggregations of events, named *complex interactions*. More recently Reason's Swiss Cheese Model [2, 3] has been considered successful in representing accidents as the result of combined failures at different levels in an organization, including unsafe acts by front-line operators and latent conditions such as weakened barriers [4, 5] and defences. Finally, other models like FRAM (Functional Resonance Accident Model) [6] or STAMP (System-Theoretic Accident Model and Processes) [4] highlighted the emergent nature of failures, which are often the result of dysfunctional interactions between different parts of the system, rather than simple malfunctions of specific components.

Compared to these theoretical advancements, there has been no comparable development in state of the art risk assessment techniques. Most of these techniques are based on a PRA approach (probabilistic risk assessment), i.e. they adopt as a central concept the well known definition *Risk=Severity x Frequency*. In such definition both the *severity* and the *frequency* are referred to the potential negative effects of the hazards which can be experienced by a certain system. Thus, in a typical safety assessment, hazards are defined as failures of one or more functions to be mitigated by reducing the frequency of their occurrence and/or the severity of their effects. The overall *level of risk* achieved by the system is the result of the aggregation of the risks identified for each specific hazard.

While this approach is theoretically appropriate for close or simple systems essentially made of hardware components, the application to complex socio-technical systems is problematic, as it relies on a linear representation of hazardous events which is inconsistent with the complex accident models mentioned above.

2.1   Assumed Linear Link Between Hazards and Their Effects

No matter which is the specific graphical notation adopted, PRA typically relies on models of accidents and incidents based on linear chains of events, representing the notion that the preceding event or condition must be present for the subsequent event to occur, i.e. if event X had not occurred than the following event Y would not have occurred [4]. These event-based models make it difficult to incorporate non-linear relationships (e.g. feedback between system components). A linear link between an identified hazard for a specific function (e.g. a technical failure or a human error) and a *final effect* of the hazard itself (e.g. a minor incident, or a severe accident) oversimplifies the relationship between a single component failure and its possible negative safety effects on a system level. I.e. it disregards the well-known notion that a failure to a single function can never be considered as the sole cause of a negative effect for the safety of a system. As the *final effect* is used as a criterion to assess the severity of a specific hazard, this considerably influences the final results of the assessment. As argued by Leveson [4], this approach -which was appropriate in process industry design (e.g. nuclear power plants)- is largely insufficient in other kind of systems in which emergent configurations of different kind or resources (humans, mechanical, procedural) are essential elements of both the correct and unsafe functioning of an organization [6].

2.2 Initiating Events in the Chain Assumed to Be Mutually Exclusive

A well known limitation of event based models (e.g. Fault Tree Analysis) is that *basic events* are usually assumed to be mutually exclusive. While this assumption simplifies the mathematics in a PRA, it may not match the reality. Leveson explained how seemingly independent failures may have common systemic causes that result in coincident failures [4]. For instance in the Bophal accident, what might have appeared as an unlikely coincidence of failures was engendered by common design and management decisions.

This methodological limitation of PRA is strictly related to the one mentioned before. Assuming the *basic events* as mutually exclusive in the determination of an accident considerably simplifies the task of modelling cause-effect configurations, thus making simpler the numerical definition of the risk associated to each failure at component level. However it can hide critical interactions between different functions or components, which are essential for identifying the appropriate mitigation means.

2.3 Functional Failures and Dysfunctional Interactions

Traditional PRAs focus on *functional failures*, i.e. on the non-performance or inability of specific components to perform their intended functions. However, the more complex safety critical systems have become, the more accidents have been determined by *dysfunctional interactions* [4]. Dysfunctional interactions happen when system elements perform as it is expected (i.e. as specified by requirements) but still the overall system behavior results to be unsafe. Accidents happen not only because a pilot deviates from a specified procedure or because a hardware component does not perform as in its specifications. Accidents may be engendered by a critical interaction among different components (electromechanical, digital, human). If the safety assessment is exclusively focussed on functions and component failures, very little insight is produced in order to mitigate the hazardous situations deriving from dysfunctional interactions.

3 Safety Assessment Methodology in Air Traffic Management

A prerequisite for performing a safety assessment based on a PRA approach is that of identifying a relationship between a set of identified failures for each specific function and a set of possible consequences. In the Air Traffic Management world, this is typically accomplished by filling in Functional Hazards Assessment (FHA) tables and by elaborating them with cause-effects propagation models, such as Fault Tree Analysis (FTA) and Event Tree Analysis (ETA).

3.1 The Assessment of Severity

SAM (Safety Assessment Methodology) [7] is the standard method for safety assessment in ATM promoted by EUROCONTROL. It is made up of three main phases: Functional Hazard Assessment (FHA), Preliminary System Safety (PSSA), System Safety Assessment (SSA) (see central column of Figure 1). The phases are in parallel with the lifecycle of the system under assessment (see left column in Figure 1). This paper is mainly focused on the first phase of the SAM, i.e. the FHA.

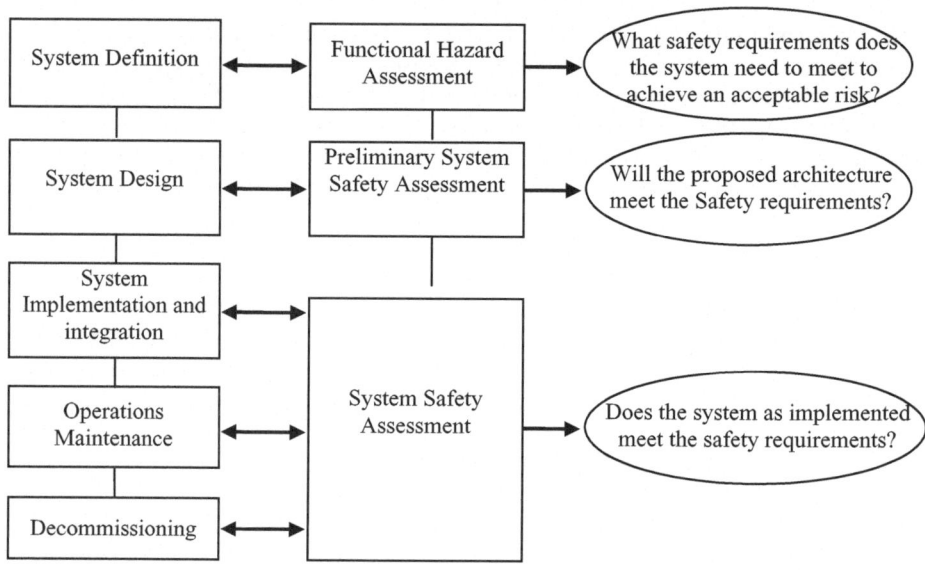

Fig. 1. The SAM Methodology

The main goal of an FHA is specifying a set of safety objectives. These are defined by following five sub-phases:

1. Identify all potential hazards associated with the system
2. Identify hazard effects on operations, including the effect on aircraft operations
3. Assess the severity of each hazard effect
4. Specify Safety Objectives, i.e. determine the maximum frequency of a hazards' occurrence
5. Assess the overall foreseen risk associated with introducing the change or new system.

Operational experts involvement is quite easily achieved in the first step (identify hazards), while the second phase (identify hazard effects) is more difficult and the third phase (assess severity) can become extremely challenging. Operational experts (typically air traffic controller and/or pilots) are supposed to identify, in collaboration with technical and safety experts, the effect of each hazard identified in phase 1. Effects are then included in textual format in a specific column of the FHA table. Subsequently the experts are required to classify each of these effects in terms of severity, by using the SAM Severity Classification Scheme. The scheme identifies 5 different Severity Classes (SC), from the most severe to the least sever:

- SC1 Accidents [most severe]
- SC2 Serious Incidents
- SC3 Major incidents
- SC4 Significant incidents
- SC5 No Immediate Effect on Safety [least severe]

In principle the same hazard can have more than one effect, based on contextual conditions. A typical example is the differentiation of the effects of the same hazard, based on traffic conditions (e.g. low vs high traffic) or weather conditions. However the FHA table should identify a specific SC for each effect, without any particular attention if two effects are produced by the same source hazard. It is to be noted that the SCs are the same adopted in the EUROCONTROL requirements on safety occurrence reporting (ESARR 2 [8]), i.e. they are used by national service providers to classify real occurrences experienced in operational air traffic control centres.

3.2 Problems with the Use of Severity Classes

According to our experience (see case studies in section 4), the severity classification scheme is not easily applied in the Safety Assessment. While SCs are fit for purpose when reporting and classifying real occurrences, they are very difficult to use when assessing the safety of "pre-operational" systems. The most problematic aspect is the assumed linear link between a specific hazard and its possible effects. A specific failure – be it a technical failure or a human error – can never be considered as the sole cause of an accident. For an accident to occur, a hazard very often combines with several other hazards and contextual conditions. However, when adopting the functional approach which is typical of PRA, hazards generally does correspond to specific failures. We deal with single failures that could at the same time cause an accident (SC1), different kinds of incidents (SC2, SC3 and SC4) or even no immediate effect on safety (SC5). This is a commonly well recognised point, as demonstrated by the emphasis safety management systems place on near-miss events collection and analysis [9]. A near-miss usually shares the same causal factors with real incidents, where mostly contextual (sometimes even fortuitous) factors determine the different outcomes (i.e. no or very limited damage in near miss). Deriving consequences from each single hazard considered in isolation neglects the above reasoning on incident dynamics. In theory, any hazard can result in a serious incident or in an accident, depending on the way it interacts with other system weaknesses.

Operational experts are normally able to provide very detailed accounts on critical situations and can give valuable insights on the possible consequences of failures or dysfunctional interactions in concrete operational contexts. However, when faced with the task of classifying a single failure in terms of the 5 SCs, they manifest uncertainties, expressed with sentences like "it depends on…it depends how…". If forced to make a choice, they generally tend to produce classifications that reflect certain assumptions about the contextual situation, or they provide a rationale justifying their answer. Neither assumption nor the rationale will be considered in the following of the assessment. The resulting classification will instead directly influence the setting of safety objectives and the definition of safety requirements.

Another spontaneous strategy is that of ranking the severity with respect to other hazards. In so doing, experts do not take into account the SC labels (accident, serious incident, etc), but rather reason on a priority ordering. The SAM guidance does acknowledge that the same hazard can have different effects and then different severities. It is specified that SCs should be assigned to the hazard effects, rather than to the hazard itself. The most commonly used method consists of identifying - based on expert judgment - the worst credible effect of each hazard, then in setting safety objectives taking into account only that effect[1].

The worst credible effect in the given environment of operation should determine the severity class leading to setting of the Safety Objective, using expert judgement. It means that somehow the probability of the hazard leading to certain effect (Pe) has been taken into account when deciding the worst credible severity of the hazard effect [10]. Even if the severity classification should not be influenced by considerations on the acceptable frequency, according to this quote one could claim that the decision on what is the worst credible effect is instead linked to considerations on the actual hazard frequency[2]. There is an implicit recommendation not to select a too much severe effect, unless its occurrence is not considered reasonably frequent. Ignoring such an implicit recommendation is likely to produce overambitious safety objectives.

4   Asking the Wrong Question to the Right People

The case studies presented in this section are both pertaining to safety assessment experiences in ATM related projects. The first case is the development of an FHA aimed at assessing the improvement of a Short Term Conflict Alert (STCA) in an European military ATC unit. The second case concerns the overall assessment process of an Airborne Separation Assurance System concept (ASAS), in the context of the European Program "Mediterranean Free Flight". These case studies provide evidence of some of the methodological limitations described in previous sections. They also document our attempts to overcome such limitations and propose an alternative approach.

Our approach is inspired by authors like Erik Hollnagel and Nancy Leveson and by their recent efforts to propose methods more in line with state of the art accident models (e.g. FRAM and STAMP). Our main strategy is that of extensively rely on the domain knowledge of expert operational personnel (e.g. controllers and pilots). The method we suggest exploits a scenario-based approach [11, 12]. The use of scenarios

---

[1] Note that the SAM guidance material proposes 4 different methods for setting safety objectives: Quantitative Method, Prescriptive Method, Criticality Method and Qualitative Method. For the sake of simplicity in this context we only refer to the last one.

[2] On the contrary, a rigorous application of the methods requires that the severity of hazard effects is assessed before. The acceptable frequencies are only established afterwards, based on a Severity x Frequency matrix.

is essential to place hazards and their possible consequences in concrete operational situations.

4.1 Case Study 1: Assessment of a New STCA for a Military Unit

Case study 1 concerns a safety assessment made in September 2006 for the introduction of an improved Short Term Conflict Alert (STCA) to be installed in a European military ATC unit. STCA is a system that assists the controller in maintaining separation between aircraft, by generating on the controller's display an alert of a potential infringement of standard separation minima. The military unit under analysis was already equipped with a modern ATC system including an STCA. However, the specific needs of the military environment (military formation flights, aerobatic maneuvers, etc.) created a large number of nuisance alert. The safety assessment was mainly focused on the safety impact of new technical solutions.

Essential part of the safety assessment was an FHA workshop, based on a number of brainstorming sessions attended by 10 people, including safety, technical and operational experts (i.e. military controllers and pilots). Main objectives were: (1) Identifying the most relevant hazards, (2) Understanding their effects on the ATM system, (3) Assessing the severity of their effects, (4) Identifying possible mitigation means.

The workshop profited from a scenario-based approach, consistently with what already made in the framework of other studies [12]. As an input to the hazard identification phase, a set of seven military related scenarios were identified, in collaboration with a controller and a technical expert. The scenarios were textual descriptions of typical operational situations representative of the military environment under analysis (an example is in Figure 2). Additional cells in the table provided information on the expected behavior of current STCA and on the technical solutions included in the improved STCA, in order to manage the specific situation.

The scenarios served to provide a description of the new system, from an operational point of view, to controllers who were not particularly familiar with STCA functioning. A second purpose was to support hazard identification brainstorming, by providing a concrete operational context. This purpose, in particular, was an attempt to integrate the functional approach, which requires starting from single functions of the system and thinking about their possible failures. The functional approach was actually maintained. However the scenarios complemented the functional perspective, as technical failures and human errors were imagined in concrete situations, allowing engineers and operational experts to derive also more complex hazards, like combination of different hazards or dysfunctional interactions.

The output of the brainstorming sessions was a list of 27 hazards, including a description of possible operational consequences and effects on safety, which were included in a typical FHA table [10, Appendix A, pp. A4-A5]. Example of hazards were: "Duplicate Mode A", "Lost Wingman", "Incorrect military formation detection", "Incorrect SSR code list input", "Controller not aware of STCA suppressed for specific aircraft", etc. In the FHA table, hazards were grouped to keep a reference to the scenario in which they were identified. According to the established method, the hazards identification phase was followed by the assessment of hazard severity and by the discussion about possible mitigation means. At the end all the results were included in the FHA table.

OS 2 - AREA TO AIRWAY	
Description	Traffic manoeuvring inside a military area next to a civil airway (ATS routes) with lateral or vertical manoeuvres.
Operational implications	Short reaction time for controllers to react if A/C penetrates civilian airspace. High speed manoeuvring, high ROC/ROD and steep turns versus steady flight profile. Aerobatics being performed both by singletons and by formation flights. Need for ATCOs to input BFL (Block Flight Levels).
STCA implications	Nuisance alerts are generated inside formations. Nuisance alerts due to excessive prediction times and high speed manoeuvring. BFL to be taken into account at the CWP Linear (any) prediction less accurate for the military traffic. If aerobatics are performed in formation, split tracks can occur.
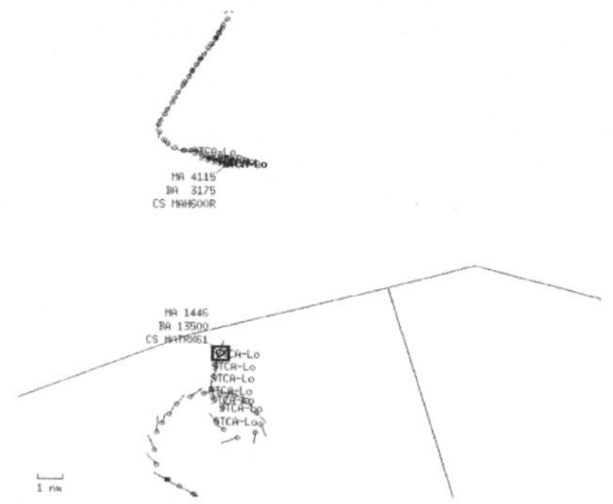	
Technical solution adopted in the new STCA	Creation of buffer zones around aerobatic areas using wider parameters as the Aircraft approaches the boundaries of the area. Use of BFL as in the current system. Dynamic activation/de-activation of STCA regions (improved FUA Level3).

Fig. 2. Example of a scenario template used during an FHA brainstorming session

4.1.1 The Decision to Give Up with Severity Classes

While in the hazard identification phase, the workshop attendees were very active in generating ideas and in providing descriptions of the possible consequences of the hazards, much more difficulties were experienced when the experts when confronted with the Severity Classification Scheme. First of all, it turned out to be difficult to identify the specific effect on safety of each hazard. Then experts stated that none of the hazards would have been the sole cause of an accident, but nearly all of them could potentially play a role in determining an accident. In addition, the categories *serious incident, major accidents, significant incidents* or *no immediate effect on safety* were considered difficult or impossible to apply. Even the safety indicators provided in the scheme (e.g. Effects on air navigation services, Exposure and Recovery) were not considered helpful, as the associated descriptions of possible hazards effects are obviously expressed in general and abstract terms: e.g "partial inability to provide or maintain a safe service" or "hazard may persist for a substantial period of time". For example defining what is a "substantial period of time" will totally depend on subjective evaluations of the specific operational circumstances experienced and will not necessary imply the risk of producing a serious accident.

The limited time available for the workshop (one day and half in total) and the feeling of being stuck with hazard classification resulted in a spontaneous solution directly proposed by some of the attendees. While both operational and technical experts were not able to classify hazards in terms of the SCs, they had no difficulties in distinguishing between *high severity* and *low severity* hazards. Furthermore they remarked the importance of establishing a priority between hazards with an immediate need for a mitigation and hazards that could have been analyzed later. In other words, they proposed to shortlist a number of candidates for the following phases. A further distinction was made between hazards the mitigation of which was considered easier and hazards requiring further study. Even though this solution could appear not rigorous in methodological terms, it highlights the strong link perceived between hazards safety effects and the phase of designing mitigation actions.

4.2 Case Study 2: Assessment of ASAS Spacing Concepts in MFF

MFF was a large project of six years duration recently concluded, sponsored by the European Union under the TenT Programme. MFF was co-ordinated by ENAV - the Italian Air Traffic Control service provider - and involved several air traffic service providers, especially from the Mediterranean area, and EUROCONTROL. The scope of MFF was to define, test and validate operational concepts and procedures for more efficient use of airspace through the delegation of some tasks related to separation assurance, relying on concepts like Free Routes, ASAS Spacing & Separation, Free Flight. It focused on the application of those procedures in the particular geographical context of the Mediterranean area. The new operational concepts and related procedures were defined in the early phases of the project [13] and their fitness-for-purpose was evaluated through a set of validation exercises, with an iterative process of concept refinement and validation. This included several cycles of Model Based Simulations, three sets of Real Time Simulations (RTS), three Safety Cases, and an extensive set of Flight Trials (FT). Cockpit simulations were used in support of both RTS and FT.

The research issue we faced in this project was mainly due to the experimental nature of procedures and applications to be assessed. The introduction of the ASAS procedures profoundly changes parts of the existing ATM system, including changes in hazardous conditions and safety issues. Given the novelty of ASAS applications, there was no previous experience of them, nor any existing system with similar characteristics. The safety assessment process was then developed to face two complementary constraints.

1. Controllers needed to be familiarized with the new procedures and applications, so that they could contribute to the safety assessment as experts.
2. No experience was available on the system behavior, so a variety of simulation exercises was set up, in order to identify potential hazardous conditions hard to anticipate in the design phase. These simulation exercises could not replicate the complexity of a real system, but still some system elements could be put in place and observed while working together.

The integration of Real Time Simulations with the safety assessment process seemed a sound solution for both of the above problems [for more information on the MFF safety assessment process and on the use of safety scenarios in RTS, please refer to 11, 12]). The key aspect of such integration was the injection of a limited sample of hazards in the simulation through the implementation of safety scenarios. The major difficulty was to reconstruct a realistic situation where the procedure and the related hazard could be analyzed from a systemic point of view, preserving all contextual factors that shape controller's behavior. Safety scenarios were then used to avoid the assessment of hazards in isolation, so that credible situations could be presented to controllers. The safety scenarios included events such as system failures, pilots and controller errors, and other operational problems (see Table 1 below for an example of scenario story board).

Table 1. Story board and actions for a safety scenarios

Hazard Identification Code: SA2		Airspace Sector: EW
Time		Events
9.48	Accomplice Pilot Action	AZA123 asks to descend to FL290 for technical reasons
9.50	If Accomplice Pilot Action	IBE3674 and AFR432 are cleared to self-separate while crossing
9.50	then Possible Event	AZA123 interfere with IBE3674 and AFR432 (self-separation on-going)

They provided at least two immediate benefits. First, they gave experts an opportunity to reason about what did not work when the system failed, thus supporting the safety analysts in clarifying some aspects of the hazards. Second the safety analysts had the opportunity to learn through the direct observation of the controller behaviour during the exercises and to obtain information directly on a series of dedicated events.

However, if we get back to the main line of reasoning of this paper, what was observed during the RTS could not be considered satisfactory as far as the severity assessment was concerned. Although the RTS context allowed making post-hoc observations of the events and not just guessing the possible effects of hazards, using the 5 severity classes resulted to be problematic. Firstly the most severe one (SC1) - corresponding to an accident- is simply not simulated in the RTS environment. The closest the simulation can get to an accident is when two aircraft pass one through the other, which in the simulated world results in no damage to any of the two. The two aircraft simply keep flying on their track after "the collision". More important, the rating on the other 4 levels was very difficult even when adopting the two basic criteria indicated to rate real occurrences, namely (i) percentage of separation infringement and (ii) whether the controller had detected the loss of separation[3]. The two criteria encounter the same drawback we mentioned in the previous section, that is they both

---

[3] The two criteria are indicated in the ESARR2 [8] and ESARR4 [10] Severity Classification Schemes from which the SAM Scheme has been derived.

address the severity of the end result of an event, which is often the product of highly specific contextual factors. In other words, safety analysts could not simply observe the RTS event and then rate the severity on the basis of the separation infringed and of the controller detection, as this would have implied rating *the factors that had produced the event in the specific RTS setting* rather than assessing *the severity of a single hazard*. Again, we tried to partially overcome this limitation by profiting of the controllers' expertise. Two workshop sessions were organised after the end of two major simulations, with the objective of reviewing the information gathered on the hazards. Hazards were presented together with the safety scenarios, so that experts could reason about the single hazards not in isolation, but bearing in mind a more realistic situation, that is in interaction with the other system elements. As in the previous case study with the STCA, experts needed to reason about concrete cases in order to draw meaningful estimate on the severity. In the MFF case, scenarios (which had been in a sense validated in the RTS) provided these concrete cases.

The lesson we draw from the MFF case is that the severity rating encountered difficulties in its application even in a case where it could be applied as a post-event classification (i.e. assessing events that were implemented in a simulation). In our opinion, these difficulties stem from the nature itself of the assessment, that is from the fact that experts are asked to assess the severity of an event as representative of a hazard, whilst experts question this very link between hazard and event. They find it hard to trace a linear link between the hazard and the event, and need to draw their estimates from more complex situations, or better said from more realistic situations.

5 An Alternative Approach to Safety Assessment

In previous sections we have presented some issues we faced in the safety assessment process, in particular those due to the severity classification scheme. In this section we would like to draw some tentative lessons learnt from the above discussion.

5.1 Assess Hazardous Situations Rather Than Single Hazards

A direct and simple link between a specific hazard and a given effect is a rare case in complex socio-technical systems. It is usually a complex configuration that jeopardizes the system defenses. However, it is almost impossible to predefine in formal terms these configurations. They can be somehow anticipated only by means of a thorough operational knowledge. Thus technical failures or human errors are better understood only if analyzed in the context of concrete operational scenarios either describing past events or envisaging future situations.

The traditional functional approach, i.e. consider individually all system functions and imagining their possible failures, is an essential starting point of all safety assessments. Nevertheless it should be always complemented by the analysis of the same events in the context of wider *hazardous situations*, which are better handled and understood by operational experts. Such an integrated approach presents at least two main advantages:

1. It gives more opportunity to identify not only the simple functional failures, but also those dysfunctional interactions which generally represent a more insidious threat for the safety of a complex system.
2. It allows the assessors to work jointly on three different aspects of a traditional safety assessment, i.e. hazards, effects and severity. The distinction between hazards and effects does not make sense from an operational point of view. What is seen as the causal factor in a certain context can be easily perceived as the consequence in a different one. With respect to the assessment of severity, critical scenarios (i.e. hazardous situations) appear as the only meaningful context to express a motivated judgment.

5.2 Prioritize Hazards Rather Than Classify Severity

A hidden assumption of functional approach methods is the need to perform an *exhaustive* assessment of all possible hazards. A corollary of such assumption is that analyzing *all* the single functions of a system and identifying *all* their potential failures will ensure that a complete assessment of risks has been performed. Nevertheless the identification of all potential hazards is far from being a viable solution for a variety of reasons.

First of all, socio-technical systems like air traffic management systems are too complex for a detailed identification of all system functions. Secondly, hazards do not derive only from failures of single functions but also from dysfunctional interactions among perfectly working functions. These cannot be identified by analyzing each function separately. In addition, due to their emerging nature, they are anyhow difficult to anticipate in pre-operational phases. Last (but not least) the time available for a safety assessment is generally limited in real situation, so an implicit prioritization is always made.

Based on these considerations, a detailed classification of each hazard in terms of the 5 SCs appears less important than a careful prioritization of what has been identified. The list of hazards can be never considered exhaustive and there is generally no time available to cover all hazards with a specific safety objective. Thus, it is of paramount importance that the most urgent hazards to be mitigated are identified, no matter which is their rating on the Risk Classification Scheme. In analogy with what has been described in Case Study 1, a subset of hazards can be classified as urgent, to make sure that fundamental design decisions are not made before these have been adequately considered. The remaining hazards - at least those which have been considered as relevant - should also be recorded, at least to make sure that they are not forgotten in following design stages.

5.3 Consider Safety Objectives and Mitigation Means Jointly

A sharp separation between safety assessment and design processes does not appear realistic. From the one end, ensuring that safety is independent from production pressures is an important requisite for the credibility of safety targets. In addition, the well known phenomenon of *risk homeostasis* [14] should be always prevented, in order to ensure that safety improvements are not automatically converted in production benefits. On the other end, looking after safety also means thinking about alternative design solutions, by considering measures on either the technical, the procedural and the training side. The same safety target can be achieved with different design solutions and with considerable variations in terms of cost and availability. Thus practical considerations suggest maintaining an adequate communication flow between safety and design at all stages of safety assessment. Separation and independence is more a requirement for different organizational functions, rather than a prescribed working method.

The need to consider jointly safety objectives and mitigation means is in contrast with traditional FHA, as the FHA is supposed to reason only in terms of abstract functions, without any speculation on how a specific function will be implemented. As for the analysis of hazards and for the identification of severity classes, the approach suggested in this paper goes in a different direction. In our opinion, if pilots and controllers' experience is essential for identifying the possible hazards effects on the system, it is hardly understandable why their expert knowledge should not be used to assess the safety benefits of various design solutions. This implies that mitigation means are considered also at the FHA level, to make sure that operational experts can actually contribute to the definition of safety objectives.

6 Conclusions

In this paper we move from the discussion of what appears to us as a fallacy in current state-of-the-art safety assessment, that is the severity assessment seems to blatantly contradict last-generation safety theories. The line of reasoning is then developed by showing the impact of such fallacy in two case studies. We also present some practical solutions we devised to mitigate the issue. We are well aware that such solutions are mostly *ad hoc* adaptations, far from representing "the solution" to the point we raised.

In our opinion the key tension we encountered in the safety assessment process is between analytical techniques and a more holistic vision. On one side, we need analytical techniques to pinpoint safety threats. On the other, these analyses "tears the system apart" and tends to overlook the fact that in reality the system elements will work together. To address the actual functioning of the system we then need more holistic techniques, to "reassemble" what we have separated for clarity's sake. Our proposal is to ground this holistic view in narrative scenarios, to show system interactions as they happen in the everyday functioning. Future research should address the tension between the two polarities – analytical *versus* holistic – and devise solutions to integrate the two perspectives. At the present moment we see the two polarities as representing a contradictory tension we have to deal with, most likely by reflecting on their complementarities rather than opting for one of the two.

Acknowledgements. The authors would like to express gratitude to the Eurocontrol SPIN Task Force representatives who promoted and supported the FHA study regarding the STCA. Special thanks are due to the ATCC Semmerzake team for hosting the FHA workshop and actively contributing to it. We would also like to thank all the colleagues of the MFF project for the fruitful collaboration on the activity. The MFF project was partially funded by the EU under the TEN-T program. The authors gratefully acknowledge the support provided to this work by the EU project "ReSIST: Resilience for Survivability in IST".

References

1. Perrow, C.: Normal Accidents: Living with High-Risk Technologies. Basic Books, 2nd edn. Princeton University Press, Princeton (1984)
2. Reason, J.T.: Human error. Cambridge University Press, Cambridge (1990)
3. Reason, J.T.: Managing the risks of organizational accidents. Ashgate Publishing Limited, Hampshire (1997)
4. Leveson, N.G.: A New Accident Model for Engineering Safer Systems. Safety Science 42(4), 237–270 (2004)
5. Leveson, N.G.: Safeware. System safety and computers. Addison Wesley Publishing Company, Reading (1995)
6. Hollnagel, E.: Barriers and accident prevention. Ashgate, Hampshire (2004)
7. EUROCONTROL, Air Navigation System Safety Assessment Methodology (SAM) (2006)
8. EUROCONTROL, ESARR 2 - EUROCONTROL Safety Regulatory Requirement. Reporting and Assessment of Safety Occurrences in ATM (2000)
9. Van der Shaaf, T.W., Lucas, D.A., Hale, A.R.: Near miss reporting as a safety tool. Butterworth-Heinemann, Oxford (1991)
10. EUROCONTROL, Air Navigation System Safety Assessment Methodology (SAM) (2004)
11. Pasquini, A., Pozzi, S., McAuley, G.: Eliciting Information for Safety Assessment. Safety Science (in press)
12. Pasquini, A., Pozzi, S.: Evaluation of Air Traffic Management Procedures - Safety Assessment in an Experimental Environment. Reliability Engineering & System Safety 89(1), 105–117 (2005)
13. Mediterranean Free Flight, MFF Operational Concepts & Requirements (2001)
14. Wilde, G.J.S.: Target Risk. Dealing with the Danger of Death, Disease and Damage in Everyday Decisions. PDE Publications, Toronto, Canada (1994)

A Context-Aware Mandatory Access Control Model for Multilevel Security Environments

Jafar Haadi Jafarian, Morteza Amini, and Rasool Jalili

Department of Computer Engineering, Sharif University of Technology, Tehran, Iran
{jafarian@ce,m_amini@ce,jalili@}sharif.edu

Abstract. Mandatory access control models have traditionally been employed as a robust security mechanism in multilevel security environments like military domains. In traditional mandatory models, the security classes associated with entities are context-insensitive. However, context-sensitivity of security classes may be required in some environments. Moreover, as computing technology becomes more pervasive, flexible access control mechanisms are needed. Unlike traditional approaches for access control, such access decisions depend on the combination of the required credentials of users and the context of the system. Incorporating context-awareness into mandatory access control models results in a model appropriate for handling such context-aware policies and context-sensitive class association mostly needed in multilevel security environments. In this paper, we introduce a context-aware mandatory access control model (CAMAC) capable of dynamic adaptation of access control policies to the context, and handling context-sensitive class association, in addition to preservation of confidentiality and integrity. One of the most significant characteristics of the model is its high expressiveness which allows us to express various mandatory access control models such as Bell-LaPadula, Biba, Dion, and Chinese Wall with it.

Keywords: Mandatory Access Control, Context-Awareness, Confidentiality, Integrity.

1 Introduction

As computing technology becomes more pervasive and mobile services are deployed, applications will need flexible access control mechanisms. Unlike traditional approaches for access control, access decisions for these applications will depend on the combination of the required credentials of users and the context and state of the system.

Unlike discretionary and role-based access control, mandatory access control models directly address multilevel security environments where information is classified based on its sensitivity; although they have been deployed in commercial sectors too.

Numerous context-aware access control models are presented in literature. Meanwhile, none of these models directly target new security requirements of multilevel environments; while some of them are applicable to such environments with considerable effort. Since mandatory access control has traditionally been used in these environments, a context-aware mandatory access control model seems the most appropriate choice in this regard.

M.D. Harrison and M.-A. Sujan (Eds.): SAFECOMP 2008, LNCS 5219, pp. 401–414, 2008.

In traditional mandatory access control models, except for some special cases, the security classes associated with entities are usually insensitive to context. However, in some systems, we may need context-sensitive association of security classes. For instance, in most intelligence agencies, the security level of documents decreases by the elapse of time. Moreover, as computing technology becomes more pervasive, applications in multilevel security domains need more flexible mandatory access control policies. Incorporating context-awareness into mandatory access control models gives rise to a flexible and expressive model suitable for management of such context-aware policies and dynamic class associations.

In this paper, we introduce CAMAC as a context-aware mandatory access control model capable of dynamic adaptation of policies with the context and handling context-sensitive class association, in addition to preserving confidentiality and integrity. In fact, CAMAC uses Bell-LaPadula and Biba properties to preserve confidentiality and integrity of information.

The rest of the paper is organized as follows. Section 2 introduces a brief survey on context-aware access control models. In Section 3, CAMAC model is formally described. In section 4, the expressiveness of CAMAC model is scrutinized. In section 5, evaluation of the model is introduced followed by our conclusion.

2 Related Work

Various mandatory access control models and policies have been introduced in literature. Bell-LaPadula [1, 2], Biba [3], Dion [4] and Chinese Wall [5] are examples of such models and policies. Bell-LaPadula and Biba constitute the infrastructure of the CAMAC model, although the definition used here is mostly based on a minimalist approach introduced by Sandhu in [6, 7].

Many researches are targeted to applying context-awareness to the RBAC model. Kumar et al. [8] proposed a context-sensitive RBAC model that enables traditional RBAC to enforce more complicated security policies dependent on the context of an attempted operation. Al-kahtani et al. [9] proposed the RB-RBAC model, performing role assignment dynamically based on users' attributes or other constraints on roles. GRBAC, Generalized RBAC, [10] incorporates three types of roles; subject roles corresponds to the traditional RBAC roles, object roles which are used to categorize objects, and environment roles to capture environmental or contextual information. Context-aware access control is achieved by employment of these role types in specification of access control policies. Zhang et al. [11] proposed DRBAC, a dynamic context-aware access control for pervasive applications. In DRBAC, there is a role state machine for each user and a permission state machine for each role. Changes in context trigger transitions in the state machines. Therefore, user's role and role's permission are determined according to the context. Georgiadis et al. [12] present a team-based access control model that is aware of contextual information associated with activities in applications. Hu et al. [13] developed a context-aware access control model for distributed healthcare applications. The model defines the notion of context type and context constraint to provide context-aware access control.

Ray et al. [14] proposed a location-based mandatory access control model by extending Bell-LaPadula model with the notion of location. In particular, every location is associated with a confidentiality level and Bell-LaPadula no read-up and no write-down properties are extended by taking confidentiality levels of locations into consideration. Based on Baldauf et al.'s classification of context-aware systems [15], location-based mandatory access control model can be categorized as a location-aware system.

3 CAMAC: A Context-Aware Mandatory Access Control Model

Through an example application enabled by a pervasive computing infrastructure in a smart building of a military environment, we discuss motivation for access control models such as ours. The building has many rooms including administration offices, campuses, etc. Sensors in the building can capture, process and store a variety of information about the building, the users, and their activities. Pervasive applications in such an environment allow military forces to access resources/information from any locations at anytime using mobile devices (PDAs) and wireless networks. While classification is still the basis for all the access control decisions, users' context information and application state should also be considered. For example, an officer can only control the audio/video equipment in a conference room if she/he is scheduled to present in that room at that time by the manager in charge. In such applications, privileges assigned to the user will change as context changes. The example above embodies many of the key ideas of the research presented in this paper. To maintain system security for such a pervasive application, we have to dynamically adapt access permissions granted to users as context information changes. Context information here includes environment of the user such as location and time that the user access the resource and system information such as CPU usage and network bandwidth. The traditional mandatory models do not directly address the requirements of such an application and although many context-aware access controls have been proposed in literature, they are not appropriate for environments where security is directly contingent upon classification. This paper aims at presenting a flexible and expressive model appropriate for multilevel security environments where classification of information is an integral property of the environment.

CAMAC is a context-aware mandatory access control model which utilizes contextual information to enhance expressiveness and flexibility of traditional mandatory access control models. Incorporation of context-awareness into the model changes traditional models in two separate ways. Firstly, contextual information can be used to define more sophisticated access control policies. As an example, an access control policy might require that for a subject to acquire a read access to an object, some timing restrictions must be satisfied. CAMAC model allows definition of such sophisticated access control policies. Secondly, the confidentiality and integrity level of entities can change based on contextual information. In traditional mandatory models, the levels initially assigned to entities are not allowed to change based on the circumstances. For instance, confidentiality level of objects might decrease as their lifetime increases (and so become accessible to less trustworthy subjects). CAMAC also allows such dynamic level association based on contextual information.

3.1 Formal Definition of CAMAC

CAMAC model can be formally described as a ten-tuple:

⟨*EntitySet, RepOf, ConfLvl, IntegLvl, λ, ω, ContextPredicateSet, ContextSet, OperationSet*⟩

in which:

- *EntitySet* is the set of all entities in the system and is composed of four sets: *User, Subject, Object* and *Environment*. *User, Subject* and *Object* are the set of all users, subjects and objects in the system respectively. *Environment* set has only one member called *environment*.

- *RepOf: Subject → User* assigns to each subject the user who has initially initiated or activated it. In other words, for $s \in Subject$, $RepOf(s)$ represents the user on behalf of whom the subject s acts.

- *ConfLvl* is a finite ordered set of confidentiality levels[1] such as $\langle c_n, c_{n-1}, ..., c_1 \rangle$ in which c_n and c_1 are the highest and lowest levels respectively. As in Bell-LaPadula model, each user, subject and object is associated with a confidentiality level. It must be noted that there exist a difference between Bell-LaPadula and CAMAC in terms of confidentiality level. While Bell-LaPadula confidentiality levels are defined by two components (a classification and a set of categories), CAMAC confidentiality levels only include the first component, i.e. classification. In section 5 we show that the second component, set of categories, is contextual information and can be easily incorporated to the model as a context type.

- *IntegLvl* is a finite ordered set of integrity levels such as $\langle i_n, i_{n-1}, ..., i_1 \rangle$ in which i_n and i_1 are the highest and lowest levels respectively. As in the Biba model, each user, subject and object is associated with an integrity level. Moreover, the above difference also applies here; i.e. CAMAC integrity levels are defined by only a classification component.

- λ is a mapping function which associates each user, subject and object with a confidentiality level: $\lambda: User \cup Subject \cup Object \rightarrow ConfLvl$

- ω is a mapping function which associates each user, subject and object with an integrity level: $\omega: User \cup Subject \cup Object \rightarrow IntegLvl$.

- *ContextPredicateSet* is the set of current *Context predicates* in the system. Each context predicate is a statement about the value of a contextual attribute. More on context predicates will come in section 3.2.

- *ContextSet* is an ordered set of *context types*. A context type is a property related to every entity or a subset of existing entities in the system. A context type $ct \in ContextSet$ can be formally described a 5-tuple:
  $ct = \langle ValueSet_{ct}, OperatorDefinerSet_{ct}, RelatorSet_{ct}, EntityTypeSet_{ct}, LURSet_{ct} \rangle$
  More details on *context types* are given in section 3.3.

- *OperationSet* is the set of all operations in the system. An operation $OPR \in OperationSet$ can be formally defined as a pair: $OPR = \langle AccessMode_{OPR}, Constraint_{OPR} \rangle$
  More details on *OperationSet* are given in section 3.5.

3.2   Context Predicate

Each *context predicate* is a predicate which represents the value for a contextual attribute. We define a context predicate $cp \in ContextPredicateSet$ as a 4-tuple:

---

[1] Since Biba uses the term 'integrity level', for Bell-LaPadula, we prefer to use the term 'confidentiality level' instead of 'security level'.

$cp = \langle en, ct, r, v \rangle$

where $en \in \{User, Subject, Object, Environment, ValueSet_{ct_1},\ldots, ValueSet_{ct_n}\}$, $ct \in ContextSet$, $r \in RelatorSet_{ct}$, $v \in ValueSet_{ct}$, and $ct_1,\ldots,ct_n \in ContextSet$. For example, $\langle John, Location, Is, Classroom \rangle$ is a context predicate and indicates the current location of subject *John*.

Management and updating context predicates is the responsibility of *Context Management Unit (CMU)*. The details on the implementation of CMU are beyond the scope of this paper, and will be explained in another paper. Context Managing Framework [16], the SOCAM project [17], CASS project [18], CoBrA architecture [19], the Context Toolkit [20] can be used as an infrastructure in implementation of *CMU*. In general, we assume that CMU updates *ContextPredicateSet* based on changes of environment, users and system and therefore the consistency and accuracy of *ContextPredicateSet* is permanently preserved.

If $\langle E, X, R, V \rangle$ is a context predicate, $X[E][R]$ will indicate the value assigned to entity E for context type X and relator R. In other words, $X[E][R] = V$. For instance if $\langle John, Location, Is, Classroom \rangle \in ContextPredicateSet$, then $Location[John][Is] = Classroom$. If such a context predicate does not exists in *ContextPredicateSet*, we will assume that $X[E][R] = \bot$ (read as null).

3.3 Context Type

Informally, *a context* is a property related to every entity or a subset of existing entities in the system. In fact, context type represents a contextual attribute of the system; e.g. time or location of entities. Formally, a context type $ct \in ContextSet$ is defined as a 5-tuple:

$$ct = \langle ValueSet_{ct}, OperatorDefinerSet_{ct}, RelatorSet_{ct}, EntityTypeSet_{ct}, LURSet_{ct} \rangle$$

More detail on each component of the context type ct is given below.

3.3.1 Set of Admissible Values: ValueSet_ct

$ValueSet_{ct}$ denotes the set of values that can be assigned to variables of context type ct. Set representation can be used to determine members of $ValueSet_{ct}$. For instance, the value set of context type *time* can be defined in the following way using *set comprehension*: $ValueSet_{time} = \{n : \mid 0 \leq n \leq 24\}$.

3.3.2 Operator Definer Set: OperatorDefinerSet_ct

$OperatorDefinerSet_{ct}$ is comprised of a finite number of functions each of which defines logical, set and other user-defined operators on the value set of context type ct. Each of these functions requires three arguments, but the types of these arguments are different among the functions. Generally speaking, each $Operator\text{-}Definer_{ct}$ determines that for two arbitrary values A and B related to $ValueSet_{ct}$ and $op \in$ a subset of *OperatorSet* whether $(A\ op\ B)$ is true or not. Since the signature of each $Operator\text{-}Definer$ function is unique, the signature must be included along the definition. The informal signature of *Operator-Definer* function is as follows:

$Operator\text{-}Definer_{ct}$: *A set of values related to* $ValueSet_{ct} \times$ *a set of operators* $\times$ *A set of values related to* $ValueSet_{ct} \rightarrow \{true, false\}$

For some context types, the specification of an *Operator-Definer* function might be complex. There exist two alternatives for definition of *Operator-Definer* function. First, it can be specified using *propositional logic* and second, it can be incorporated into model using an external module. The detail is omitted due to lack of space.

3.3.3 Set of Admissible Relators: RelatorSet$_{ct}$

RelatorSet$_{CT}$ represents the set of admissible relators for context type *CT*. For instance, for context type *location*, *RelatorSet$_{location}$* can be defined as follows:

$$RelatorSet_{location} = \{Is, Entering, Leaving\}$$

3.3.4 Set of Admissible Entity Types: EntityTypeSet$_{ct}$

EntityTypeSet$_{ct}$ denotes the set of entity types related to context type *ct*. In addition, the value set of other context types can be included in *EntityTypeSet$_{ct}$* and it simply means that a context type might express a property about a value of another context type. In fact, *EntityTypeSet$_{ct}$* is a subset of the set {*Subject, Object, Environment, Value-Set$_{ct_1}$,... , ValueSet$_{ct_n}$*}. As an example, context type location represents a property which is only related to users, subjects and objects and therefore:

EntityTypeSet$_{Location}$ = {*User, Subject, Object*}

As another example, consider a context type *locationlvl* which associates a confidentiality level with each value of context type *location*. Then:

$$EntityTypeSet_{locationlvl} = \{ValueSet_{location}\}$$

3.3.5 Level Update Rules: LURSet$_{ct}$

Each *level update rule* (LUR) describes how confidentiality or integrity levels of users, subjects and objects are updated based on their contextual values for context type *ct*. Informally, a *LUR* $\in$ *LURSet$_{ct}$* is a state machine in which confidentiality or integrity levels represent states and 'conditions on contextual values' corresponds to transitions. When a contextual value of context type *ct* related to an entity changes, the conditions are evaluated and entity's (confidentiality or integrity) level is updated based on the result of evaluation.

LURSet$_{ct}$ denotes a set which itself is comprised of two sets of LURs: confidential level update rule set or *C-LURSet$_{ct}$* and integral level update rule set or *I-LURSet$_{ct}$*.

C-LURSet$_{ct}$ includes confidential level update rules of type ct (*C-LUR$_{ct}$*). A *C-LUR$_{ct}$* specifies how confidentiality level of entities is updated based on changes in context predicates of type ct.

The confidential level update rules of *C-LURSet$_{ct}$* are generally divided into four categories. The first, second and third categories includes *C-LUR$_{ct,USR}$*, *C-LUR$_{ct,SBJ}$*, and *C-LUR$_{ct,OBJ}$* respectively. Each of these rules defines a level update rule for confidentiality level of users/subjects/objects based on changes in their contextual value for context type *ct*. The fourth category includes a group of *C-LURs* in the form of *C-LUR$_{ct,en}$*. Each of these *LURs* defines a level update rule for confidentiality level of a special entity. For instance, *C-LUR$_{ct,en}$* defines how confidentiality level of an entity *en* changes based on its contextual value for context type *ct*. It is evident that if *C-LURSet$_{ct}$* contains a specialized *C-LUR* for an entity, it overrides the general *C-LURs* defined in other categories. Notice that inclusion of these categories in *C-LURSet$_{ct}$* is optional and *C-LURSet$_{ct}$* might be even empty.

I-LURSet$_{ct}$ includes integral level update rules of context type *ct* (*I-LUR$_{ct}$*). An *I-LUR$_{ct}$* specifies how integrity level of entities is updated based on changes in context predicates of type *ct*. The integral level update rules of *I-LURSet$_{ct}$* are generally divided into four categories as defined for *C-LURSet$_{ct}$*. As above, inclusion of these categories in *I-LURSet$_{ct}$* is optional and *I-LURSet$_{ct}$* might be even empty.

Confidential/Integral Level Update Rule: C-LUR$_{ct}$, I-LUR$_{ct}$. As mentioned earlier, each LUR is simply a state machine. Also, LURs are divided into two categories: C-LURs and I-LURs. For a C-LUR, ConfLvl denotes the set of states and for an I-LUR, IntegLvl constitutes this set. The transitions, on the other hand, are simply some conditions on contextual values of entities for context type ct.

For an *LUR* to act in a correct way, we need to store the previous confidentiality/integrity levels of an entity, before applying that *LUR* to it. The reason for such need will be explained later. Specifically, we need two extra variables for every pair of (entity, context type). For a pair like (*en, ct*), these variables are represented by $\lambda_{ct}(en)$ and $\omega_{ct}(en)$ and are initialized in the following way:

$$\forall ct \in ContextSet \; \forall en \in (Subject \cup Object \cup User). \lambda_{ct}(en) = \lambda(en) \wedge \omega_{ct}(en) = \omega(en)$$

Each transition is composed of a set of statements each of which is a conjunction of two conditions: one on contextual value and one on previous confidentiality/integrity levels. The transition takes place if all conditions of all statements are evaluated to true. For instance, suppose in a *C-LUR$_{ct}$* the following transitions is defined:

$$\{(Is, \geq 10, (=,TS)),(Is, \leq 20,(=,TS))\}$$

This transition takes place if the following statement is evaluated to true:

$$(ct[en][Is] \geq 10 \wedge \lambda_{ct}(en) = TS) \wedge (ct[en][Is] \leq 20 \wedge \lambda_{ct}(en) = TS)$$

Furthermore, the second condition is optional and can be equal to $(\perp, \perp)$; since sometimes there is no restriction on the previous confidentiality/integrity level.

Due to lack of space, the formal definition of a level update rule is omitted here. Instead, an example is used to clarify the concept. Assume *ConfLvl* = $\langle TS,S,C,U \rangle$. Fig. 1 shows *C-LUR$_{Age,OBJ}$* that describes how objects' confidentiality level is updated based on their Age.

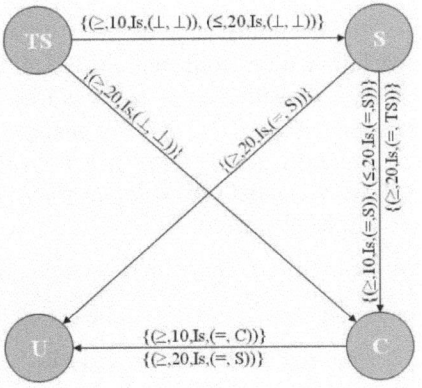

Fig. 1. Confidential level update rules of context type *Age* for objects: *C-LUR$_{Age,OBJ}$*

C-LUR$_{Age,OBJ}$ simply specifies that the confidentiality level of an object decreases every decade with the restriction that the confidentiality level of an object can never decrease more than two levels. In particular, assume a document named *Doc* is 10 to 20 years old and $\lambda_{Age}(Doc) = \lambda(Doc) = S$. When *C-LUR$_{Age,OBJ}$* is applied to *Doc* for the first time, the transition $(\{(Is,\geq,10,(=,S)), (Is,\leq,20, (=,S))\})$ is evaluated to true and therefore, $\lambda_{Age}(Doc) = S$, $\lambda(Doc) = C$. In other words, the above transition denotes the following conditional statement:

$$(Age[Doc][Is] \geq 10 \wedge \lambda_{Age}(Doc) = S) \wedge (Age[Doc][Is] \leq 20 \wedge \lambda_{Age}(Doc) = S)$$

As long as the age of *Doc* is between 10 and 20, application of *C-LUR$_{Age,OBJ}$* on *Doc* causes no change in levels, since none of the transitions from state *C* to *U* are evaluated to true. When its age is changed to above 20, the transition $(\{(\geq,20,Is,(=,S))\})$ is evaluated to true and the following assignments takes place:

$$\lambda_{Age}(Doc) = C, \lambda(Doc) = U$$

Algorithms for Applying LURs to Entities. In this section, algorithms for applying LURs to entities are presented. To reduce the complexity, we propose two algorithms: one for C-LURs and one for I-LURs.

> *Apply-CLUR (ct ∈ ContextSet, I ∈ C-LURSet$_{ct}$, e ∈ EntitySet\ {environment}){*
> $\lambda_{ct}(e) = \lambda(e)$
> *For each state s in ConfLvl*
> > *For each transition from $\lambda(e)$ with label $\{(co_1,v_1,r_1,P_1),...,(co_n,v_n,r_n,P_n)\}$*
> > *to s in I*
> > > *if ((P$_1$ = (⊥,⊥) AND Operator-Definer$_{ct}$(ct[e][r$_1$], co$_1$, v$_1$)) OR*
> > > *(P$_1$ = (do$_1$,l$_1$) AND Operator-Definer$_{ct}$(ct[e][r$_1$],co$_1$,v$_1$) AND*
> > > $\lambda_{ct}(e)$ *do$_1$ l$_1$))*
> > > *AND*
> > >
> > > *...*
> > > *AND*
> > > *if ((P$_n$ = (⊥,⊥) AND Operator-Definer$_{ct}$(ct[e][r$_n$], co$_n$, v$_n$)) OR*
> > > *(P$_n$ = (do$_n$,l$_n$) AND Operator-Definer$_{ct}$(ct[e][r$_n$], co$_n$, v$_n$) AND*
> > > $\lambda_{ct}(e)$ *do$_n$ l$_n$))*
> > > > $\lambda(e) = s$
>
> *}*

In order to preserve the confidentiality level of entity before being changed, $\lambda(e)$ is assigned to $\lambda_{ct}(e)$. Next, each transition from state $\lambda(e)$ to all other states is evaluated. If the result of evaluation for a transition to a state *s* is true, *s* is assigned to $\lambda(e)$. If none of the transitions is evaluated to true, $\lambda(e)$ is not changed.

Furthermore, for every statement (co_i,v_i,r_i,P_i) of a transition, if $P_i = (⊥,⊥)$, then only the first condition will be evaluated (*Operator-Definer$_{ct}$(ct[e][r$_i$], co$_i$, v$_i$)*). But if $P_i \neq (⊥,⊥)$ both conditions will be evaluated:

$$Operator\text{-}Definer_{ct}(ct[e][r_i], co_i, v_i) \text{ } AND \text{ } \lambda_{ct}(e) \text{ } do_i \text{ } l_i$$

To apply *C-LUR$_{ct}$* to an entity *en*, *Apply-CLUR* will be called in the following way:

$$Apply\text{-}CLUR(ct,C\text{-}LUR_{ct},en)$$

Since the algorithm for applying *I-LUR*s to entities has minor changes compared to *Apply-CLUR* (λ substituted with ω and *ConfLvl* substituted with *IntegLvl*), we omit the details here.

The Reason for Storing Previous Levels of Entities. As mentioned above, we need two extra variables for each pair of (e, ct) where e $\in$ EntitySet \ {environment} and ct $\in$ ContextSet: one for storing previous confidentiality level of the entity before being changed by one of C-LURs of ct and one for storing its previous integrity level before being changed by one of I-LURs of ct. They are represented by $\lambda_{ct}(en)$ and $\omega_{ct}(en)$ respectively.

Since, *LUR*s are applied to entities on special occasions, for a change in context, it is impossible to find out whether an *LUR* has already been applied to an entity or not. In order words, when a change occurs in context, there must be a way to recognize whether this change has already been considered or not. These extra variables are needed to for this matter. Further detail on this issue is omitted due to lack of space.

An Algorithm for Updating Levels of an Entity. UpdateEntityLevels updates the confidentiality and integrity levels of a specific entity (passed to it as an argument) based on the appropriate LURs of all context types in ContextSet.

> *UpdateEntityLevels(e $\in$ EntitySet\{environment}, ET $\in$ {USR,SBJ,OBJ}){*
> *for each context type ct $\in$ ContextSet in order*
>   *if C-LUR$_{ct,e}$ $\in$ C-LURSet$_{ct}$*
>   *Apply-CLUR(ct, C-LUR$_{ct,e}$, e)*
>   *else if C-LUR$_{ct,ET}$ $\in$ C-LURSet$_{ct}$*
>   *Apply-CLUR(ct, C-LUR$_{ct,ET}$, e)*
>   *if I-LUR$_{ct,e}$ $\in$ I-LURSet$_{ct}$*
>   *Apply-ILUR(ct, I-LUR$_{ct,e}$, e)*
>   *else if I-LUR$_{ct,ET}$ $\in$ I-LURSet$_{ct}$*
>   *Apply-ILUR(ct, I-LUR$_{ct,ET}$, e)*
> *}*

In this algorithm, *ET* represents the type of Entity. *USR*, *SBJ*, and *OBJ* represent *User*, *Subject* and *Object* sets respectively. The *LUR*s of context types are applied based on the ordering defined by *ContextSet*; i.e. the first element of the ordered set is applied first and so forth. For each context type *ct*, it first checks if there is a specific *C-LUR* defined for entity *e* (*If C-LUR$_{ct,e}$ $\in$ C-LURSet$_{ct}$*) and if so, the *C-LUR* is applied to the entity. Otherwise, it checks if there is a general *C-LUR* based on the type of entity (*else if C-LUR$_{ct,ET}$ $\in$ C-LURSet$_{ct}$*) to be applied to it. The same procedure is adopted for *I-LUR*s.

3.4 Operations

3.4.1 AccessRightSet$_{opr}$

The set of access rights in CAMAC model is comprised of *read* and *write*. In CAMAC, every operation, based on what it carries out, includes a subset of these modes; e.g. if it only does an observation of information and no alteration, it only includes *read* and so on. *AccessRightSet$_{opr}$* is a subset of the set {*read*, *write*} which denotes access right set of the operation.

3.4.2 Constraint$_{opr}$

Each operation includes a *constraint* which denotes the prerequisite conditions that must be satisfied before the operation is executed. For *opr* ∈ *OperationSet*, this constraint is represented by *Constraint$_{opr}$* and is mainly composed of *condition blocks*. There exist three types of condition blocks: Confidential condition blocks (*C-CB*), Integral condition blocks (*I-CB*) and Contextual condition blocks (*Cxt-CB*). In defining each condition block, we make use of variable *USR, SBJ* and *OBJ* to represent user, subject and object respectively. Use of these variables allows us to define generic constraints. Next, we define different types of condition blocks and later a grammar for derivation of constraints is presented.

Confidential Condition Block (C-CB). A confidential condition block is defined as a triple ⟨λ$_1$, op, λ$_2$⟩ *in which* λ$_1$, λ$_2$ ∈ *ConfLvl and op* ∈ *DomOperatorSet. For instance* ⟨λ(SBJ), ≥, λ(OBJ)⟩ *is a C-CB denoting the simple security property of Bell-LaPadula.*

Integral Condition Block (I-CB). An integral condition block is defined as a triple ⟨ω$_1$, op, ω$_2$⟩ *in which* ω$_1$, ω$_2$ ∈ *IntegLvl and op* ∈ *DomOperatorSet. For instance* ⟨ω(SBJ), ≥, ω(OBJ)⟩ *is an I-CB denoting the integrity *-property of Biba.*

Contextual Condition Block (Cxt-CB). A contextual condition block is defined as a triple ⟨Value$_1$, op, Value$_2$⟩$_{ct}$ *in which Value$_1$,Value$_2$ ∈ ValueSet$_{ct}${.element}, ct ∈ ContextSet and op* ∈ *OperatorSet. The subscript ct determines that operator definer functions of context type ct must be used to evaluate this Cxt-CB. Instances of Cxt-CB are* ⟨Time[environment][Is], <,9⟩$_{Time}$ *and* ⟨Age[SBJ][Is],>,Age[OBJ][Is]⟩$_{Age}$.

A Grammar for Derivation of Constraints. Constraints are built using the following unambiguous grammar:

$$Constraint \rightarrow Constraint \vee C1$$
$$Constraint \rightarrow C1$$
$$C1 \rightarrow C1 \wedge C2$$
$$C1 \rightarrow C2$$
$$C2 \rightarrow (Constraint)$$
$$C2 \rightarrow Cxt\text{-}CB | C\text{-}CB | I\text{-}CB.$$

For example, for an operation named *GenerateReport* the following constraint may be defined using the above grammar:

Constraint$_{GenerateReport}$ = (⟨λ(SBJ),≥,S⟩) ∨ (⟨λ(SBJ),=,C⟩ ∧ ⟨Time[environment][Is], ≥, 6⟩$_{Time}$ ∧ ⟨Time[environment][Is], ≤, 12⟩$_{Time}$)

Definition of operations finalizes specification of elements of CAMAC model. Next we consider how requests are authorized in CAMAC.

3.5 Authorization of Action

A subject's request to access an object is represented by an *action*. Formally, an action *A* is a triple ⟨s, o, opr⟩ in which s ∈ *Subject*, o ∈ *Object* and *opr* ∈ *Operation*. Furthermore the user of an action is the user on behalf of whom the subject is acting; i.e. *u = RepOf(s)*. The algorithm *AuthorizeAction* handles authorization of actions.

$AuthorizeAction(A = \langle s, o, opr \rangle)$
{

$\quad u = RepOf(s)$
$\quad Constraint_A = Constraint_{opr}$
$\quad UpdateEntityLevels(u, USR)$
$\quad UpdateEntityLevels(s, SBJ)$
$\quad UpdateEntityLevels(o, OBJ)$
$\quad \lambda(s) = GLB(\lambda(s), \lambda(u))$
$\qquad\quad \omega(s) = GLB(\omega(s), \omega(u))$
$\quad if\ Read \in AccessRightSet_{opr}$
$\qquad\quad Constraint_A \quad = \quad Constraint_A \quad \wedge \quad (\langle\langle\lambda(SBJ),\geq,\lambda(OBJ)\rangle \quad \wedge$
$\langle\omega(OBJ),\geq,\omega(SBJ)\rangle\rangle$
$\quad if\ Write \in AccessRightSet_{opr}$
$\qquad\quad Constraint_A \quad = \quad Constraint_A \quad \wedge \quad (\langle\langle\lambda(SBJ),\leq,\lambda(OBJ)\rangle \quad \wedge$
$\langle\omega(OBJ),\leq,\omega(SBJ)\rangle\rangle$

$\quad$ *Assign u, s, o to USR, SBJ, OBJ in Constraint$_A$ respectively*
$\quad$ *return Evaluate(Constraint$_{opr}$)*

}

Upon occurrence of an action, initially the confidentiality and integrity levels of user, subject and object of an action must be updated. As mentioned in section 3.3, *UpdateEntityLevels* algorithm updates the levels of an entity using all the applicable *LUR*s of all context types. Calling the algorithm for user, subject and object takes care of these updates. Since the confidentiality and integrity levels of a subject must be dominated by the corresponding levels of its user, after updating levels of user and subject, the following assignments seems indispensable:

$$\lambda(s) = min(\lambda(s), \lambda(u)),\ \omega(s) = min(\omega(s), \omega(u))$$

After level updates are done, the constraint of the action must be evaluated. Constraint of an action A is represented by $Constraint_A$ and is initially equal to operation constraint. Before evaluation takes place, the corresponding confidentiality and integrity constraints must be added to the constraint of action based on access right set of the operation. In other words, if *read* $\in AccessRightSet_{opr}$, simple security property of Bell-LaPadula and simple integrity property of Biba must be added to $Constraint_A$

$read \in AccessRightSet_{opr}.Constraint_A = Constraint_A \wedge (\langle\langle\lambda(SBJ),\geq,\lambda(OBJ)\rangle \wedge \langle\omega(OBJ),\geq,\omega(SBJ)\rangle\rangle$

Also, if *write* $\in AccessRightSet_{opr}$, *-property of Bell-LaPadula and integrity *-property of Biba must be added to $Constraint_A$.

$write \in AccessRightSet_{opr}.Constraint_A = Constraint_A \wedge (\langle\langle\lambda(OBJ),\geq,\lambda(SBJ)\rangle \wedge \langle\omega(SBJ),\geq,\omega(OBJ)\rangle\rangle$

At last, u, s and o are assigned to USR, SBJ and OBJ respectively and the constraint is evaluated using a parser, operator definer functions of context types, and dominance relationship. If the result of evaluation is true, the action is granted and otherwise denied.

4 CAMAC Expressiveness

Various mandatory concepts can be expressed using CAMAC. In this paper, due to lack of space, we only express set of categories with it, while some famous models and policies such as Dion and Chinese Wall can conveniently be expressed by the model.

The confidentiality levels in the original Bell-LaPadula model are defined by two components: a classification and a set of categories. On the other hand, as defined in section 3.1 the confidentiality levels of CAMAC model consists of the first component and the set of categories is simply ignored. The same statement holds for integrity levels of Biba. We intend to show that the set of categories is inherently a contextual concept and can be simply modeled as a context type. Here, we take confidentiality levels into consideration. The set of categories for integrity levels can be modeled in a similar way.

The set of categories is a subset of a non-hierarchical set of elements and the elements of this set depend on considered environment and refer to the application area to which information pertains or where data is to be used. A classic example of this set is $\{Nato, Nuclear, Crypto\}$ which denotes the categories in which the classification of the confidentiality level is defined. We define a context type $C\text{-}Category$ as follows:

$C\text{-}Category = \langle ValueSet_{C\text{-}Category}, OperatorDefinerSet_{C\text{-}Category}, RelatorSet_{C\text{-}Category}, EntityType\text{-}Set_{C\text{-}Category}, LURSet_{C\text{-}Category}\rangle$

- $ValueSet_{C\text{-}Category} = \{P(\{Nato, Nuclear, Crypto\})\}$
- $OperatorDefinerSet_{C\text{-}Category}$

    $\{$

    $\qquad Operator\text{-}Definer_{C\text{-}Category} (A \in ValueSet_{C\text{-}Category}, o \in OperatorSet, B \in ValueSet_{C\text{-}Category})\{$

    $\qquad\qquad (A = \{Nato\} \wedge B = \{Nato, NuClear\} \wedge o = '\subseteq') \vee ....$

    $\qquad \}$

    $\}$

- $RelatorSet_{C\text{-}Category} = \{Is\}$
- $EntityTypeSet_{C\text{-}Category} = \{User, Subject, Object\}$
- $LURSet_{C\text{-}Category} = \{C\text{-}LURSet_{C\text{-}Category}, I\text{-}LURSet_{C\text{-}Category}\}$
    - $C\text{-}LURSet_{C\text{-}Category} = \emptyset, I\text{-}LURSet_{C\text{-}Category} = \emptyset$

Now the constraints of all operations in $OperationSet$ are changed in the following way:

$\forall\, opr \in OperationSet \mid read \in AccessRightSet_{opr}.$
$Constraint_{opr} = (Constraint_{opr}) \wedge (\langle C\text{-}Category[OBJ][Is], \subseteq, C\text{-}Category[SBJ][Is]\rangle_{C\text{-}Category})$

$\forall\, opr \in OperationSet \mid write \in AccessRightSet_{opr}.$
$Constraint_{opr} = (Constraint_{opr}) \wedge (\langle C\text{-}Category[SBJ][Is], \subseteq, C\text{-}Category[OBJ][Is]\rangle_{C\text{-}Category})$

Assume $opr \in OperationSet$ and $read \in AccessRightSet_{opr}$. Based on definition, a confidentiality level $L_1 = (c_1, s_1)$ is higher or equal to (dominates) level $L_2 = (c_2, s_2)$ if and only if the following relationships are valid: $c_1 \geq c_2, s_1 \supseteq s_2$

Notice that an action $A = \langle s, o, opr\rangle$ is authorized if the following condition blocks are true: $\langle \lambda(s), \geq, \lambda(o)\rangle, \langle C\text{-}Category[s][Is], \supseteq, C\text{-}Category[o][Is]\rangle$

These condition blocks denote aforementioned relationships and since both of them must be satisfied for an action including *opr* to be authorized, it has the same effect as incorporating set of categories in confidentiality levels.

5 Evaluation and Conclusion

CAMAC model could be evaluated and compared with other mandatory models on plenty of basis: authorization time complexity, complexity of policy description, support for context-awareness, expressiveness and security objective. Here we only consider time complexity of authorization due to lack of space.

One important metric would be the computational time needed to authorize an action.

It can be shown that for the computational time to be polynomial, the maximum of time complexities of all *Operator-Definer* functions, must be polynomial. This assumption may not be necessarily true in all cases. Specifically, if the function is added as an external module to the system, there is no guarantee in this regard.

In this paper, we explained the need for a context-aware mandatory access control model and presented CAMAC as a model which satisfies such a need. CAMAC model utilizes context-awareness to provide dynamicity and context-sensitivity of levels to enable specification of sophisticated mandatory policies. In addition, various mandatory controls can be incorporated into the CAMAC model. Bell-LaPadula and Biba strict integrity policy are the inbuilt part of the model and other Biba policies, Chinese Wall policy and Dion can be appended to the model using context types. Also, an amalgamation of mandatory policies can be used simultaneously. For instance, Bell-LaPadula, Biba strict integrity policy, and Chinese Wall Policy can all be deployed at once.

References

1. Bell, D.E., LaPadula, L.J.: Secure Computer System: Unified Exposition and Multics Interpretation. Technical Report MTR-2997 Rev. 1. MITRE Corporation (1976)
2. Bell, D.E., LaPadula, L.J.: Secure Computer Systems: Mathematical Foundations. Technical Report MTR-2547. MITRE Corporation (1976)
3. Biba, K.: Integrity Considerations for Secure Computer Systems. In: Corporation, M. (ed.): Technical Report MTR-3153, Bedford, MA (1977)
4. Dion, L.C.: A Complete Protection Model. In: IEEE Symposium on Security and Privacy, Oakland, CA, pp. 49–55 (1981)
5. Brewer, D.F.C., Nash, M.J.: The Chinese Wall Security Policy. In: IEEE Symposium Research in Security and Privacy, pp. 215–228. IEEE CS Press, Los Alamitos (1989)
6. Sandhu, R.S.: Lattice-Based Access Control Models. IEEE Computer 26(11), 9–19 (1993)
7. Sandhu, R.S., Samarati, P.: Access Controls: Principles and Practice. IEEE Communications 32 (9), 40–48 (1994)
8. Kumar, A., Karnik, N., Chafle, G.: Context Sensitivity in Role Based Access Control. ACM SIGOPS Operating Systems Review, 53–66 (2002)
9. Al-Kahtani, M.A., Sandhu, R.: A Model for Attribute-Based User-Role Assignment. In: 18th Annual Computer Security Applications Conference, pp. 353–364. IEEE Computer Society Press, Las Vegas (2002)

10. Covington, M., Moyer, M., Ahamad, M.: Generalized role-based access control for securing future applications. In: 23rd National Information Systems Security Conference, Baltimore, MD, USA (2000),
    http://csrc.nist.gov/nissc/2000/proceedings/toc.pdf
11. Zhang, G., Parashar, M.: Context-aware dynamic access control for pervasive applications. In: Communication Networks and Distributed Systems Modeling and Simulation conference, San Diego (2000)
12. Georgiadis, C.K., Mavridis, I., Pangalos, G., Thomas, R.K.: Flexible Team-based Access Control Using Contexts. In: Sixth ACM Symposium on Access Control Models and Technologies, pp. 21–27. ACM Press, Chantilly (2001)
13. Hu, J., Weaver, A.C.: A Dynamic, Context-Aware Security Infrastructure for Distributed Healthcare Applications. In: First Workshop on Pervasive Privacy Security, Privacy, and Trust, Boston, MA, USA (2004), http://www.pspt.org/techprog.html
14. Ray, I., Kumar, M.: Towards a location-based mandatory access control model. Computers & Security 25, 36–44 (2006)
15. Baldauf, M., Dustdar, S.: A Survey on Context-aware Systems. Technical report TUV-1841-2004-24. Distributed Systems Group, Technical University of Vienna (2004)
16. Korpipää, P., Mäntyjärvi, J., Kela, J., Keränen, H., Malm, E.-J.: Managing Context Information in Mobile Devices. IEEE Pervasive Computing 2 (3), 42–51 (2003)
17. Gu, T., Pung, H.K., Zhang, D.Q.: A Middleware for Building Context-Aware Mobile Services. In: IEEE Vehicular Technology Conference, Milan, Italy, vol. 5, pp. 2656–2660 (2004)
18. Fahy, P., Clarke, S.: CASS: Middleware for Mobile, Context-Aware Applications. In: Workshop on Context Awareness at MobiSys., Boston, pp. 304–308 (2004)
19. Chen, H., Finn, T., Joshi, A.: Using OWL in a Pervasive Computing Broker. In: Workshop on Ontologies in Open Agent Systems, AAMAS 2003, Melbourne, Australia, pp. 9–16 (2003)
20. Dey, A.K., Salber, D., Abowd, G.D.: A Conceptual Framework and a Toolkit for Supporting the Rapid Prototyping of Context-Aware Applications. Human-Computer Interaction (HCI) Journal 16(2-4), 97–166 (2001)

Formal Security Analysis of Electronic Software Distribution Systems

Monika Maidl[1], David von Oheimb[1] Peter Hartmann[2],
and Richard Robinson[3]

[1] Siemens Corporate Technology, Otto-Hahn Ring 6, 80200 München, Germany
{monika.maidl,david.von.oheimb}@siemens.com
[2] Landshut University of Appl. Sciences,
Am Lurzenhof 1, 84036 Landshut, Germany
peter.hartmann@fh-landshut.de
[3] Boeing Phantom Works, P. O. Box 3707, MC 7L-70,
Seattle, WA 98127-2207, USA
richard.v.robinson@boeing.com

Abstract. Software distribution to target devices like factory controllers, medical instruments, vehicles or airplanes is increasingly performed electronically over insecure networks. Such software often implements vital functionality, and so the software distribution process can be highly critical, both from the safety and the security perspective. In this paper, we introduce a novel software distribution system architecture with a generic core component, such that the overall software transport from the supplier to the target device is an interaction of several instances of this core component communicating over insecure networks. The main advantage of this architecture is reduction of development and certification costs. The second contribution of this paper describes the validation and verification of the proposed system. We use a mix of formal methods, more precisely the AVISPA tool, and the Common Criteria (CC) methodology, to achieve high confidence in the security of the software distribution system at moderate costs.

1 Introduction

1.1 Network Enabled Software Distribution

In recent years, computer systems that support industrial applications, energy management and distribution, transportation systems, medical and many other applications started to use network interconnections for a range of communication needs. One such need is the distribution of software to devices in the field, in particular to allow for software updates. If such software is used to implement critical functionality that can affect the safety of people or valuable property, the software distribution process itself becomes highly critical. In other words, networked software distribution makes the safety and/or security of a system dependent upon securing communication over potentially insecure channels, facing threats like corruption, injection, diversion, replay, and disclosure of the software payload.

M.D. Harrison and M.-A. Sujan (Eds.): SAFECOMP 2008, LNCS 5219, pp. 415–428, 2008.

Various methods can be used to ensure security properties of networked systems. However, methods typically used in software development, such as testing, do not work well for security properties due to the severe consequences of subtle errors or small oversights. After all, security properties have to hold in the presence of attackers who actively try to exploit any weaknesses. A better approach to assess security of systems is to work with a well-designed catalog of requirements that is based on a broad range of experience. Certification according to Common Criteria, as discussed in the next section, falls into this category. Another proven approach is to use exhaustive search as offered by formal methods, in our case by model checking.

1.2 Security Certification

For assessing the security of a system, i.e., assuring that the system implements countermeasures for all relevant security threats, the Common Criteria (CC) [5] is one of the most advanced and widely accepted methodologies. The aim of an evaluation according to the CC is to systematically and objectively demonstrate that the countermeasures are sufficient and correctly implemented. The first step is to produce a specification called Security Target (ST). It defines the Target of Evaluation (TOE) which is the software, firmware and/or hardware component(s) to be evaluated, identifies threats the TOE is exposed to, derives objectives to cover the threats, states functional requirements to implement the objectives, and demands assurance requirements. The Security Target can be an instance of a generic Protection Profile (PP) which specifies the evaluation of a class of systems. We have defined such PPs for an Airplane Asset Distribution System (AADS) and its core component [7].

The CC predefined Evaluation Assurance Levels (EALs) range from 1 to 7 and determine the rigor and depth of the analysis process. Evaluation at high assurance levels, i.e., EAL5-EAL7, requires high effort for the design and implementation and also for the CC evaluation. For example, EAL6 requires a semiformally verified design based on a formal security model, and EAL7 requires full formal verification.

In [8] we have determined the assurance levels that must be met by a distribution system for airplane software. Given the high criticality of some airplane software, according to the NSA, EAL6 is recommended for safety-relevant threats, whereas EAL4 is shown sufficient for threats on airline business. In general, the distribution of software controlling safety-critical processes will require a high assurance level.

Usually CC certifications are applied to single strongly confined IT components, not to whole distributed systems consisting of several interacting entities. This is done mainly in order to limit the evaluation effort. The component-wise certification of complex systems also gives flexibility for the assembly of the overall system: components may be developed and certified individually, even by different partners.

On the other hand, we face the composition problem: the threats and vulnerabilities at system level may be different from the ones at component level. Therefore, whether the security objectives of the overall system are met as a consequence of the security properties of the individually certified components is a question to be addressed separately. The latest version 3.1 of the CC provides a first step to address this problem by providing composed assurance package (CAP) evaluations. However, CAP evaluations cannot achieve a high evaluation assurance level.

1.3 Model Checking

As mentioned above, high assurance calls for formal analysis. Tool-supported formal methods range from automatic model checkers to powerful theorem provers. In the last years, several tools targeted for the verification of security protocols, i.e. protocols that are based on the use of cryptographic measures, have been developed and proven very successful. Among those, the AVISPA tool [1,2] offers a front-end and several model checkers. In its design special care has been given to offer easy use even in an industrial setting. It has been applied to many protocols, mainly of the IETF. Other tools for verifying correctness of security protocols are ProVerif [3], based on resolution theorem proving, and LySa [4], which is based on static analysis.

1.4 Our Contributions

Based on our experience with software distribution for avionics, automotive, and healthcare equipment, we define a generic system architecture for a Software Distribution System (SDS). We simplify the system design and its certification by defining a generic core component, the Software Signer Verifier (SSV), instances of which are used at every node of the system. The overall SDS from the software supplier to the target device is essentially an interaction of several SSV instances.

For a cost-efficient and still rigorous assessment of the distributed SDS, we propose a *hybrid approach*, based on the Common Criteria and on formal methods, that takes advantage of the architecture outlined above and addresses the composition problem for CC-high assurance as mentioned in Section 0. We analyze and specify the security requirements for the SSV and for the overall SDS with Protection Profiles like [7]. Assuming that the involved SSV components are certified, we use the AVISPA tool to formally specify and model check that the overall SDS protocol fulfils the security objectives at system level.

The main contributions of this paper are the system architecture for a SDS, its formal model as an abstract security protocol, and the validation of its system-level security properties.

2 System Architecture of the Software Distribution System

2.1 Threats and Security Objectives for a SDS

In [8] we have presented a threat analysis and security objectives for an Airplane Asset Distribution System (AADS). We can generalize those threats to more general software distribution systems as follows:

Corruption. The contents of software items could be altered or replaced.

Injection. The target device's configuration could be affected by invalid software items created by the attacker and installed on the target device.

Diversion. Software items could be diverted to an unsuitable destination, e.g. by disturbing the execution of other software at that destination.

Wrong version. A mismatch between the target's intended and actual configuration could be caused by replaying outdated versions or by forging version numbers.

Disclosure. The attacker can get hold of the software item contents without having a license, or reengineer functionality in order to help manipulating software.

The last threat was not included in [8] because it is not needed in the AADS context. Yet in general, confidentiality might be necessary, e.g. to protect intellectual property.

Based on the threats described above we derive a set of security objectives that must be met by the SDS:

Authenticity. Every software item accepted must originate from a genuine supplier.

Integrity. For every software item accepted at a target, its identity and contents must not have been altered on the way—it must be exactly the same as at the supplier.

Confidentiality. If required, software items must be kept secret from the entry point of the SDS (at the supplier) until reaching the target device.

Correct Destination. A target device must accept and receive only software items for which it is the true destination intended by the target operator.

Correct Version. A target device must accept software items only in the latest version approved by the target operator.

Note that the first three requirements are stated *end-to-end*, i.e. they are properties stretching from the initial source of software assets to their final destination. In contrast, *hop-by-hop* properties refer to the transport of assets between adjacent entities, for instance that in each step the integrity of an asset is preserved.

2.2 SDS Architecture

On the way from the software *supplier* to the target device, software items may be handled at intermediate entities: software *distributors* or OEMs might receive the software items from the supplier, and send it to the *target operator*, who bears responsibility for the safe operation of the *target device*, and has the authorization to send software there. So the software distribution process consists of several hops, and the SDS stretches over the IT systems related to the process at each of these entities.

Fig. 1 shows the overall flow of software items. Simpler scenarios are possible, e.g. where the operator coincides with the distributor or even with the supplier.

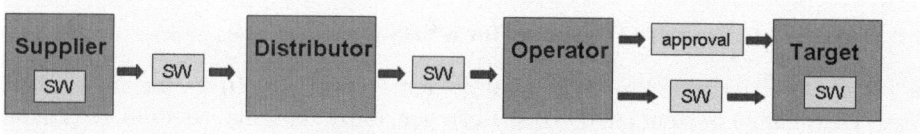

Fig. 1. A typical Software Distribution System

For every transportation step, the software item has to be protected against the threats listed above. Digital signatures and encryption using public key technology are the fundamental security mechanisms used to implement protection for the SDS. Signatures are generated by applying the private key of the sender to the contents, or rather to the hash (which is a cryptographic checksum) of the contents. The recipient applies the corresponding public key, compares the result with the contents which have been received in the clear, and if there are no differences, the receiver can be sure that the contents have not been modified during transport and that only the owner of the private key could have produced this signature. If in addition confidentiality is required, the sender encrypts the signed message with the public key of the receiver. Only the owner of the corresponding private key can decrypt and hence read the contents, not an attacker intercepting it.

The intermediaries might just store and forward the software, or perform some local processing, such as including owner specific license keys and setting target specific software parameters. In any case, the intermediary has to check the signature of the previous entity and might add a new signature.

As the target operator is responsible for its target devices, he has the special task of managing the software configurations on the devices, i.e. deciding which software versions may be installed on which targets. This may take the form of an explicit *installation approval statement* that is sent by the operator to a target, and authorizes the installation of the software item with a suitable version at the specific target instance. We do not specify how installation approval statements are transported securely from the operator to the target. This can be done for example in an out of band communication, or in a protected separate message, or it can be included in the distributed software package.

The target device verifies the integrity and authenticity of the software item using the signature of the operator and checks, using the approval statement of the operator, whether it is an approved recipient of the software item with the given version. Airplane software distribution typically uses an out of band process for the installation approval: the airplane operator (i.e., the airline) issues installation orders in the form of a work order on paper, to be executed by a mechanic. Similar processes might apply in software distribution systems if target devices are located in the vicinity of the operator. For other SDS, administration of the target device should be automatic under remote control of the operator.

We structure the SDS into several instances of a signature application component called Software Signer Verifier (SSV), which is responsible for applying digital signatures on software items before transmitting them, and for verifying signatures on software items received from other entities in the distribution process. For different nodes involved in the software distribution, the SSV can be developed and certified independently or one and the same SSV product can be used at all nodes.

2.3 SSV: The SDS Core Component

Each node in the above distribution chain runs an instance of the SSV, i.e. the SDS core component. The SSV instances are used for:

Introducing unsigned software into the SDS by digitally signing and optionally encrypting it and making it available for other SSV instances.

Verifying the signature on software received from other SSV instances (after decrypting it if needed) and checking the authenticity and authorization of the sender.

Approving the software by adding a signature and optionally re-encrypting the software and making it available to further SSV instances.

Delivering software out of the SDS after successfully verifying it.

Introduction of software into the SDS typically takes place at the supplier, yet may take place also at intermediate entities, while software delivery happens at the software target. All SSV instances except at a supplier verify incoming software. Adding a new signature will be done usually at SSV instances located at distributors and operators after some local processing of the software, such as adding license information or by performing a quality inspection. Such processing is performed within the local environment of the respective. Fig. 2 shows the SSV in its environment including the flow of software.

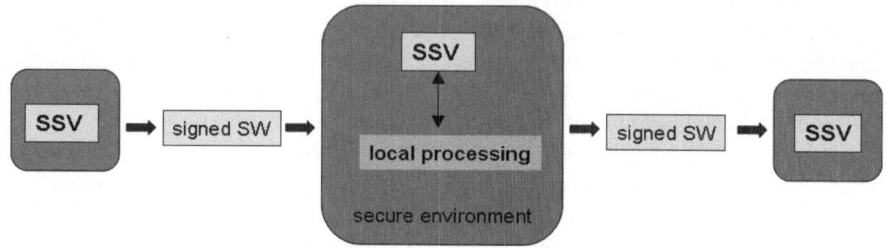

Fig. 2. The generic Software Signer Verifier and its environment

3   Security Assessment of the SDS

3.1   Assumptions on the Operational Environment

Not all assurance issues related to software distribution can be covered by the security assessment of the SDS at reasonable costs. For example the reliability of the Public Key Infrastructure (PKI), which is used to provide keys and certificates for asset protection, is considered out of scope. According to the CC methodology, such aspects are collected as assumptions on the operational environment of the assessed system. The assumptions on the SSV environment are the following:

SSV protection. The SSV instances are protected against direct manipulation and misuse. The SSVs run on a hardened operating system (OS), user access is possible only locally and restricted by effective access control mechanisms. Keys, certificates and other critical data are protected against manipulation and disclosure. Authorized personnel are assumed to be trustworthy.

Secure local environment. The SSV instances and their underlying OS run in a secure local environment, which may contain processing facilities for performing local operations on software items. An adequately configured firewall ensures that the SSV, its underlying OS and the local environment are not compromised through network access.

Reliable PKI. It is assumed that the PKI used to certify keys used by the SSVs is trustworthy and properly managed. Revocation information is issued regularly and immediately after revocation of a signing key.

Target configuration enforcement. The local environment of the target SSV checks whether the installation of received software items is authorized by an approval statement of the target operator. Depending on the system design, this assumption can be relaxed, e.g. the SSV itself might perform such checks.

3.2 Certification of the SSV

The SSV is a component for which a security target may be produced according to the CC methodology. The document [7] is a Protection Profile (PP) for the SSV in the special case of airplane software distribution; however this PP can easily be adapted to handle the very generic case discussed in this paper. The PP specifies the security objectives of integrity and authenticity and – if required – of confidentiality at the component level, and hence after successful CC certification of the SSV instances, there will be sufficient evidence that the security mechanisms of SSVs achieve these security objectives under the assumptions on the SSV environment stated above. The remaining security objectives of correct destination and correct version and end-to-end integrity and authenticity will be covered at system level by the formal analysis described in the next sections.

3.3 The Protocol for End-to-End Software Distribution

In order to assess the correctness of the SDS at system level, we consider the interaction between the SSV instances located at the different nodes. As the interaction consists of exchanging cryptographically secured messages, we have chosen the form of a cryptographic protocol analysis.

First we present the protocol in the common Alice-Bob notation. The different nodes are abbreviated as follows: SUP software supplier, DIS software distributor, OP target operator, TD target device and CA certificate authority. For each node N, the associated private key is denoted by inv(KN).

In the first step, the supplier SSV imports assets from its local environment. In every further step, the SSV at the respective node receives a signed asset and checks the signature. Except in the last step, the SSV adds its approval signature, encrypts the whole message if needed, and sends the new message to the next SSV instance.

```
1. SUP - {Asset.{h(Asset).DIS}_inv(KSUP).CertSUP}_KDIS -> DIS
2. DIS - {Asset.{h(Asset).DIS}_inv(KSUP).CertSUP
              .{h(Asset).OP }_inv(KDIS).CertDIS}_KOP  -> OP
3. OP  - {Asset.{h(Asset).DIS}_inv(KSUP).CertSUP
              .{h(Asset).OP }_inv(KDIS).CertDIS
              .{h(Asset).TD }_inv(KOP ).CertOP}_KTD   -> TD
```

We shortly explain the constructs used in Alice-Bob notation:

A - M -> B	means message M sent from A to B
Asset	means a software item including its identity
M.N	means the concatenated contents of M and N
h(M)	means the hash value of content M
{M}_inv(K)	means content M signed with private key K
{M}_K	means content M encrypted with public key K

As usual when producing a signature, not the asset itself is signed but only its hash value. Note that the signature also includes the identity of the intended receiver. The sender's certificate, which ties the sender's identity together with its public key, is also included into the message. The certificates are self-signed or signed by a certificate authority (CA) that confirms the identity of the certificate holder.

In the SDS protocol, signatures are applied in parallel: every SSV keeps the old signatures and adds its own. However, each SSV only checks the signature applied by its immediate predecessor, but not the signatures applied in the steps before, as it is not assumed that an SSV has a trust relationship with all previous nodes. For example, the target device trusts its operator, but is not configured to know all potential suppliers. For signatures with self-signed certificates, the check consists in looking up the public key in a locally stored set of authorized senders. For instance, the target device typically knows the public key of its operator. For CA-signed certificates, the CA key has to be contained in a locally stored set of public keys of trusted CAs. We assume that the two locally stored sets of trusted public keys are managed by a trustworthy administrator.

We do not model installation approvals explicitly. Instead, we model part of the approval information by including the identity of the intended target device in the asset signature applied by the operator.

3.4 Security Properties

For the SDS protocol, we formally validate the authenticity of the asset origin and the integrity and confidentiality during asset transport. More precisely, we show that

(1) assets accepted by the target device have indeed been sent by the supplier,
(2) assets accepted by the target have not been modified during transport,
(3) asset authenticity and integrity also hop-by-hop, i.e. from any SSV instance to the next, in particular between the operator and target device, and
(4) assets remain secret among the SSVs.

Clearly the security objectives of *authenticity, integrity* and *confidentiality*, stated in Section 0, are covered by (1), (2), and (4). Further, when sending a message, every sender includes the name of the intended receiver in the signature, and the receiving SSV checks whether it is the intended destination, so together with (3), the objective *correct destination* is also satisfied. In other words, the signature of the operator containing the name of the target device models part of the installation approval statement for the asset. The remaining part of the installation approval statement, namely the version information, is not contained in our model. The corresponding security objective of *correct version* is covered by the *target configuration enforcement* assumption, i.e. that version checking is done by the SSV local environment.

Hence the formal analysis, presented in the next section, implies that our formal model of the SDS architecture satisfies the security objectives at the system level. As the implementation details of the SSVs at the different nodes are covered by CC certification, we gain substantial confidence in the overall security of the SDS.

4   Formal Analysis of the SDS Protocol

The Alice-Bob notation, showing only message exchanges, is not detailed and precise enough for any thorough analysis. It leaves important processing steps implicit, in particular the checks an agent performs to accept a message and the parts of received messages and other state information the agent uses to construct further messages. The specification language of the AVISPA tool, HLPSL, offers constructs to express all steps involved in the message exchange in a precise, declarative way. Agents are defined generically as a role, of which multiple instances may exist in a given system or scenario. The behavior of a role is specified as a set of state transitions. During such a transition, an agent receives and checks messages before sending new messages, which then can be received in a transition by another agent.

Instead of individually modeling all roles, i.e. supplier, distributor, operator and target device, we use the fact that all run an instance of the SSV component. Hence we can specify a parameterized role, called SSV, which is then instantiated multiple times to represent the overall SDS protocol.

Figure 3 shows the header declaration for the SSV role. The parameters are used to configure the different instances, e.g. Import is true if signed assets may be received. The parameter KeySet holds a set of public keys that acts as authorization information: software items signed with a key in this set are accepted. For instance, the target device only accepts software items signed by its operator. Alternatively, signed software items can be sent together with a CA-signed certificate, and are accepted if the public key of the CA is contained in KCASet.

The local variables of the SSV include the variable State, which acts as a program counter, and others that are mainly used to hold values received in messages.

```
role softwareSignerVerifier(
        SND,RCV: channel(dy),
        SessN: nat,       % session number, needed just for technical reasons
        SUP,TD: agent,    % supplier and target, just for expressing asset_end_to_end
        Import,Export: bool,% Import is true if a signed asset is expected,
                          % Export is true if a signature has to be added .
        SSV, NextSSV: agent,
        KSSV,KNextSSV: public_key,   % public key of this SSV and the one to
                                     % which it sends messages
        CertSSV: {agent.public_key
                  }_inv(public_key),  % certificate for the private key inv(KSSV)
        KCASet: public_key set,       % set of accepted CA certificates
        KeySet: public_key set        % set of public keys of authorized senders
)
local
        State: nat,
        Asset: text,
        Msg,X,PrevSigs: message,
        KCA,KprevSSV: public_key,
        Cert: {agent.public_key}_inv(public_key),
        PrevSSV: agent

init
        State := 0
```

Fig. 3. Header and local variables of the SSV role

There are five transition rules, presented in Figure 4. The first covers the case that an asset is imported from the local environment (in unsigned form). The second and third rules cover the reception of a signed part, authorized either by a CA-signed certificate or by a public key contained in the internal key set. The remaining two rules describe what the SSV does with the received asset: either forward it in signed form to the next one, or consume it.

```
transition

introduceNew.
     State   = 0 /\ Import = false /\ RCV(start)
 =|> State':= 1 /\ Asset' := new() /\ PrevSigs' := nil
     /\ secret(Asset',asset,{})

importCASignedCert.
     State   = 0 /\ Import = true
     /\ RCV({Asset'.PrevSigs'}_KSSV)
     /\ PrevSigs' =
            X'.({h(Asset').SSV.SessN}_inv(KprevSSV').Cert')
     /\ Cert' = {PrevSSV'.KprevSSV'}_inv(KCA')
     /\ in(KCA',KCASet) % check if CA is in the accepted CA set
 =|> State':= 1
     /\ wrequest(SSV,PrevSSV',asset_hop_by_hop,Asset')

importSelfSignedCert.
     State   = 0 /\ Import = true
     /\ RCV({Asset'.PrevSigs'}_KSSV)
     /\ PrevSigs' =
            X'.({h(Asset').SSV.SessN}_inv(KprevSSV').Cert')
     /\ Cert' = {PrevSSV'.KprevSSV'}_inv(KprevSSV')
     /\ in(KprevSSV',KeySet) % check if signing key acceptable
 =|> State':= 1
     /\ wrequest(SSV,PrevSSV',asset_hop_by_hop,Asset')

send.
     State   = 1 /\ Export = true /\ RCV(start)
 =|> State':= 2
     /\ SND({Asset.PrevSigs.({h(Asset).NextSSV.SessN}_inv(KSSV)
            .CertSSV)}_KNextSSV)
     /\ witness(SSV,NextSSV,asset_hop_by_hop,Asset)
     /\ witness(SUP,TD      ,asset_end_to_end,Asset)

final.
     State   = 1 /\ Export = false /\ RCV(start)
 =|> State':= 2 /\ wrequest(TD,SUP,asset_end_to_end,Asset)
```

Fig. 4. Transitions of the SSV role

We explain the second transition in more detail. A transition is divided into a condition part in which a message may be received and checked, and an action part in which a message may be sent. Variables can occur in a transition in primed or unprimed form, where the unprimed from refers to the value of the variable *before* the transition, whereas the primed form refers to the value of the same variable *after* the transition.

Variables can obtain a new value once during a transition, either by assignments, written in the action part, or by pattern matching in the condition part, typically during reception of a message. For example, `State = 0` means the condition that the variable `State` has the value zero, while `State':= 1` means that the variable `State` is assigned a new value: one. The expression `{Asset'.PrevSigs'}_KSSV` means that a message that must be encrypted with the key `KSSV` is received, the first part of which is stored in the variable `Asset`, and the second part is stored in `PrevSigs`. The next line specifies the constraint that the second part of the message has a specific form, namely `X'.({h(Asset').SSV.SessN}_inv(KprevSSV').Cert')`. As `Asset` has already been assigned a value in this transition, in this way it is checked whether the hash value of the asset is correct. Furthermore, the name of the receiving agent must be the identity of the current `SSV` The public key with which the signature can be validated is stored in `KprevSSV`. Next the certificate is validated: It has to contain the identity of `KprevSSV`, and has to be signed by a CA whose public key is contained in `KCASet`. As a by-product of these checks, the SSV learns the identity of the sender, stored in the variable `PrevSSV`. As explained above, the SSV checks only the signature applied by the direct sender. This is modeled by using the variable `X'` for the signatures that are not handled and by not performing verification on `X'`

Fig. 5 shows the composed role called `session`, which ties together the instantiations of the SSV needed for the end-to-end transport of one asset. Each instantiated SSV is configured with its own parameters. For instance, the eighth parameter is the name of the agent, i.e. SUP in the first instantiation, DIS in the second and so on.

```
role session(SND,RCV: channel(dy),SessN: nat,
     SUP,DIS,OP,TD: agent,
     KSUP,KDIS,KOP,KTD,KCA: public_key,
     SUPCert,DISCert,OPCert,TDCert:
                            {agent.public_key}_inv(public_key),
     SUPKeySet,DISKeySet,OPKeySet,TDKeySet: public_key set)
def=

composition
   softwareSignerVerifier(SND,RCV,SessN,SUP,TD,false,true,
                        SUP,DIS,KSUP,KDIS,SUPCert,{KCA},SUPKeys)
/\ softwareSignerVerifier(SND,RCV,SessN,SUP,TD,true,true,
                        DIS,OP ,KDIS,KOP ,DISCert,{KCA},DISKeys)
/\ softwareSignerVerifier(SND,RCV,SessN,SUP,TD,true,true,
                        OP ,TD ,KOP ,KTD ,OPCert ,{KCA},OPKeys)
/\ softwareSignerVerifier(SND,RCV,SessN,SUP,TD,true,false,
                        TD,none,KTD,knone,TDCert ,{}    ,TDKeys)
end role
```

Fig. 5. Specification of the session role

The last part of the model specifies the environment, including initializing channels and other parameters, defining the initial knowledge of the attacker, and starting three different sessions of the protocol, for instance a session between supplier `sup1` with a CA-signed certificate `Sup1Cert`, distributor `dis`, operator `op`, and target device `td`.

```
role environment() def=

local
        SND,RCV: channel(dy),
        SUP1Cert,SUP2Cert,DISCert,OPCert,TDCert:
                        {agent.public_key}_inv(public_key),
        SUPKeys,DISKeys,OPKeys,TDKeys: public_key set

const
        sessN1,sessN2,sessN3: nat,
         sup1, sup2, sup3, dis, op, td      : agent,
        ksup1,ksup2,ksup3,kdis,kop,ktd,kca: public_key,
        asset_hop_by_hop,asset_end_to_end,asset: protocol_id

init
        SUP1Cert := {sup1.ksup1}_inv(kca   ) /\
        SUP2Cert := {sup2.ksup2}_inv(ksup2) /\ % self-signed
         DISCert := {dis .kdis }_inv(kca   ) /\
          OPCert := {op  .kop  }_inv(kop   ) /\ % self-signed
          TDCert := {td  .ktd  }_inv(ktd   ) /\ % self-signed, unused
        SUPKeys  := {}              /\        % unused
        DISKeys  := {ksup2, ksup3} /\        % ksup3 is unused
        OPKeys   := {}             /\
        TDKeys   := {kop}

intruder_knowledge = { sup1, sup2, sup3, dis, op, td,
                       ksup1,ksup2,ksup3,kdis,kop,ktd,kca}

composition
        session(SND,RCV,sessN1, sup1, dis, op, td,
                          ksup1,kdis,kop,ktd,kca,
                          SUP1Cert,DISCert,OPCert,TDCert,
                          SUPKeys ,DISKeys,OPKeys,TDKeys)
    /\  session(SND,RCV,sessN2, sup2, dis, op, td,
                          ksup2,kdis,kop,ktd,kca,
                          SUP2Cert,DISCert,OPCert,TDCert,
                          SUPKeys ,DISKeys,OPKeys,TDKeys)
    /\  session(SND,RCV,sessN3, sup2, dis, op, td,
                          ksup2,kdis,kop,ktd,kca,
                          SUP2Cert,DISCert,OPCert,TDCert,
                          SUPKeys ,DISKeys,OPKeys,TDKeys)
end role
```

Fig. 6. Specification of the environment and session role instances

In order to validate or falsify the security goals specified for the system, the model checker enumerates (essentially) all message exchanges possible for the given model applying the usual Dolev-Yao attacker model [6], which assumes an intruder capable of controlling the whole network traffic. He can intercept and take apart messages (as far as he knows the secret keys required to decrypt them) and learn their contents, construct new messages out of the material known to him, and send them to any party.

As stated in the previous section, the security properties checked for the SDS protocol are authenticity, integrity and confidentiality. These properties are specified in HLPSL by adding annotations, as shown in Fig. 4. For instance, the annotation witness(SSV,NextSSV,asset_hop_by_hop,Asset) asserts that agent SSV has sent to agent NextSSV the value Asset, while the corresponding annotation *wrequest(SSV,PrevSSV',asset_hop_by_hop,Asset')* expresses that the agent SSV expects that the agent `PrevSSV'` has sent the value `Asset'`. If during the model checker run, a *wrequest* event is not matched by a previous *witness* event with the same identifier (in this case, *asset_hop_by_hop*) such that the values of sender, receiver and asset correspond, an attack has been found. The confidentiality goal is expressed by another annotation: `secret(Asset',asset,{})`. An attack against the confidentiality of the value `Asset'` is found if during the model checker run this value becomes part of the evolving intruder knowledge, which the model checker keeps track of. The overall system goals and system run are activated as follows:

```
goal
        weak_authentication_on asset_hop_by_hop
        weak_authentication_on asset_end_to_end
        secrecy_of asset
end goal
environment()
```

The AVISPA tool offers several model checkers as back-ends, which we have used to validate the SDS protocol, i.e. to check the specified security properties. We have performed the analysis on the protocol with and without encryption of messages, and in both cases, no attack has been found.

5 Conclusions and Future Work

We have proposed an architecture for a security-critical software distribution system, in particular the use of a generic component that is instantiated at different points of the SDS. For assessing the security of our design, we have composed two approaches, namely CC certification and formal analysis in the form of model checking. While the CC methodology is strong in systematically covering the secure implementation of a confined IT-product, it does not offer cost-efficient support for the assessment of a system composed of several instances of a generic component with a high assurance level. On the other hand, the automatic state exploration done by model checking is restricted to relatively small systems, like high-level security protocols, due to the exponential size of the state space of formal models, and dealing with implementation details requires the use of abstractions. Hence by assessing the implementation of the core component, the SSV, with the CC methodology and by formally analyzing the overall SDS protocol at high level, we combine the two methodologies according to their strengths, and gain substantial confidence in the overall security of the SDS. Apart from its role in the security assessment, the process of writing a formal model helps removing the inconsistencies and omissions usually present in a design speci-fied in natural language. Moreover, having a formal model of the SDS protocol is valuable in itself, as it provides a highly precise documentation.

As further work, we plan to extend the formal model and include full configuration management with explicit installation instructions and configuration reports. We also have formally modeled aspects of the PKI underlying our software distribution sys-tem, in particular certificate initialization, and we plan to continue this work.

References

1. Armando, A., von Oheimb, D., et al.: The AVISPA Tool for the Automated Validation of Internet Security Protocols and Applications. In: Etessami, K., Rajamani, S.K. (eds.) CAV 2005. LNCS, vol. 3576. Springer, Heidelberg (2005)
2. AVISPA project homepage (2005), http://www.avispa-project.org/
3. Blanchet, B.: From Secrecy to Authenticity in Security Protocols. In: Hermenegildo, M., Puebla, G. (eds.) SAS 2002. LNCS, vol. 2477, pp. 342–359. Springer, Heidelberg (2002)
4. Bodei, C., Buchholtz, M., Degano, P., Nielson, H.R., Nielson, F.: Static validation of security protocols. Journal of Computer Security 13(3), 347–390 (2005)
5. Common Criteria, http://www.commoncriteriaportal.org/
6. Dolev, D., Yao, A.: On the Security of Public-Key Protocols. IEEE Transactions on Information Theory 29(2), 198–208 (1983)
7. Hartmann, P., Tappe, J., von Oheimb, D.: Asset Signer Verifier Protection Profile, Available upon request (2008)
8. Robinson, R., Li, M., Lintelman, S., Sampigethaya, K., Poovendran, R., von Oheimb, D., Bußer, J., Cuellar, J.: Electronic Distribution of Airplane Software and the Impact of Information Security on Airplane. In: Saglietti, F., Oster, N. (eds.) SAFECOMP 2007. LNCS, vol. 4680, pp. 28–39. Springer, Heidelberg (2007)

The Advanced Electric Power Grid: Complexity Reduction Techniques for Reliability Modeling

Ayman Z. Faza, Sahra Sedigh, and Bruce M. McMillin

Missouri University of Science and Technology, Rolla, MO, 65409-0040, USA
Tel: +1(573)341-7505; Fax: +1(573)341-4532
{azfdmb,sedighs,ff}@mst.edu

Abstract. The power grid is a large system, and analyzing its reliability is computationally intensive, rendering conventional methods ineffective. This paper proposes techniques for reducing the complexity of representations of the grid, resulting in a mathematically tractable problem to which our previously developed reliability analysis techniques can be applied. The IEEE118 bus system is analyzed as an example, incorporating cascading failure scenarios reported in the literature.

Keywords: Reliability, complexity reduction, power grid.

1 Introduction

Analyzing the reliability of the power grid is a computationally intensive task, due to its scale and complexity and large number of interconnected components. For a system of n components, where each component can either fail or function, the number of system states is 2^n, increasing exponentially with the number of components. As n approaches 30, analyzing the matrices representing those states becomes a cumbersome and difficult task. This problem can be alleviated by choosing smaller systems for analysis, in hope that the insights gained can be generalized to larger systems. However, smaller systems rarely experience the failure modes that affect large real-life systems, and hence the information gained by analyzing them is of limited value. One of the main goals of any reliability model is to capture the effects of cascading failures on the overall system reliability. Small systems rarely experience cascading failures, which necessitates the analysis of large systems, where such failures do occur.

The objective of our work is to quantitatively evaluate the reliability of the power grid, and we use the IEEE118 bus system as our case study. As its name suggests, the system has 118 buses. A total of 186 transmission lines connect the buses to each other, making conventional analysis methods insufficient. In this paper, we develop a method for breaking down the large system into a number of smaller subsystems, which simplifies the analysis. We analyze each subsystem separately and aggregate the results obtained from each subsystem into the original large system. In doing so, we develop methods to evaluate the overall system reliability based on the Markov chain Imbeddable Structures

M.D. Harrison and M.-A. Sujan (Eds.): SAFECOMP 2008, LNCS 5219, pp. 429–439, 2008.

(MIS) technique discussed in our previous work [1], and introduce additional techniques to evaluate the reliability of the subsystems created.

A load flow solution is used to simulate the IEEE118 bus system. Line outages or contingencies can be added to the simulator by changing the parameters of the system, and the effects of the contingencies on the functionality of the system can be observed. The results obtained using this simulator provide useful information about failures in the system and areas most likely to cause damage to the grid.

The main contribution of this paper is the introduction of a method to reduce the computation complexity of representations of the grid used in analyzing reliability, and the application of that method to the IEEE118 bus system as an example. In addition, we produced simulation results that identify the specific lines in the example system that could cause cascading failures, and discussed them within the context of the reliability of the IEEE118 bus system.

In this analysis, our goal is to get closer to an accurate evaluation of the reliability of the power grid, which will help identify reliability bottlenecks, eventually resulting in a more robust and survivable grid.

The rest of the paper is organized as follows. Section 2 provides a summary of related literature, and Section 3 defines in detail the problem being addressed. Section 4 provides an analysis of the grid reliability at the system level, while Section 5 explains the techniques used to evaluate the reliability of each subsystem, and explains the results obtained from simulation. Section 6 completes our analysis by showing how to aggregate the data from all the cascade scenarios, and Section 7 concludes the paper and describes future plans for the research.

2   Related Work

The reliability of the power grid is the topic of several research studies. In this section, we present the studies that are most closely related to our work.

Laprie et al. present a study in [2], where they analyze interdependencies among the electric power infrastructure and the information infrastructures supporting its management, control and maintenance. The paper develops a qualitative model that captures these interactions for various types of failures, including cascading, escalating and common-cause failures. Despite conceptual similarities, the qualitative nature of their work significantly differentiates it from our quantitative approach.

Several papers investigate the 2003 blackout in North America, during which 50 million people were left without power. [3]. Notable examples include [4], which also describes in detail other significant blackouts in the United States and Europe. This study also provides a general model that describes the causes and typical sequence of events in a cascading failure leading to a blackout, and suggests means of mitigating such risks. In [5], Stefanini et al. provide another analysis of the causes of blackouts, and observe that the interaction between the information and power networks could lead to such failures when not properly controlled.

A number of other studies describe efforts to model and estimate the reliability of the power grid. In [6], Chassin et al. describe the grid as a Barabasi-Albert network, which is a scale free network characterized by a power-law connectivity probability, and propose a model for failure propagation in the grid. Another study related to this topic is presented in [7], which derives the reliability of the power grid from component attributes such as failure rate, outage time, and unavailability.

In [8], Kazemi et al. present a reliability assessment for an automated power distribution system, in which control equipment is added to isolate earth faults and short-circuit faults from the rest of the system. This procedure restores functionality to some parts of the system that were originally affected by those faults but not directly involved in them. The behavior of the system when such faults occur, and the probabilities of the faults occurring were used as factors that determine the failure rates and repair times of the system components.

A project of particular relevance to our work is described in [9], where Walter et al. develop a modeling tool called OpenSESAME, which can be used to model fault-tolerant, highly available systems. The model can be modified to incorporate component dependencies that lead to failure propagation and common-cause failures. The complexity of our system, however, limits our ability to use this tool. In another study, presented in [10], a modeling framework is developed that aims to capture the interdependencies between the electric infrastructure and the information technology-based control system supporting it. The authors define the states of their system and analyze its behavior, while identifying the major challenges to refining the modeling framework and proposing approaches for meeting these challenges.

In the area of system reduction in terms of reliability, Shooman et al. propose several techniques to reduce complex systems [11]. Such techniques include delta-Y transformation, edge factoring and polygon-chain transformations. While these techniques are quite useful in reducing the complexity of communication networks for the purpose of reliability estimation, they are insufficient for analysis of the power grid. The reduction methods assume that a system is functioning as long as it stays connected with at least one link. While this may be true for communication networks, it is rarely sufficient for the power grid, where in addition to a connected network there should be enough capacity to handle the power flow in the system. This renders the techniques proposed in [11] inadequate for our purposes.

3   Problem Statement

A number of cascading failures have been documented in [12]. In our approach, we take one cascading scenario and aggregate into subsystems all parts of the system not involved in the cascade, while leaving intact the lines involved in the cascade. This significantly reduces the size of the problem, and helps us evaluate the reliability of the grid using the methods we developed in [1]. The reliability of each subsystem is then evaluated separately, and the results are used in evaluation of the reliability of the system as a whole. In order to illustrate this approach, we use a cascading failure scenario as an example. The cascading

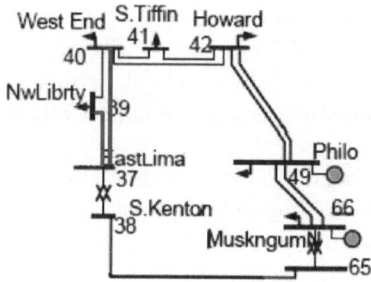

Fig. 1. Line outage (37-39) (Reprinted from [1])

scenario selected occurs due to the failure of line (37-39) in the IEEE118 bus system, a portion of which is depicted in Figure 1. As explained in [1] and [13], the cascade occurs when line (37-39) fails. This leads to an overload in line (37-40), which fails as a result and causes lines (40-42) and (40-41) to fail subsequently.

To simplify representation of the IEEE118 bus system, we aggregate into subsystems all lines and buses that are not directly involved in the specific (37-39) cascade. Figure 2 shows the original IEEE118 bus diagram, with the subsystems highlighted as boxes around parts of the large system. The subsystems were

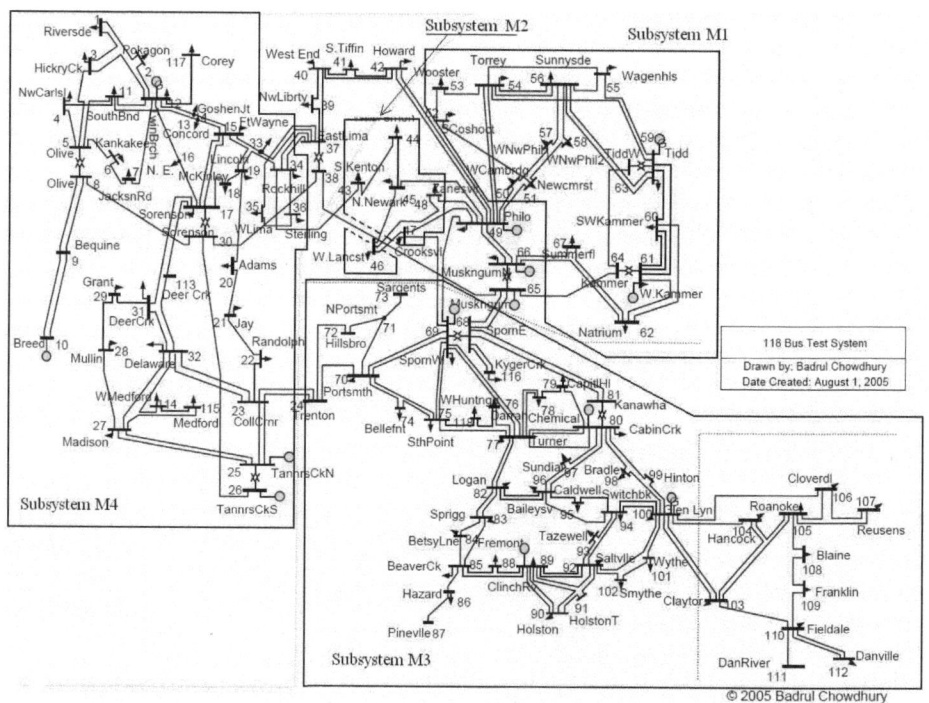

Fig. 2. Highlighting the subsystems of the IEEE118 bus system

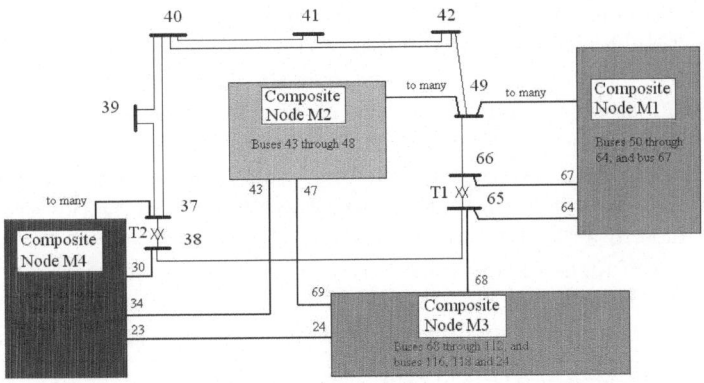

Fig. 3. Simplified IEEE118 bus system diagram

formed based on minimum number of cuts; i.e., borders of the subsystems were drawn at locations where the number of lines that would connect the subsystem to the rest of the system is minimum. In Figure 3, the equivalent system is shown after replacing the subsystems with nodes bearing the names M1 through M4. Having carried out the reduction, the system now appears much smaller. Instead of 186 lines and 118 buses, the system now has 4 large nodes (or subsystems), 9 regular nodes representing the buses, 21 lines, and 2 transformers. The resulting topology is shown in Figure 4, in which the buses and subsystems are replaced by small and large nodes, respectively, and the transmission lines and transformers are represented by lines. This aggregation brings the total number of components to 36, which is a significant reduction of the original 304 components. The number of system states is therefore reduced significantly from 2^{304} to 2^{36}. Further reduction can be carried out by assuming that the buses do not fail. This assumption is justified because the power generators usually have enough backup units to cover for the failed ones, and the main sources of failures in the grid come from the transmission lines. This reduces the number of components of interest to 27, and the overall number of states to 2^{27}. With this reduction in mind, we can apply the MIS technique to the system to evaluate its reliability.

4   Evaluation of System Level Reliability

A detailed introduction to the MIS technique can be found in [1] or [14]; however, we briefly review the important aspects of the method in this section. The reliability of a system of n components can be found using Equation 1 below.

$$R_n = (\mathbf{\Pi_0})^T (\prod_{l=1}^{n} \mathbf{\Lambda}_l) \mathbf{u} \ , \tag{1}$$

where Π_0 denotes a vector of probabilities, and

$$\Pi_0 = [\Pr(Y_0 = S_0), \Pr(Y_0 = S_1), ..., \Pr(Y_0 = S_N)]^T \ , \tag{2}$$

where $\Pr(Y_0 = S_i)$ is the probability of the system initially being in state S_i. Λ_l represents the state transition probabilities of the system as a function of l.

Each element $p_{ij}(l)$ in the matrix Λ_l represents the probability that the system would switch from state S_i to another state S_j due to the failure of component l. Finally, the vector $\mathbf{u}$ is a vector of length equal to the number of states, in which each element has a value of 1 if the corresponding state is considered a "good" state for the system, and 0 otherwise. A system is in a "good" state if it is functioning. It is in a "bad" state if the system has failed.

At this point, the focus of our work turns to identifying the states in which the system is considered to be "good" or functional, and the states at which the system is considered to have failed. The following assumptions will be made.

- If any of the subsystems M1, M2, M3 or M4 fails, the entire system will fail.
- If any two or more lines fail simultaneously, the system will fail.
- If line (37-39) fails, a cascading failure will occur and the system will fail.

The case where the failure of line (37-39) causes a cascading effect is documented in the literature [12]. Simulation was used to identify other cases that could lead to system failure. The results of this simulation show that failures in the system occur when any of the lines (34-37), (38-65), (42-49), (37-40), or (41-42) fails. Details of the cascades that occur as a result are summarized in Table 1 below.

In summary, the good states are the states in which all subsystems M1 through M4 are functioning, and all lines in the first column of Table 1 are functioning. Of the remaining lines, any individual line can fail without jeopardizing the good state; i.e., two or more lines are not allowed to fail independently at the same time. While this case is unlikely to occur, it should be stressed that it is a bad state.

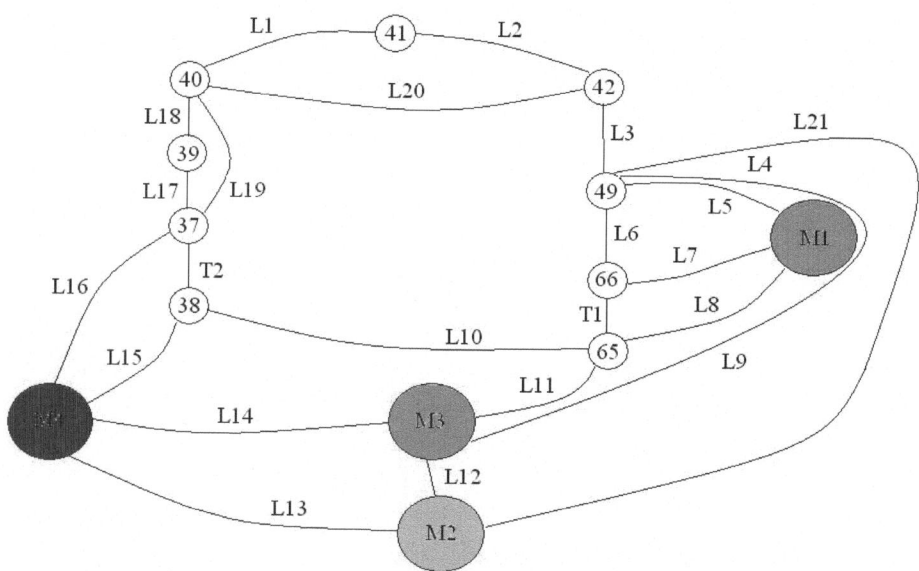

Fig. 4. The IEEE118 bus system as a general node and line diagram

Table 1. Cascading failures

	Stage 1	Stage 2	Stage 3	Stage 4	Stage 5
1	34-37	35-36	43-44	failure	
2	37-39	37-40	40-42	40-41	failure
3	38-65	49-69	47-69	65-68	failure
4	42-49	40-41	40-42	failure	
5	37-40	37-39	40-41	failure	
6	41-42	40-41	failure		

Identification of the good states enables population of the values in vector **u**. However, the matrix Λ_l is yet to be populated. As mentioned before, the values $p_{ij}(l)$ in the matrix Λ_l represent the probability that the system would switch from state S_i to another state S_j due to the failure of component l. For example, assume S_1 is the state where all components are functioning properly, and S_2 is the state in which everything is operational other than line (37-40). The probability of switching from state S_1 to state S_2 is therefore equal to the probability that line (37-40) will fail. It is important to notice that the probability of interest p_{12} is the probability that line (37-40) fails under non-overloading conditions, as explained below.

We know from the cascading failure of line (37-39) that line (37-40) becomes overloaded when (37-39) fails. This leads to the following interpretation of the probability of failure of line (37-40).

$$\Pr(\overline{37-40}) = \Pr(\overline{37-40}|NOV) * \Pr(NOV) + \Pr(\overline{37-40}|OV) * \Pr(OV) \quad (3)$$

where $\Pr(\overline{37-40}|OV)$ and $\Pr(\overline{37-40}|NOV)$ represent the probability that line (37-40) fails under overloading and non-overloading conditions, respectively. $\Pr(OV)$ and $\Pr(NOV)$ represent the probability of overload and no overload, respectively.

The cascading scenario indicates that line (37-40) overloads when line (37-39) fails; hence, $\Pr(OV) = \Pr(\overline{37-39})$.

Returning to the original purpose of this analysis, the value we needed is the probability that line (37-40) fails, provided that all other lines stay functional, including line (37-39). This means that line (37-39) cannot fail concurrently with line (37-40); therefore, the probability the we need is the probability that line (37-40) fails under non-overloading conditions; i.e., $\Pr(\overline{37-40}|NOV)$.

Similarly, the remaining probabilities used to populate Λ_l represent the probabilities of lines failing under non-overloading conditions.

5   Reliability Evaluation of the Reduced Subsystems

Having developed a method to evaluate the reliability of the reduced system, we now turn to determining the probability of failure for the subsystems being reduced. Each of the subsystems is a group of buses interconnected through a

Table 2. Summary of line failures causing subsystem failures

Subsystem 1	Subsystem 2	Subsystem 3	Subsystem 4	Remaining Lines
64-65	47-69	69-70	4-5	34-37
49-51	48-49	82-92	26-30	37-39
53-54	45-46	69-70	69-70	38-65
55-59	45-49	68-69	5-8	42-49
	48-49	68-116	17-30	37-40
		71-73	68-69	41-42
		85-86	8-9	
		86-87	9-10	
		110-111	12-117	
		110-112	2-12	
		68-69	3-5	
		69-70	5-6	
		69-75	5-11	
		74-75	11-13	
		75-118	22-23	
		76-77	25-27	
		77-78	31-32	
		77-80	34-6	
		79-80	35-37	
		80-98		
		83-85		
		85-89		
		94-95		
		99-100		
		100-103		
		100-106		
		103-110		
		105-107		
		106-107		

number of transmission lines. As in the analysis of the previous section, our main concern is with line failures, we assume that buses do not fail. A failure in a line within a subsystem can cause a failure of the entire subsystem. In some cases, this failure occurs directly due to the failure of the line, and in other cases it is due to a cascade initiated by the failure of the line.

Using the simulation, we performed a complete test to evaluate the effect of single line failures on the functionality of the system. Some of the lines were found to be safe; i.e., the network still functions properly after the failure of these lines, while in other cases the system was found to fail. Table 2 shows the lines that can cause a failure in the subsystem when they fail.

The data generated by the simulation can now be used to develop an equation representing the failure probability of each subsystem. Since subsystem failure can occur due to the failure of any of the lines in Table 2, the failure probability of Subsystem m can be described as shown in Equation 4 below.

$$\text{Pr}(\text{Subsystem } m \text{ fails}) = \text{Pr}(\text{Subsystem } m \text{ fails due to failure in line 1})$$
$$+\text{Pr}(\text{Subsystem } m \text{ fails due to failure in line 2})$$
$$+... \tag{4}$$

This can be translated into Equation 5 below, which can be subsequently reduced to Equation 6.

$$\text{Pr}(\text{Subsystem } m \text{ fails}) =$$
$$\text{Pr}(\text{Subsystem } m \text{ fails|line 1 fails}) * \text{Pr}(\text{line 1 fails}) \tag{5}$$
$$+\text{Pr}(\text{Subsystem } m \text{ fails|line 2 fails}) * \text{Pr}(\text{line 2 fails})$$
$$+...$$

$$\text{Pr}(\text{Subsystem } m \text{ fails}) =$$
$$\sum_{\text{all lines } i} \text{Pr}(\text{Subsystem } m \text{ fails|line } i \text{ fails}) * \text{Pr}(\text{line } i \text{ fails}) \tag{6}$$

Note that the probability that a subsystem fails as a result of the failure of a certain line can be either '1' or '0'. If the line is one that can cause system-level failure, then the probability is equal to '1', otherwise it is '0'.

In a more general approach, we would also need to investigate the effect of concurrent failures of two or more lines. For simplicity, at this stage of the research, we will assume that such an event is highly unlikely. Hence, the probability of concurrent line failures is omitted from the equations.

Further Analysis of the results obtained
Table 3 presents further analysis of the results obtained through simulation.

Table 3. Analysis of the simulation results

Subsystem	M1	M2	M3	M4	Remaining Lines
Total number of lines causing cascades	4	5	30	19	6
Overall number of lines in the subsystem	29	8	74	56	19
Percentage of lines causing cascades	13.8%	62.5%	40.5%	33.9%	31.6%

The table shows the percentage of lines in a subsystem that can lead to cascading failures. Note that subsystem M1 is the most stable, with less than 14% of the lines causing cascading failures. On the other hand, subsystems M2 and M3 have the highest percentage of lines causing cascades, with 62.5% and 40.5%, respectively. These numbers will help direct our future efforts towards the less stable areas, i.e., the areas with a large percentage of line failures that could lead to cascades.

6   Aggregation of Data from All Cascade Scenarios

In the previous sections of this paper, the reliability of the grid was evaluated by aggregating parts of the system around a particular cascading failure scenario;

namely, the failure of line (37-39). For the general case, a similar approach can be taken to iteratively evaluate the system reliability, with the aggregations centered around a different scenario in each iteration. For each case, the MIS technique can be used to evaluate the system-level reliability, and the probability of failure of each subsystem can be evaluated separately. After obtaining system reliability evaluations from all different scenarios, we can aggregate them by averaging all values.

Averaging the results, however, can be inaccurate, and a more general approach to aggregating the data would be in the form of a statistical expectation. In other words, not all cascades are equally likely to happen, and this needs to be reflected in the equation. Equation 7 below captures that effect.

$$\text{System Reliability} = \sum_{\text{all cascades, } i} R_i * Rel_i \tag{7}$$

where R_i is the reliability of the system as seen by the cascading failure i, and Rel_i is the relative probability that the cascading failure i will occur. Rel_i can be found as follows.

$$Rel_i = \frac{\text{Pr(cascade } i \text{ occurs)}}{\sum_{\text{all cascades } j} \text{Pr(cascade } j \text{ occurs)}} \tag{8}$$

7   Conclusions and Future Work

The electric power grid is a large system with numerous interconnected components. In this paper, we have presented a method to reduce the complexity of representations of the power grid, in order to simplify the evaluation of its reliability. We selected a particular cascading scenario reported in the literature, and reduced the size of the grid surrounding it by replacing large areas of the system with equivalent nodes. Then, using the MIS technique, we evaluated the reliability of the grid at the system level. In order to complete the analysis, the probability of failure of each of the subsystems had to be calculated. This was carried out by analyzing the transmission lines in each subsystem and determining the conditions under which the subsystem would fail.

We used simulation to gain insight into system behavior in the event of a line failure. With this simulation, we were able to identify a number of cascading failure scenarios in the power grid, which were then used to evaluate the failure probabilities of the subsystems previously defined. After we developed equations to evaluate the reliability of the system, a method was devised that would present a more accurate estimate of the grid reliability based on the different cascading scenarios that could take place.

Future extensions to this research will focus on determining confidence intervals for these reliability estimates. The results obtained at this stage are useful in understanding the operation of the grid, and can help in identifying critical areas in the system, which in turn facilitates efforts in hardening the grid through redundancy and intelligent control of the system.

The complexity reduction and reliability models developed will also be applied to other critical infrastructures, in which the flow problem resembles that of the power grid. Examples of such infrastructures include the ground transportation system and the aviation system. Our ultimate goal is to help build more reliable infrastructures in the future, and reduce interruptions and failures in existing infrastructures.

References

1. Faza, A., Sedigh, S., McMillin, B.: Reliability Modeling for the Advanced Electric Power Grid. In: Saglietti, F., Oster, N. (eds.) SAFECOMP 2007. LNCS, vol. 4680, pp. 370–383. Springer, Heidelberg (2007)
2. Laprie, J.C., Kanoun, K., Kaaniche, M.: Modelling interdependencies between the electricity and information infrastructures. In: Saglietti, F., Oster, N. (eds.) SAFE-COMP 2007. LNCS, vol. 4680, pp. 54–67. Springer, Heidelberg (2007)
3. U.S. Canada Power System Outage Task Force: Final report on the August 14, 2003 blackout in the United States and Canada: Causes and recommendations. Technical report (April 2004)
4. Pourbeik, P., Kundur, P., Taylor, C.: The anatomy of a power grid blackout - root causes and dynamics of recent major blackouts. IEEE Power and Energy Magazine 4(5), 22–29 (2006)
5. Stefanini, A., Masera, M.: The security of networked infrastructures and the role of information and communication technologies: lessons from recent blackouts. In: International Workshop on Complex Networks and Infrastructure Protection (March 2006)
6. Chassin, D.P., Posse, C.: Evaluating North American electric grid reliability using the barabasi-albert network model. Physica A 55, 667–677 (2005)
7. Billinton, R., Jonnavithula, S.: A test system for teaching overall power system reliability assessment. IEEE Transactions on Power Systems 11(4), 1670–1676 (1996)
8. Kazemi, S., Fotuhi-Firuzabad, B.R.: Reliability assessment of an automated distribution system. IET Generation, Transportation and Distribution 1(2), 223–233 (2007)
9. Walter, M., Siegle, M., Bode, A.: OpenSESAME-the simple but extensive, structured availability modeling environment. Reliability Engineering and System Safety (2007)
10. Chiaradonna, S., Lollini, P., Giandomenico, F.D.: On a modeling framework for the analysis of interdependencies in electric power systems. In: Proceedings of the 37th Annual IEEE/IFIP International Conference on Dependable Systems and Networks (DSN 2007) (2007)
11. Shooman, A.M., Kershenbaum, A.: Methods for communication-network reliabilty analysis: Probabilistic graph reduction. In: Proceedings of the annual Reliability and Maintainability Symposium (1992)
12. Chowdhury, B.H., Baravc, S.: Creating cascading failure scenarios in interconnected power systems. In: IEEE Power Engineering Society General Meeting (June 2006)
13. Lininger, A., McMillin, B., Crow, M., Chowdhury, B.: Use of max-flow on FACTS devices. In: North American Power Symposium (2007)
14. Kuo, W., Zuo, M.J.: Optimal Reliability Modeling, Principles and Applications, pp. 164–171. John Wiley and Sons, Inc., Hoboken (2003)

Automating the Processes of Selecting an Appropriate Scheduling Algorithm and Configuring the Scheduler Implementation for Time-Triggered Embedded Systems

Ayman K. Gendy and Michael J. Pont

Embedded Systems Laboratory,
University of Leicester,
University Road, Leicester LE1 7RH, UK
{akg14,M.Pont}@le.ac.uk

Abstract. Predictable system behaviour is a necessary (but not sufficient) condition when creating safety-critical and safety-related embedded systems. At the heart of such systems there is usually a form of scheduler: the use of time-triggered schedulers is of particular concern in this paper. It has been demonstrated in previous studies that the problem of determining the task parameters for such a scheduler is NP-hard. We have previously described an algorithm ("TTSA1") which is intended to address this problem. This paper describes an extended version of this algorithm ("TTSA2") which employs task segmentation to increase schedulability. We show that the TTSA2 algorithm is highly efficient when compared with alternative "branch and bound" search schemes.

Keywords: Safety-related embedded systems, automatic code generation, scheduler, time triggered.

1 Introduction

Developers creating software for use in the majority of "desktop" applications face a very different set of challenges from those creating embedded software. For example, the time interval within which the desktop system responds to a command may vary significantly without causing a major problem: by contrast, even small variations in timing behaviour (milliseconds or much less) in embedded systems may prove life threatening in (for example) an industrial, automotive or medical system.

There are two common approaches used in building real-time embedded systems: event-triggered (ET) and time-triggered (TT). The ET approach may prove cost effective in cases where the system must handle many aperiodic and sporadic events ([4], [5]), since the conversion of such events to periodic events may reduce the system utilisation. On the other hand *"time-triggered systems are to be preferred with respect to fault tolerance"* [4] and are also considered as best match for supporting safety critical applications ([1], [2], [3], [4]). In addition, it is widely recognised that *"Very safety critical systems, like X-by-wire require fault-tolerance and redundancy. The implementation of such systems probably will fail without the framework of time-triggered architectures"* [5]. For these reasons, this paper focuses on systems with a TT architecture.

M.D. Harrison and M.-A. Sujan (Eds.): SAFECOMP 2008, LNCS 5219, pp. 440–453, 2008.
© Springer-Verlag Berlin Heidelberg 2008

In most TT designs, an "offline" (also known as "pre-runtime", or "static") schedule is considered the best choice ([6], [7], [12], [14], [3]). It has been demonstrated in previous studies that the problem of testing the schedulability and determining the scheduler and task parameters for a set of tasks for such a system is NP-hard ([7], [8], [9], [10]). As part of an effort to address these problems we previously introduced a novel two-stage heuristic search technique ("TTSA1") which is intended to support the configuration of TT schedulers for use with resource-constrained embedded systems which employ a single processor [11]. In this paper, we extend our TTSA1 technique. Our goal is to show that, with appropriate (static) task execution behaviour, tasks may be cleanly segmented, allowing an increase in schedulability, while meeting the constraints of a set of periodic tasks for use with reliable embedded systems.

The remainder of this paper is organised as follows. In Section 2, we review previous work in scheduler design and selection. In Section 3, we introduce and describe a modified scheduling algorithm ("TTSA2") which is used to automate the process of scheduler selection and configuration. In Section 4, we describe the process used to assess the TTSA2 algorithm and present the results obtained from this assessment. Finally, in Section 5, we discuss the results, present our conclusions and make some suggestions for future work.

2 Related Work

In this section we review previous work in this area.

2.1 Scheduling Safety Critical Resource-Constrained Embedded Systems

A wide range of software architectures can be used for real-time systems, ranging from a simple scheduler to a full real-time operating system (RTOS).

For resource-constrained embedded systems, which have a very limited memory and CPU performance, a simple form of "time triggered co-operative" (TTC) – a form of cyclic executive – scheduler ([2], [8], [12], [14], [15]), *"which has low run-time overhead"*[14], is often used. For safety-critical applications which have hard real-time constraints, such as low jitter requirements, TTC architectures demonstrate very low levels of task jitter [16], and can maintain their low-jitter characteristics even when techniques such as dynamic voltage scaling (DVS) are employed to reduce system power consumption [17].

2.2 Time Triggered Co-operative Scheduler (TTC)

The TTC implementation discussed in this paper executes each task in a predefined time intervals which is derived from a scheduler "tick". The scheduler tick is usually signalled by an interrupt associated with the (periodic) overflow of a hardware timer. At each tick the status of each task is updated and tasks which are due to run are dispatched. The processor is then often placed in an "idle" mode, where it will remain until the next tick (in order to reduce the system power consumption).

2.3 Time Triggered Hybrid Scheduler (TTH)

Despite some attractive features, a TTC solution is not always appropriate. For example a TTC system cannot respond to a critical external event while executing specific task: this presents problems if the required response time is shorter than the worst case execution time, "WCET", of any of the system tasks [18].

In these circumstances, the TTC architecture can be replaced with a fully pre-emptive architecture (for example, a rate monotonic or the earliest deadline first, architecture [22]). Such an approach provides flexibility (and, possibly, portability), but it will also tend to increase the system complexity and overhead when compared to pre-run-time scheduling ([6], [7]).

In some designs the system responsiveness can be increased while maintaining the minimal resource requirements, by allowing a limited level of pre-emption in the system. This can be done by employing what we call a "time-triggered hybrid" (TTH) scheduler ([3], [19]), and others have called a "multi-rate executive with interrupts" [20]. The TTH can be seen as a rate-monotonic scheduler that supports a single, short, high priority, pre-empting task, and a collection of co-operative tasks (which have equal priority which is less than that of the pre-empting task).

The pre-empting task may be used for periodic data acquisition, typically by means of an analogue-to-digital converter or similar device. Such requirements are common in, for example, control systems [13], and applications which involve data sampling and Fast-Fourier transforms (FFTs) or similar techniques: see, for example, the work by Schlindwein et al. [21].

2.4 Scheduler Design and Configuration

When implementing a TTC or TTH scheduler, the system designer has to determine some parameters (including the length of the tick interval, the order in which tasks must be dispatched, and the initial delay - or phase - of each task). Inappropriate choice of these parameters may affect systems reliability (by violating task constraints) or lead to unnecessarily high levels of task jitter and / or to increased system power consumption. It has been demonstrated in previous studies that the problem of determining these parameters is NP-hard ([7], [8], [9], [10]).

In order to cope with this challenge, schedulability analysis and scheduler design have been studied extensively over many years: see, for example, work by Liu and Layland [22] through to work by Xu [7]. Researchers have proposed solutions based on simulated annealing techniques [9], constraint programming heuristics [30], branch and bound algorithm ([28]), and genetic algorithms [29].

None of the work summarised above relates directly to TTC / TTH architectures: instead, most previous studies have tended to focus on "conventional" RT operating systems (e.g. VxWorks: [29]). Such operating systems exceed the resource require-ments available in the types of processor considered in this study.

2.5 The TTSA1 Scheduling Algorithm

In an effort to support creation of TTC / TTH designs we have previously introduced a novel two-stage heuristic search technique, "TTSA1", which is intended to support the configuration of these time-triggered schedulers for use with resource-constrained embedded systems which employ a single processor [11].

As noted above the TTSA1 algorithm helps to automate the process of both scheduler selection and configuration. If a suitable scheduler is identified for a given task set, TTSA1 attempts to determine the suitable scheduler parameters (the tick interval) and task parameters (such as the task order and task offset). In determining these parameters, TTSA1 aims to ensure that: (i) task constraints are met; (ii) power consumption is "as low as possible"; (iii) a fully co-operative scheduler architecture is employed whenever possible.

The input to TTSA1 is a list of task specifications and constraints. The algorithm tests the schedulability of the given task set, first using the TTC scheduler. If the task set cannot be scheduled with this scheduler, the process is repeated using the TTH scheduler. The algorithm calculates a suitable tick interval, along with the task order and the required offset value for each task if all the tasks are schedulable; otherwise a list of the schedulable tasks is generated.

To achieve this result, TTSA1 begins by sorting the tasks according to two criteria: a) task precedence, b) task deadline, laxity, period, WCET, or jitter. It is then assumed that the first task will run with zero offset and the algorithm tries to find a suitable offset for the second task, using the longest possible tick interval. If such an offset is identified (and the constraints of both tasks are met), a third task is added to the system and the process is repeated. We carry on in this way until all tasks have been scheduled (if this proves possible).

3   The TTSA2 Scheduling Algorithm

In this section, we describe a modified version of the TTSA1 algorithm ("TTSA2"). TTSA2 employs task segmentation to increase the number of task sets which can be scheduled.

3.1   Overview

Despite its attractive features, the TTSA1 algorithm fails to find a suitable schedule for a set of tasks in some cases. For example assume that for a given set all tasks have deadlines equal to their periods. Assume also that this set includes two short tasks (Task S1 and Tasks S2), and at least one long task (Task L).

Table 1. Task specifications for task set that cannot be scheduled with TTC/TTH

Task	WCET (ms)	Deadline (ms)	Period (ms)
A	10	50	50
B	1	10	10
C	1	10	10

The TTSA1 algorithm fails to find a suitable schedule for this set if:

Deadline (S1) < WCET (S1) + WCET (L)

and

Deadline (S2) < WCET (S2) + WCET (L)

Table 2. Task specifications for task set that can be scheduled with TTC/TTH

Task	WCET (ms)	Deadline (ms)	Period (ms)
SA_1	5	45	50
SA_2	5	50	50
SB_1	1	10	10
SC_1	1	10	10

For example Task B and / or Task C shown in Table 1 will miss their deadlines every time Task A runs if these three tasks are scheduled using TTC / TTH. To overcome this situation, while still using a TTC / TTH architecture, long task(s) can be divided into multiple short tasks ([3],[8], [16]): for example Task A can be divided into two segments, Segment SA_1 and Segment SA_2, as shown in Table 2.

As previously indicated, testing the schedulability of a set of tasks and finding a suitable scheduler for them (if any) is an NP-hard problem. The problem becomes more complex if some parts of some tasks are required to exclude parts of other tasks in the set. For example, it may be that Segment SA_2 in Task A excludes Segment SB_3 and Segment SC_2 in Task B and Task C respectively, and Segment SB_1 in Task B excludes Segment SC_1 in Task C.

In this section we extend our previous TTSA1 algorithm to deal with such situations. We call the resulting algorithm TTSA2. The input to TTSA2 is a list of task specifications and constraints, including critical-section boundaries.

The TTSA2 algorithm tests the schedulability of the given task set, first using the TTC scheduler, if possible, otherwise it will try the TTH, considering each task as a single segment. If the task set cannot be scheduled the process is repeated after dividing one or more long tasks into two or more shorter tasks. The algorithm calculates a suitable tick interval, the task order, the smallest number of task segments along with the required offset value for each task and task segment if all the tasks are schedulable; otherwise a list of the schedulable tasks and task segments is generated.

To achieve this result, TTSA2 begins (like TTSA1) by sorting the tasks according to two criteria: a) task precedence, b) task deadline, laxity, period, WCET, or jitter. It is then assumed that the first task will run as one segment with zero offset and the algorithm tries to find a suitable offset for the second task (in one segment), using the longest possible tick interval (the greatest common divisor of the task periods). If such an offset is identified (and the constraints of both tasks are met), a third task is added to the system and the process is repeated. We carry on in this way until all tasks have been scheduled (if this proves possible). If a schedule cannot be found at any stage the last task added to the design is removed and divided into two segments. After adding the segmentation overhead and updating the segment deadlines (as explained in the next subsections), the search proceeds (Fig. 1).

Please note that this search process is not exhaustive, and might be described as "best characteristics first" approach: for example, it starts with a long tick interval (which is known to reduce power consumption) and it gradually reduces the tick interval until it matches the timing needs of the application (if ever). We proceed iteratively, stopping the search when we have identified the first workable solution. We assume that - because we have begun the search with "best characteristics" - any schedule identified will represent a good (but not necessarily completely optimal) solution.

```
START
Arrange tasks;
// Common divisors of task periods in descending order
GCD[t] = {GCD1, GCD2, ..., GCDm}, t=1, 2,,.., m;
Sched_Strategy = {TTC, TTH};

// First check schedulability using TTC strategy
Sched_Strategy_Index = 1;
DO{
   Tick_index = 1;
   DO{
     Tick_Interval = GCD[Tick_index];
     Add the first task as one segment with zero offset;
     i = 1; Sched_Tasks = 1;
     DO{
       i++;
       Add the next task, one segment at a time;
       start segment with zero offset;
       DO{
         Sched[i] = Check_Sched(i, Tick_index,
                              Sched_Strategy_Index);
         IF (Sched[i] = TRUE)
           { Sched_Tasks ++ ;}
         ELSE
           {
           Increment offset of the last added segment;
           Add the segmentation overhead;
           }
         } WHILE((constraints violated) and
                 (offset<=Period));
       } WHILE (i < total number of tasks);
       IF ( Sched_Tasks = total number of tasks )
           {
           Print task offsets, tick interval,
           scheduler type;
           EXIT;
           }
       ELSE
           { Tick_index++;}
       } WHILE (Tick_index <= m )
   Sched_Strategy++;         // Try the TTH strategy
   } WHILE (Sched_Strategy_Index <= 2);
Print list of scheduled and unscheduled tasks;
END
```

Fig. 1. Pseudo code for the TTSA2 algorithm

3.2 Adjusting the Segment Deadline

If Task T is divided into n segments, ST_1, ST_2.., ST_n, then the TTSA2 algorithm calculates the deadline of each segment as follows:

$$\text{Deadline } (ST_n) = \text{Deadline } (T) . \tag{1}$$

$$\text{Deadline } (ST_{i-1}) = \text{Deadline } (ST_i) - \text{WCET } (ST_i), \text{ where } i = n, n\text{-}1, n\text{-}2...,2 . \tag{2}$$

Please notice the deadline of Segment SA_1 in Table 2 as an example.

To be able to divide long tasks into multiple short tasks accurate information about the task WCET and the points at which the task can / cannot be pre-empted (the critical sections boundaries) must be specified. This can be done using techniques such as the "single path programming paradigm" ([24], [25]) or code balancing techniques [23].

3.3 Adding Segmentation Overhead

If a task is divided into two or more segments, the TTSA2 algorithm takes segmentation overhead into account. This overhead represents the time needed to save the context of the current segment and the time needed to restore this context when the next segment becomes ready to run. The time required for saving the context (*Context_Saving_overhead*) may not be the same as that required for loading the context (*Context_Loading_overhead*).

If Task T is divided into n segments, ST_1, ST_2.., ST_n, then the TTSA2 algorithm updates the segments WCETs to reflect this overhead, as follows:

$$\text{WCET } (ST_1) = \text{WCET } (ST_1) + \textit{Context\_Saving\_overhead} . \tag{3}$$

$$\text{WCET } (ST_i) = \textit{Context\_Loading\_overhead} + \text{WCET } (ST_i) + \\ \textit{Context\_Saving\_overhead}, \text{ where } i = 2, 3..., n\text{-}1. \tag{4}$$

$$\text{WCET } (ST_n) = \textit{Context\_Loading\_overhead} + \text{WCET } (ST_n) . \tag{5}$$

4 The Effectiveness of the TTSA2 Scheduling Algorithm

In this section the complexity and the effectiveness of the TTSA2 is evaluated. We compare the performance of the TTSA2 with that of the "branch and bound" algorithm (BaB), a standard benchmark which has been used previously to test the effectives of heuristic algorithms[26].

4.1 Algorithm Complexity

Assume we have a set of n independent tasks and that each consists of s segments. Investigating the schedulability of these tasks by means of a BaB algorithm requires testing n paths, each of length $n!$, this has a complexity of $O(n.n!)$ which is "*computationally intractable and cannot be used in practical systems when the number of tasks is high*"[27].

As noted elsewhere [11], the offset of each task can be any value in the range [0, Period], in ticks. Taking all possible offset combinations (t^n), where t is the period, and considering each task as set of s segments, each may has different offset, the complexity will increase to $O(t^{n.s} .n!)$.

By contrast, the TTSA2 algorithm does not try all paths. In addition, it does not change the task or / and segment offset of a given task once it has been added successfully to the schedule (that is, added without causing violation of the constraints of any of the tasks and segments which have been included in the schedule previously). The complexity of this algorithm is $O(n.s.t)$.

4.2 Algorithm Performance

An empirical test was carried out to explore the performance of the TTSA2 algorithm. The procedure and the results of this test are discussed in this section.

4.2.1 The Test Tools

The chosen hardware platform was an NXP (formerly Philips) LPC2129 micro-controller running on a small evaluation board. The LPC2129 is based on an ARM7TDMI core and is typical of modern (low cost) embedded processors. The measurements of the scheduler overhead and the segmentation overhead were carried out using this microcontroller. This overhead was taken into account while performing the scheduling selection and configurations.

To compare the performance of the TTSA2 with that of the BaB a simple (custom) scheduler simulator was executed on a desktop PC.

4.2.2 Task Constraints

The constraints considered in this study are described in this section.

Jitter calculation
As far as we are concerned in this paper, a task's jitter is the variation in the interval between the start times of the task. The starting time of each task is recorded so that the jitter statistics can be estimated. In the experiment discussed in the present paper, the upper bound of task jitter is (pseudo) randomly generated according to Equation 6.

$$0 \leq \text{Jitter} \leq P(i) \text{, where } P(i) \text{ is the period of Task } i. \tag{6}$$

Precedence
If it is required that Task A precedes Task B, then, in any tick, Task B is allowed to start its execution only after Task A completes its execution (e.g. see [28]).

In the current study, the precedence relation between any two tasks, A and B, is (pseudo) randomly generated iff

$$P(A) = P(B) . \tag{7}$$

and

$$P(A) \geq (\text{WCET}(A) + \text{WCET}(B)) . \tag{8}$$

Exclusion
If it is required that Segment SA_2 in Task A excludes Segment SB_3 in Task B, then, at any tick, Segment SA_2 is not allowed to pre-empt Segment SB_3 and vice versa [28].

In the current study some tasks segments in each set were (pseudo) randomly selected to have an exclusion relation between them.

Distance
The distance relation between any two tasks, A and B, can be defined as the minimum distance between the completion of Task A and the start of Task B [29].

In the current study the precedence relation between two tasks was (pseudo) randomly generated according to Equation 9.

$$0 \leq \text{Distance(A, B)} \leq P(A) - (WCET(A) + WCET(B)) . \quad (9)$$

given that:

$$P(A) = P(B) \text{ and } P(A) \geq (WCET(A) + WCET(B)) \quad (10)$$

Latency

The latency relation between any two tasks, A and B, can be defined as the maximum distance between the completion of Task B and the start of Task A [29].

In the current study the latency relation between two tasks was (pseudo) randomly generated as follows:

If there are no distances constraint between Task A and Task B then:

$$(WCET(A) + WCET(B)) \leq \text{Latency(A, B)} \leq \text{Max}(P(A), P(B)) \quad (11)$$

given that:

$$\text{Max}(P(A), P(B)) \geq (WCET(A) + WCET(B)) . \quad (12)$$

Otherwise:

$$(WCET(A) + WCET(B) + \text{Distance}(A,B)) \leq \text{Latency(A, B)} \leq P(A) . \quad (13)$$

4.2.3 Dataset Used

To explore the effectiveness of this algorithm, 1000 sets of tasks were (pseudo) randomly generated. Each set consisted of 3, 4 and 5 tasks specified by WCETs, deadlines and periods. These specifications were generated according to the following criteria:

$$0 < WCET(i) \leq 2000 \, \mu s . \quad (14)$$

$$WCET(i) < P(i) \leq 10000 \, \mu s . \quad (15)$$

$$WCET(i) \leq D(i) \leq P(i) . \quad (16)$$

In order to simplify the calculations, task periods were (pseudo) randomly generated at multiples of 1 ms (constrained by (15)). Task constraints of precedence, exclusion, distance, latency, and upper bound of jitter were also (pseudo) randomly generated and were in line with the findings from previous studies (e.g. see [28], [29]) and are (pseudo) randomly generated constrained by Equation 6 – Equation 16.

4.2.4 Results (Small Task Sets)

We tested the effectiveness of the TTSA2 algorithm when scheduling small sets of tasks (each contains 3, 4, or 5 tasks) and compared the results with those from the TTSA1 and the BaB algorithms. The results obtained from the BaB algorithm with / without using task segmentation are recorded as BaB1 and BaB2 respectively.

Fig. 2 to Fig. 4. show the number of task sets that was found to be schedulable using TTSA1, TTSA2, BaB1, and BaB2. Please note that the results obtained by combining the (unique) results from TTSAx-EDF, TTSAx-LLF, TTSAx-Jitter,

TTSAx-RM, and TTSAx-SJF are shown in these figures as TTSAx-ALL, where x equals 1 or 2 for TTSA1 and TTSA2. The number of trials until each algorithm identified the set of tasks as schedulable/unschedulable and the total time are also shown in Table 3.

Table 3. Number of trials and the total time

3-task set	TTC				TTH			
	TTSA1	TTSA2	BaB1	BaB2	TTSA1	TTSA2	BaB1	BaB2
Minimum number of trials	2.00E+00	2.00E+00	2.00E+00	2.00E+00	2.00E+00	2.00E+00	2.00E+00	2.00E+00
Maximum number of trials	2.70E+02	5.80E+02	2.97E+03	4.66E+06	1.65E+02	3.30E+02	2.83E+03	1.99E+05
Average number of trials	2.46E+01	4.63E+01	3.51E+02	1.75E+04	1.59E+01	3.04E+01	1.97E+02	2.08E+03
Total number of trials	2.46E+04	4.63E+04	3.51E+05	1.75E+07	1.59E+04	3.04E+04	1.97E+05	2.08E+06
Total time (s)	4.00E+00	4.00E+00	3.90E+01	8.49E+02	2.50E+00	2.50E+00	2.50E+01	1.17E+02

4-task set	TTC				TTH			
	TTSA1	TTSA2	BaB1	BaB2	TTSA1	TTSA2	BaB1	BaB2
Minimum number of trials	3.00E+00	3.00E+00	3.00E+00	3.00E+00	3.00E+00	3.00E+00	3.00E+00	3.00E+00
Maximum number of trials	2.65E+02	5.30E+02	7.23E+04	1.60E+08	2.65E+02	5.30E+02	4.21E+04	2.78E+07
Average number of trials	3.49E+01	7.01E+01	5.45E+03	5.30E+05	2.49E+01	4.99E+01	3.24E+03	1.13E+05
Total number of trials	3.49E+04	7.01E+04	5.44E+06	5.29E+08	2.49E+04	4.99E+04	3.23E+06	1.13E+08
Total time (s)	3.50E+00	4.50E+00	2.22E+02	1.73E+04	2.50E+00	4.50E+00	1.11E+02	4.17E+03

5-task set	TTC				TTH			
	TTSA1	TTSA2	BaB1	BaB2	TTSA1	TTSA2	BaB1	BaB2
Minimum number of trials	4.00E+00	4.00E+00	4.00E+00	4.00E+00	4.00E+00	4.00E+00	4.00E+00	4.00E+00
Maximum number of trials	1.40E+02	3.14E+02	1.03E+06	1.21E+09	1.02E+02	2.26E+02	1.33E+06	1.98E+08
Average number of trials	4.04E+01	8.58E+01	8.84E+04	2.04E+07	2.93E+01	6.34E+01	5.41E+04	5.09E+06
Total number of trials	4.04E+04	8.58E+04	8.84E+07	2.04E+10	2.93E+04	6.34E+04	5.41E+07	5.09E+09
Total time (s)	2.50E+00	6.00E+00	3.69E+03	2.94E+06	2.50E+00	3.50E+00	5.26E+03	3.79E+05

From the results obtained it was noted that:

- TTSA2 found a suitable scheduler for more sets than TTSA1.
- Because TTSA2 tries to find a suitable (TTC or TTH) scheduler using the lowest number of task segments, the results obtained from TTSA1 are found to be a subset of the complete list of valid schedules identified by TTSA2. This means that all the schedulers identified by both TTSA1 and TTSA2 have the same scheduling overhead and power consumption.
- The results obtained from TTSA1 and TTSA2 (when overheads are taken into account) are found to be a subset of the complete list of valid schedules identified by BaB1 and BaB2, respectively. In addition, although TTSA1 and TTSA2 test the schedulability using a subset of all the possible offset combinations, they produce results which are similar to those obtained with the BaB1 and BaB2 methods.
- The criteria used for adding the tasks to the TTSA1 and TTSA2 have an impact on the schedulability of the set (different criteria may give different results).

- Combining results from the variations of TTSA1 and variations of TTSA2 together gives results which are very close to those obtained from the BaB1 and BaB2 respectively while requiring a much lower number of trials, and hence less time (see Table 3).

4.2.5   Results (Large Task Set)

The results shown in Fig. 2. to Fig. 4. consider a maximum of 5 tasks. This is not an unrealistic number for the resource-constrained systems we are concerned with in this paper. However, this task set does not fully test the algorithm. In order to explore the performance of TTSA2 on larger problems, 1000 new data sets were created. Each data set consisted of 50 tasks, each with a maximum execution time of 2 ms and

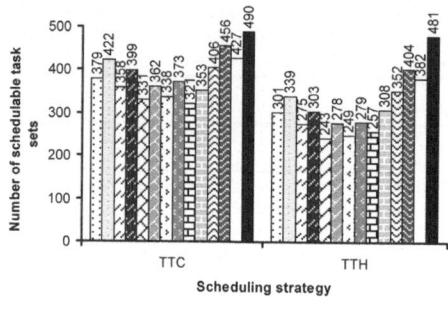

Fig. 2. Number of Scheduled task sets (3 interdependent tasks in each set)

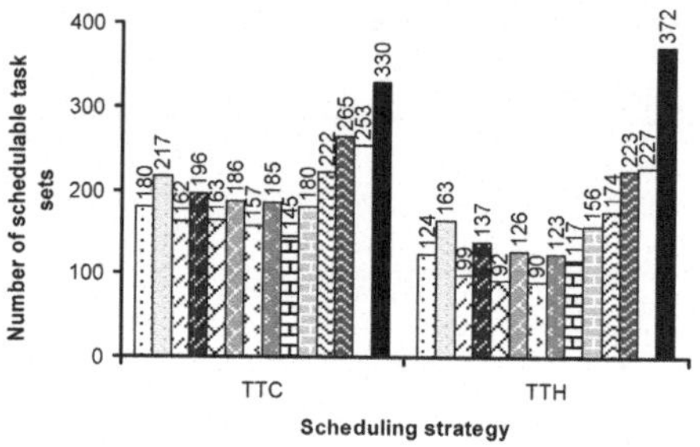

Fig. 3. Number of Scheduled task sets (4 interdependent tasks in each set)

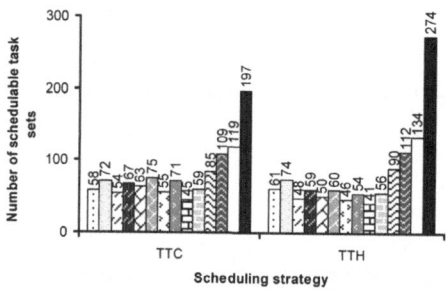

Fig. 4. Number of Scheduled task sets (5 interdependent tasks in each set)

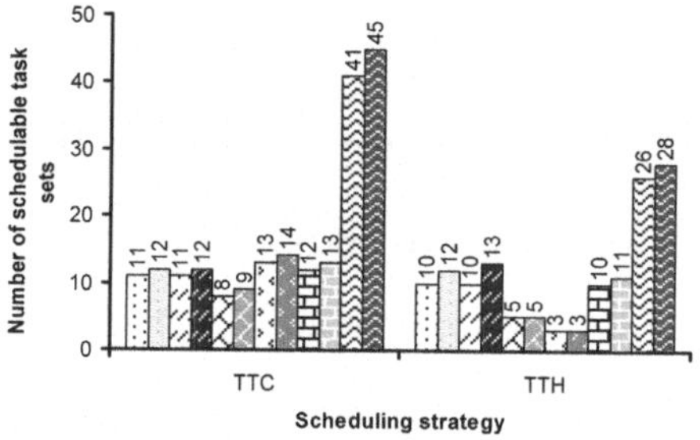

Fig. 5. Number of Scheduled task sets (50 interdependent tasks in each set)

TTSA1-EDF   TTSA2-EDF   TTSA1-LLF   TTSA2-LLF
TTSA1-Jiiter   TTSA2-Jiiter   TTSA1-RM   TTSA2-RM
TTSA1-SJF   TTSA2-SJF   TTSA1-ALL   TTSA2-ALL
BnB1   BnB2

maximum period of 200 ms. The task sets were (pseudo) randomly created according to the constraints described previously. To reduce the length of the major cycle, task periods were (pseudo) randomly generated as a multiple of 20 ms.

The results from this test are shown in Fig. 5. It took approximately 1 minute to complete the schedulability test for one set of 50 tasks using TTSA2-EDF, and a total of approximately 6 minutes to complete the test for TTSA2-All. It was not possible to complete this search using a BaB approach as this would have required performing a huge number of trials.

5   Discussion and Conclusions

It has been previously demonstrated that using offline, or pre-runtime, scheduling helps in reducing the complexity of inspecting and verifying the timing behaviour of safety critical embedded systems ([6], [7]).

In this paper we introduced a new offline scheduling algorithm, TTSA2, which helps to automate the process of determining the parameters required to schedule a given set of tasks in a resource-constrained embedded system employing a TTC or TTH architecture. The TTSA2 algorithm tries to find a suitable scheduler for the set of tasks by dividing each task into two or more segments. While searching for a workable scheduler the proposed scheduling algorithm ensures that the CPU power consumption is "as low as possible" (by choosing the longest possible tick interval), and the task constraints as well as individual segment constraints are met (by adjusting the segment offsets, tick interval, and task orders), using the lowest number of segments. If the tasks cannot all be scheduled (for example, if some timing constraints of one or more tasks cannot be met), a list of the schedulable/unschedulable tasks is generated. The algorithm improves on the performance of both a BaB search and a previous version of this algorithm (TTSA1).

The algorithm can be used as part of a tool for automatic code generation for safety–critical, resource-constrained embedded systems. Using such a tool will not only reduce the time and effort required to develop such systems but will reduce the probability of scheduling errors, which may cause serious damage (see [31] for an example), as well.

In the current work we assume that a task can be divided into two or more segments only at certain points of time (the critical segment boundaries). Future work needs to be done to determine more efficient way to find the points of time at which task can be divided. In addition, other methods are needed to explore ways for choosing the pre-empting task in the TTH scheduler, in order to improve the results.

References

1. Domaratsky, Y., Perevozchikov, M.: Highly dependable time-triggered operating system-static scheduling approach and effective run-time implementations, Dedicated Systems Magazine, pp. 77–84 (October-December 2000)
2. Kopetz, H.: Real-Time Systems, Design Principles for Distributed Embedded Applications. Kluwer Academic, Dordrecht (1997)
3. Pont, M.J.: Patterns For Time-Triggered Embedded Systems. Addison-Wesley, Reading (2001)

4. Scheler, F., Schröder-Preikschat, W.: Time-Triggered vs. Event-Triggered: A matter of configuration? In: Dulz, W., Schröder-Preikschat, W. (eds.) MMB Workshop Proceedings (GI/ITG Workshop on Non-Functional Properties of Embedded Systems Nuremberg, pp. 107–112. VDE Verlag, Berlin (2006) ISBN 978-3-8007-2956-2
5. Albert, A.: Comparison of Event-Triggered and Time-Triggered Concepts with Regard to Distributed Control Systems, Embedded World 2004, Nuremberg, WEKA Verlag, pp. 235–252 (2004)
6. Xu, J.: Making Software Timing Properties Easier to Inspect and Verify. IEEE Software 20(4), 34–41 (2003)
7. Xu, J., Parnas, D.L.: Priority Scheduling Versus Pre-Run-Time Scheduling. Int. Journal of Time-Critical Systems 18, 7–23 (2000)
8. Baker, T.P., Shaw, A.: The Cyclic Executive Model and Ada. Real-Time Systems 1(1), 7–25 (1989)
9. Tindell, K., Burns, A., Wellings, A.: Allocating Hard Real-Time Tasks: An NP-Hard Problem Made Easy. Real-Time Systems 4(2), 145–165 (1992)
10. Baruah, S.K.: The Non-Preemptive Scheduling of Periodic Tasks Upon Multiprocessors. Real-Time Systems 32(1-2), 9–20 (2006)
11. Gendy, A.K., Pont, M.J.: Automatically Configuring Time-Triggered Schedulers for Use with Resource-Constrained, Single-Processor Embedded Systems. IEEE Trans. on Industrial Informatics 4(1), 37–46 (2008)
12. Gangoiti, U., Marcos, M., Estévez, E.: Using Cyclic Executives for Achieving Closed Loop Co-Simulation. In: Proc. of the Joint 44th IEEE Control and Decision Conference and European Control Conference CDC-ECC 2005, Sevilla, pp. 3790 –4785 (2005)
13. Buttazzo, G.C.: Rate monotonic vs. EDF: Judgement day. Real-Time Systems 29(1), 5–26 (2005)
14. Huang, C., Chang, L., Kuo, T.: A Cyclic-Executive-Based QoS Guarantee over USB. In: IEEE 9th Real-Time and Embedded Technology and Applications Symposium, Toronto, Canada, May 27-30, 2003, pp. 88–95 (2003)
15. Burns, A.: Generating Feasible Cyclic Schedules. Control Engineering Practice 3(2), 151–162 (1995)
16. Locke, C.D.: Software Architecture for Hard Real-Time Applications: Cyclic Executives Vs. Fixed Priority Executives. Real-Time Systems 4(1), 37–52 (1992)
17. Phatrapornnant, T., Pont, M.J.: Reducing Jitter in Embedded Systems Employing A Time-Triggered Software Architecture and Dynamic Voltage Scaling. IEEE Transactions on Computers (Special Issue on Design and Test of Systems-On-a-Chip) 55(2), 113–124 (2006)
18. Allworth, S.T.: An Introduction to Real-Time Software Design. Macmillan, London (1981)
19. Maaita, A., Pont, M.J.: Using Planned Pre-Emption to Reduce Levels ff Task Jitter in a Time-Triggered Hybrid Scheduler, UK Embedded Forum, Birmingham, UK, University of Newcastle (2005)
20. Kalinsky, D.: Context Switch. Embedded Systems Programming 14(1), 94–105 (2001)
21. Schlindwein, F.S., Smith, M.J., Evans, D.H.: Spectral Analysis of Doppler Signals and Computation of the Normalized First Moment in Real Time. Using a Digital Signal Processor, Medical & Biological Engineering & Computing 26, 228–232 (1988)
22. Liu, C.L., Layland, J.W.: Scheduling Algorithms for Multiprogramming in a Hard Real-Time Environment. Journal of the ACM 20(1), 40–61 (1973)
23. Gendy, A.K., Pont, M.J.: Towards a Generic Single-Path Programming Solution with Reduced Power Consumption. In: Proceedings of the ASME 2007 International Design Engineering Technical Conferences & Computers and Information in Engineering Conference (IDETC/CIE 2007), Las Vegas, Nevada, USA, September 4-7 (2007)

24. Puschner, P., Burns, A.: Writing Temporally Predictable Code. In: Proc. 7th IEEE International Workshop on Object-Oriented Real-Time Dependable Systems, pp. 85–91 (January 2002)
25. Puschner, P.: The Single-Path Approach Towards WCET-Analysable Software. In: Proc. IEEE International Conference on Industrial Technology, pp. 699–704 (December 2003)
26. Cucu, L., Sorel, Y.: Non-Preemptive Multiprocessor Scheduling for Strict Periodic Systems with Precedence Constraints. In: Proc. 23rd Annual Workshop of the UK Planning and Scheduling Special Interest Group, PLANSIG 2004, Cork, Ireland (December 2004)
27. Buttazzo, G.: Hard Real-Time Computing Systems: Predictable Scheduling Algorithms and Applications. Kluwer Academic, Dordrecht (1997)
28. Xu, J., Parnas, D.L.: Scheduling Processes with Release Times, Deadlines, Precedence and Exclusion Relations. IEEE Transactions on Software Engineering 16(3), 360–369 (1990)
29. Sandström, K., Norström, C.: Managing Complex Temporal Requirements in Real-Time Control Systems. In: 9th IEEE Conf. Engineering of Computer-Based Systems. IEEE, Sweden (2002)
30. Ekelin, C., Jonsson, J.: Evaluation of Search Heuristics for Embedded System Scheduling Problems. In: Proc. Int. Conf. Principles and Practice of Constraint Programming, Paphos, Cyprus, pp. 640–654 (2001)
31. Reeves, G.: What Really Happened on Mars?, – Authoritative Account (1997), http://research.microsoft.com

Author Index